Y0-AAF-348

LIBRARY OF THE UNIVERSITY OF CALIFORNIA
Berkeley

MATH/STAT.

CONTEMPORARY MATHEMATICS

Titles in this Series

LIBRARY
DEC 19 1985
UNIVERSITY OF CALIFORNIA
BERKELEY

Titles in this Series

Volume

21 **Topological methods in nonlinear functional analysis,** S. P. Singh, S. Thomeier, and B. Watson, Editors

22 **Factorizations of $b^n \pm 1$, $b = 2, 3, 5, 6, 7, 10, 11, 12$ up to high powers,** John Brillhart, D. H. Lehmer, J. L. Selfridge, Bryant Tuckerman, and S. S. Wagstaff, Jr.

23 **Chapter 9 of Ramanujan's second notebook—Infinite series identities, transformations, and evaluations,** Bruce C. Berndt and Padmini T. Joshi

24 **Central extensions, Galois groups, and ideal class groups of number fields,** A. Fröhlich

25 **Value distribution theory and its applications,** Chung-Chun Yang, Editor

26 **Conference in modern analysis and probability,** Richard Beals, Anatole Beck, Alexandra Bellow and Arshag Hajian, Editors

27 **Microlocal analysis,** M. Salah Baouendi, Richard Beals and Linda Preiss Rothschild, Editors

28 **Fluids and plasmas: geometry and dynamics,** Jerrold E. Marsden, Editor

29 **Automated theorem proving,** W. W. Bledsoe and Donald Loveland, Editors

30 **Mathematical applications of category theory,** J. W. Gray, Editor

31 **Axiomatic set theory,** James E. Baumgartner, Donald A. Martin and Saharon Shelah, Editors

32 **Proceedings of the conference on Banach algebras and several complex variables,** F. Greenleaf and D. Gulick, Editors

33 **Contributions to group theory,** Kenneth I. Appel, John G. Ratcliffe and Paul E. Schupp, Editors

Contributions to Group Theory

EDITORIAL BOARD

R. O. Wells, Jr., managing editor
Jeff Cheeger
Adriano M. Garsia
Kenneth Kunen
James I. Lepowsky
Johannes C. C. Nitsche
Irving Reiner

1980 *Mathematics Subject Classification.* Primary 01A70, 18G40, 20C05, 20D30, 20E05, 20E06, 20E07, 20E22, 20E26, 20F05, 20F06, 20F10, 20F28, 20F36, 20G15, 20H10, 20J05, 30F10, 57M05, 57S25.

Library of Congress Cataloging in Publication Data
Main entry under title:

Contributions to group theory.
(Contemporary mathematics; v. 33)
Papers published in honor of Roger Lyndon on his sixty-fifth birthday.
Bibliography: p.
1. Groups, Theory of—Addresses, essays, lectures. 2. Lyndon, Roger C.—Addresses, essays, lectures. I. Appel, Kenneth I., 1932– . II. Ratcliffe, John G., 1948– . III. Schupp, Paul E., 1937– . IV. Lyndon, Roger C. V. Series.
QA171.C683 1984 512′.22 84-18454
ISBN 0-8218-5035-0 (alk. paper)

Copying and reprinting. Individual readers of this publication, and nonprofit libraries acting for them, are permitted to make fair use of the material, such as to copy an article for use in teaching or research. Permission is granted to quote brief passages from this publication in reviews provided the customary acknowledgement of the source is given.

Republication, systematic copying, or multiple reproduction of any material in this publication (including abstracts) is permitted only under license from the American Mathematical Society. Requests for such permission should be addressed to the Executive Director, American Mathematical Society, P. O. Box 6248, Providence, Rhode Island 02940.

The appearance of the code on the first page of an article in this volume indicates the copyright owner's consent for copying beyond that permitted by Sections 107 or 108 of the U. S. Copyright Law, provided that the fee of $1.00 plus $.25 per page for each copy be paid directly to Copyright Clearance Center, Inc., 21 Congress Street, Salem, Massachusetts 01970. This consent does not extend to other kinds of copying, such as copying for general distribution, for advertising or promotion purposes, for creating new collective works or for resale.

Copyright © 1984 by the American Mathematical Society
Printed in the United States of America
All rights reserved except those granted to the United States Government
This volume was printed directly from author prepared copy.
The paper used in this book is acid-free and falls within the guidelines
established to ensure permanence and durability.

CONTEMPORARY
MATHEMATICS

Volume 33

Contributions to Group Theory

Kenneth I. Appel, John G. Ratcliffe
and Paul E. Schupp, Editors

AMERICAN MATHEMATICAL SOCIETY
Providence · Rhode Island

72719850

Math Sep lac

SD 12/19/85 RD

QA171
C6831
1984
MATH

Papers dedicated to

ROGER C. LYNDON

on the occasion
of his sixty-fifth birthday.

TABLE OF CONTENTS

Preface

This volume consists of five short articles on Roger Lyndon and his contributions to mathematics, and twenty-seven invited research papers. The research articles involve topics in combinatorial group theory and closely related areas. Several of the articles fall into subfields of combinatorial group theory in which much of the initial work was done by Lyndon. Most of the rest are in areas in which he has made major contributions. It is a tribute to Lyndon's mathematical breadth that papers covering such a wide array of topics are all closely related to the work he has done.

The articles on Lyndon and his work include a biographical essay by Kenneth Appel and expository articles by Saunders Mac Lane, John Ratcliffe, Jerome Keisler, and Paul Schupp. Mac Lane describes the results in Lyndon's doctoral thesis and explains how they fit into the early history of spectral sequences. Ratcliffe describes Lyndon's fundamental work in cohomology of groups in the early part of his career. Keisler discusses Lyndon's work in logic, especially his fundamental results in model theory in the mid 1950s. Schupp then describes Lyndon's work in group theory over the past twenty years.

The volume grew out of a desire by the editors to honor our teacher, Roger Lyndon, on the occasion of his sixty-fifth birthday. It has been a labor of love for us to gather together papers of solid mathematical interest dedicated to Roger Lyndon by his students, colleagues and friends, all of whom have shown great enthusiasm for this project.

We would like to thank all of the contributors and all the people who have made this collection possible. All the papers in this volume have been referred.

Kenneth I. Appel
John G. Ratcliffe
Paul E. Schupp

Contemporary Mathematics
Volume 33, 1984

ROGER C. LYNDON: A BIOGRAPHICAL AND PERSONAL NOTE

Kenneth I. Appel

We who are privileged to be Roger Lyndon's students know him as a genuine "Doctor-father," a man who patiently guided us in our first efforts at producing mathematics, and insisted that we learned to present our arguments in decent English. Although we have worked in many areas of mathematics since leaving Ann Arbor, he has continued to be interested in our mathematical work and has often made helpful suggestions. Because of his essential modesty and his enthusiasm for discussing our work rather than his own, some of us did not initially appreciate the extent and depth of his contributions to mathematics. Each of us knew of the profound work that he had done in our thesis specialties -- logic, homological algebra, many areas of group theory -- but we later discovered that people who were not familiar with his work in our own particular specialties considered him a mathematician of the first rank for his work in their areas of expertise.

Roger Lyndon, who will always be associated with the Midwest because of his distinguished career at the University of Michigan, was born on December 18, 1917, at Calais, Maine, almost the easternmost point of the continental United States. His family lived in the small nearby town of Eastport, where his father, Percy Lyndon, was a Unitarian minister. His paternal grandfather had come to America from England at the age of two. Lyndon's maternal grandfather came to Maine from Louisiana and owned a busy wharf on the Bay of Fundy. When Roger was two years old, his mother, Ann Aymar Milliken Lyndon, died. In the next fifteen years Percy, Roger and his sister lived in various towns in Massachusetts and New York. After graduation

© 1984 American Mathematical Society
0271-4132/84 $1.00 + $.25 per page

from the Derby School, Lyndon enrolled at Harvard in 1935, intent on studying literature and becoming a writer.

Lyndon had always enjoyed mathematics. He soon discovered that he could get good grades in mathematics with no effort, while having to memorize the complete works of Shakespeare and a number of obscure poets seemed forbidding. Thus he completed his undergraduate degree in mathematics in 1939. This was the Harvard of G. D. Birkhoff, Mac Lane, Quine, Huntington, M. H. Stone, H. Whitney, J. L. Coolidge, G. Birkhoff, and L. Loomis -- as always, a major center of mathematics in America. Upon graduation, after having spent most of his life in New England, Lyndon decided to see what the rest of the country was like and spent almost a year working in a bank in Albuquerque. Then he returned to Harvard to earn a Master's degree in 1941.

Lyndon taught at Georgia Tech during the academic year 1941-42. After the war began he participated in a program for teaching navigation to Navy pilots. In 1942 he returned to Harvard to teach in the Navy V-12 program and remained to receive a Ph.D. in 1946. Lyndon's first mathematical interest was logic. He studied Quine's system and discovered that it contained a major error but never published this result. After becoming interested in relational algebras he finally decided to write a thesis in homological algebra. The thesis was a brilliant piece of mathematics; it anticipated much of the development of spectral sequences. Saunders Mac Lane, who was Lyndon's thesis adviser, describes this work in detail in another article in this volume. Alfred Tarski arrived in Cambridge during Lyndon's final year at Harvard and Lyndon attended Tarski's course. This was the beginning of a long personal and professional friendship that led to Lyndon's work in model theory a decade later.

After two years in London with the Office of Naval Research, Lyndon felt that to work in cohomology he should study topology further, and he went to work at Princeton University, the center of American topology at that time. The Princeton faculty then included Lefshetz, Steenrod, Artin, Church, Bochner,

Tucker, Spencer, and Fox. While at Princeton he was greatly influenced by Ralph Fox, whose course in knot theory he attended. The first semester of Fox's course covered the group theoretic background for knot theory and Lyndon became much more interested in the group theory than the knot theory. During that time he had close contact with K. Reidemeister who visited for a year, and S. Eilenberg whom he had known at Harvard and who came frequently to Princeton to collaborate with Steenrod. After five years as an instructor and assistant professor at Princeton, Lyndon joined the faculty of the University of Michigan as an assistant professor in 1953. He has remained at Michigan ever since except for visiting positions at the University of California, Berkeley (1956-57), the Institute for Defense Analyses, Princeton (1959-60), Queen Mary College, London (1960-61 and 1964-65), Morehouse College, Atlanta (1969), Université de Montpellier, France (1974-75), and Université de Picardie, France (1980-81). He was promoted to associate professor in 1956 and professor in 1959.

During his years on the Michigan faculty Lyndon has been the thesis adviser of twelve Ph.D. students, of whom I was privileged to have been the second. The diversity of topics of the theses of his students serves as well as his own bibliography to illustrate the breadth of his mathematical interests. These topics include decision problems, model theory, algorithmic problems, cohomology, combinatorial group theory, and group rings (the names of Lyndon's students and their thesis titles are listed separately).

I fondly remember my experiences working as a thesis student in the years 1957-59. My conversations with Lyndon often took place in the newly established commons room, which was supplied with a large coffee percolator. As a "morning person," I would want to discuss my work with Lyndon as soon as he arrived at Angell Hall. It soon became clear to me that no mathematical discussion was possible until he had consumed at least two cups of coffee, so during my last two years of graduate work I was the unofficial coffee preparer for the commons room.

Lyndon produces elegant mathematics and thinks in terms of broad and deep ideas. Paul Schupp and John Ratcliffe, who have styles similar to his, tell me that as students they had a very easy time communicating their ideas to him and understanding his. I tend to work from the combinatorial aspects of particular problems without thinking of abstract structural considerations. While I was a graduate student, many of my conversations with Lyndon proceeded as follows. I would announce proudly that I could prove a lemma and write it on the blackboard, and he would copy it down. I would then proceed through a detailed if rough hewn proof. When I had finished, he would admit that he had not followed much of what I had said. A day later, working from the statement of my lemma and a few recollections of my methods, he would show me an extremely elegant proof. It often took me quite some time to understand what he had done, and I learned a good bit of mathematics in the process.

I have become fascinated with several problems that he has attacked in his papers and the methods he has developed to work on them. This interest had led me to study these problems further and proceed a bit further along the path. Thus he has continued to inspire my work in the more than twenty years since I have left Ann Arbor. I am sure that this is the case with several of his other students. In reading Lyndon's papers, I am impressed by the mathematical power that he uses, even in the proof of specialized results. I once asked him whether there was a thread common to his diverse work in so many fields of mathematics; he replied that he felt that the problems on which he worked had all been combinatorial in nature. In agreeing with this method of classifying his work, one would certainly have to put him in the very first rank of those who have used combinatorial techniques in the last forty years.

The articles by Keisler, Mac Lane, Ratcliffe, and Schupp in this volume help us to appreciate Lyndon the mathematician. To appreciate Lyndon the man, one should read his article "Saunders Mac Lane as a shaper of mathematics and mathematicians" [52]. In that tribute to Mac Lane, Lyndon provides a catalogue of his own personal values and evidence of his own modesty that would be hard to match here.

A tall man of distinguished appearance with a deep voice, Lyndon is the kind of person who quickly makes people feel comfortable working with him. Mathematical conversations with him have the same effect that locker room orations of a great football coach have on team members -- one leaves with the enthusiasm to work on a level far above one's ability and a feeling that it might be possible to do so, forgetting, perhaps because of Lyndon's self-effacing manner, that he brings to similar tasks an extraordinary talent and energy that few others possess.

Ph.D. Students of Roger C. Lyndon

		Thesis Title
Gerald O. Losey	1958	Group Rings and Dimension Subgroups
Kenneth I. Appel	1959	Two Investigations on the Borderline of Logic and Algebra
Calvin C. Elgot	1960	Decision Problems of Finite Automata Design and Related Arithmetic
Paul E. Schupp	1966	On Dehn's Algorithm and the Conjugacy Problem
Charles S. Holmes	1967	Projectivities of Free Products
Arthur G. Conn	1969	Continuous Model Theory and Set Theory
Nancy E. King	1969	Real Length Functions in Groups
Gerald E. Meike	1969	A Decision Procedure for the Prefix Class
Norman J. Frisch	1970	Automorphisms of the Fundamental Group of an Orientable 2-Manifold
John H. Remmers	1971	Some Algorithmic Problems for Semigroups: A Geometric Approach
Ian M. Chiswell	1973	On Groups Acting on Trees
John G. Ratcliffe	1977	The Theory of Crossed Modules with Applications to Cohomology of Groups and Combinatorial Homotopy Theory (jointly directed by James Kister)
Libo Lo	1984	On the Computational Complexity of the Theory of Abelian Groups

Publications by Roger C. Lyndon

Books

1. Notes on Logic, Van Nostrand, Princeton, N.J. (1966), 97 pp.

2. Word Problems: Decision Problems and the Burnside Problem in Group Theory, Editor (with W. W. Boone and F. B. Cannonito), Studies in Logic, Vol. 71, North-Holland, Amsterdam (1973), 646 pp.

3. Combinatorial Group Theory, Ergebnisse der Math. 89 (1977), Springer Verlag, Heidelberg, 339 pp. (With P. E. Schupp).

4. Théorie des Groupes: Groupes et Géométrie. Notes par A. Boidin, A. Fromaget, et R. Lyndon. Bound polycopied, Université de Picardie, 1980/1981, 245 pp.

Articles

1. The Zuse computer, Math. Tables and Other Aids to Computation II, 20 (1947), 355-359.

2. The cohomology theory of group extensions, Duke Math. J. 15 (1948), 271-292.

3. New proof for a theorem of Eilenberg and Mac Lane, Ann. Math. 50 (1949), 731-735.

4. The representation of relational algebras, Ann. Math. 51 (1950), 707-729.

5. Cohomology theory of groups with a single defining relation, Ann. Math. 52 (1950), 650-665.

6. Identities in two valued calculi, Trans. Amer. Math. Soc. 71 (1951), 457-465.

7. Two notes on nilpotent groups, Proc. Amer. Math. Soc. 3 (1952), 579-583.

8. On the Fouxe-Rabinovich series for free groups, Port. Math. 12, Fasc. 3 (1953), 115-118.

9. Identities in finite algebras, Proc. Amer. Math. Soc. 5 (1954), 8-9.

10. On Burnside's problem, Trans. Amer. Math. Soc. 77 (1954), 202-215.

11. On Burnside's problem, II, Trans. Amer. Math. Soc. 78 (1955), 329-332.

12. Commutator subgroups of free groups, Amer. J. Math. 72 (1955), 929-937.

13. A nonspecial operator Lie ring, Bull. Amer. Math. Soc. 61 (1955), 519.

14. A theorem of Friedrichs, Mich. Math. J. 3 (1956), 27-29.

15. The representation of relation algebras, II, Ann. Math. 63 (1956), 294-307.

16. Free differential calculus, IV, The quotient groups of lower central series, Ann. Math. 68 (1958), 81-95. (With K. T. Chen and R. H. Fox).

17. An interpolation theorem in the predicate calculus, Pac. J. Math. 9 (1959), 129-142.

18. Properties preserved under homomorphism, Pac. J. Math. 9 (1959), 143-154.

19. Properties preserved in subdirect products, Pac. J. Math. 9 (1959), 155-164.

20. Properties preserved under algebraic constructions, Bull. Amer. Math. Soc. 65 (1959), 287-299.

21. The equation $a^2b^2 = c^2$ in free groups, Mich. Math. J. 6 (1959), 89-95.

22. Existential Horn sentences, Proc. Amer. Math. Soc. 10 (1959), 994-998.

23. Burnside groups and Engel rings, Proc. Symp. Pure Math. I, Providence (1959), 4-14.

24. Equations in free groups, Trans. Amer. Math. Soc. 96 (1960), 445-457.

25. Groups with parametric exponents, Trans. Amer. Math. Soc. 96 (1960), 518-533.

26. Relation algebras and projective geometries, Mich. Math. J. 8 (1961), 21-28.

27. Equivalence of elements under automorphisms of a free group, Mimeo., Queen Mary College (1961). (With P. J. Higgins).

28. Metamathematics and algebra: an example, Proc. International Congress for Logic, Methodology and Philosophy of Science, Stanford (1962), 143-150.

29. Dependence and independence in free groups, J. Reine Angew Math. 10 (1962), 148-173.

30. The equation $a^M = b^N c^P$ in a free group, Mich. Math. J. 9 (1962), 289-298. (With M. P. Schutzenberger).

31. Free bases for normal subgroups in free groups, Trans. Amer. Math. Soc. 108 (1963), 526-537. (With D. E. Cohen).

32. Length functions in groups, Math. Scand. 12 (1963), 209-234.

33. Grushko's theorem, Proc. Amer. Math. Soc. 16 (1965), 822-826.

34. Dependence in groups, Colloq. Math. 14 (1966), 275-283.

35. Equations in free metabelian groups, Proc. Amer. Math. Soc. 17 (1966), 728-730.

36. On Dehn's algorithm, Math. Annalen (1966), 208-228.

37. A maximum principle for graphs, J. Comb. Theory 3 (1967), 34-37.

38. Groups of elliptic linear fractional transformations, Proc. Amer. Math. Soc. 18 (1967), 1119-1124. (With J. L. Ullman).

39. Pairs of real 2-by-2 matrices that generate free products, Mich. Math. J. 15 (1968), 161-166. (With J. L. Ullman).

40. Groups generated by two parabolic fractional transformations, Canad. J. Math. 21 (1969), 1388-1403. (With J. L. Ullman).

41. On the Freiheitssatz, J. London Math. Soc. (2), 5 (1972), 95-101.

42. Commutators as products of squares, Proc. Amer. Math. Soc. 39 (1972), 267-272. (With Morris Newman).

43. Geometric methods in the theory of abstract infinite groups, Proc. Colloq. on Permutations, Paris, V (July, 1972), 9-14.

44. Two notes on Rankin's book on the modular groups, J. Australian Math. Soc. 16 (1973), 454-457.

45. On products of powers in groups, Proc. Amer. Math. Soc. 40 (1973), 419-420. (With T. McDonough and Morris Newman).

46. On products of powers in groups, II, Comm. Pure Appl. Math. 26 (1973), 781-784.

47. Equivalence of elements under automorphisms of a free group, J. London Math. Soc. 8 (1974), 254-258. (With P. J. Higgins).

48. On non-Euclidean crystallographic groups, Proc. Second International Conf. Theory of Groups, Canberra (August 1973), Lecture Notes in Mathematics 372 (1974), Springer Verlag, 437-442.

49. A remark about permutations, J. Comb. Theory 18 (1975), 234-235. (With W. Feit and L. Scott).

50. On the combinatorial Riemann-Hurwitz formula, Convegno sui gruppi infiniti, Rome (December 1973), Symposia Math. 17 (1976), 435-439.

51. Quadratic equations in free products with amalgamation, Houston J. Math. 4 (1978), 91-103.

52. Saunders Mac Lane as a shaper of mathematics and mathematicians, Saunders Mac Lane Selected Papers, Springer Verlag (1979), 515-517.

53. Cancellation theory in free products with amalgamation, Word Problems II, The Oxford Book. Studies in Logic 95 (1979), 8 pp.

54. Commutators in free groups, Canadian Math. Bull. 24 (1981), 101-106. (With M. J. Wicks).

55. (Abstract of above). Commutateurs dans les groupes libres, in Proc. of Septième Ecole de Printemps d'Informatique Theorique, Jougne, France, 1979, 1 p.

56. Equations in groups, Boletim Soc. Brasiliero de Mathematica 11 (1980), 79-102.

57. (Abstract of above). Equations in groups, in Proc. of 12° Coloquio Brasiliero de Mathematica, Pocos de Caldas, Brasil, 1979, ca. 4 pp.

58. Proof of the fundamental theorem of algebra, Amer. Math. Monthly 88 (1981), 253-256. (With J. L. Brenner).

59. Equations over cyclic groups, Publ. du Laboratoire Informatique-Théorique et Programmation, Univ. Paris, VII, No. 81-41, pp. 25.

60. Nonparabolic subgroups of the modular group, J. Algebra 77 (1982), 311-322.

61. Permutations and cubic graphs, Pacific J. Math. 104 (1983), 285-315.

62. The orbits of the product of two permutations, European J. Combinatorics 4 (1983), 279-293. (With J. L. Brenner).

63. Maximal nonparabolic subgroups of the modular group, Math. Annalen 263 (1983), 1-11. (With J. L. Brenner).

64. Infinite Eulerian Tesselations, Discrete Math. 46 (1983), 111-132. (With J. L. Brenner).

65. A theorem of G. A. Miller on the order of the product of two permutations I, Jñānābha 14 (1984) (to appear). (With J. L. Brenner).

66. A theorem of G. A. Miller on the order of the product of two permutations II, Indian J. Math. (to appear). (With J. L. Brenner).

67. A theorem of G. A. Miller on the order of the product of two permutations III, Pure Applied Math. Sci. (to appear). (With J. L. Brenner).

68. Doubly Eulerian trails on rectangular grids, J. Graph Theory (to appear). (With J. L. Brenner).

69. Some uses of coset graphs, Proc. Groups-Korea-1983 (to appear).

70. Representations of permutations by words (submitted). (With J. Mycielski).

71. The word problem for geometrically finite groups (submitted). (With W. J. Floyd and A. H. M. Hoare).

Contemporary Mathematics
Volume 33, 1984

SPECTRAL COMPLICATIONS IN COHOMOLOGY COMPUTATIONS

Saunders Mac Lane

1. <u>The Background</u>. Roger Lyndon's doctoral thesis, written in 1946, was addressed to the problem of computing the cohomology groups of a group extension in terms of the cohomology groups of the factors of that extension. It turned out that in many cases one could not completely compute the whole cohomology group. Now we know that the difficulty lies in the additional invariants presented by the spectral sequence of that group extension. Lyndon's thesis included a partial anticipation of the notion of a spectral sequence. This note is a commentary on that anticipation, combined with observations as to why spectral sequences must arise in the situation of extensions and in the corresponding situations for fiber spaces.

At that time, the cohomology of groups was new. Eilenberg and Mac Lane [6], stimulated by a seminal paper of Hopf [10], had provided an algebraic formula to determine the cohomology groups of a space $K(G,1)$ with fundamental group G and vanishing higher homotopy groups. For such a space, the cohomology groups with coefficients in an abelian group A depended only upon A and the fundamental group G. They were thus called the cohomology groups of the group G:

$$H^n(G,A) = H^n(K(G,1),A) \ .$$

This dependence on G could be exhibited by taking a minimal singular complex of the "Eilenberg-Mac Lane" space $K(G,1)$. This complex has one zero cell, one 1-cell $[x]$ for each element x of the fundamental group G and n-cells $[x_1,\ldots,x_n]$, one for each list of n elements $x_i \in G$. These cells may be pictured by a singular simplex of $K(G,1)$ in which each vertex is

© 1984 American Mathematical Society
0271-4132/84 $1.00 + $.25 per page

mapped to the base point, while the edge from vertex $i-1$ to vertex i represents the element $x_i \in G$. Since all the higher homotopy groups vanish, the interior of the simplex can be filled in. These cells constitute the (reduced) "bar-resolution" $B(G)$. A singular n-cochain f for the group G is then a function $f\colon G^n \to A$, while its coboundary δf, calculated as the alternating sum of the $n+2$ faces on the boundary of a simplex $[x_1,\ldots,x_{n+1}]$, is given by the formula

$$\delta f(x_1,\ldots,x_{n+1}) = f(x_2,\ldots,x_{n+1}) + \sum_{i=1}^{n} (-1)^i f(x_1,\ldots,x_i x_{i+1},\ldots,x_{n+1}) + (-1)^{n+1} f(x_1,\ldots,x_n) \,. \tag{1}$$

For $n = 2$, the equation $\delta f = 0$ states that a 2-cocycle f is just a "factor set" for a central group extension of A by G, so that H^2 is the abelian group of all such extensions:

$$H^2(G,A) = \mathrm{Extcent}(G,A) \,.$$

Such group extensions (and the corresponding "crossed product algebras" were long known, as were the not necessarily central extensions of an abelian group A by the group G, as presented by a short exact sequence

$$0 \to A \to E \to G \to 1 \,.$$

In such cases conjugation makes the abelian group A into a left G-module; that is, a module over the integral groups ring $\mathbb{Z}(G)$. The higher dimensional cohomology groups $H^n(G,A)$ were also defined in this case, simply replacing the first term on the right-hand side of (1) by $x_1 f(x_2,\ldots,x_n)$, and so using the action of x_1 on A.

Beyond dimension 2, very few of these cohomology groups had been computed. For a free group F it was known that $H^n(F,A) = 0$ when $n \geq 2$, and the cohomology groups $H^n(G,A)$ had been computed when G is cyclic. Even the calculation in this case for large n was an elaborate one, as one may see by noting the Eilenberg-Mac Lane calculation made for this case and presented in §1 of Eilenberg [5]. The calculation started with the bar resolution $B(G)$ for G cyclic and reduced it term by term to a standard form suggested by the known treatment of cyclic group extensions. The quicker method of choosing a suitable special resolution of the (trivial) G-module $\mathbb{Z}$ was not then known -- such a study of resolutions in general apparently first appeared in 1948 in Cartan [3] and in 1949 in Eilenberg-Mac Lane [7]; and it was only then recognized that the reduced bar resolution $B(G)$ could be obtained by taking a free resolution of the trivial G-module $\mathbb{Z}$ and then "dividing out" the action of the group G.

Roger Lyndon had been an undergraduate student at Harvard, where he had studied logic with W. V. Quine. After leaving Harvard he found (but did not publish) the Rosser paradox in Quine's Mathematical Logic (first edition). He returned to Harvard for graduate work. For a period, he hoped to write a thesis extending the Stone representation theorem from Boolean algebras to algebras of relations. When the project seemed to progress too slowly, I suggested to Roger that he might instead take up the needed calculation of the cohomology groups of groups: It should be possible to determine this cohomology for a finitely generated abelian group G and perhaps for a group G with a given normal subgroup Γ -- in other words, for a group G which is given as an extension of Γ by $Q = G/\Gamma$. At about that time, a favorite new tool in topology (Kelly-Pitcher [11]) was the long exact cohomology sequence for a space and a subspace. Perhaps (I do not recall) I hoped that there could be a similar long exact sequence for group cohomology, involving the evident homomorphisms

$$H^n(Q,A) \to H^n(G,A), \quad H^n(G,A) \to H^n(\Gamma,A), \tag{2}$$

now called inflation and restriction, and induced respectively by the projection $G \to Q$ and the inclusion $\Gamma \to G$. If that was what I expected, I was wrong, as Lyndon demonstrated; wrong because what is really there is not a long exact sequence but a spectral sequence. Lyndon also did complete the suggested calculation for the cohomology of G with G a finite abelian group.

2. <u>The case of a direct product</u>. For a group G which is a direct product $G = Q \times \Lambda$, the cartesian product of the corresponding Eilenberg-Mac Lane spaces $K(Q,1) \times K(\Lambda,1)$ is readily seen to be a space $K(Q \times \Lambda,1)$. The well known Künneth formulas then allow the calculation of homology groups of the product space from homology groups of the factors. More explicitly, the Eilenberg-Zilber theorem (not published till 1950) shows that the singular chain complex of the product space is chain-equivalent to the tensor product of the chain complexes of the two factors. This equivalence yields a chain equivalence for the reduced bar constructions

$$B(G) \simeq B(Q) \otimes B(\Lambda) , \tag{3}$$

where the right-hand side is the tensor product of the two chain complexes with the usual boundary formula, while chains in dimension n are a direct sum of tensor products:

$$B_o(Q) \otimes B_n(\Lambda) + B_1(Q) \otimes B_{n-1}(\Lambda) + \ldots + B_n(Q) \otimes B_o(\Lambda) .$$

An n-dimensional cochain of this complex with values in the abelian group A is then a sum $f_o + \ldots + f_n$, where $f_p : Q^p \times \Lambda^q \to A$ and $q = n-p$ is called the "complementary" degree. If we write this cochain as $f = (f_o, \ldots, f_n)$ its coboundary is

$$\delta f \equiv (\delta'' f_o, \delta' f_o \pm \delta'' f_1, \ldots, \delta' f_{n-1} \pm \delta'' f_n, \delta' f_n) \tag{4}$$

where the choice of signs need not concern us, but where $\delta' f_p$ denotes the coboundary of f_p computed from the variables in Q only, while $\delta'' f_p$ is computed only on the variables in Λ. This analysis suggests that we consider in succession for $p = 0,1,\ldots,n$ those cocycles $f = (0,\ldots,0, f_p,\ldots,f_n)$ which have an f_p as the first non-zero term. Let $F^p \subset H^n(Q \times \Lambda, A)$ be the group of all those homology classes represented by such a cocycle. This defines a descending sequence of subgroups

$$H^n(G,A) = F^o \supset F^1 \supset F^n \supset \dots \supset F^n \supset 0 , \tag{5}$$

called a (descending) filtration of the cohomology group. Examine the quotient group F^p/F^{p+1}. Each cocycle f in filtration F^p has leading term a function $f_p : Q^p \times \Lambda^q \to A$ with

$$\delta'' f_p = 0 , \quad \delta' f_p = \pm \delta'' f_{p+1} \tag{6}$$

for some $f_{p+1} : Q^{p+1} \times \Lambda^{q-1} \to A$. The function f_p can be reinterpreted as a function $Q^p \to C^q(\Lambda,A)$ from p arguments in Q into the q-dimensional cochains of Λ. The first condition of (6) asserts that the values of this function are cocycles, in $Z^q(\Lambda,A)$, so that f_p thus determines a map

$$c: Q^p \to H^q(\Lambda,A) \tag{7}$$

and thus a p dimensional cochain of Q. The second condition of (6) then states that this cochain is a cocycle, which gives a cohomology class

$$\mathrm{cls}(c) \in H^p(Q,H^q(\Lambda,A)) .$$

The resulting correspondence $\mathrm{cls}(f) \mapsto \mathrm{cls}(c)$ might thus lead to a homomorphism

$$F^p/F^{p+1} \to H^p(Q,H^q(\Lambda,A)) . \tag{8}$$

This map may not be onto; also it may not be well-defined, since replacing f by a cohomologous cycycle may change the class of c.

Lyndon pursued the argument, although he did not use the Eilenberg-Zilber theorem for the product of spaces, preferring to use a direct algebraic argument giving a cochain equivalent of the reduction (4). In some cases the possible difficulties noted above for (8) can be avoided. In this way he proved:

Theorem 1. If Q, Λ and A are finitely generated abelian groups, with $G = Q \times \Lambda$ operating trivially on A, then the cohomology groups of the product can be written as a direct sum

$$H^n(Q \times \Lambda, A) \cong \sum_{p=o}^{n} H^p(Q, H^{n-p}(\Lambda, A)) . \tag{9}$$

For a general group extension, however, one cannot get a direct sum on the right in (9) but only a filtration of H^n in which the successive factor groups are not the groups $H^n(Q, H^{n-p}(\Lambda, A))$ but suitable subquotients (subgroups of quotient groups) of these groups.

3. The Complications. The difficulties with (8) above come from two sources. First, each cocycle $c \in Z^p(Q, H^q(\Lambda, A))$ determines a function $f_p: Q^p \times \Lambda^q \to A$ with $\delta'' f_p = 0$ and for which there is an f_{p+1} with $\delta' f_p = \delta'' f_{p+1}$, as in (6). However, this does not necessarily mean that f_p is the leading term of a cocycle $(f_p, f_{p+1}, \ldots)$ for F^p. If this is to be the case there must be, besides f_{p+1}, a sequence of

$$f_{p+2}, f_{p+3}, \ldots \text{ with}$$

$$\delta' f_{p+1} = \pm \delta'' f_{p+2} , \ \delta' f_{p+2} = \pm \delta'' f_{p+3}, \ldots \ . \tag{10}$$

Hence the usable cocycles c represent only a subgroup of $H^p(Q, H^q(\Lambda, A))$ -- that subgroup for which all of (10) is possible.

On the other hand, the lead term f_p in the cocycle in F^p can be altered by coboundaries, such as the coboundary δ_g of a cochain $g = (g_o, g_1, \ldots, g_{n-1})$. Then f_p is replaced by $f_p + \delta' g_{p-1} + \delta'' g_p$, where moreover the terms before p in δg vanish:

$$\delta'' g_o = 0 \ , \ \delta' g_o = \pm \delta'' g_1, \ldots, \ \delta' g_{p-2} = \pm \delta'' g_{p-1} \ . \tag{10'}$$

If f_p is replaced just by $f_p + \delta'' g_p$, the cochain c of (6) is unchanged. If f_p is replaced by $f_p + \delta' g_{p-1}$ where $g_{p-2} = 0$ and so $\delta'' g_{p-1} = 0$, then the cochain c is changed but the cohomology class $\text{cls}(c)$ is unchanged.

But when additional g_i with $i < p-1$ enter, the class of c may be changed, hence the correspondence $\mathrm{cls}(f_p) \mapsto \mathrm{cls}(c)$ carries the group F^p/F^{p-1} not into $H^p(Q,H^q(\Lambda,A))$ but into some one of its quotient groups -- dividing by a suitable kernel to record the possible change in $\mathrm{cls}(c)$ caused by the presence of such g_i.

All told, the best that can result is a homomorphism of F^p/F^{p-1} into a subquotient of $H^p(Q,H^q(\Lambda,A))$, plus a proof that this homomorphism is an isomorphism. The calculation of F^p/F^{p-1} thus remains in part ambiguous. This ambiguity, illustrated here for the tensor product complex $B(Q) \otimes B(\Lambda)$, is likely to occur in the corresponding calculation of the cohomology of any tensor product of two chain complexes and also of any "double complex" C_{pq} with commuting boundary operations $\partial' : C_{p,q} \to C_{p-1,q}$ and $\partial'' : C_{p,q} \to C_{p,q-1}$.

4. Cohomology of Group Extensions. By analogous means Lyndon considered the cohomology of a group extension

$$1 \to \Gamma \to G \to Q \to 1 \tag{11}$$

with coefficients in a G-module A. He showed that the cohomology $H^q(\Gamma,A)$ is naturally a Q-module, so that the cohomology groups $H^p(Q,H^p(\Gamma,A))$ with these operators are defined. He then proved

Theorem 2. For the cohomology of a group extension (11) there is for each dimension n a canonical sequence of subgroups

$$H^n(G,A) = F^o \supset F^1 \supset F^2 \supset \ldots \supset F^n \supset 0 ,$$

one for each $p = 0,\ldots,n$ and an isomorphism (where $q = p-n$)

$$F^p/F^{p-1} \cong \text{subquotient of } H^p(Q,H^q(\Gamma,A)) . \tag{12}$$

For $p = 0$, the right-hand side is just a subgroup; for $p = n$, it is a quotient group.

In fact, the isomorphism (12) for $p = 0$ is essentially the restriction homomorphism, while for $p = n$ it is essentially the inverse of the inflation (see Mac Lane [23], Prop. XI 10.2).

The proof of this result cannot be achieved by calculations with a tensor product of complexes. Instead, Lyndon constructed a filtration F^p of the cocycles of the bar resolution $B(G)$, stating that a cocycle f belongs to F^p if and only if one has

$$f(x_1\lambda_1,\dots,x_p\lambda_p,x_{p+1},\dots,x_n) = f(x_1,\dots,x_p,\dots,x_n)$$

for all elements $x_i \in G$ and $\lambda_i \in \Gamma$; in other words, F^p consists of just those cocycles f which in the first p arguments depend only on Q, not on G. (Today [2] we use the same filtration, extended also to cochains.) To handle this filtration, he chose representatives w of Q in G; that is, he chose a function $w: Q \to G$ which is a section of the projection $G \to Q$. He then defined a cochain f to be <u>normal</u> if and only if, for all $0 \leqq r < s < n$, it satisfies the condition

$$f(w_1,\dots,w_r,\lambda_{r+1},\dots,\lambda_s,w_{s+1},x_{s+2},\dots,x_n) = 0$$

for all representatives w_i, for all $\lambda_j \in \Gamma$, and for all $x_k \in G$. An elaborate calculation then proved that every cochain is cohomologous to a normal cochain. The filtration for these normalized cochains,

$$F^o \supset F^1 \supset F^2 \supset \dots \supset F^p \supset \dots \supset F^n \supset 0 ,$$

then leads gradually to a proof of the main Theorem 2. This argument using normalization might be viewed as a wide extension of the earlier Eilenberg-Mac Lane calculation for the cohomology of a cyclic group -- which again used a representative of a generator of the cyclic group.

Lyndon then used this result and these techniques to calculate $H^n(G,A)$ in many special cases -- more than meeting the original objectives of the thesis. However, he found the general Theorem 2 somewhat unsatisfactory.

In his thesis [19] he says:

"(i) This factorization no longer gives $H^n(G,A)$ as a direct product (sum) but only as the result of a succession of group extension (of F_p/F_{p+1} by F_{p+1}/F_{p+2}, etc.) . . . since the nature of these extensions is not fully specified, the group composition for $H^n(G,A)$ is not fully determined by that of the $H^p(G,H^q(\Gamma,A)$."

"(ii) The factors obtained are no longer the groups $H^R(G,H^q(\Gamma,A))$ themselves, but the derivative groups (the subquotients) . . . ."

When Lyndon came to publish the results of his thesis, the published paper [20] gave all the particular calculation, with proofs, stated the Theorem 2 above (there Theorem 4') but omitted its proof, with the words

"Thus analysis of the present type, although it has proven useful in particular instances, can hardly be expected to yield any strong general theorems."

The thesis problem had been: Calculate some cohomology groups $H^n(G,A)$. This led inevitably to a Theorem (Theorem 2) which is clearly <u>not</u> a calculation. It seemed to be a disappointing outcome, except for the case of direct products. (In 1975 Beyl [1] proved that the subquotients are not needed when G is the direct product $Q \times \Lambda$, operating trivially on A.)

5. <u>Leray and Spectral Sequences</u>. In the fall of 1947, Jean Leray became a Professor at the College de France: For his inaugural series of lectures 1947-48 he presented his new ideas on the spectral and the filtered ring of a continuous application -- in other words on spectral sequences and the spectral sequence of a filtered ring. There had been a préliminary publication [13, 14] in 1946 and a definitive publication [16] in 1950. Leray dealt with the cohomology <u>ring</u> of a fibre space $E \to B$ and its relation to the cohomology rings of the base B and the fibre F. The projection of the fibre space naturally induces a filtration $F^o \supset \ldots \supset F^p \supset \ldots$ on the cohomology of E, and so inevitably leads to calculations like those forced on Lyndon. It so happens that I was in Paris in the late fall of 1947 and

heard the first of Leray's lectures (and talked with him at some length). I did not recognize the connection with Lyndon's work. Indeed, I did not at that time recognize the deep connection between topological fibrations and algebraic ones such as the projection $G \to Q$ of a group extension or the corresponding fibrations $L(\pi,n) \to K(\pi,n)$ which arose in the study of the general Eilenberg-Mac Lane spaces.

Leray had explicitly recognized that more than calculation was necessary. He saw that it was plainly impossible to express the cohomology ring of a fibre space E directly in terms of the cohomology rings of base and fibre. One must proceed, more indirectly, by spectral sequences. In Paris, Henri Cartan took up the idea with enthusiasm, Koszul published a note [12] codifying the algebraic aspects of spectral sequence and Borel and Serre took interest. Shortly after the publication in 1948 of Lyndon's results, Serre took note of them, wrote a note [26] in the <u>Comptes Rendus</u> constructing a spectral sequence for the cohomology of a group extension, saying "One notes the analogy between this proof and that used by Lyndon to demonstrate the Theorem 4' of his paper." (This is our Theorem 2 above; it is the one which Lyndon did not demonstrate in his published paper.)

6. <u>What is spectral</u>? What does the spectral sequence add to Theorem 2 above? It provides a mechanism which shows <u>in part</u> how the subquotients involved can be determined. We had observed above that subgroups are necessary in order to take into account the successive functions $f_{p+1}, f_{p+2}, \ldots$ which may be required to make the "leading term" f_p into a complete cocycle of $Q \times \Lambda$. We also noted that the quotients are necessary in order to take into account the coboundary terms $g_{p-1}, g_{p-2}, \ldots$ which may modify this leading term f_p. If one takes these <u>two</u> changes into account step by step and together one constructs (c.f. <u>Homology</u> [23], Chap. XI) a whole sequence of subquotients

$$H^p(Q,H^q(\Lambda,A)) = E_2^{p,q} > E_3^{p,q} > E_4^{p,q} > \ldots \tag{13}$$

where $E > E'$ means that E' is a subquotient of E. <u>Moreover</u> one constructs for each $r = 2,3,\ldots$ a "differential" which is a homomorphism

$$d_r : E_r^{p,q} \to E_r^{p+r,q-r+1} \tag{14}$$

with the usual property $d_r d_r = 0$ of a differential. We will not trouble to record here the exact (and well-known) formulas (see e.g. Mac Lane [24] except to say that each d_r is essentially induced by the original coboundary operator δ on cochains. Moreover, and this is the crux, one shows that for each r, the next subquotient $E_{r+1}^{p,q}$ is the cohomology $H(E_r, d_r)$; that is, can be expressed as kernel modulo image in the (non-exact) sequence

$$E_r^{p-r,q+r-1} \xrightarrow{d_r} E_r^{p,q} \xrightarrow{d_r} E_r^{p+r,q-r+1}$$

of differentials d_r. In other words the successive intermediate terms E_r are <u>all</u> of the form "cocycles modulo coboundaries." This allows for a "homology like" expression of the final subquotient -- since, for given values of p and q, only a finite number of steps are required in (13).

Thus Lyndon's Theorem 2 becomes (a little) more like a calculation when the whole of the spectral sequence is included.

The rest of the story is well known. Serre's famous thesis (1951) showed that in topology the spectral sequences could be used very effectively to calculate homotopy groups. Shortly thereafter Hochschild and Serre [9] constructed several spectral sequences for group extensions, using both Lyndon's filtration, a modification of Lyndon's filtration and a different filtration obtained by using more general resolutions. In particular, when the coefficient module A is a ring the cohomology cup product gives $H^*(G,A)$ a ring structure which carries over to the whole spectral sequence. In particular, the differentials d_r behave properly for this ring structure, which makes them a more powerful means of partial calculation. For some years it was unclear whether the several different filtrations used by Hochschild-Serre really gave the same spectral sequences, or simply different spectral sequences which "look alike" in the sense that they have the same terms E_2 and the same limit. It has now been definitely proved (Evens [8], Beyl [2] and in not yet published work of Donald W. Barnes) that they are all the same. It is the Lyndon-Hochschild-Serre sequence.

It is still true that the spectral sequence does not "calculate" $H^n(G,A)$. As Cartan was fond of saying, it gives first the terms $H^n(Q,H^q(\Lambda,A))$ -- no, just a little bit less (E_3) -- no, again just a little bit less, and so on. The differentials which express this are themselves subtle invariants of the situation, as exhibited in a few cases (other would be of interest) in a recent paper by Ratcliffe [25]. In the meantime, cohomology led Lyndon in [21] and [22] to the cohomology of groups with one defining relation and then the many other problems in combinatorial group theory, while books on spectral sequences [18] can be quite extensive.

REFERENCES

[1] Beyl, F. R. Abelian groups with a vanishing homology group. Journ. Pure Appl. Alg. 7 (1976), 175-193.

[2] Beyl, F. R. The spectral sequence of a group extension. Bull. Sci. Math. 2e series 105 (1981), 407-434.

[3] Cartan, H. Sur la cohomologie des espaces où opère un groupe. Notions Algébriques préliminaries. CR Acad. Sci. Paris 226 (1948), 148-150.

[4] Cartan, H. Séminaire Henri Cartan 3e Année 1950/51 Cohomologie des Groupes, Suite Spectrale, Faisceaux. Sec. Math. 11 Rue Pierre Curie Paris, 1957.

[5] Eilenberg, S. Topological methods in abstract algebra. Cohomology Theory of Groups Bull. A.M.S. 55 (1949), 3-37.

[6] Eilenberg, S. and Mac Lane, S. Relations between the homology and homotopy groups of spaces. Ann. of Math. 46 (1945), 480-509.

[7] Eilenberg, S. and Mac Lane, S. Homology of spaces with operators II. Trans. A.M.S. 65 (1949), 49-99.

[8] Evens, L. The spectral sequence of a group extension stops. Trans. A.M.S. 212 (1975), 269-277.

[9] Hochschild, G. P. and Serre, J. P. Cohomology of group extensions. Trans. A.M.S. 74 (1953), 110-134.

[10] Hopf, H. Fundamentalgruppe und zweite Bettische Gruppe. Comment. Math. Helv. 14 (1941/42), 257-309.

[11] Kelley, J. L. and E. Pitcher. Exact homomorphism sequences in homology. Theory. Ann. of Math. (2) 48 (1947), 682-709.

[12] Koszul, J-L. Sur les opérations de dérivation dans un anneau. C. R. Acad. Sci. Paris 225 (1947), 217-219.

[13] Leray, J. L'anneau d'homologie d'une représentation. C. R. Acad. Sci. Paris 222 (1946), 1366-1368.

[14] Leray, J. Structure de l'anneau d'homologie d'une représentation. C. R. Acad. Sci. Paris 222 (1946), 1419-1422.

[15] Leray, J. L'homologie filtrée, Topologie Algébrique, pp. 61-82, Colloques Int. du Centre Nat. de la Recherche Sci. No. 12, CNRS, Paris, 1949.

[16] Leray, J. L'anneau spectral et l'anneau filtré d'homologie d'un espace localement compact et d'une application continue. J. Math. Pures Appl. (9), 29 (1950), 1-139.

[17] Leray, J. P. L'homologie d'un espace fibré dont la fibre est connexe. Journal de Math. Pures et Appliqués (9), 29 (1950), 169-213.

[18] Lubkin, S. Cohomology of completions. North Holland Mathematical Studies 42, 802 p. North Holland Publishing Co., Amsterdam, New York, 1980.

[19] Lyndon, R. C. The cohomology theory of group extensions. Ph.D. Thesis, Harvard University, May 18, 1946.

[20] Lyndon, R. C. The cohomology theory of group extensions. Duke Math. J. 15 (1948), 271-292.

[21] Lyndon, R. C. New proof of a theorem of Eilenberg and Mac Lane. Ann. of Math. (2), 50 (1949), 731-735.

[22] Lyndon, R. C. Cohomology theory of groups with a single defining relation. Ann. of Math. (2), 52 (1950), 650-665.

[23] Mac Lane, S. Homology. Springer-Verlag, 422 pp. Berlin, Göttingen, Heidelberg, 1963.

[24] Mac Lane, S. Origins of the cohomology of groups. L'Enseignement Math. 24 (1978), 1-29.

[25] Ratcliffe, J. G. On the second transgression of the Lyndon-Hochschild-Serre spectral sequence. J. of Alg. 61 (1979), 593-598.

[26] Serre, J. P. Cohomologie des extensions de groupes. C. R. Acad. Sci. Paris 23 (1950), 653-646.

DEPARTMENT OF MATHEMATICS
UNIVERSITY OF CHICAGO
CHICAGO, ILLINOIS 60637

Contemporary Mathematics
Volume 33, 1984

LYNDON'S CONTRIBUTION TO COHOMOLOGY OF GROUPS

John G. Ratcliffe

Roger Lyndon's contribution to cohomology of groups began with his 1946 Harvard Ph.D. thesis in which he laid the groundwork for the development of the spectral sequence relating the cohomology of a group extension to the cohomology of its factors. This spectral sequence has come to be called the Lyndon-Hochschild-Serre spectral sequence in honor of Lyndon's thesis and the fundamental paper of Hochschild and Serre [6] in which the spectral sequence was first defined. The LHS-spectral sequence is a basic tool for computation and Lyndon used its machinery in his thesis to compute the cohomology of a finitely generated abelian group. In another essay in this volume, Lyndon's thesis advisor, Saunders Mac Lane, describes Lyndon's thesis and its historical significance in more detail.

Lyndon's next contribution to cohomology of groups was his 1948 Annals of Mathematics paper [7] in which he gave a short proof of the cup product reduction theorem of Eilenberg and Mac Lane. This paper is an early example of Lyndon's ability to give short, elegant proofs of difficult, complicated theorems. Other examples are his proof of Grushko's Theorem [10], Dehn's algorithm [11], and the Lyndon-Higgins proof of J. H. C. Whitehead's algorithm [5].

In 1950 Lyndon published his famous Annals of Mathematics paper "Cohomology theory of groups with a single defining relation" [8]. In this fundamental paper he computed the cohomology of a one-relator group. Although the formulas given for the cohomology of a one-relator group are beautiful in themselves, the most significant contribution of the paper is not the end

© 1984 American Mathematical Society
0271-4132/84 $1.00 + $.25 per page

result but the method introduced to make the computation. Lyndon's method was to construct a resolution that is as simple as possible and is completely natural.

Let $(x_1,\ldots,x_m; r_1,\ldots,r_n)$ be a finite presentation for a group G. Lyndon constructed an exact sequence of free G-modules

$$\cdots \longrightarrow F_i \xrightarrow{\partial_i} F_{i-1} \longrightarrow \cdots \longrightarrow F_2 \xrightarrow{\partial_2} F_1 \xrightarrow{\partial_1} F_0$$

such that F_0 has rank one, F_1 has rank m, F_2 has rank n, the homomorphism ∂_1 has the matrix $(x_i - 1)$, and ∂_2 has the matrix $(\partial r_i/\partial x_j)$ defined by the Fox derivatives $\partial r_i/\partial x_j$. This sequence is called the Lyndon resolution and it has a natural topological interpretation.

Lyndon tells the story that he was a young instructor at Princeton when he presented his cohomology theory of a one-relator group in a seminar. Kurt Reidemeister, who was visiting the Institute for Advanced Study, was in the audience. After the seminar, Reidemeister explained to Lyndon that his resolution could be given a topological interpretation in terms of incidence matrices of a cellular chain complex.

Today this is a very familiar construction. First, start with the cell complex K^2 which models the presentation of G. Thus, K^2 has one zero-cell, m one-cells, and n two-cells attached by spelling out the relators. Then attach higher dimensional cells to K^2 to obtain an aspherical cell complex K. The cellular chain complex of the universal cover of K is naturally isomorphic to the Lyndon resolution.

Because of this natural topological interpretation, Lyndon's resolution has become an extremely useful tool in the computation and interpretation of low-dimensional homology and cohomology of groups and topological spaces. For example, one can show that the Alexander polynomial is a knot invariant by using the Lyndon resolution to compute the first homology of the universal abelian cover of the complement of a knot.

Lyndon gratefully acknowledged Reidemeister's topological insight on page 651 of his paper. Even though he first saw his resolution algebraically, Lyndon immediately appreciated the topological interpretation of what he had done. In fact, using geometric methods to interpret algebraic problems has been a recurring theme in his work from that point on.

Let $(x_1,\ldots,x_m; r)$ be a one-relator presentation for a group G with r non-trivial. Without loss of generality we may assume that r is cyclically reduced. Then there is a maximal integer q such that $r = s^q$, and the element s is unique. The form of the presentation of G determines the start of the Lyndon resolution

$$F_2 \xrightarrow{\partial_2} F_1 \xrightarrow{\partial_1} F_0 .$$

In order to define ∂_3, one only needs to choose a set of generators for the kernel of ∂_2. The main theorem of Lyndon's cohomology theory of a one-relator group states that the kernel of ∂_2 is generated by the element $s-1$. This is a restatement of Lyndon's Identity Theorem. The proof is by induction on the length of r and follows the general outline of Magnus's proof of the Freiheitssatz.

Lyndon then defines ∂_i for $i > 2$ as follows. If $q = 1$ then $s = 1$ in G, whence ∂_2 is a monomorphism and we may take $F_i = 0$ and $\partial_i = 0$ for all $i > 2$. If $q > 1$, then we may take F_3 to have rank one and define ∂_3 to be left multiplication by $s-1$. This gives exactness at F_2. An easy argument shows that the kernel of ∂_3 is generated by $t = 1 + s + \ldots + s^{q-1}$. Thus we may take F_4 to have rank one and define ∂_4 to be left multiplication by t. This gives exactness at F_3. Another easy argument shows that the kernel of ∂_4 is generated by $s-1$. Now we may take F_i to have rank one for all $i > 2$ and define $\partial_i = \partial_{i-2}$ for $i > 4$.

A beautiful application of Lyndon's Identity Theorem is the following topological theorem of Cockcroft [3].

Theorem: A finite two-dimensional cell complex K^2 with only one two-cell is aspherical if and only if the attaching map of the two-cell does not represent a proper power in the fundamental group of K^1.

This theorem generalizes the classical theorem that a closed surface is aspherical if and only if it is not a sphere or projective plane.

Cockcroft's Theorem follows almost immediately from the Identity Theorem and the topological interpretation of Lyndon's resolution. Let K^2 be a finite two-dimensional cell complex with only one two-cell. The cell structure of K^2 determines a one-relator presentation $(x_1,\ldots,x_m; r)$ for the fundamental group of K^2; moreover, the class of the attaching map of the two-cell in the fundamental group of K^1 corresponds to r. Now the process of attaching higher dimensional cells to K^2 to obtain an aspherical cell complex corresponds to the process of extending Lyndon's resolution beyond F_2. Since we can have $F_i = 0$ for all $i > 2$ if and only if r is not a proper power, K^2 is aspherical to begin with if and only if the attaching map of the two-cell does not represent a proper power in the fundamental group of K^1.

Cockcroft's theorem is actually equivalent to Lyndon's Identity Theorem. In [4], Dyer and Vasquez gave a topological proof of Cockcroft's Theorem and derived Lyndon's Identity Theorem as a corollary.

The ideas in Lyndon's fundamental paper were seminal to his later work in group theory (described in another paper in this volume by Paul Schupp). The Identity Theorem and the topological interpretation of his resolution led naturally to small cancellation theory. His combinatorial proof of the Identity Theorem developed his interest in combinatorial group theory and led naturally to his work on the theory of equations in groups.

From this point on, Lyndon's research in group theory moved towards combinatorial group theory; however, he continued to use homological techniques to study groups. In a 1955 American Journal of Mathematics paper [1], M. Auslander and Lyndon used the cup product reduction theorem to study normal subgroups of free groups. In a 1958 Annals of Mathematics paper [2],

K. T. Chen, R. Fox, and Lyndon derived an algorithm to compute the quotients of the successive terms of the lower central series of a finitely presented group. Lyndon applied this algorithm in his work on Burnside's problem [9].

Eventually, homological techniques became just one tool in a powerful arsenal of techniques Lyndon developed to study groups.

REFERENCES

1. M. Auslander and R. C. Lyndon, Commutator subgroups of free groups, Amer. J. Math. 77 (1955), 929-931.
2. K. T. Chen, R. H. Fox, and R. C. Lyndon, Free differential calculus, IV. The quotient groups of the lower central series, Ann. Math. 68 (1958), 81-95.
3. W. H. Cockcroft, On two-dimensional aspherical complexes, Proc. London Math. Soc. 4 (1954), 375-384.
4. Eldon Dyer and A. T. Vasquez, Some small aspherical spaces, J. Austral. Math. Soc. 16 (1973), 332-352.
5. P. J. Higgins and R. C. Lyndon, Equivalence of elements under automorphisms of a free group, J. London Math. Soc. 8 (1974), 254-258.
6. G. Hochschild and J-P. Serre, Cohomology of group extensions, Trans. Amer. Math. Soc. 74 (1953), 110-134.
7. R. C. Lyndon, New proof for a theorem of Eilenberg and Mac Lane, Ann. Math. 50 (1949), 731-735.
8. Roger C. Lyndon, Cohomology theory of groups with a single defining relation, Ann. Math. 52 (1950), 650-665.
9. R. C. Lyndon, On Burnside's Problem, Trans. Amer. Math. Soc. 77 (1954), 202-215.
10. R. C. Lyndon, Grushko's Theorem, Proc. Amer. Math. Soc. 16 (1965), 822-826.
11. R. C. Lyndon, On Dehn's algorithm, Math. Annalen 166 (1966), 208-228.

DEPARTMENT OF MATHEMATICS
UNIVERSITY OF ILLINOIS
URBANA, ILLINOIS 61801

Contemporary Mathematics
Volume 33, 1984

LYNDON'S RESEARCH IN MATHEMATICAL LOGIC

H. Jerome Keisler

Roger Lyndon has made pioneering contributions to two areas of mathematical logic, algebraic logic and model theory.

Algebraic logic began with the study of Boolean algebras in the nineteenth century. More recently, relation algebras, cylindric algebras, and polyadic algebras have been introduced as algebraic approaches to logics with individual variables and quantifiers. Lyndon worked on the earliest of these, relation algebras, which correspond to first order logic with binary relations and individual variables but no quantifiers. A relation algebra was defined by Tarski in 1946 as a Boolean algebra with two extra operations, composition and inverse, and one extra constant, the identity relation, which satisfies a set of five natural axioms. A relation algebra is representable if it is isomorphic to a set of relations such that the boolean and relation algebraic operations have the natural meanings. The representable relation algebras are the intended interpretations of the axioms. By the Stone representation theorem, every Boolean algebra is representable. On the other hand, Bjarni Johnson gave an example of a non-representable relation algebra. The main questions which Lyndon considered in his three papers in relation algebras of 1950, 1956, and 1961 are: "Which relation algebras are representable?" "Are there natural examples of non-representable relation algebras?"

In his first paper in 1950, "The representation of relation algebras," Lyndon gave a characterization of the finite representable relation algebras. In fact, he found an explicit set C of sentences of first order logic such

© 1984 American Mathematical Society
0271-4132/84 $1.00 + $.25 per page

that a finite relation algebra is representable if and only if it is a model of C. He also gave an example of a finite relation algebra which is not representable. Later, Tarski gave a set of sentences (in fact, equations) in first order logic which characterizes the set of all representable relation algebras. In a sequel to his 1950 paper, published in 1956, Lyndon showed that Tarski's result could not be extended to strongly representable relation algebras, that is, relation algebras representable in such a way that the identity element is a union of atoms. He proved that the set of strongly representable relation algebras cannot be characterized by a set of universal sentences of first order logic. Lyndon continued his work on relation algebras in the 1961 paper "Relation algebras and projective geometries." Each projective geometry G induces a relation algebra A(G) in a natural way. He proved that A(G) is representable whenever G has dimension greater than two, and provided elegant characterizations of the representability of A(G) when G has dimension one or two. If G is of dimension two, A(G) is representable if and only if G is Desarguesian, and if G is of dimension one, A(G) is representable if and only if there is a projective plane with the same order as G. This yields some very natural examples of relation algebras which are not representable. The simplest example is A(G) where G is a line with seven points.

Lyndon's most important contributions to mathematical logic are in the area of model theory, particularly the two 1959 papers, "An interpolation theorem in the predicate calculus," and "Properties preserved under homomorphism." The first paper contains the celebrated Lyndon interpolation theorem, an improvement of the Craig interpolation theorem which keeps track of positive and negative occurrences. The result states that if S and T are first order sentences such that S implies T, then there is a sentence M (a Lyndon interpolant of S and T) such that S implies M, M implies T, every relation symbol which occurs positively in M occurs positively in both S and T, and every relation symbol which occurs negatively in M occurs negatively in both S and T. The theorem itself belongs to the area known as proof theory, but it has important applications to model theory. The best known application is the Lyndon homomorphism theorem, in the second paper. This theorem, in its simplest form, states that a sentence of first order logic is preserved under homomorphic images if and only if it is equivalent to a positive sentence (a sentence whose only connectives are "and" and "or"). This result is perhaps the most striking example of a family of results in model theory known as "preservation theorems." It is easy to show by induction on sentences that every positive sentence is preserved under homomorphic images. There are many familiar examples of this phenomenon in algebra, such as the axioms for groups, rings, and lattices. The other direction of the theorem is a deep result which Lyndon proved using his interpolation theorem. He also showed, by a method which has later proved to be widely applicable to other preservation problems, that it is undecidable whether a given first order sentence is equivalent to a positive sentence.

Two other papers published in 1959, "Properties preserved under subdirect products" and "Existential Horn sentences," contain preservation theorems which concern products of algebras. Both papers go back to the result of A. Horn that every sentence of a certain form is preserved under direct products. These sentences, called Horn sentences, are obtained as follows. A basic Horn formula is a formula of the form $A \rightarrow B$ where A is a finite conjunction of atomic formulas and B is either atomic or negated atomic. A sentence obtained from basic Horn formulas by conjunctions and quantifiers is called a Horn sentence. Chang and Morel had shown that the converse of Horn's result is false by giving an example of a sentence which is preserved under direct products but is not equivalent to a Horn sentence. In the paper on subdirect products, Lyndon introduced the notion of a special Horn sentence, and proved that a first order sentence is preserved under subdirect products if and only if it is equivalent to a special Horn sentence. In the paper on existential Horn sentences, Lyndon proved that a first order sentence is preserved under extensions and direct products if and only if it is equivalent to a Horn sentence all of whose quantifiers are existential. Thus the converse of Horn's result is true for existential sentences. (Much progress has been made in the area since 1959. For example, the reviewer proved a preservation theorem for reduced products and Horn sentences, and J. Weinstein proved a preservation theorem for direct products and a larger class of sentences.)

Lyndon has also written some excellent expository articles in mathematical logic. The paper "Properties preserved under algebraic constructions" is based on an invited lecture to the meeting of the American Mathematical Society in Evanston in November 1958, and contains a survey of preservation theorems in model theory. "Notes in Logic," 1966, is a monograph aimed at the advanced undergraduate or beginning graduate level. It very quickly introduces the reader to some of the central ideas in logic, particularly model theory and proof theory.

DEPARTMENT OF MATHEMATICS
UNIVERSITY OF WISCONSIN
MADISON, WISCONSIN 53706

Contemporary Mathematics
Volume 33, 1984

SOME ASPECTS OF LYNDON'S WORK ON GROUP THEORY

Paul E. Schupp

There is a footnote after the title of the paper in which John Milnor [1968] introduced the idea of the rate of growth of a finitely generated group. The footnote says "The connection of this idea with random walks on topological groups was pointed out to the author by R. C. Lyndon." I think that this footnote is a good illustration of Lyndon's broad range of interests, his insightful grasp of very diverse areas of mathematics and his clarity of vision in seeing the ramifications of an idea. I would like to discuss some of Lyndon's numerous contributions to group theory. A logical starting point, both chronologically and in terms of its importance, would be Lyndon's contributions to cohomology theory but these are discussed in the articles by Mac Lane and Ratcliffe.

One of Lyndon's major contributions to infinite group theory was to the development of "small cancellation theory." When Max Dehn [1912] first posed the word and conjugacy problems for finitely generated groups he solved these problems for fundamental groups of closed orientable surfaces. A feature of the standard presentation $G_k = \langle a_1, b_1, \ldots, a_k, b_k; [a_1, b_1] \ldots [a_k, b_k] \rangle$ of the fundamental group of the orientable surface of genus k is that it has a single defining relator r with the property that if s is any cyclic conjugate of r or r^{-1} with $s \neq r^{-1}$ then there is very little cancellation in forming the product rs. V. A. Tartakovskii [1949] initiated an algebraic study of groups having a presentation satisfying a similar "small cancellation" condition. Tartakovskii's work was clarified by Greendlinger [1960] and

© 1984 American Mathematical Society
0271-4132/84 $1.00 + $.25 per page

J. Britton [1956] investigated a similar situation but dealt with quotients of free products instead of quotients of free groups. The subject remained incredibly complicated, however, and the cited papers of Tartakovskii, Greendlinger and Britton present a total of one hundred fifty pages of very detailed and difficult argument.

Lyndon's paper [1966c] "On Dehn's Algorithm" completely transformed the foundations of the subject. It is typical of much of Lyndon's work that his paper was simultaneously a considerable advance forward, a great simplification and a connecting back to earlier work. Dehn solved the word problem for a surface group G_k by making use of the fact that the entire Cayley group Γ of G_k is the dual graph of a regular tessellation of the hyperbolic plane and thus Γ is embedded in the hyperbolic plane. A non-trivial freely reduced word w is equal to the identity in G_k if and only if w is the label on a closed path δ in Γ. Dehn calculated the curvature around the curve δ and concluded that w must contain more than half of a cyclic permutation of the defining relator or its inverse. This fact gives Dehn's Algorithm for solving the word problem. If a non-trivial word w can be written $w \equiv w_1 t w_2$ where there is a cylic permutation $s \equiv tu$ of the relator or its inverse and $|t| > |u|$ then $w = w' \equiv w_1 u^{-1} w_2$ in G_k and w' is a shorter word. Finally, w is equal to the identity in G_k if and only if w is reducible to the empty word by a sequence of such replacements.

The Cayley graph of a presentation $G = \langle X;R \rangle$ is a complicated object and is planar only in very special cases. From now on we shall assume that all sets R of defining relators are "symmetrized", that is, if $r \in R$ then r is cyclically reduced and all cyclic permutations of $r^{\pm 1}$ are also in R. Lyndon's approach starts with the concept of a "cancellation diagram" which is a "local slice" of the Cayley graph that is small enough to always be planar. If a word w represents the identity in $G = \langle X;R \rangle$ then there is a finite, planar, connected and simply connected diagram M with each edge labelled by an element of the free group $F = \langle X \rangle$ and such that reading the label around the boundary of each region of M yields an element of R while the label on the boundary of the entire diagram M is the word w. Lyndon arrived at the concept of this geometric framework after thinking about the work of Tartakovskii and Greendlinger. Interestingly, it turns out that van Kampen [1933] had had the same idea thirty-three years earlier but his paper was completely neglected until it was used by C. M. Weinbaum [1966] about the same time as Lyndon's paper. Ironically, van Kampen's paper occurs immediately next to his very famous paper in which he gives a proof of the Seifert-van Kampen Theorem.

The geometric framework of cancellation diagrams immediately makes clear the meaning of the small cancellation hypotheses. The hypotheses directly impose conditions on the degrees of interior regions and vertices in the diagrams constructed. It turns out that there are three minimal hypotheses which allow one to obtain results -- the hypotheses being in one-to-one correspondence with the three regular tessellations of the Euclidean plane. Lyndon's proof that under a sufficiently strong hypothesis Dehn's Algorithm solves the word problem is completely in the spirit of Dehn's original argument. Given a cancellation diagram one computes a certain discrete sum which is the exact analogue of the curvature around the boundary. From the fact that this sum cannot be too small one concludes that a non-trivial word which is equal to the identity must contain more than half of a defining relator. The ideas involved have now come full circle with the proof in the general abstract algebraic case reduced to the combinatorial principles underlying Dehn's original geometric argument.

After Lyndon's paper, small cancellation theory grew rapidly with the geometric approach serving as the foundation of the subject. The scope of the theory was widened in two ways. On the one hand, questions concerning conjugacy, torsion and commuting elements involved considering more general types of diagrams -- not necessarily simply connected or even planar. On the other hand, "small cancellation quotients" of groups with natural length functions, such as free products with amalgamation and HNN extensions, were considered. Applications have ranged from very general embedding theorems to settling questions about knot groups.

One aspect of the Dehn's Algorithm paper which went largely unnoticed until Lyndon himself returned to the subject in [LS 177] was the connection between the idea of diagrams on spheres and the structure of the relation module for a presentation. The crucial point is the fact that if every spherical diagram for a presentation is "reducible" -- that is, it can be broken down into simpler spherical diagrams -- then the relation module is a direct sum of permutation modules involving certain canonical finite cyclic subgroups. (It should be noted that the argument given in [LS 1977] requires some elaboration. See Chiswell, Collins and Huebschmann [1981].)

This train of thought looks back to the work on cohomology where Lyndon determined the structure of the relation module of a one-relator group. The Dehn's Algorithm paper shows that every spherical diagram is reducible for a small cancellation presentation. One interesting consequence of the connection between spherical diagrams and relation modules is the following. A theorem of Serre gives a classification of torsion elements in groups whose cohomology in dimensions higher than two is the same as the cohomology of a direct sum of cyclic groups. It follows that the only elements of finite order than can occur in small cancellation groups are conjugates of the roots of the relators.

Lyndon calls a presentation $\langle X;R\rangle$ "aspherical" if every spherical diagram for the presentation is reducible. Now this definition is not the same as saying that the standard 2-complex $K(X;R)$ which realizes $\langle X;R\rangle$ is topologically aspherical. If a presentation $\langle X;R\rangle$ is "Lyndon aspherical" then $K(X;R)$ is topologically aspherical but there are counterexamples to the converse direction. Lyndon's idea has recently attracted attention from topologists interested in the long-standing problem of J. H. C. Whitehead which asks whether or not every subcomplex of an aspherical 2-complex is also aspherical. Lyndon's concept has given a method of constructing many aspherical 2-complexes all of whose subcomplexes are aspherical and has suggested methods for analyzing the kinds of homotopy involved in showing that a singular complex is trivial. For a detailed discussion of these ideas see [CCH 1981].

These considerations are also related to the Cohen-Lyndon Theorem [CL 1963] showing that the normal closure N is a free group F of a single relator r has a free basis consisting of conjugates of r. A similar result holds for the normal closure of a subset R satisfying a suitable small cancellation condition.

Before leaving the subject of cancellation diagrams mention must be made of their use in Lyndon's paper [1972] "On the Freiheitssatz" in which he gave an essentially new proof of the famous Magnus Freiheitssatz which states that, in a one relator group G, if one excludes a generator occurring in the defining relator then the remaining generators freely generate a free subgroup of G. Lyndon's proof is very elegant and the arguments are reminiscent of arguments from complex function theory. In the diagram approach the Freiheitssatz becomes a "maximum modulus principle." One establishes this principle by induction on the number of regions in the diagram and reduction to the previous case is achieved by using a certain "potential function" defined on the diagram.

Connecting a rather abstract algebraic setting to classical arguments involving analysis or topology is a persistent theme in Lyndon's work. The work on Fuchsian complexes [1974b, 1976] clarifies the essentially combinatorial nature of several arguments about Fuchsian groups, particularly the Riemann-Hurwitz formula. This work is in turn somewhat related to the general investigation of cohomology and Euler characteristics of groups. (See Chiswell [1976].) The papers of Lyndon and Ullman [1967, 1968, 1969] on the generation of a free group or a free product by a pair of unimodular matrices with complex entries give a considerable refinement of the well-known result that the group $G(\lambda)$ of matrices generated by $A = \begin{pmatrix} 1 & \lambda \\ 0 & 1 \end{pmatrix}$ and $B = \begin{pmatrix} 1 & 0 \\ \lambda & 1 \end{pmatrix}$, where $\lambda \in \mathbb{C}$ and $|\lambda| \geq 2$, is freely generated by A and B.

The recent work of Lyndon and Brenner [BLa,b,c] is also in the same general vein. An element of the modular group M is "parabolic" if it has a single fixed point. A non-parabolic subgroup of M is one containing no parabolic elements while a parabolic subgroup is one with all non-trivial elements parabolic. If P is a maximal parabolic subgroup of M and S is a complement of P in M then S is a maximal non-parabolic subgroup. B. H. Neumann [1933] and later C. Tretkoff [1975] studied such groups and showed that they are associated with a triple (Ω, A, B) where A and B are permutations of a countably infinite set Ω and $A^2 = B^3 = 1$ while $C = AB$ is transitive on Ω. Lyndon and Brenner associate a graph $\Gamma(\Omega, A, B)$ with each such triple. The graph Γ is always cuboid (all vertices have degree greater than or equal to three) and is cubic (all vertices have degree exactly three) if neither A nor B has a fixed point in Ω. Furthermore, Γ has an Eulerian path, that is, a path traversing each directed edge exactly once. Conversely, they show that each pair consisting of a connected cuboid graph together with an Eulerian path arises from some such triple (Ω, A, B) which is essentially unique if neither A nor B has a fixed point.

Brenner and Lyndon are able to conclude that if the modular group is written $M = PS$ as in the situation above (P maximal parabolic, S maximal non-parabolic) then S is a free product of r_2 cyclic groups of order two, r_3 cyclic groups of order three and r_∞ infinite cyclic groups where r_2 and r_3 are respectively the number of fixed points of A and B, r_∞ is the Betti number of Γ, and $r_2 + r_3 + r_\infty$ is infinite. They also give a new proof of a result given by G. A. Miller in 1900. If a, b and c are integers such that $2 \leq a \leq b \leq c$ then there is a finite set Ω with $|\Omega| \leq c + 2$ and permutations A and B of Ω such that A, B and $C = AB$ have orders a, b and c respectively.

Lyndon [1963] investigated a common unified approach to the very similar cancellation arguments used to prove the Nielsen-Schreier and Kurosh Subgroup Theorems. He defined a "length function" $|x| \to x$ on a group G to be an integer-valued function satisfying certain axioms. By adding further axioms he was able to completely characterize those length functions which arise from free groups and free products. (Lyndon's [1965] proof of Grushko's Theorem uses similar cancellation arguments although the abstract formulation in terms of length functions is not used.) The current usage is to call a function a length function if it satisfies only the following two of Lyndon's axioms:

A.2. $$|x^{-1}| = |x|$$

A.4. $d(x,y) < d(x,z)$ implies $d(y,z) = d(x,z)$

where $d(x,y) = \frac{1}{2}(|x| + |y| - |xy^{-1}|)$ gives a measure of the "cancellation" occurring in the product xy^{-1}. Interestingly, the equivalence of groups with a length function and groups acting on trees was established by Chiswell [1976a]. The method of groups acting on trees is an easier way of proving subgroup theorems but has so far not yielded some of the important information obtainable from Nielsen type reduction arguments. For a detailed discussion of these points see Hoare [1981]. Some investigation has also been made of real-valued length functions and groups acting on tree-like spaces.

An immediate consequence of Nielsen's argument is that the automorphism group of a finitely generated free group $F = \langle X \rangle$ is generated by the elementary Nielsen automorphisms. A very important result about automorphisms of free groups is a theorem of J. H. C. Whitehead [1936] describing an algorithm which, when given two elements u and v (or two n-tuples of elements) decides if v is the image of v under some automorphism of F. Whitehead's original proof used very difficult topological arguments. A combinatorial but still very difficult proof was given by Rapaport [1958]. However, Whitehead's Theorem is almost completely known through the Higgins-Lyndon [1974] proof which was available only in mimeo notes for many years but was finally published in 1974. Joan Birman has called this proof "a triumph of notation over nature."

A Whitehead automorphism of F is an automorphism which either permutes the elements of $X^{\pm 1}$ or, for some fixed $a \in X^{\pm 1}$, takes each $x \in X^{\pm 1}$ to one of x, xa, $a^{-1}x$, or $a^{-1}xa$. We wish to consider "cyclic words," that is, given a cyclically reduced word w, we may freely replace w by any of its cyclic permutations. The key conclusion of Whitehead's investigation is that if u and v are cyclic words with $|u| \geq |v|$ or v is the image of u under some automorphism of F then there is a sequence of Whitehead automorphisms $\tau_1, \ldots, \tau_n$ where, for each $2 \leq i \leq n$, one has $|u\tau_1 \ldots \tau_{i-1}| \leq |u\tau_1 \ldots \tau_i|$ and strict inequality holds unless $|u\tau_1 \ldots \tau_i| = |v|$. The Higgins-Lyndon proof forms the basis of the further work of McCool [1975] who proves that the stabilizer $A_S \subseteq \mathrm{Aut}(F)$ of a finite subset $S \subseteq F$ is finitely presented.

During the 1976 Oxford Conference on decision problems in algebra, V. I. Remeslennikov gave a talk on which he included the following compliment to Lyndon, who was sitting in the audience. Remeslennikov began a sentence "Last winter, in our seminar at Novosibirsk, it was my pleasure to give a report on the very elegant proof of Higgins and Lyndon of Whitehead's Theorem which says that ..." and then proceeded to state the theorem. I was translating the talk but my facility for remembering Russian sentences was inadequate and by the time the sentence was finished I had forgotten the first half -- thus omitting this tribute to the proof.

Some mention must certainly be made of two of Lyndon's contributions to the theory of varieties of algebras. In "Two Notes on Nilpotent Groups" he showed that the set of laws of any finitely generated nilpotent group is finitely based. In "Identities in finite algebras" [1954] he constructed a seven element algebra with one binary operation and one constant whose laws are not finitely based.

A subject which has long interested Lyndon is that of solutions of equations in groups. The paper with Schützenberger [1962] shows that if (a,b,c) is a solution of the equation $x^m y^n z^p = 1$ in a free group where

$m,n,p \geq 2$ then a, b and c are all powers of a common element. A related result [1973] is that if $w(x_1,\ldots,x_n)$ is a word which is quadratic in the sense that each x_i occurs exactly twice in w and $(a_1,\ldots,a_n)$ is a solution of $w = 2$ in a free group then the rank of the subgroup generated by $a_1,\ldots,a_n$ cannot exceed the greatest integer in $n/2$. Probably his most well-known result on equations in free groups is his explicit description [1960] of all solutions of an equation in a single variable, say $w(x) = 1$, where w contains only one variable x and generators from a fixed free group F, in terms of parametric words. This result cannot be extended to two generator groups for Appel [1969] has shown that there are two variable equations in free groups whose solutions do not admit any parametric description.

I began by quoting a footnote to Milnor's paper [1968]. Gromov [1981] has recently proven Milnor's conjecture that a finitely generated group with polynomial growth is nilpotent by finite. At the 1981 Oberwolfach group theory meeting Hyman Bass gave a lecture on Gromov's proof. Sitting in the back row, Lyndon at one point said, "That construction reminds me of ultraproducts."

Van den Dries and Wilkie [1984] have since given a proof using non-standard analysis. It strikes me as a typical Lyndon remark.

REFERENCES

1. K. I. Appel, On two-variable equations in free groups. Proc. Amer. Math. Soc. 21 (1969), 179-181.
2. J. L. Brenner and R. C. Lyndon, Nonparabolic subgroups of the modular group. J. Algebra 77 (1982), 311-322.
3. J. L. Brenner and R. C. Lyndon, Maximal nonparabolic subgroups of the modular group. Math. Ann. (to appear).
4. J. L. Brenner and R. C. Lyndon, A theorem of G. A. Miller on the order of the product of two permutations. (preprint).
5. J. L. Britton, Solutions of the word problem for certain types of groups I and II. Proc. Glasgow Math. Assoc. 3 (1956), 45-54, 68-90.
6. I. M. Chiswell, Abstract length functions in groups. Proc. Camb. Phil. Soc. 80 (1976), 451-563.
7. I. M. Chiswell, Euler characteristics of groups. Math. Zeit. 147 (1976), 1-11.
8. I. M. Chiswell, D. J. Collins, and J. Huebschmann, Aspherical group presentations. Math. Zeit. 178 (1981), 1-36.

9. D. E. Cohen and R. C. Lyndon, Free bases for normal subgroups of free groups. Trans. Amer. Math. Soc. 180 (1963), 528-537.

10. M. Dehn, Uber unendliche diskcontinuerliche Gruppen. Math. Ann. 71 (1912), 116-144.

11. P. J. Higgins and R. C. Lyndon, Equivalence of elements under automorphisms of a free group. J. London Math. Soc. 8 (1974), 254-258.

12. A. H. M. Hoare, Nielsen methods in groups with a length function. Math. Scad. 48 (1981), 153-164.

13. M. Greendlinger, Dehn's algorithm for the word problem. Comm. Appl. Math. 13 (1960), 641-677.

14. M. Gromov, Groups of polynomial growth and expanding maps. Pub. IHES. 53 (1981), 53-73.

15. R. C. Lyndon, Identities in finite algebras. Proc. Amer. Math. Soc. 5 (1954), 8-9.

16. R. C. Lyndon, Equations in free groups. Trans. Amer. Math. Soc. 96 (1960), 445-457.

17. R. C. Lyndon, Length functions in groups. Math. Scand. 12 (1963), 209-234.

18. R. C. Lyndon, Grushko's theorem. Proc. Amer. Math. Soc. 16 (1965), 822-826.

19. R. C. Lyndon, On Dehn's algorithm. Math. Ann. 166 (1966), 208-228.

20. R. C. Lyndon, On the Freiheitssatz. J. London Math. Soc. 5 (1972), 95-101.

21. R. C. Lyndon, On products of powers in groups. Comm. Pure Appl. Math. 26 (1973), 781-784.

22. R. C. Lyndon, On non-Euclidean crystallographic groups. In: Proc. Conf. Canberra 1973, Springer Lecture Notes in Math. 372, 437-442, Berlin-Heidelberg-New York, Springer, 1974.

23. R. C. Lyndon, On the combinatorial Riemann-Hurwitz formula. In: Convegi sui gruppi infiniti Rome 1973, 435-439, New York and London, Academic Press, 1976.

24. R. C. Lyndon and P. E. Schupp, Combinatorial Group Theory. Ergebnisse der Mathematik und ihrer Grenzgebiete 89, Berlin-Heidelberg-New York, Springer, 1977.

25. R. C. Lyndon and M. P. Schützenberger, The equation $a^M = b^N c^P$ in a free group. Mich. Math. J. 9 (1962), 289-298.

26. R. C. Lyndon and J. L. Ulmman, Groups of elliptic linear fractional transformations. Proc. Amer. Math. Soc. 18 (1967), 1119-1124.

27. R. C. Lyndon and J. L. Ullman, Pairs of real 2-by-2 matrices that generate free products. Mich. Math. J. 15 (1968), 161-166.

28. R. C. Lyndon and J. L. Ullman, Groups generated by two parabolic linear fractional transformations. Canad. J. Math. 21 (1969), 1388-1403.

29. J. Mc Cool, Some finitely presented subgroups of the automorphism group of a free group. J. Algebra 35 (1975), 205-213.

30. J. Milnor, A note on curvature and fundamental groups. J. Diff. Geometry 2 (1968), 1-7.

31. B. H. Neumann, Uber ein gruppentheoretische-arithmetisches Problem. Sitzungsber. Preuss. Akad. Wiss. Phys. Math. Kl. 10 (1933).

32. E. S. Rapaport, On free groups and their automorphisms. Acta. Math. 99 (1958), 139-163.

33. V. A. Tartakovskii, The sieve method in group theory. Math. Sb. 25 (1949), 3-50.

34. V. A. Tartakovskii, Application of the sieve method to the solution of the word problem for certain types of groups. Math. Sb. 25 (1949), 251-274.

35. V. A. Tartakovskii, Solution of the word problem for groups with a k-reduced basis for $k > 6$. Izv. Akad. Nauk. SSSR Ser. Math. 13 (1949), 483-494.

36. C. Tretkoff, Nonparabolic subgroups of the modular group. Glasgow Math. J. 16 (1975), 90-102.

37. L. van den Dries and A. Wilkie, A non-standard analysis and Gromov's solution of Milnor's Conjecture. J. Algebra, to appear.

38. E. R. van Kampen, On some lemmas in the theory of groups. Amer. J. Math. 55 (1933), 268-273.

39. C. M. Weinbaum, Visualizing the word problem, with an application to sixth groups. Pacific J. Math. 16 (1966), 557-578.

40. J. H. C. Whitehead, On equivalent sets of elements in a free group. Ann. Math. 37 (1936), 782-800.

DEPARTMENT OF MATHEMATICS
UNIVERSITY OF ILLINOIS
URBANA, ILLINOIS 61801

Contemporary Mathematics
Volume 33, 1984

IA-AUTOMORPHISMS OF FREE PRODUCTS OF ABELIAN GROUPS

S. Andreadakis

To R. C. Lyndon on his 65th birthday

The group of automorphisms of a free product of groups has been described in terms of generators and relations by Fuchs-Rabinowtiz [3] and Golovin and Sadovskii [4] many years ago. Here we consider a free product of a finite number of abelian groups none of which is infinite cyclic and give a set of generators for the group of IA-automorphisms of this group, a result which is analogous to a result of Magnus [5] concerning the automorphisms of a free group.

Let $G = A_1 * \ldots * A_n$ be a free product of n abelian groups. If G' is the derived group of G, then the quotient group G/G' is isomorphic to the direct product $A_1 \times A_1 \times \ldots \times A_n$ of the given groups $A_1, A_2, \ldots, A_n$. Since G' is a characteristic subgroup of G, each automorphism of G induces in a natural way an automorphism on G/G' and thus there exists a mapping

$$\phi\colon \mathrm{Aut}(A_1 * \ldots * A_n) \to \mathrm{Aut}(A_1 \times \ldots \times A_n)$$

which is a homomorphism. The kernel $\ker\phi = K(G)$ consists of all the automorphisms of G which induce the identity on $G/G' \cong A_1 \times \ldots \times A_n$. These autmorphisms have been called IA-automorphisms by S. Bachmuth [2]. It has been proved in [1] that for $n = 2$ the group $K(G)$ coincides with the group of inner automorphisms of G. Our main result is the following:

© 1984 American Mathematical Society
0271-4132/84 $1.00 + $.25 per page

THEOREM 1. <u>If none of the abelian groups</u> $A_1, A_2, \ldots, A_n$ <u>is infinite cyclic, then</u> $K = K(G)$ <u>is generated by the automorphisms</u> $k(i,j,a)$, $i \neq j$ $a \in A_j$ <u>where</u>

$$k(i,j,a): x \to a^{-1}xa \quad \text{for every } x \in A_i$$
$$y \to y \quad \text{for every } y \in A_r,\ r \neq i .$$

Proof: Let $\alpha \in K$. Let A_i^α be the image of A_i under α. Since A_i^α is not infinite cyclic it follows from Kurosh's subgroup theorem that A_i^α is a conjugate of a subgroup of some A_j, $j = 1,2,\ldots,n$. But A_i coincides with its normalizer in $G = A_i * \ldots * A_n$ and therefore the same must hold for A_i^α since $G = A_i^\alpha * \ldots * A_n^\alpha$. Hence A_i^α coincides with a conjugate of A_j for some j. Now $x^\alpha = x (\text{mod } G')$ and since $G/G' \cong A_1 \times \ldots \times A_n$ the subgroup A_i^α is conjugate to A_i. Consequently there exists $g_i \in G$ such that $x^\alpha = g_i^{-1}xg_i$ for every $x \in A_i$. In conclusion there exist $g_1, g_2, \ldots, g_n \in G$ such that

$$x^\alpha = g_i^{-1}xg_i \quad \text{for every } x \in A_i \text{ and each } i = 1,2,\ldots,n .$$

For each automorphism $\alpha \in K$ we can define a length by

$$\ell(\alpha) = \sum_{i=1}^{n} \ell(g_i) ,$$

where $\ell(g_i)$ is the length of g_i as an element of the free product $G = A_1 * \ldots * A_n$.

Let now $\overline{K}$ be the subgroup generated by the $k(i,j,a)$'s. Clearly $\overline{K} \leqq K$. We have to show that $\overline{K} = K$. Suppose on the contrary that $\overline{K} \neq K$ and let $\alpha \in K, \alpha \notin \overline{K}$ with $\ell(\alpha)$ minimal. Then $\ell(\alpha) > 0$ and

$$x^\alpha = g_i^{-1}xg_i \quad \text{for } x \in A_i ,\ i = 1,2,\ldots,n .$$

We may assume, by renumbering the factors of the free product if necessary, that there exists a $k \geqq 0$ such that $g_1 = g_2 = \ldots = g_k = 1$ and $g_i \neq 1$ for $i > k$. Clearly $k < n$. Let now a be an element of A_i for $i > k$. Then

$$a = g_{i_1}^{-1}a_{i_1}g_{i_1}g_{i_2}^{-1}a_{i_2}g_{i_2} \ldots g_{i_m}^{-1}a_{i_m}g_{i_m} , \tag{*}$$

where $a_{i_\rho} \in A_{i_\rho}$, $a_{i_\rho} \neq 1$, $\rho = 1,2,\ldots,m$ and $i_\rho \neq i_{\rho+1}$, $\rho = 1,\ldots,m-1$. Clearly $m \geqq 2$. The above representation of a is unique as a representation of a in the free product $G = A_1^\alpha * \ldots * A_n^\alpha$. If we now replace in (*) the g's with their representation as elements of the free product $A_1 * \ldots * A_n$, then the right-hand side in (*) will reduce to a. Therefore there must be cancellations.

We may suppose that for each ρ the element g_{i_ρ} does not begin with an element of A_{i_ρ}. We may also suppose that if $g_{i_\rho} \neq 1$, then g_{i_ρ} does not end in an element of A_s, for $s = 1,2,\ldots,k$ otherwise by premultiplying α by $k(i_\rho,s,b)$ with suitable b we could reduce $\ell(\alpha)$ by 1. Hence in (*) the cancellations will start in the products $g_{i_\rho} g_{i_{\rho+1}}^{-1}$. Because of the above assumptions there must be products in which one of the g_{i_ρ}, $g_{i_{\rho+1}}$ is trivial or is entirely cancelled by the other. Suppose that in $g_{i_\rho} g_{i_{\rho+1}}^{-1}$ the first is cancelled (or it is trivial) (respectively the second is cancelled). Then $g_{i_{\rho+1}} = g'_{i_{\rho+1}} g_{i_\rho}$ (resp. $g_{i_\rho} = g'_{i_\rho} g_{i_{\rho+1}}$) with no cancellations in the product $g'_{i_{\rho+1}} g_{i_\rho}$ (resp. $g'_{i_\rho} g_{i_{\rho+1}}$). If $g'_{i_{\rho+1}} = 1$ (resp. $g'_{i_\rho} = 1$), then the cancellation will stop there. This cannot happen for all pairs $g_{i_\rho}, g_{i_{\rho+1}}$ and therefore there must exist a pair such that $g'_{i_{\rho+1}} \neq 1$ and $g'_{i_{\rho+1}}$ ends in a_{i_ρ} (resp. $g'_{i_\rho} \neq 1$ and g'_{i_ρ} ends in $a_{i_{\rho+1}}^{-1}$). Thus $g_{i_{\rho+1}} = g''_{i_{\rho+1}} a_{i_\rho} g_{i_\rho}$ (resp. $g_{i_\rho} = g''_{i_\rho} a_{i_{\rho+1}}^{-1} g_{i_{\rho+1}}$) and there are no cancellations in the product $g''_{i_{\rho+1}} a_{i_\rho} g_{i_\rho}$ (resp. $g''_{i_\rho} a_{i_\rho}^{-1} g_{i_{\rho+1}}$). We now take $\gamma = k(i_{\rho+1}, i_\rho, a_{i_\rho}^{-1})$ (resp. $\gamma = k(i_\rho, i_{\rho+1}, a_{i_{\rho+1}})$ and the product $\gamma\alpha = \beta$. Then

$$x^\beta = (x^\gamma)^\alpha = (a_{i_\rho} x a_{i_\rho}^{-1})^\alpha = g_{i_\rho}^{-1} a_{i_\rho} g_{i_\rho} g_{i_{\rho+1}}^{-1} x g_{i_{\rho+1}} g_{i_\rho}^{-1} a_{i_\rho}^{-1} g_{i_\rho} =$$

$$= g_{i_\rho}^{-1} g''^{-1}_{i_{\rho+1}} x g''_{i_{\rho+1}} g_{i_\rho} \quad \text{for} \quad x \in A_{i_{\rho+1}}$$

and

$$y^\beta = (y^\gamma)^\alpha = y^\alpha = g_s^{-1} y g_s \quad \text{for} \quad y \in A_s,\ s \neq i_{\rho+1}$$

$$(\text{resp.} \quad x^\beta = (x^\gamma)^\alpha = (a_{i_{\rho+1}}^{-1} x a_{i_{\rho+1}})^\alpha =$$

$$= g_{i_{\rho+1}}^{-1} a_{i_{\rho+1}}^{-1} g_{i_{\rho+1}} g_{i_\rho}^{-1} x g_{i_\rho} g_{i_{\rho+1}}^{-1} a_{i_{\rho+1}} g_{i_{\rho+1}} =$$

$$= g_{i_{\rho+1}}^{-1} g_{i_\rho}''^{-1} x g_{i_\rho}'' g_{i_{\rho+1}} \quad \text{for} \quad x \in A_{i_\rho}$$

and

$$y^\beta = (y^\gamma)^\alpha = y^\alpha = g_s^{-1} y g_s \quad \text{for} \quad y \in A_s,\ s \neq i_\rho) .$$

But

$$\ell(g_{i_{\rho+1}}'' g_{i_\rho}) < \ell(g_{i_{\rho+1}}'' a_{i_\rho} g_{i_\rho}) = \ell(g_{i_{\rho+1}})$$

$$(\text{resp.} \quad \ell(g_{i_\rho}'' g_{i_{\rho+1}}) < \ell(g_{i_\rho}'' a_{i_{\rho+1}}^{-1} g_{i_{\rho+1}}) = \ell(g_i))$$

and therefore $\ell(\beta) < \ell(\alpha)$. Thus $\beta \in \overline{K}$ and therefore $\alpha = \gamma^{-1}\beta \in \overline{K}$. This contradiction shows that $K = \overline{K}$ and the proof of the Theorem 1 is complete.

COROLLARY. *If* $n = 2$, *then* $K(G)$ *is the group of inner automorphisms of* G.

The corollary has been proved, in the general case, in [1]. Theorem 1 is not valid if one or more of the $A_1, \dots, A_n$, $n > 2$, is infinite cyclic as the result of Magnus shows [5]. For example, if A_1 is infinite cyclic with a as generator, then the automorphism $a \to a[b,c]$ and $x \to x$, $x \in A_i$, $i \neq 1$, $b \in A_j$, $c \in A_k$, $1 \neq j \neq k \neq 1$, $b \neq 1$, $c \neq 1$ belongs to K, though it does not belong to the subgroup generated by the k's as described in the Theorem 1. We can conjecture that it is enough to add to the set of generators automorphisms like the above or which amounts to the same thing that in the general case the group $K(G)$ is normally generated in $\text{Aut } G$ by all the automorphisms of the form $k(i,j,a)$. On the other hand Theorem 1 is no longer valid if the number n of factors is infinite. For example if n is infinite and A_1 is a factor then for fixed $a \in A_1$ the mapping $x \to x$ for every $x \in A_1$ and $y \to a^{-1}ya$ for every $y \in A_i$ and $i \neq 1$ gives an element of $K(G)$ which obviously does not belong to the group generated by the $k(i,j,a)$'s. The set of generators needed for $K(G)$ can clearly be reduced. For example for fixed j we need only take the automorphisms $k(i,j,a)$ where a runs over a generating set of A_j. Note that for fixed i and j the $k(i,j,a)$'s generate a group isomorphic to A_j.

The image of Aut G under the homomorphism ϕ defined above can also be described explicitly. Namely

THEOREM 2. <u>If the non isomorphic groups among the abelian groups</u> $A_1, A_2, \ldots, A_n$ <u>none of which is infinite cyclic are</u> $B_1, B_2, \ldots, B_k$ <u>each appearing</u> $n_1, n_2, \ldots, n_k$ <u>times respectively then the image of</u> $\text{Aut } G = \text{Aut}(A_1 * \ldots * A_n)$ <u>in the homomorphism</u>

$$\phi: \text{Aut}(A_1 * \ldots * A_n) \to \text{Aut}(A_1 \times \ldots \times A_n) \text{ is the group}$$

$$A^*(G) = (\text{Aut } B_1 \wr S_{n_1}) \times \ldots \times (\text{Aut } B_k \wr S_{n_k}) ,$$

<u>where</u> S_{n_i} <u>is the symmetric group on</u> n_i <u>letters and</u> $\text{Aut } B_i \wr S_{n_i}$ <u>is the usual wreath product, for</u> $i = 1,2,\ldots,k$.

Proof: It follows from the proof of Theorem 2 that if $\alpha \in \text{Aut } G$, then for given i we have $A_i^{\alpha} = g^{-1}A_j g$ for some $g \in G$ and some $j \in \{1,2,\ldots,n\}$. But this shows that the groups A_i, A_j are isomorphic and therefore $\alpha\phi$ permutes the isomorphic factors among themselves. On the other hand a set of automorphisms $\{\alpha_1, \alpha_2, \ldots, \alpha_n\}$ with $\alpha_i \in \text{Aut } A_i$, $i = 1,2,\ldots,n$ can easily be extended to an automorphism of $G = A_1 * \ldots * A_n$ which agrees with α_i on A_i, $i = 1,2,\ldots,n$. The rest follows now easily.

Using Theorems 1 and 2 one can now describe fairly well the group Aut G as an extension of $K(G)$ by $A^*(G)$.

REFERENCES

[1] S. Andreadakis, On semicomplete groups, J. London Math. Soc. 44 (1969), 361-364.

[2] S. Bachmuth, Automorphisms of free metabelian groups, Trans. Amer. Math. Soc. 118 (1965), 93-104.

[3] D. I. Fuchs-Rabinowitz, On the groups of automorphisms of free products, I.II. Mat. Sb. 8 (1940), 265-276; and 9 (1941), 183-220.

[4] O. N. Golovin and L. E. Sadovskii, On the automorphism groups of free products, Math. Sb. 4 (1938), 505-514.

[5] W. Magnus, Über n-dimensionale Gittertransformationen, Acta. Math. 64 (1935), 353-367.

DEPARTMENT OF MATHEMATICS
UNIVERSITY OF ATHENS
ATHENS 15781, GREECE

Received July 8, 1982

Contemporary Mathematics
Volume 33, 1984

ON ARTIN GROUPS AND COXETER GROUPS OF LARGE TYPE[1]

Kenneth I. Appel

1. Introduction

The purpose of this paper is to extend the results obtained in [1] to Artin and Coxeter groups of large type. First, fundamental definitions will be recalled.

A Coxeter matrix $\underset{\approx}{M}$ over a set I is a symmetric matrix with entries $m_{ij} \in \mathbb{N} \cup \{\infty\}$ where $m_{ii} = 1$ for $i \in I$ and $m_{ij} \geq 2$ for $i \neq j \in I$. The Artin group G defined by $\underset{\approx}{M}$ is the group with generating set $\{a_i : i \in I\}$ and for each pair $i \neq j$ with $m_{ij} < \infty$ a defining relation of the form

$$a_i a_j a_i \ldots = a_j a_i a_j \ldots$$

saying that the alternating string of a_i's and a_j's of length m_{ij} beginning with a_i is equal to the alternating string of length m_{ij} beginning with a_j. Associated with the Artin group G is the Coxeter group $\overline{G}$ obtained by setting the squares of the generators equal to the identity so that the defining relations become $a_i^2 = 1$, $i \in I$, and $(a_i a_j)^{m_{ij}} = 1$ for $i \neq j$, $m_{ij} < \infty$. An Artin or Coxeter group is of <u>large type</u> if $m_{ij} \geq 3$ for $i \neq j$, and is of <u>extra-large type</u> if $m_{ij} \geq 4$ for all $i \neq j$.

---

[1]Dedicated to Professor Roger Lyndon on the occasion of his sixty-fifth birthday. This study was supported by NSF grant MCS 790512. The work followed from the collaboration with Paul Schupp that led to [1]. The author would like to thank Professor Schupp for many helpful discussions and his generous decision that his participation in the extended results reported in this paper was not sufficient for co-authorship.

© 1984 American Mathematical Society
0271-4132/84 $1.00 + $.25 per page

In [1], an analysis of cancellation diagrams for two generator Artin and Coxeter groups plus the use of techniques of small cancellation theory yielded theorems on the generalized word problem, the conjugacy problem, and torsion as well as a "Freiheitssatz" for finitely generated Artin and Coxeter groups of extra-large type. The most important tools in that analysis will be recalled in the next section. Here an extension of these tools and a few new techniques are used to prove comparable theorems for groups of large type. The theorems of [1] are implied by the corresponding theorems of this paper with the exception that Theorem 4 of this paper does not specify the forms of conjugating elements as do Theorems 4' and 4" of [1]. However [1] is organized to make it possible for the reader unfamiliar with small cancellation theory to follow in detail while this paper is somewhat more technical and assumes familiarity with small cancellation theory or with [1].

The results obtained are the following.

<u>Theorem</u> 1. Let G be an Artin or Coxeter group of large type. If $J \subseteq I$ then G_J has a presentation defined by the Coxeter matrix M_J and the generalized word problem for G_J in G is solvable. If $J,K \subseteq I$ then $G_J \cap G_K = G_{(J \cap K)}$.

<u>Theorem</u> 2. An Artin group of large type is torsion free.

<u>Theorem</u> 3. Let G be an Artin group of large type. Then the set $\{a_i^2, i \in I\}$ freely generates a free subgroup of G.

<u>Theorem</u> 4. An Artin or Coxeter group of large type has solvable conjugacy problem.

As in [1] all groups are assumed to be finitely presented. It is easy to check that Theorems 1, 2, and 3 extend directly to recursively presented groups. Somewhat surprisingly, Theorem 4 requires an additional hypothesis before it can be extended to recursively presented groups.

In what follows, when it is necessary to refer to a lemma of [1] we shall use a prefix 1. Thus Lemma 1.4 will mean Lemma 4 of [1]. To avoid confusion the numbering of lemmas in this paper will begin with 13 since [1] has 12 lemmas. Those lemmas of [1] to which reference is made will be restated here.

2. Small Cancellation Theory and Material from [1]

In this paper we must extend some of the work that leads to Results III, IV and V of [1]. The basic tool with which we begin is Theorem 3.1 of [11] which states that if M is an arbitrary map and $1/p + 1/q = \frac{1}{2}$ then

$$(*) \quad p(Q-h) = \Sigma^{\cdot}(p/q + 2 - d(v)) + \Sigma^{\circ}(p - d(v)) + p/q\,\Sigma(q - d(D)) + p/q(V^{\cdot} - E^{\cdot}).$$

Here Q is the number of components of M, h is the number of holes, $V^{\cdot}$ is the number of boundary vertices of M and $E^{\cdot}$ is the number of edges in a boundary cycle of M. The symbols $\Sigma^{\cdot}$ and Σ° indicate sums over boundary and interior vertices respectively.

Note that in our diagrams and annuli (which are similar to those in [1]) $V^{\cdot} = E^{\cdot}$. Thus, if we use the notation $a \dot{-} b$ to mean the greater of $a - b$ and 0 we obtain the following equations from setting p equal to three and four. (Note that $a - b = (a \dot{-} b) - (b \dot{-} a)$.) For simply connected diagrams with no vertices of degree one:

$$(1) \quad 2\Sigma^{\cdot}(5/2 \dot{-} d(v)) = 2\Sigma^{\cdot}(d(v) \dot{-} 5/2) + 2\Sigma^{\circ}(d(v) - 3) + \Sigma(d(D) - 6) + 6$$

$$(2) \quad \Sigma^{\cdot}(3 \dot{-} (d(v))) = \Sigma^{\cdot}(d(v) \dot{-} 3) + \Sigma^{\circ}(d(v) - 4) + \Sigma(d(D) - 4) + 4\,.$$

Equations (1) and (2) are displayed so that the left side is the number of boundary vertices of degree two. Thus we shall often write N_2 instead of the left sides of these equations.

If M is a non-empty map such that each interior vertex of M has degree at least p then if all regions of M have degree at least q, M is defined as a [p,q] map, if all interior regions of M have degree at least q we will call M a (p,q) map. (We will often call a [3,6] map a "C(6) diagram" to emphasize the applicable cancellation condition.) Following [2], given a map M we define a dual map M^* as follows. Pick a point v_i^* in each region D_i of M. The v_i^* are vertices of M^*. If D_1 and D_2 are regions of M, $D_1 \neq D_2$ but having an edge e in common, an edge e^* is drawn from v_1^* to v_2^* crossing e but no other edges of M or edges of M^* already constructed. If a region D_i of M has an edge e such that D_i lies on both sides of e a loop is drawn at v_i^* crossing e but no other edges. The regions of M^* are the regions of the resulting graph that contain interior vertices of M.

We prove a result similar to the curvature formula of [2].

Result VI. Let A be a connected (3,6) annulus with no edges that do not lie on regions and no regions that lie on both boundaries. Then

$$\Sigma^{\bullet}_A(4 - i(D)) = \Sigma^{\circ}_A(d(D) - 6) + 2\Sigma^{\circ}_A(d(v) - 3) .$$

Proof. Let A^* be the dual of A. Each boundary vertex v^* of A^* corresponds to a boundary region D of A and each edge of v^* corresponds to an interior edge on D. Each interior vertex of A^* corresponds to an interior region of A of the same degree. Each region of A^* corresponds to an interior vertex of A. But A^* is a [6,3] diagram so

$$\Sigma^{\bullet}_{A^*}(4 - d(v)) + \Sigma^{\circ}_{A^*}(6 - d(v)) + 2\Sigma_{A^*}(3 - d(D)) = 0 .$$

Hence the result follows from the correspondence described.

In [1] the concept of strips in diagrams that satisfy C(4) and T(4) was introduced. In a connected simply-connected diagram satisfying C(4) and T(4) a singleton strip was defined as a simple boundary region (i.e., one whose boundary has connected intersection with the boundary of the diagram) of interior degree at most one while a compound strip consisted of consecutive simple boundary regions of which the first and last had degree two and the others had degree three. The group $G_{ij} = \langle a_i, a_j; \underset{\sim}{R}_{ij} \rangle$, where $\underset{\sim}{R}_{ij}$ is the symmetrized set obtained from defining relator $r_{ij} = a_i a_j a_i \ldots \bar{a}_j \bar{a}_i \bar{a}_j$ if $2 \leq m_{ij} < \infty$ and $\underset{\sim}{R}_{ij}$ is empty if $m_{ij} = \infty$, was a crucial tool in the study of more general Artin groups. It was observed that the boundary label of any region in a cancellation diagram for such a G_{ij} has exactly two places where the exponent on the generators change sign. In a minimal diagram, if such a sign change takes place on an interior edge then it must take place at a vertex. If it occurs on the boundary of the diagram them if it does not occur at a vertex of degree three or more it is convenient to define a boundary vertex of degree two at the point of change. Such vertices were called <u>separating</u> <u>vertices</u> of the corresponding regions in [1]. Separating vertices were used to prove the following lemma.

<u>Lemma</u> 1.3. The set $\mathcal{R}_{ij}$ satisfies the small cancellation conditions C(4) and T(4).

The following useful observation was made about regions D_1 and D_2 in a minimal $\mathcal{R}_{ij}$ diagram.

<u>Lemma</u> 1.4. If D_1 and D_2 have a common edge incident to a vertex v of M then v is not a separating vertex for both D_1 and D_2.

Defining $|w|$ as the length of a word w and $\|w\|$ as its syllable length, a combination of the above ideas was used to prove another important lemma.

<u>Lemma</u> 1.7. Let $w = 1$ in G_{ij} and write $w = w_1 w_2$.

a) If $\|w_1\| \leq m_{ij}$ then $|w_1| \leq |w_2|$.

b) If $\|w_1\| < m_{ij}$ then $|w_1| < |w_2|$.

Next, the following lemma shows that if we restrict to words on generators indexed by a subset J of I and group with only these relators and generators with subscripts in J we get the expected local property.

<u>Lemma</u> 1.9. Let G be an Artin or Coxeter group of large type. Let w be a word on A_j. If $w = 1$ in G then $w = 1$ in G_J.

3. Diagrams for Two Generator Artin Groups

In this section, M will always stand for a minimal connected, simply connected $\mathcal{R}_{ij}$-diagram with cyclically reduced boundary label. Recall that if D is a region of M its boundary level is the relator r_{ij} or its inverse. We write $r_{ij} = x\tilde{x}$ where x consists of an alternation of generators a_i and a_j of length m_{ij} and $\tilde{x}$ is a similar alternation of $\bar{a}_j$ and $\bar{a}_i$. x and $\tilde{x}$ are the labels on boundary paths between separating vertices of regions corresponding to r_{ij} while $\tilde{x}^{-1}$ and x^{-1} play the same role with respect to r_{ij}^{-1}. By Lemma 1.4 the label on any interior edge of M must be a <u>proper</u> subword of $x, x^{-1}, \tilde{x}$ or $\tilde{x}^{-1}$. It is convenient to require the same condition on boundary edges of M so that regions of M all have degree at least four. We do so as follows. If D is a region both of whose separating vertices lie on the boundary of a diagram M with at least two regions, let v be the first of these separating vertices in counterclockwise order. We require that the edge e of $\partial D \cap \partial M$ incident on v have a label consisting of a single generator. This is accomplished by adding an additional vertex called a <u>B-vertex of M</u>, in $\partial D \cap \partial M$. A separating vertex of a region D that is a boundary vertex of M of degree two is called an S-vertex of M. Note that every boundary vertex of M either is a B-vertex or an S-vertex or lies on an interior edge of M. If v and v^* are boundary vertices of M we write $\delta(v,v^*)$ for the part of the boundary path of M in the counterclockwise direction from v to v^*. If v and v^* are boundary vertices of M of degree two such that no other vertex of degree two lies on $\delta(v,v^*)$, v and v^* are called a consecutive pair of degree two boundary vertices of M.

By Equation (2), the number of boundary vertices of M of degree two is

$$N_2 = \Sigma_M^{\cdot}(d(v) - 3) + \Sigma_M^{\circ}(d(v) - 4) + \Sigma_M(d(D) - 4) + 4 .$$

By Lemma 1.3 and the B-vertex condition each summand is non-negative.

If starting at some vertex on ∂M, in one counterclockwise circuit of ∂M we encounter degree two boundary vertices in the following order: first

σ_1 S-vertices then β_1 B-vertices, next σ_2 S-vertices . . . and finally β_k B-vertices (where all σ_i and β_i are strictly positive), then we say that M has a boundary sequence $\sigma_1, \beta_1, \ldots, \beta_k$.

<u>Lemma</u> 13. If M has a boundary sequence $\sigma_1, \beta_1, \ldots, \beta_k$ then the number of degree two boundary vertices of M is at least $\Sigma_{i=1}^{k}(\sigma_1 + \beta_i - 2) + 4$.

<u>Proof</u>. Let v and v^* be a consecutive pair of degree two boundary vertices of M. Let the vertices on $\delta(v,v^*)$ in counterclockwise order be v_0 (= v), $v_1, \ldots, v_m$ (= v^*). Suppose that v and v^* are both of the same type (either B-vertices or S-vertices). Then, from the definition of B-vertex it is evident that $m \geq 2$ and $v_1, \ldots, v_{m-1}$ have degree at least three. Let D_i, $i = 1, \ldots, m$ be the region that lies on the edge joining v_{i-1} and v_i. Now if v_i is a vertex of degree three, by Lemma 1.4, it cannot be a separating vertex for both D_i and D_{i+1}. On the other hand it cannot be a non-separating vertex for both D_i and D_{i+1} for then the part of the boundary label of M at v_i would consist of a generator and its inverse, contradicting the assumption that M has cyclically reduced boundary label. Thus if v and v^* are of the same type it cannot be the case that all the other vertices on $\delta(v,v^*)$ have degree three.

Now, by Equation (2), the number of boundary vertices of M of degree two is at least four plus the number of boundary vertices of degree greater than three plus the sum of the excesses over four of the degree of the regions of M. To prove the lemma we need only show that to each consecutive pair of degree two boundary vertices of the same type we can attribute either a distinct boundary vertex of degree greater than three or a distinct excess in the degree of a region.

If v and v^* form a consecutive pair of S-vertices then the reasoning above shows that a boundary vertex of degree greater than three lies in $\delta(v,v^*)$ and since the paths $\delta(v_1,v_1^*)$ and $\delta(v_2,v_2^*)$ associated with different consecutive pairs of boundary vertices of degree two can only intersect at their endpoints such attributed vertices must be distinct. If

v and v^* form a consecutive pair of B-vertices then $\delta(v,v^*)$ must contain either a vertex of degree higher than three or an edge joining a pair of vertices v_{i-1},v_i neither of which is a separating vertex of D_i. But the degree of D_i is just two plus the number of its non-separating vertices and so this condition forces an (attributable) increase in the degree of D_i. (Note that the same region may play the role of D_i for more than one such pair v,v^* but then its degree is at least four plus the number of such pairs.) □

Corollary. M must have at least two S-vertices and not all its S-vertices can be consecutive.

Proof. $N_2 = \Sigma_{i=1}^k(\sigma_i + \beta_i) \geq \Sigma_{i=1}^k(\sigma_i + \beta_i - 2) + 4 = N_2 - 2k + 4$, hence $k \geq 2$. □

Suppose v is an interior vertex of degree four that lies on four regions, say D_1,D_2,D_3,D_4 (in counterclockwise order) of degree 4. Let M be the diagram consisting of D_1,D_2,D_3,D_4 and their incident edges and vertices. (In following the discussion it will be convenient to examine Figure 1 which illustrates the situation in full generality.) Each region lies on exactly one vertex of degree two of ∂M and these vertices alternate in type. Without loss of generality we may assume that D_1 and D_3 lie on S-vertices and D_2 and D_4 lie on B-vertices. Let qs be the label

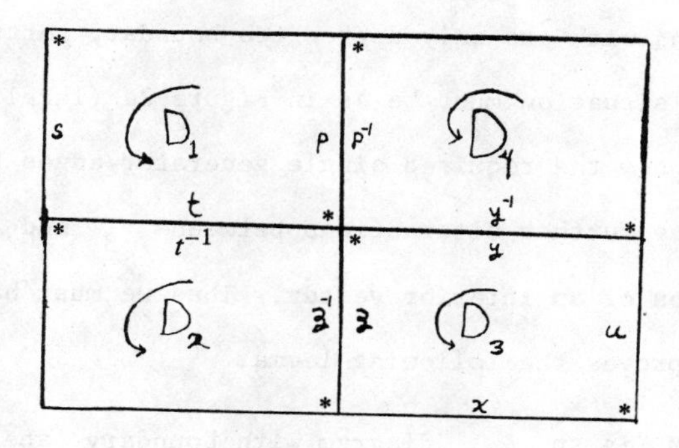

Figure 1

on $\partial D_1 \cap \partial M$ with q ending at the S-vertex of D_1 and let xu be the label on D_3 with x ending at the S-vertex of D_3. (Note that *'s in Figure 1 represent separating vertices.)

<u>Lemma</u> 14. In the circumstances described above $|s| + |x| = |q| + |u| = m_{ij}$.

<u>Proof</u>. From Figure 1 we see that $|x| + |z| = m_{ij}$ (since the label on a path joining separating vertices has length m_{ij}). Similarly $|z| + |t| = m_{ij} = |t| + |s|$, whence

$$|x| + |s| = (|x| + |z|) - (|z| + |t|) + (|t| + |s|) = m_{ij} .$$

The proof that $|q| + |u| = m_{ij}$ is similar.

Suppose we know that the boundary label of M has precisely $2m_{ij}$ syllables and $m_{ij} \geq 3$. We know that there is a syllable change at each S-vertex of M and that each B vertex signals the fact that one region of M has two separating vertices on the boundary whence there are $m_{ij}-1$ syllable changes along the boundary between them. Now since M has at least two B-vertices and at least two S-vertices, all $2m_{ij}$ syllable changes are accounted for by these alone and there can be no further degree two boundary vertices. Thus all regions must have degree four, all interior vertices must have degree four and all other boundary vertices must have degree three. Now suppose M had an interior vertex. Then M must be a "rectangle" composed of degree four regions with its only degree two boundary vertices at the "corners." Thus the situation must be as in Figure 2a (labelled for the case $m_{ij} = 3$). But now the required single generator edges between v_1 and v_2, and Lemma 14 force further alternations between v_3 and v_4, contradicting the assumption of an interior vertex. Thus we must have the situation in Figure 2b. This proves the following lemma.

<u>Lemma</u> 15. If M is an $\mathcal{R}_{ij}$ diagram with boundary label of $2m_{ij}$ syllables then that label is one of $a_i a_j a_i^n \bar{a}_i \bar{a}_j \bar{a}_i^n$, $a_i^n a_j a_i \bar{a}_j^n \bar{a}_i \bar{a}_j$ or their inverses for some $n \geq 1$.

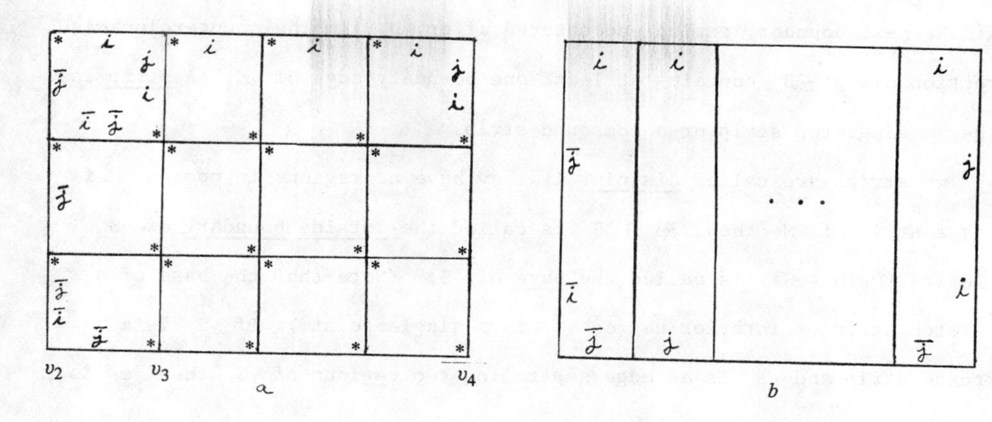

Figure 2

4. Strips in C(6) Diagrams

Lemma 1.7 that shows that if M is an R_{ij} diagram with boundary label st then the fact that $\|s\| \geq \|t\|$ implies $|s| \geq |t|$ was proved using the idea of a strip. The arguments for the various theorems about Artin and Coxeter groups of extra-large type make heavy use of the existence of Dehn regions. For groups of large type, to replace Dehn regions, we use arguments based on strips. Here the strips will be defined for [3,6] diagrams in a manner very similar to those defined for [4,4] diagrams in [1].

Let M be a connected, simply connected diagram satisfying C(6). A <u>singleton strip</u> of M is a simple boundary region D of M with $i(D) \leq 1$ or with $i(D) = 2$ and such that the next boundary region encountered in the counterclockwise direction along ∂M contains at least one boundary edge of M. A <u>compound strip</u> of M is a subdiagram S of M consisting of regions $D_1,\ldots,D_n$, $n \geq 2$ which are all simple boundary regions of M successively encountered in the given order in the counterclockwise directions along ∂M such that $i(D_1) = i(D_n) = 3$ and $i(D_k) = 4$ for $1 < k < n$, such that for $1 < k \leq n$, D_{k-1} and D_k meet along a single interior edge of S, and such that the next boundary region encountered after D_n in the counterclockwise direction along ∂D contains at least one boundary edge of M. A strip is either a singleton strip or a compound strip.

Two strips are called disjoint if they have no regions in common. If S is a strip of M then $\partial M \cap \partial S$ is called the outside boundary of S while $\partial S \cap \partial(M - S)$ is called the base of S. (Note that the base of a singleton strip of interior degree 0 is a single point.) If S is a compound strip and e is an edge separating two regions of S then e is called an interior edge of S.

Lemma 16. Let M be a connected, simply connected diagram with more than one region. If M satisfies $C(6)$ then M contains at least two disjoint strips and if M has no singleton strips then M has at least three disjoint strips.

Proof. The proof is by induction on the number of regions of M. First, suppose that the boundary of M is a simple closed curve and that every boundary region of M is a simple boundary region. From [11] we know that $\Sigma^*_M[4 - i(D)] \geq 6$ (where the sum is taken over simple boundary regions of M) and we examine how this can happen. Beginning at some boundary vertex we traverse the boundary of M in counterclockwise order, at each boundary region computing the partial sum $s = \Sigma^*[4 - i(D)]$ over those simple boundary regions encountered so far. Since M has more than one region and the boundary of M is a simple closed curve, M has no region with interior degree zero. A region D of interior degree one or two is a singleton strip, the former contributes three to s while the latter contributes two.

If D is a boundary region with $i(D) = 3$ then $[4 - i(D)] = 1$. If $i(D) = 4$ then $[4 - i(D)]$ makes no contribution to s. If $i(D) = 4 + k$, $k > 0$ then $[4 - i(D)]$ decreases s by k and can be thought of as cancelling the most recent contributions of regions of interior degree less than four.

Suppose that M has no singleton strips. There must be at least six regions of degree three whose contributions have not been cancelled. Suppose that D_1 is the first such region. Then, perhaps after encountering some regions of degree four, we must find a region D_n with interior degree three before encountering any region of interior degree five or greater. Furthermore, again due to the uncancelled status of D_1, D_n cannot be followed by a region with no boundary edges (for such a region would have interior degree at least six). Thus D_1, D_n and the regions between them form a strip. Since $s \geq 6$ we see that there must be at least three disjoint strips in this case. Listing all of the possibilities for singleton strips and their degrees we see that the conclusions of the lemma hold.

Now, still assuming that the boundary of M is a simple closed curve we suppose that there is a boundary region E of M with $\partial E \cap \partial M$ disconnected. Then $M-E$ has at least two components, say C_1 and C_2. The diagrams $J_i = C_i \cup E$ each have more than one region. By induction, each J_i has two disjoint strips. At most one of the strips in each J_i can involve E so that each J_i contributes at least one strip to M so M has two disjoint strips and at least four strips.

Finally, if the boundary of M is not a simple closed curve then M has two subdiagrams M_1 and M_2 that are extremal disks in the sense that the boundary of each M_i is a simple closed curve and M_i is connected to the rest of M by a single vertex, v_i. Each M_i is either a single region, and thus a singleton strip or contains more than one region and satisfies the inductive conclusion. Attachment at v_i can make at most one strip of M_i not encountered consecutively in the boundary of M. Thus M satisfies the conclusion of the lemma. □

To replace the argument based on Dehn regions for the word problem for groups of extra-large type, we need the following lemma.

<u>Lemma</u> 17. Let w be a non-trivial freely reduced word such that $w = 1$ in G and let M be a minimal $\mathcal{R}$-diagram for w. Then there is a

factorization $w = w_1 s w_2$, and a strip S of M with label s on its outside boundary and label b on its base (so that $s = b$ in G), such that the result $w' = w_1 b w_2$ of replacing s by b in w has shorter syllable length, i.e., $\|w'\| < \|w\|$.

Proof. If M has only one region then w is a conjugate of a defining relator r. Replacing r by 1 yields $w' = 1$ and, since M consists only of a singleton strip with interior degree zero the result holds. If M has more than one region then M has at least two strips (in the $C(6)$ sense) and the outside boundary label, say s, of one of the strips, S, is a subword of w. By looking at the interior degrees of regions in a $C(6)$ strip and noting that the label on an interior edge cannot be longer than one syllable, it is easy to see that the syllable length of the label s on the outside boundary of S exceeds the syllable length of the label b on the base of S by at least two. If $\|s\| \geq \|b\| + 3$ then replacing s by b yields w' with $\|w'\| < \|w\|$ even if both the first and last syllables of s are part of larger syllables of w.

Now $\|s\| \geq \|b\| + 3$ unless the label r on the last region D_n of S has a syllable change at the last vertex v on $\partial D_n \cap \partial M$.

Since we are dealing with a $C(6)$ strip, the region E meeting D_n along the edge b_n of ∂D_n incident to v has a boundary ege f on ∂M that follows the last edge e of $\partial D_n \cap \partial M$. The label on f must involve a generator different from that in the label $\varphi(e)$ for otherwise E and D_n would be of the same type, violating the minimality of M. In this situation, the last syllable of s is not part of a larger syllable of w, so replacing s by b still yields w' with $\|w'\| < \|w\|$. □

We now show how Lemma 17 gives an algorithm for the word problem for G. We will define the notation (s,b) for a pair of possible labels for the outside boundary and base of a strip in an $\mathcal{R}$-diagram (see Figure 3).

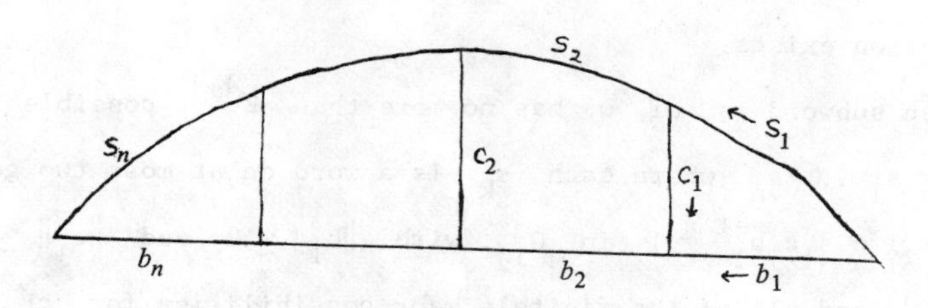

Figure 3

Define a <u>reduction pair</u> (for G) to be a pair (s,b) of words, with $|s| \geq |b|$, satisfying one of the following two conditions.

(i) s and b involve only two generators, say a_i and a_j, $\|b\| \leq 2$ and $sb^{-1} = 1$ in G_{ij}.

(ii) s and b have factorizations $s = s_1 \ldots s_n$, $b = b_1 \ldots b_n$, each $\|b_h\| \leq 2$, there exist $c_1, \ldots, c_n$ with each $\|c_i\| \leq 1$, and each of $s_1c_1b_1^{-1}$, $s_hc_hb_h^{-1}c_{h-1}^{-1}$, $h = 2,\ldots,n = 1$, $s_nb_n^{-1}c_{n-1}^{-1}$ is a relator of some type (i_ℓ, j_ℓ).

Suppose that w is a freely reduced word on A of the form $w = w_1sw_2$ where (s,b) is a reduction pair and $\|w_1bw_2\| < \|w\|$. The process of replacing s by b in w and freely reducing the resulting word is called an <u>extended</u> <u>R-reduction</u>. A freely reduced word w is <u>strongly</u> <u>R-reduced</u> if there is no extended R-reduction applicable to w. A cyclically reduced word is <u>strongly cyclically R-reduced</u> if there is no extended R-reduction applicable to any cyclic conjugate of w.

<u>Lemma</u> 18. Let G be an Artin group of large type. If w is a freely reduced non-trivial word such that $w = 1$ in G then there is an extended R-reduction applicable to w. If G is recursively presented then there is an effective procedure which, given any word w, calculates a strongly R-reduced word w' with $w = w'$ in G.

<u>Proof</u>. The first statement is immediate from Lemma 17. To prove the second part it suffices to show that there is an effective procedure which,

given w, finds an extended $\underset{\approx}{R}$-reduction applicable to w or concludes that no such reduction exists.

Any given subword s of w has no more than $2^{|s|}$ possible factorizations $s = s_1 \dots s_n$ where each s_k is a word on at most two generators. By Lemma 1.7, if $s_1 c_1 b_1^{-1} = 1$ in G_{ij}, with $\|b_1\| \leq 2$ and $\|c_1\| \leq 1$ then $|c_1 b_1^{-1}| \leq |s_1|$ and all of the finitely many possibilities for c_1 and b_1 can be listed. Thus all reduction pairs that could be applicable to w can be found and tested. If replacement of some s by the corresponding b yields a word of shorter syllable length the procedure can be repeated with the new word. A word w' such that no replacement reduces syllable length is strongly $\mathcal{R}$-reduced. □

Next we modify Lemma 1.10 to replace Dehn regions by strips.

Lemma 19. Let G be an Artin or Coxeter group of large type. If v is a strongly $\mathcal{R}$-reduced word that represents an element of G_J then v is a word on A_J.

Proof. Choose u from among all words on A_J such that $v = u^{-1}$ in G so that the $\mathcal{R}$-diagram M for uv has as few regions as possible, and such that u is shortest among words yielding diagrams with this number of regions. Then u must be freely reduced for if it were not then identification of adjacent inverse edges would provide a diagram with the same number of regions and shorter u. If the diagram M has a spike, it must occur such that one of u, v begins at the vertex of degree one on the spike and the other ends there. Thus the label on the spike is a word on A_J and if the diagram with spikes removed is not minimal then neither is M. Similarly if there are no spikes and (say) v has a terminal subword that is inverse to an initial subword of u, identification of inversely labelled adjacent edges provides a diagram that itself must satisfy our minimality condition. Thus, we may assume that uv is cyclically reduced. Suppose that M has at least one region. Then M has a boundary cycle $\alpha\beta$ where α and β have labels u and v respectively. A region D labelled by an element of $\mathcal{R}_J$ cannot have its boundary label intersect that of M in a subword of α, for deleting the region D from M would result in a diagram M^* with fewer regions and boundary label u^*v with u^* a word on A_J contradicting the minimality of choice of u.

Suppose v were not a word on A_J. If M consisted of a single region then u would have to consist of a single syllable and so v could not be strongly $\mathcal{R}$-reduced. Thus we need only consider the case in which M has at least two strips. If S is a strip of M let $\sigma = \partial S \cap \partial M$. Now σ must contain edges of both α and β. For if $\sigma \subseteq \beta$ then v would not be strongly $\mathcal{R}$-reduced. Also σ can have at most one edge in α since an end region D of a strip satisfies $i(D) \leq 3$ and two consecutive edges in α would force D to be labelled by a relator from $\mathcal{R}_J$ thus, as noted above, contradicting the minimal choice of u. If S is a singleton strip with $i(S) = 1$ then S has all but two of its edges in α and removal of S would constitute an $\mathcal{R}$-reduction. Hence no such singleton strip occurs and Lemma 16 guarantees that M has at least three disjoint strips. But only two strips can contain edges from both α and β. This contradiction shows that if v is not a word on A_J then M cannot have any regions, and thus u is v. □

Now Theorem 1 follows from Lemmas 1.9 and 19 since given a word w we can effectively calculate a strongly $\mathcal{R}$-reduced word v with $v = w$ in G. Now v does not contain any generators not contained in w and $v \in G_J$ only if v is a word on A_J.

Next, we prove Theorem 2 which states that an Artin group of large type is torsion free.

<u>Proof</u>. Suppose that G is an Artin group of large type and w is a word on the generators of G, $w \neq 1$ and $w^n = 1$ in G for some $n > 1$. Then w has a strongly cyclically $\mathcal{R}$-reduced conjugate u such that u is

non-empty and u has order n. Furthermore $|u| \leq |w|$. If such a w exists assume that one is chosen with u as short as possible. We may assume that u involves at least three generators for if u involved only two generators, say a_i and a_j then $u^n = 1$ in the one relator group G_{ij}. But a one-relator group whose relator is not a proper power is torsion free.

Let M be a minimal $\mathcal{R}$-diagram for u^n. M must have more than one region for if M has only one region we would not have three generators in u. Thus M has strips; let S be one such. If S is a singleton strip then the label s on $\partial M \cap \partial S$ involves only two generators and hence is a subword of a cyclic conjugate of u. In this situation, replacing s by the label on the base of S is a strong $\mathcal{R}$-reduction, contradicting the assumption that u is strongly cyclically $\mathcal{R}$-reduced. Thus M has no singleton strip and S must be a compound strip.

Let u^* be the cyclic conjugate of u that begins at the beginning of the label s on $\partial M \cap \partial S$. Then u^* must be a proper subword of s or the same type of reduction will apply. Now we examine the structure of S somewhat further. Let S consist of h regions. For $1 \leq k \leq h$ let $N_k = D_1 \cup \ldots \cup D_k$ consist of the first k regions of S ordered in the counterclockwise sense. Let s_k be the label on $\partial N_k \cap \partial M$ and let b_k be the label on $\partial N_k \cap \partial(M - N_k)$. The observation above on reduction makes it clear that if s_k is a subword of u^* then $|s_k| \leq |b_k|$. Let $f \leq h$ be such that if $k \leq f$ then s_k is a proper initial subword of u^*. For $1 \leq i \leq h$, let p_i be the label on $\partial D_i \cap \partial M$ and let q_i^{-1} be the label in D_i on $\partial D_i \cap \partial(M - S)$. For $1 \leq i < h$, let t_i be the label, in D_i, on $\partial D_i \cap \partial D_{i+1}$. Thus, for example, if $h > 2$, the label on D_2 is $p_2 t_2 q_2^{-1} t_1^{-1}$ (since we read a boundary label in the counterclockwise direction) (see Figure 4). Note that $s_k = p_1 \ldots p_k$ while, for $k < h$, $b_k = q_1 \ldots q_k t_k^{-1}$. Now we claim that for $1 \leq k \leq h$, $|s_k| \geq |b_k|$. The proof is by induction. For $k = 1$ we note that $i(D_1) = 3$, whence b_1 has three syllables and by Lemma 1.7, $|s_1| \geq |b_1|$. Now for $1 < k < h$, $i(D_k) = 4$, thus $q_k t_k^{-1}$ has three

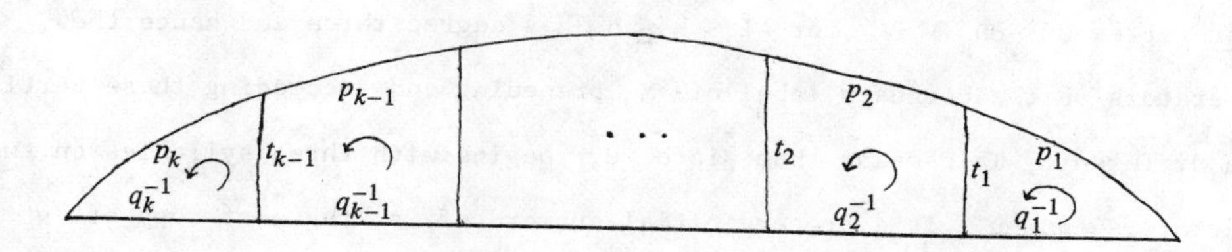

Figure 4

syllables and $|t_{k-1}p_k| \geq |q_k t_k|$. But $|s_k| = |s_{k-1}| + |p_k|$ while $|b_k| = |b_{k-1}| - |t_{k-1}| + |q_k| + |t_k|$ so the increase in $|s_k|$ is at least that in $|b_k|$ and the claim is proved. Thus, for $k \leq f$, $|s_k| = |b_k|$. This means that $m_{ij} = 3$ for the pairs a_i, a_j involved in the labels on regions $D_1, \ldots, D_f$. For if $m_{ij} > 3$ for region D_k then $d(D_k) \geq 8$ and since $i(D_k) \leq 4$ the reasoning above shows that $|s_k| > |b_k|$. But each $D_k \in S$ has at least two syllables in s, whence all relators represented in these D_k must already be represented in those D_k for $k \leq f$, so all regions in S correspond to relators with $m_{ij} = 3$. Thus (writing i for a_i, etc.) by Lemma 15 all have the form $iji^n\bar{j}\bar{i}\bar{j}^n$ or $i^n jij^{-n}\bar{i}^{-n}\bar{j}$.

Now consider D_k, $1 < k \leq f$. For these values of k, D_k has a six syllable label and $i(D_k) = 4$. If $\|p_k\| > 2$ then p_k must share some syllable of the label of D_k with either t_{k-1}^{-1} or t_k. But since $|t_{k-1}| + |p_k| = |q_k| + |t_k|$ as shown above, the last letter of p_k cannot be the same as the first letter of t_k (or, from the form of the relator, the equality could not hold). Thus t_k must be a full syllable of the label of D_k. Note also that the syllable containing t_{k-1}^{-1} and the syllable t_k, since they are precisely three syllables apart in the label of D_k, are of opposite sign (i.e., one is a power of a generator, the other is a power of the inverse of the other generator). Furthermore they are of the same length, hence if t_{k-1}^{-1} is not a full syllable of the label of D_k then $|t_k| > |t_{k-1}|$, otherwise $|t_k| = |t_{k-1}|$. In particular if for any k, $1 < k \leq f$, p_k has more than two syllables then $|t_k| > |t_1|$. It can also be seen that each syllable t_k (as part of the label D_k) has the same sign.

Now by the definition of compound strip, every vertex of M that is the last vertex of $\partial D_k \cap \partial M$ for $1 \leq k \leq h$ has degree three and hence the generators on the boundary label of M preceding and succeeding these vertices are different. This means that since u^* begins with three syllables on two generators and u^* is a proper initial subword of s, the next copy of u^* must begin at the initial letter of the label of some D_k, $k \leq h$. Thus u^* recommences at D_{f+1} and $p_{f+1} = p_1$.

Now, we first consider the case that $h = f + 1$, i.e., the second copy of u^* begins with the label on the last region of the strip. Now $p_1 = p_n$ and each consists of three full syllables of the label of the corresponding region. Now $|s_f| = |b_f|$. Hence u^* is equal in G to b_f. Now we will show that b_f is not cyclically reduced and hence has a shorter cyclic conjugate which is also a proper n-th power.

To show that b_f is not cyclically reduced we begin by noting that the first and last generators in p_1 are the same, whence the first generator of b_f is the same as the generator for the syllable t_{h-1} (from the form of the relators as dictated by Lemma 15). Now the sign on the last generator of p_1 is the same as that on the last generator of p_n, hence the same as that on t_{n-1} and opposite that of the first generator of p_1. But now, from the form of the relators it is immediate that the sign of the first generator in b_f is opposite to that of the last generator and b_f is not cyclically reduced.

Now suppose $h > f + 1$. Then, since each p_k has length at least two, for $f + 1 < k \leq n$, $p_k = p_{k-f}$ by the periodicity of the boundary label u^n. Now p_{f+1} has three syllables since $p_{f+1} = p_1$ so, as noted above, $|t_k| > 1$ for $k \geq f + 1$ and $|t_{f+1}| > |t_f|$. From the form of the relator we now see that only the first syllable of p_{f+1} (and hence of p_1) can have length greater than 1. This implies that in each D_k, $k \leq f$, if any syllable has length greater than one it is the syllable containing t_{k-1}^{-1}. On the other hand consider p_n. Its last syllable has length equal to that of t_{h-1}. If $|t_{h-1}| = 1$ then we contradict the existence of three syllables in p_f. If $|t_{j-1}| > 1$ then the last syllable of p_{h-f} has length greater than one since $p_{h-f} = p_h$, again a contradiction. □

To prove Theorem 3 for groups of large type we proceed as follows.

We will say that an Artin group G has the <u>separated (single) syllable</u> property if whenever w is a cyclically reduced word with $w = 1$ in G_{ij} and w has a factorization $w = uv$ where $\|u\| = 2$ then v has a syllable that is neither its first nor its last syllable consisting of a single generator with exponent ± 1.

<u>Lemma</u> 20. Let G be an Artin group with the separated single syllable property. Let H be a subgroup of G generated by the set $S_I = \{a_i^2 | i \in I\}$. Then H is freely generated by S_I.

<u>Proof</u>. If H were not freely generated by S_I then some cyclically reduced word w all of whose syllables have length at least two would be equal to 1 in G. Let M be a minimal $\mathcal{R}$-diagram for such a word w. M certainly cannot consist of a single $\mathcal{R}_{ij}$-region since that region (as a G_{ij} diagram) has the separated syllable property and hence has a single generator syllable. Hence M must have a strip S. If S is a singleton strip then $i(S) \leq 2$ and the label on the edge(s) of the base of S plays the role of u in the separated syllable property, whence $\varphi(\partial S \cap \partial M)$ must contain a single generator syllable that is not an end syllable (and hence is not a part of a longer syllable of the boundary label of M). Thus we may assume that S is a compound strip. Let D_n be the last region of S in the counterclockwise direction along ∂M. D_n has three interior edges in M and, by the definition of compound strip, lies on two boundary vertices of degree three. Now there must be a syllable change in a cyclically reduced boundary label at each degree three boundary vertex since the two regions at the vertex share a common edge. Thus, applying the separated syllable property to D_n as a region of a G_{ij} diagram, we see that even if a single generator syllable is an end syllable of the label of $\partial D_n \cap \partial M$ (which could

happen since there are three interior edges) it cannot be part of a longer syllable in the label w of ∂M.

Now we must show that two generator Artin groups of large type have the separated syllable property. To do so we examine properties of a purported counterexample. It is relatively straightforward to show (see the proof of Theorem 3 of [1]) that if $m_{ij} \geq 4$ then any connected, simple connected $\mathcal{R}_{ij}$-diagram M has the separated syllable property. Now suppose that M is a connected, simply connected $\mathcal{R}_{ij}$-diagram that does not have the separated syllable property. We can then be sure that $m_{ij} = 3$.

Each B-vertex of M, since it lies on a boundary path joining separating vertices of a region has a label including parts of three syllables of the boundary label of M, the middle one of which must have precisely one generator occurrence. The separated syllable property could only be false if all single generator syllables occurred among four consecutive syllables of the boundary label of M. This means that M must have precisely two B-vertices (since it must have at least two or all S-vertices would be consecutive, something shown impossible as a result of Lemma 13).

Now suppose that there are more than two S-vertices in M. In the proof of Lemma 13 we showed that there is a vertex of degree greater than three on ∂M on the path between consecutive S-vertices such that one unit of its degree is attributed to the pair of consecutive S-vertices. Suppose that M contained any regions of degree greater than four or interior vertices of degree greater than three or other boundary vertices of degree greater than three except those boundary vertices of degree four due to attributed units. We see by Lemma 13 that k, the number of disjoint consecutive strings of B-vertices, would be forced to be more than two, contradicting the assumption that M has precisely two B-vertices. Thus, we can assume that no such vertices or regions exist and that M consists of "rectangles joined at vertices of degree four."

Let v_1 and v_3 be the two B-vertices of M and let v_2 be the single S-vertex on $\delta(v_1,v_3)$. Note that to insure the separated syllable property all edges on $\delta(v_1,v_3)$ between the separating vertices of the regions on which v_1 and v_3 lie must bear labels consisting of a single generator occurrence, else the isolated syllables at the B-vertices would be more than two syllables apart and the separated syllable property would hold (see Figure 5a).

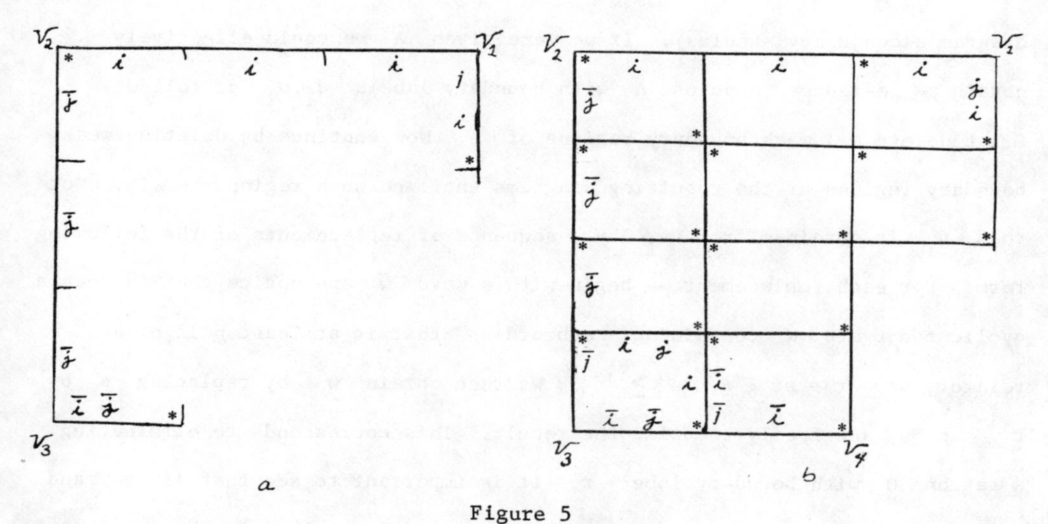

Figure 5

Next we note that the requirement that there be at most two B-vertices forces M to be a subdiagram of the "rectangle" with three corners at v_1,v_2,v_3. (One such example is shown in Figure 5b.) But now it is immediate that either v_4, the S-vertex following v_3 lies on the same region as v_3 forcing an immediate syllable change or else Lemma 14 forces two syllable edges on $\delta(v_3,v_4)$ hence the last syllable preceding v_4 must be a single generator syllable, demonstrating the separated syllable property for M. Thus the theorem is proved. □

5. The Conjugacy Theorem

To prove Theorem 4 it is necessary to refine our tools somewhat further. Let A be an annulus testifying to the conjugacy of its boundary labels w_1 and w_2 in an Artin group G of large type. If D is a boundary region of

A then D is called a <u>weak boundary region</u> of A if the intersection of ∂D with one boundary b of A is a consecutive part of b with label of length at least half of the length of the boundary label of D. A is called e-<u>reduced</u> if both of its boundary labels are cyclically reduced and has no weak boundary regions.

Suppose v_1 and v_2 are the labels on the inner and outer boundaries, respectively of some annular cancellation diagram $\tilde{A}$ (we will call such a diagram a conjugacy annulus). If we were given $\tilde{A}$ we could effectively obtain an e-reduced annulus A with boundary labels u_1, u_2 as follows. First delete all weak boundary regions of A. Now continue by deleting weak boundary regions of the resulting diagrams until no such regions remain. Note that u_1 is obtained from v_i by a sequence of replacements of the following form. For each replacement we begin with a word $\tilde{w}$ and notice that $\tilde{w}$ has a cyclic conjugate W containing a subword s that is at least half of a relator, say $r = st \in \mathcal{R}$, $|s| \geq |t|$. We then obtain w^* by replacing s by t^{-1} in w and freely reducing the result. This corresponds to eliminating a region D with boundary label r. It is important to see that if v_1 and v_2 are the labels on the boundaries of a conjugacy annulus $\tilde{A}$ then we can effectively find all possible pairs (U_1, U_2) of words that could be the boundary labels of e-reduced annuli obtainable from $\tilde{A}$.

Lemma 1.7 tells us that if $r = st \in \mathcal{R}$ and $|s| \geq |t|$ the number of syllables in s is at least m_{ij}, since $\|s\| \geq \|t\|$. Hence s involves both generators in r. Thus if D is a region involved in the e-reduction process it must have a label r from some $\mathcal{R}_{ij}$ such that a_i and a_j occur in v_1 or v_2. Furthermore $|r| \leq 2\max(|v_1|, |v_1|)$. Thus, given the Coxeter matrix M of G it is easy to list the finite collection of relators that can serve as labels in the process. But then the collection of words u_i obtainable from v_i $(i = 1,2)$ by such successive substitutions is a subset of the collection of words of length at most $|v_i|$ on the generators of v_i and hence is finite and may be completely listed in a straightforward manner.

Now suppose we provide a procedure to determine of cyclically reduced words u_1 and u_2 if they are boundary words of an e-reduced annulus (and hence are conjugate in G). Then, to determine whether or not arbitrary words v_1 and v_2 are conjugate we need only list all pairs (u_1,u_2) obtainable from cyclically reduced conjugates of (v_1,v_2) as above and see if any pair is the pair of boundary words of an e-reduced annulus. Since v_1 and v_2 can only be conjugate if one of the pairs (u_1,u_2) passes the test we would have a procedure to test the hypothesis that v_1 is conjugate to v_2. For this reason we need only consider e-reduced annuli in what follows.

Let A be an e-reduced conjugacy annulus without self-identified regions for an Artin group of large type. Further suppose that A is special in the sense that no regions lies on both boundaries of A. Since all regions have degree at least six if a boundary region D satisfied $i(D) < 4$ then, by Lemma 1.7, A could not be e-reduced. Thus $i(D) \geq 4$ for every region of A.

Since A is a (3,6) annulus Result VI yields

$$\Sigma^{\cdot}(4 - i(D)) = \Sigma^{\circ}(d(D) - 6) + 2\Sigma^{\circ}(d(v) - 3)$$

but the left side is non-positive and the terms on the right are non-negative; hence each interior vertex has degree three, each interior region has degree six and each boundary region has interior degree four. (Note that this implies that every boundary region has at least two boundary edges.)

Next, from Equation (*) of Section 2, we obtain

$$\Sigma^{\cdot}(\tfrac{5}{2} - d(v)) + \Sigma^{\circ}(3 - d(v)) + \tfrac{1}{2}\Sigma^{\circ}(6 - d(D)) + \Sigma^{\cdot}(6 - d(D)) = 0$$

and combining this with the results above, we obtain

$$\Sigma^{\cdot}(\tfrac{5}{2} - d(v)) + \tfrac{1}{2}\Sigma^{\cdot}(6 - d(D)) = 0 .$$

Now consider the set B of boundary vertices of degree greater than two. Each of these occurs on two boundary regions and each boundary region lies on two such vertices so $|B|$ is both the number of such vertices and the number of boundary regions. Furthermore each boundary region D has interior degree four and hence has $(d(D) - 5)$ boundary vertices of degree two. Thus we have

$$\Sigma^{\cdot}(\frac{5}{2} - d(v)) = -\frac{1}{2}\Sigma^{\cdot}(d(D) - 5) + \sum_B(\frac{5}{2} - d(v))$$

whence

$$-\frac{1}{2}\Sigma^{\cdot}(d(D) - 5) + \sum_B(\frac{5}{2} - d(v)) + \frac{1}{2}\Sigma^{\cdot}(6 - d(D)) = 0$$

so

$$\sum_B(\frac{5}{2} - d(v)) = -\frac{1}{2}\Sigma^{\cdot} 1 = -\frac{B}{2}$$

and all boundary vertices have degree at most three.

Next note that since A is e-reduced and $i(D) = 4$ for each boundary region no boundary region can have degree greater than or equal to eight. (Note that degree seven is possible by splitting a syllable between a pair of edges but no relator involved can have more than six syllables.)

Lemma 15 tells us that a six syllable relator has the form of $a_i a_j a_i^n \bar{a}_j \bar{a}_i \bar{a}_j^n$ or its inverse. Thus all regions have such boundary labels. Now consider the regions lying on the inner boundary of A. Each of these regions has interior degree four so the vertices on these regions that do not lie on the inner boundary of A determine a closed path P of edges such that every second vertex on P shares an edge with a vertex on the inner boundary of A. Since we have assumed that no region lies on both boundaries of A, P consists entirely of interior edges. We examine the sub-annulus B of A consisting of all regions lying on edges of P. Since all interior vertices of A have degree three B is of the form of the annulus of Figure 6 (where boundary vertices are not shown).

Now suppose the region marked I has label involving generators a_i and a_j and region II has label involving a_i and a_k (with the common edge having label a power of a_i. Then it is easy to see that region III has label involving a_j and a_k, whence region IV has label involving a_i and a_j, etc. Indeed there must be 3m regions on each boundary of B, for some integer m,

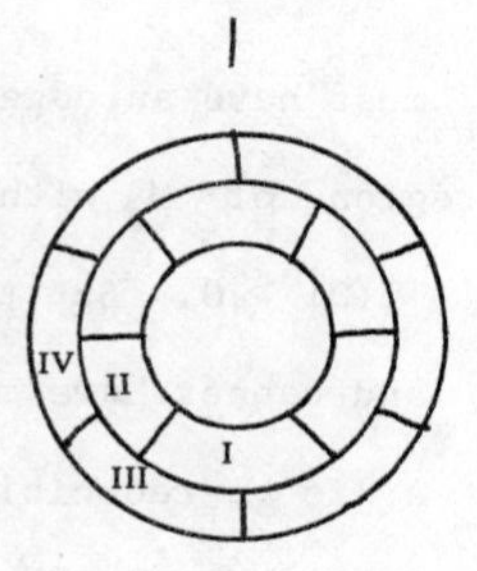

Figure 6

and all regions involve relators on a_i, a_j and a_k of the form prescribed by Lemma 15.

If the outer path, Q, is not the outer boundary of A then the conditions on degrees and interior degrees force it to consist entirely of interior edges and vertices, whence A consists of two or more full layers of 3m regions each for some $m \geq 1$, and all regions are labelled by relators involving two of three specified generators.

Lastly, we note that if A is an annulus with the minimal number of regions for its boundary words it can have no more than six layers of 3m regions each where its boundary words have length 6m. Thus, given u_1 and u_2 it is straightforward to test for the existence of such an A. (Actually one can describe the possible pairs (u_1,u_2) in considerably more detail but this is not necessary for our purposes.)

Now suppose that an e-reduced annulus A without self-identified regions has a region D that lies on both boundaries of A. Then we claim that every region must lie on both boundaries. For suppose not. Let M be a subdiagram of A that contains a region that does not lie on both boundaries of A, contains no region that lies on both boundaries of A and has edges in common with regions D_1 and D_2 (possibly $D_1 = D_2$ and the edges are distinct) that lie on both boundaries of A.

Now M is a simply connected (3,6) diagram, hence by Result II, $\Sigma^*_M(4 - i(D)) \geq 6$. But since A is e-reduced if D is a simple boundary region of M then if $\partial D \cap \partial M = \partial D \cap \partial A$ we must have $4 - i(D) \leq 0$ whereas if $\partial D \cap \partial M \neq \partial D \cap \partial A$ then D must have an edge in common with D_1 or D_2. Thus only those (at most two) regions of M with boundary edges on both ∂A and ∂D_1 or ∂D_2 can have $4 - i(D) > 0$. But then, since such regions can lie on only one boundary of A and cannot have more edges on the boundary of A than interior to A (since A is e-reducible), $\Sigma_M^*(4 - i(D)) \leq 4$, a contradiction.

Now given u_1 and u_2 it is easy to determine if there is an e-reduced conjugacy annulus for G all of whose regions lie on both boundaries that has u_1 and u_2 as boundary labels.

Now consider an e-reduced annulus A that has at least one self-identified region. Let B be the submap of A consisting of those regions that are not self-identified. We will show that B is empty.

First, suppose that a non-empty submap C of B is bounded by two self-identified regions, E and F of A. If no vertex of E lies on F then C has only two boundary vertices of degree two, namely those on the interior paths of E and F respectively. But then C violates Equation (1). Suppose E and F have common vertices. Then C is a union of edge disjoint simply connected diagrams. Each of these lies on at most two vertices common to E and F and at most two vertices on interior edges of E and F, hence has at most four boundary vertices of degree two. But for a non-empty $(3,6)$ diagram, six such vertices are required, thus C is empty.

Next, suppose C is a non-empty submap of B consisting of all regions outside the outermost self-identified region, E. First suppose E has no vertex on the boundary of A. Then C is an annulus with no vertex on both its boundaries. If C has a region that lies on both of its boundaries then, since C is an e-reduced annulus, every region lies on both boundaries. But then since C is not self-identified it must have a region with no vertex of degree two on its interior boundary, hence it must have only three edges interior to A. This contradicts e-reducibility. If no region of C lies on both boundaries of C then, since some region of C on its inner boundary has interior degree at least five, C is not e-reduced, whence A is not e-reduced.

Last, suppose C is a union of edge disjoint simply connected diagrams. Then, let M be one such subdiagram. By Result II, $\Sigma_M^*(4 - i(D)) \geq 6$. But each simple region of M can have at most one edge on E. Thus either M has a region of interior degree at most three with no edge on E or a region of interior degree at most one with a single edge on E. In either case A is not e-reduced, a contradiction.

The identical reasoning applies to the inner boundary of E.

Thus annuli with self-identified regions may be treated in exactly the same manner as in the case of Artin or Coxeter groups of extra-large type in [1] and the proof of Theorem 4 is complete.

If one examines the proof of Theorems 1, 2 and 3 one discovers that no essential use is made of the fact that G is finitely presented -- the hypothesis that G is recursively presented would permit us to find all needed m_{ij} and the fact that all necessary computations can be done in the subgroup G_J generated by the generators that occur in boundary labels is all that is needed. Theorem 4 is not true for recursively presented groups for the reason that one needs to know of a pair a_i, a_j of generators of G whether they are joined by an odd path in the Coxeter graph Γ of G.

There exist infinite recursive presentations for which the existence of odd paths connecting vertices in Γ is undecidable. The following elegant example is due to Carl Jockusch. Let I be the natural numbers and let S be a recursively enumerable, non-recursive set enumerated by recursive function f. Let p be an odd integer, $p \geq 5$. If i and j are both even let $m_{ij} = p$. If exactly one is odd, say $i = 2s$ and $j = 2t + 1$, let $m_{ij} = p$ if $f(s) = t$ and let $m_{ij} = \infty$ otherwise. If i and j are both odd, let $m_{ij} = \infty$. The function m_{ij} is certainly recursive. Choose odd integers h and k, $h = 2a + 1$, $k = 2b + 1$. There is an odd path connecting

h and k if and only if both a and b are elements of S. But, since S is not recursive this question cannot be effectively decided.

If we add the hypothesis that the "odd path problem" is solvable for recursively presented G of large type then the conjugacy problem is solvable.

REFERENCES

1. Appel, K. I. and Schupp, P. E., Artin Groups and Infinite Coxeter Group, Inventiones Math. 72 (1983), 201-220.
2. Birman, J., Braids, Links, and Mapping Class Groups (Princeton, N.J.: Princeton University Press, 1974).
3. Bourbaki, N., Groups et algèbres de Lie, Chapitres 4, 5, et 6, Eléments de Mathématique XXXIV (Paris: Hermann, 1968).
4. Brieskorn, E., Die Fundamentalgruppe des Raumes der Regulären Orbits einer endlichen komplexen Spielsgruppe, Inventiones Math. 12 (1971), 57-61.
5. Brieskorn, E., and Saito, K., Artin-gruppen und Coxeter-gruppen, Inventiones Math. 17 (1972), 245-271.
6. Coxeter, H. S. M., The complete enumeration of finite groups of the form $R_i^2 = (R_iR_j)^{k_{ij}} = 1$, J. London Math. Soc. 10 (1935), 21-25.
7. Deligne, P., Les immeubles des groupes de tresses généralisés, Inventiones Math. 17 (1972), 273-302.
8. Garside, F. A., The braid group and other groups, Quart. J. Math., Oxford, 2 Ser. 20 (1969), 235-254.
9. Len, V. Ya, Artin's braids and their connections with groups and spaces, in Etoge Nauke e Texneke, Algebra, Topologeya, Geometreya, Tom 17.
10. Lyndon, R. C., on Dehn's algorithm, Math. Ann. 166 (1966), 208-228.
11. Lyndon, R. C. and Schupp, P. E., Combinatorial Group Theory, Ergebnisse der Mathematik und ihrer Grenzgebeite, 1977 (Berlin: Springer-Verlag).
12. Schupp, P. E., A survey of small cancellation theory, in Word Problems (Amsterdam: North Holland, 1973), pp. 569-589.
13. Tits, J., Le problème des mots dans les groupes de Coxeter, Instituto Nazionale di Alta Mathematica, Symposia Mathematica 1 (1968), 175-185.
14. Tits, J., Normalisateurs de tores I. Groupes de Coxeter étendus, J. Algebra 4 (1966), 96-116.

DEPARTMENT OF MATHEMATICS
UNIVERSITY OF ILLINOIS
1409 W. GREEN
URBANA, ILLINOIS 61801

Received April 14, 1983

Contemporary Mathematics
Volume 33, 1984

THE FINITE GENERATION OF AUT(G), G-FREE METABELIAN OF RANK ≥ 4

Seymour Bachmuth* and Horace Y. Mochizuki*

(To Roger C. Lyndon on his 65th birthday)

Let $G(n)$ denote the free metabelian group of rank n. Thus, $G(n) = F(n)/F(n)''$ where $F(n)$ is the free group of rank n and $F(n)''$ is the second derived group of $F(n)$. It is known that Aut $G(2)$ is finitely generated whereas Aut $G(3)$ is not finitely generated. However, contrary to expectations, we can now announce

<u>Theorem</u> 1: For any integer n larger than 3, $\mathrm{Aut}(G(n))$ is finitely generated.

Actually, a much stronger result holds, of which Theorem 1 is a consequence.

<u>Theorem</u> 2: The natural homomorphism $\mathrm{Aut}(F(n)) \to \mathrm{Aut}(G(n))$ is a surjection for $n = 2$ and $n \geq 4$. In other words, each automorphism of $G(n)$ is induced by an automorphism of $F(n)$ for $n = 2$ and $n \geq 4$.

Nielsen [7] showed that $\mathrm{Aut}(F(n))$ can be generated by four automorphisms, and B. H. Neumann [6] reduced the number of generators to two for $n \geq 4$ (cf. [5], Sec. 3.5 or [4], page 43).

A comparison of Theorems 1 and 2 with known results suggests a relationship between the sequence of groups

$$\mathrm{Aut}(G(n)),\ n = 2,3,\dots ,$$

*Research supported by NSF Grant No. MCS 81-01904.

© 1984 American Mathematical Society
0271-4132/84 $1.00 + $.25 per page

and

$$GL(n-1, R(n-1)), \ n = 2,3,\dots ,$$

where $R(n) = \mathbb{Z}[x_1,x_1^{-1},\dots,x_n,x_n^{-1}]$, the Laurent polynomial ring in n indeterminants and their inverses over the ring of rational integers $\mathbb{Z}$. We observe that $Aut(G(n))$ and $GL(n-1, R(n-1))$ play analogous roles. Theorem 1 completes the story which tells us that for $n = 2$ and $n \geq 4$, both $Aut(G(n))$ and $GL(n-1, R(n-1))$ are finitely generated whereas $Aut(G(3))$ and $GL(2, R(2))$ are not finitely generated. (For the case $GL(2, R(2))$ see [2]; for $Aut(G(3))$ see [1]; and for $GL(n, R(n))$, $n \geq 3$, see [8].) However, Theorem 2 makes clear a much deeper relationship between the sequences $Aut(G(n))$ and $GL(n-1, R(n-1))$, $n = 2,3,\dots$. To understand this relationship, we must define the common designation of "tame" element in groups in both sequences.

The subgroup of $GL(n, R(n))$ generated by the elementary matrices and the invertible diagonal matrices is denoted by $GE(n, R(n))$. An element of $GL(n, R(n))$ is called <u>tame</u> if it is in $GE(n, R(n))$ and <u>non-tame</u> otherwise.

In $Aut(G(n))$, an element is called <u>tame</u> if it is the image of an element of $Aut(F(n))$ under the natural homomorphism $Aut(F(n)) \to Aut(G(n))$ and <u>non-tame</u> otherwise.

Thus, in both sequences of groups, the tame elements are just the obvious elements which these groups possess. In this context what Theorem 2 says is that for $n \geq 4$, each of $GL(n-1, R(n-1))$ and $Aut(G(n))$ consists entirely of tame elements (see [8] for the case of the GL's). By contrast, for both $GL(2, R(2))$ and $Aut(G(3))$, any set of generators must contain an infinite number of non-tame elements ([2] and [1]).

Because of certain similarities between the automorphism groups of free metabelian groups and the automorphism groups of polynomial rings, one should consider a third sequence of groups

$$Aut_K(P(n)), \ n = 2,3,\dots ,$$

where $Aut_K(P(n))$ is the automorphism group of the ring $P(n)$ of polynomials in n (commuting) indeterminates over the field K. As in the groups $Aut(G(n))$, $Aut_K(P(n))$ contains some obvious automorphisms which can be designated as the "tame" elements of $Aut_K(P(n))$. These are rather precise analogues of the tame elements of $Aut(G(n))$, and we refer the reader to [3] for definitions and a full discussion of this third sequence $Aut_K(P(n))$, $n = 2,3,\ldots$. In closing, we cannot resist the speculations that, as in the two previous sequences, for $n \neq 3$, $Aut_K(P(n))$ consists only of tame elements, whereas $Aut_K(P(3))$ contains "mostly" non-tame elements. Knowing Roger's interest in the automorphism problems described here, we are pleased to offer these conjectures as a sixty-fifth birthday present.

REFERENCES

[1] S. Bachmuth, H. Y. Mochizuki, The non-finite generation of Aut(G), G-free metabelian of rank 3, Trans. Amer. Math. Soc. 270 (1982), 693-700.

[2] S. Bachmuth, H. Y. Mochizuki, $E_2 \neq SL_2$ for most Laurent polynomial rings, Amer. J. Math. 104 (1982), 1181-1189.

[3] S. Bachmuth, H. Y. Mochizuki, GL_n and the automorphism groups of free metabelian groups and polynomial rings, Groups-St. Andrews 1981, edited by Campbell and Robertson, Cambridge University Press, 1982, 160-168.

[4] R. C. Lyndon, P. E. Schupp, Combinatorial Group Theory, Berlin-Heidelberg-New York, Springer-Verlag, 1977.

[5] W. Magnus, A. Karrass, D. Solitar, Combinatorial Group Theory, New York, Wiley, 1966.

[6] B. H. Neumann, Die Automorphismengruppe der freien Gruppen, Math. Ann. 107 (1932), 367-386.

[7] J. Nielsen, Die Isomorphismengruppe der freien Gruppen, Math. Ann. 91 (1924), 169-209.

[8] A. A. Suslin, On the structure of the special linear group over polynomial rings, Isv. Akad. Nauk. 11 (1977), 221-238.

DEPARTMENT OF MATHEMATICS
UNIVERSITY OF CALIFORNIA, SANTA BARBARA
SANTA BARBARA, CALIFORNIA 93106

Received April 21, 1982

Contemporary Mathematics
Volume 33, 1984

TWO-GENERATOR GROUPS III

J. L. Brenner, R. M. Guralnick, and James Wiegold

Dedicated to our dear friend R. C. Lyndon

§1. Introduction. The main theme of this note is the

PROBLEM 1. What groups can be generated by two conjugates of a single element?

This appears to be a well-known question for finite simple groups. It is clear that alternating and symmetric groups have generating sets of this sort, but as far as we know no other big classes of finite groups have been investigated from this point of view. For simplicity let us call a group a C-group if it can be generated by two conjugates of one element. In §2 we shall use some elementary calculations to prove that the Mathieu groups M_{11} and M_{12} are C-groups; in §3 we do the same for the groups PSL(2,q).

A related question is one called to our attention by Joachim Neubüser:

PROBLEM 2. Prove that every finite 2-generator simple group G has two ordered pairs of generators, (a,b) and (c,d), such that there is no automorphism α of G with $a^{\alpha} = c$, $b^{\alpha} = d$.

All known finite non-abelian simple groups do have this property, but a uniform elementary proof seems out of reach at present. Call a two-generator group G a J-group if it has two pairs of generators as in the formulation of Problem 2. The connection mentioned above is the following result. Note that there is no assumption as to finiteness or simplicity or the like.

© 1984 American Mathematical Society
0271-4132/84 $1.00 + $.25 per page

THEOREM 1.1. Every non-trivial C-group is a J-group.

Proof. Suppose that $G = \langle a, a^x \rangle$ for some a, x, with $a \neq 1$. Then $G = \langle a,x \rangle = \langle a,ax \rangle$. We claim that there is no automorphism α of G with $a^\alpha = a$, $x^\alpha = ax$. Suppose there were. Then $(a^x)^\alpha = a^{ax} = a^x$, and so α is the identity mapping since it fixes the generators a, a^x. But then $ax = x^\alpha = x$, $a = 1$, a contradiction.

Clearly, a necessary condition for a group G to be a C-group is that $d(G) \leq 2$, $d(G/G') \leq 1$. This shows that not every J-group is a C-group, since for example the dihedral group D_8 is a J-group. Moreover, the conditions are not sufficient, as the following example shows:

EXAMPLE 1.2. Let G be a finite two-generator nonabelian simple group, and n the largest positive integer such that $d(G^n) = 2$, where now G^m means the m-th direct power of G. Then G^n is not a J-group and thus not a C-group either.

To see this, observe that there is just one normal subgroup N of the free group F of rank 2 such that $F/N \cong G^n$, namely the intersection of all M with $F/M \cong G$ (see P. Hall [6]). Thus every pair of ordered pairs of generators of G^n can be connected by an automorphism as required. A concrete example is A_5^{19}. Similar observations and examples can be made for higher generating numbers.

The example given here is insoluble, and the question arises as to the soluble case. We believe that every (finite?) soluble group G with $d(G) = 2$ and $d(G/G') = 1$ is a C-group, but we have been able to prove this only for the metabelian case. (This assertion is a corollary of the following lemma.)

LEMMA 1.3. Let G be a metabelian group with $d(G) = k$, $d(G/G') = 1$. Then G can be generated by k conjugates of a single element.

PROOF. Using Tietze transformations we can find a generating set $\{a, x_2, x_3, \ldots, x_k\}$ for G with all x_i in G'. Then G' is the normal closure of all commutators $[a,x_2], \ldots, [a,x_k]$. But every element of G is of the form ga^{λ} for suitable $g \in G'$, $\lambda \in \mathbb{Z}$; thus, as G' is abelian, every conjugate of $[a,x_i]$ is of the form $[a,x_i]^{a^{\lambda}}$. This means that $G = \langle a,[a,x_2],\ldots,[a,x_k]\rangle$, since the group on the right contains G' and thus $x_2,\ldots,x_k$. As $[a,x] = a^{-1}a^x$ for all x, we have our assertion.

Our attempts to extend 1.3 to higher solubility lengths, even for finite groups, have failed. We believe it should be possible! Note that every finite group G with G/G' cyclic is the normal closure of one element [9, Lemma 3.1], so it is generated by conjugates of that element; our difficulty arises in proving one can do it with $d(G)$ such conjugates. (See §5.)

In the first paper of this series [2], we introduced the concept of <u>spread</u> of a finite group, and we repeat it here. A two-generator group G is said to have spread r if for any non-trivial elements $x_1,x_2,\ldots,x_r$ in G there exists an element y in G such that $G' = \langle x_1,y\rangle = \ldots = \langle x_r,y\rangle$. For example, A_{2n}, $n \geq 4$ has spread 4 but not spread 5, A_{2n+1} has spread 3 (and probably has spread tending to infinity with n); $PSL(2,q)$ has large spread. We conjectured that every finite simple group has spread 1; the concept has the following connection with the others dealt with here:

LEMMA 1.4. <u>Let</u> G <u>be a non-cyclic group of spread</u> 1 <u>containing an involution</u>. <u>If</u> G <u>is not a</u> J-<u>group</u>, <u>then</u> $G \cong V_4$.

PROOF. Let a be an involution and b a mate for it, $G = \langle a,b\rangle$. Then b has order 2 since there is an automorphism α with $a^{\alpha} = b$, $b^{\alpha} = a$. Similarly $G = \langle a,ab\rangle$ and so ab has order 2.

By Feit-Thompson, simple finite groups of spread 1 are J-groups. But then, they are probably C-groups anyhow.

In §4 we show how to prove that M_{11} and M_{12} have spread 1. They probably have much higher spread, but we do not feel the necessary calculations would be worth pursuing. (See §5.)

§2. M_{11} and M_{12} are C-groups

There are several ways of proving this. We choose the one which is most elementary, viz.; exhibition of suitable generators.

Firstly, M_{11} can be generated by the permutations

$$\alpha = (1,2,\dots,11)$$

$$\beta = (2,5,6,10,4)(3,9,11,8,7)$$

$$\gamma = (3,11,9,7)(4,10,5,6)$$

(see Coxeter and Moser [4, p. 99]). Now $\langle \alpha, \alpha^{\gamma} \rangle$ is clearly primitive, and it contains the commutator

$$[\alpha, \gamma] = (1,10)(3,11,6,9,7,4,8,5) ,$$

which is of order 8. Sims' list [8, pp. 169-183] asserts that the primitive proper subgroups of M_{11} are of order 11,22,55,660; none of these contains an element of order 8, so then $\langle \alpha, \alpha^{\gamma} \rangle = M_{11}$.

For M_{12} we use the Sims generators. We have $M_{11} = \langle a_9, c_9^2, c_9 d_9, c_{10}, d_{11} \rangle$, $M_{12} = \langle M_{11}, c_{12} \rangle$, where

$$a_9 = (1,2,3)(4,5,6)(7,8,9)$$

$$c_9^2 = (2,4,3,7)(6,9,8,5)$$

$$c_9 d_9 = (2,9,3,5)(4,6,7,8)$$

$$c_{10} = (1,10)(4,7)(5,6)(8,9)$$

$$d_{11} = (4,8)(5,9)(6,7)(10,11)$$

$$c_{12} = (4,7)(5,8)(6,9)(11,12) .$$

Next, we calculate the element e given by

$$e = c_9^2 a_9 c_{10} d_{11} = (1,2,7,3,5,6,8,9,4,11,10) .$$

We shall show that $\langle e, c_{12}^{-1} e c_{12} \rangle = M_{12}$. Firstly it is transitive since $c_{12}^{-1} e c_{12}$ moves 12, and e moves 1,2,...,11; and primitive since it is even doubly transitive (e is an 11-cycle). Now we make the computations

$$c_{12}^{-1}ec_{12}e = (1,7,12)(2,11,10)(3,9,6)(4,5,8)$$

$$(c_{12}^{-1}ec_{12}e^4)^2 = (1,8,6)(3,5,7)(10,11,12) .$$

Now it is clear from Sims' list that the Sylow 3-subgroup of every primitive subgroup of M_{12} other than M_{12} itself is cyclic of order 3; since this is manifestly false for $\langle e, c_{12}^{-1}ec_{12}\rangle$, this group is M_{12}.

§3. PSL(2,q) is a C-group

For $q \leq 5$ everything is clear; $PSL(2,2) \cong S_3$, $PSL(2,3) \cong A_4$, $PSL(2,5) \cong A_5$, so we may assume that $q > 5$. Our strategy is always to find a C-subgroup of PSL(2,q) strictly containing a dihedral subgroup known to be maximal. We content ourselves with stating which elements work, and giving brief reasons why. Convenient references are [5] and [7]. There are three cases:

I. $q \equiv 1 (\text{mod } 4)$. Let θ be a primitive element of F_q, write $r = \frac{1}{4}(q-1)$, $a_1 = \text{diag}(\theta, \theta^{-1})$, $a_2 = a_1^r$. Then a_2 has order 2 mod the centre of SL(2,q). With $w = \frac{1}{2}(\theta + \theta^{-1})$ and

$$b = \begin{pmatrix} 1 & \frac{1}{2}(w-1) \\ 1 & \frac{1}{2}(w+1) \end{pmatrix} ,$$

we claim that $\langle a_1, b^{-1}a_1b\rangle = SL(2,q)$. This follows since it contains $\langle a_2, b^{-1}a_2b\rangle$, which modulo the centre is dihedral and maximal. [Note trace $a_2^{-1}b^{-1}a_2b = -2w$.]

II. $q \equiv 3 (\text{mod } 4)$. Let a_1 be any element of order $\frac{1}{2}(q+1)$ in PSL(2,q); set $r = \frac{1}{4}(q+1)$ and $a_2 = a_1^r$. Then a_2 lies in a dihedral group $\langle a_2, b^{-1}a_2b\rangle$ of order $q+1$ not containing a_1, so that $\langle a_1, b^{-1}a_1b\rangle = PSL(2,q)$.

III. $q \equiv 0 (\text{mod } 4)$. Here we let $a_3 = \text{diag}(\theta, \theta^{-1})$, where θ is a primitive element in F_q. Set $a_1 = \begin{pmatrix} 1 & 1 \\ -1 & 0 \end{pmatrix}$, $a_2 = a_1^2$. Then $\langle a_1, a_3^{-1}a_1a_3\rangle$ contains the dihedral subgroup $\langle a_2, a_3^{-1}a_2a_3\rangle$ properly, and so it is PSL(2,q).

§4. M_{11} and M_{12} have spread 1.

We show that every nontrivial element $x \in G = M_{11}$ $[M_{12}]$ has a mate y of order 11 such that $\langle x,y\rangle = G$.

LEMMA 4.1. Let M_{11} $[M_{12}]$ be represented in the usual way as a permutation group on 11 [12] symbols. Let T be any 11-subgroup of M_{11} $[M_{12}]$. Then T is contained in a unique transitive subgroup H isomorphic to $PSL(2,11)$.

PROOF. Clearly T is contained in one such subgroup H. Suppose (for a contradiction) that $T < H_1 \cap H_2$, where $H_1 \cong H_2 \cong PSL(2,11)$, $H_1 \neq H_2$. From [3a] we know that $H_2 = H_1^g$ for some g in M_{11} $[M_{12}]$. But then, both T and T^g are contained in the 11-Sylow subgroup of H_2. Thus $T = T^{gk}$ for some k in H_2. From this, $gk \in N(T) < H_2$. ($N(T)$ is the normalizer of T in M_{11} $[M_{12}]$.) Thus $g \in H_2$, so that $H_1 = H_2$.

THEOREM 4.2. M_{11} has spread 1. In fact if $1 \neq x \in M_{11}$, then y exists in M_{11}, $y^{11} = 1$, with $M_{11} = \langle x,y\rangle$.

PROOF. The conjugacy class Γ of the subgroups H of M_{11} that are isomorphic to $PSL(2,11)$ has trivial intersection: $\cap H = 1$. Let $1 \neq x \in M_{11}$ be given, and find $H_o \cong PSL(2,11)$ such that $x \notin H_o$. Choose $y \in H_o$ of order 11. Then $K = \langle x,y\rangle$ is transitive. Moreover, if $L \in \Gamma$, then the lemma shows that $K \not\leq L$, since $x \notin H_o$, and if $L \neq H_o$, then $y \notin L$. The list of maximal subgroups on page 235 of [3a] shows that $K = M_{11}$.

REMARK 4.3. We have another proof of Theorem 4.2 using Sims' tables and direct calculation.

THEOREM 4.4. M_{12} has spread 1. In fact if $1 \neq x \in M_{12}$ then y exists in M_{12}, $y^{11} = 1$, with $M_{12} = \langle x,y\rangle$.

PROOF. As in the proof of Theorem 4.2, let Γ be the conjugacy class of the maximal subgroups of M_{12} that are isomorphic to $PSL(2,11)$. Choose $H_o \in \Gamma$ such that $x \notin H_o$. Since H_o is a transitive subgroup of M_{12}, there exists a permutation y of period 11 in H_o such that x,y do not fix a common point (in either permutation representation). Then $K = \langle x,y\rangle$ will be doubly transitive. If $K \neq M_{12}$, the list of maximal subgroups of M_{12} in [3a] shows that $K \leq L$ for some L in Γ, $L \simeq PSL(2,11)$. But on the one hand, $L \neq H_o$ (since $x \notin H_o$), and on the other hand, $y \notin L$ if $L \neq H_o$ (Lemma 4.1). This contradiction shows that $K = M_{12}$.

REMARK 4.5. A refinement of the argument in the proofs of 4.2, 4.4 shows that M_{11}, M_{12} have spread 2. They probably have somewhat greater spread.

REMARK 4.6. Using Sims' tables, 4.4 can be proved by direct calculation.

§5. Postscript.

This article was submitted in 1981. The authors have since obtained proofs of the following; these results are unpublished.

5.1 Every sporadic finasig has a pair of conjugate generators. The same holds for $PSL(n,q)$.

5.2 There is a class C in $PSL(n,q)$ (if $q > n+1$) such that $CC = PSL(n,q)$.

5.3 All five Mathieu groups have spread ≥ 2.

5.4 Further results have been obtained in connection with Lemma 1.3.

5.5 The conjecture on page 84, line 9 has been settled negatively. Stewart, Heineken, Lennox, and Wiegold have found a 2-generator group of solubility length 3 not generated by two conjugates of a single element.

5.6 J. Neubüser has computed the structure constants for all squares of classes in each of the 26 sporadic simple groups. In that way he showed that every one of these groups has a class C such that CC covers the group. In particular, every element in each of these groups is a commutator.

REFERENCES

1. H. R. Brahana, "Pairs of generators of the known simple groups whose orders are less than one million," Ann. of Math. (2) 31 (1930), 529-549.

2. J. L. Brenner and James Wiegold, "Two-generator groups I," Michigan Math. J. 22 (1975), 53-64.

3. J. L. Brenner and James Wiegold, "Two-generator groups II," Bull. Austral. Math. Soc. 22 (1980), 113-124.

3a. J. H. Conway, "Three Lectures on Exceptional Groups," in Finite Simple Groups, edited by M. Powell and G. Higman, Academic Press, New York, 1971.

4. H. S. M. Coxeter and W. O. J. Moser, Generators and relations for discrete groups, Ergebnisse der Math. und ihrer Grenzgebiete, Springer-Verlag, Berlin-Göttingen-Heidelberg, 1967.

5. L. E. Dickson, Linear Groups, Dover Publications, New York, 1958.

6. P. Hall, "The Eulerian functions of a group," Quart. J. Math. Oxford 7 (1936), 134-151.

7. B. Huppert, Endliche Gruppen, Springer-Verlag, Berlin-Heidelberg-New York, 1967.

8. C. C. Sims, "Computational methods in the theory of permutation groups," in Computational Problems in Abstract Algebra (Proc. Conf. Oxford, 1967).

9. James Wiegold, "Growth sequences of finite groups," J. Austral. Math. Soc. XVII (1974), 133-141.

10 PHILLIPS ROAD
PALO ALTO, CALIFORNIA 94303 USA

Received October 25, 1980

UNIVERSITY OF SOUTHERN CALIFORNIA
LOS ANGELES, CALIFORNIA 90089

UNIVERSITY COLLEGE
CARDIFF CF1 1XL
WALES

Contemporary Mathematics
Volume 33, 1984

THE SUBGROUP SEPARABILITY OF FREE PRODUCTS OF TWO FREE GROUPS WITH CYCLIC AMALGAMATION

A. M. Brunner, R. G. Burns and D. Solitar

0. Introduction

In line with terminology introduced by Mal'cev [12], we shall say that a group G has the property of finite separability of its finitely generated subgroups (or, more briefly, is subgroup separable) if for each finitely generated subgroup H of G and each $g \in G - H$, there is a subgroup of finite index in G containing H and still avoiding g, or, equivalently, if there is a homomorphism of G to a finite group such that the image of g remains outside the image of H.

One reason for interest in this property is that it bears the same relationship to the generalized word problem (or "embedding problem") as the weaker property of residual finiteness bears to the word problem: the proof that a finitely presented, residually finite group has soluble word problem adapts readily to yield that a finitely presented, subgroup separable group has soluble generalized word problem.

Before stating our main result (as in the title), we briefly sketch in its background. In [7] M. Hall, Jr. showed that free groups are subgroup separable. (Cf. Nielsen's result in [13] that a finitely generated free group has soluble generalized word problem.) In Romanovskiĭ's paper [14] (and independently in [4]) M. Hall's result, together with the basic idea of his proof, were generalized to show that free products of subgroup separable groups are subgroup separable. Finally Scott [15] showed, using geometrical methods,

© 1984 American Mathematical Society
0271-4132/84 $1.00 + $.25 per page

that the orientable surface groups (and so all finitely generated Fuchsian groups) are subgroup separable. Since a non-abelian surface group is a free product of two free groups with a rather special kind of cycle amalgamated, our main result, which we now state, significantly improves on Scott's result.

THEOREM 1. Any free product of two free groups with a cyclic amalgamation is subgroup separable.†

We note that this strengthens also G. Baumslag's result in [1] that such amalgamated products are residually finite, and lends further support to his statement in [2] that these groups (which he calls "cyclically pinched one-relator groups") are "very well-behaved." (Note that by a result of J. L. Dyer they are conjugacy separable.)

We prove Theorem 1 by establishing general sufficient conditions (given in the lemma in §2) on groups A and B for any amalgamated product $(A*B; U = V)$ with U (and V) infinite cyclic, to be subgroup separable; thus the result we prove is actually more general than indicated by Theorem 1, although admittedly the conditions are rather stringent. The proof is essentially a generalization of M. Hall's proof for free groups (with a modification due to Romanovskiǐ). Thus where M. Hall required a converse of the Schreier subgroup theorem for free groups, we shall need a converse of the Karrass-Solitar subgroup theorem for arbitrary amalgamated products $(A*B; U = V)$. This converse may conceivably prove useful in other contexts.

We thank C. Y. Tang for a germinal remark, and E. J. Mayland for his interest and encouragement.

1. Converse of the subgroup theorem for $(A*B; U = V)$

As mentioned above, this converse will play a role in the proof of Theorem 1 analogous to that played respectively by M. Hall's converse of the

---

†It follows (see the earlier remark) that such groups have soluble generalized word problem. This result is however a particular case of a theorem of S. G. Ivanov, "The embedding problem for a free product of groups with an amalgamated subgroup," Sib. Mat. Zh. 16, No. 6 (1975), 1155-1171.

Schreier subgroup theorem in his result in [7], and by the converse of the Kurosh subgroup theorem for free products in [4] and [14]. (The latter converse was first proved by Dey [6].)

We first state the Karrass-Solitar subgroup theorem, following, with minor changes, [9] and [5]. Thus write $G = (A*B; U = V)$ and let T_A, T_B be left transversals containing 1 for $U = V$ in A,B respectively. Then G is characterized by the fact that each of its elements g has a unique normal form

$$g = t_1 \ldots t_n u ,$$

where $n \geq 0$; $t_1,\ldots,t_n \in (T_A \cup T_B) - \{1\}$; for $i = 1,\ldots,n-1$, t_i,t_{i+1} are not both from T_A nor both from T_B; and $u \in U$. We call u the U-<u>ending</u> of g. If $t_n \in T_A$, then we call $t_n u$ the A-<u>ending</u> of g; otherwise u is the A-<u>ending</u> of g (and similarly with B replacing A). Now let H be a subgroup of G. Then, as in [9] (or [5]), one can construct a <u>cress</u> for H in G; this is a pair (C_A,C_B) of right transversals for H in G satisfying the following three conditions. (To facilitate the statement of these conditions, we shall adopt, following [9], the notation $d_A e_A e_U$ for a typical element $g = t_1 \ldots t_n u \in C_A$, where $e_U = u$, and if $t_n \in A$, then $d_A = t_1 \ldots t_{n-1}$, $e_A = t_n$; while if $t_n \in B$, then $d_A = t_1 \ldots t_n$, and $e_A = 1$. Thus $e_A e_U$ is the A-ending of g. We denote a typical element of C_B similarly by $d_B e_B e_U$.) The three conditions defining a cress (C_A,C_B) for H in G are as follows:

I. (a) Whenever $d_A e_A e_U \in C_A$, then also $d_A e_A$, $d_A \in C_A$ (and similarly for C_B).

(b) If $g = t_1 \ldots t_n u \in C_A \cup C_B$, and $t_n \in A - \{1\}$, then $g \in C_A$; i.e. any element of $C_A \cup C_B$ with A-ending outside U belongs to C_A (and similarly for C_B).

(c) If $g = t_1 \ldots t_n u \in C_A \cup C_B$ and $gu^{-1} \in C_A \cap C_B$, then $g \in C_A \cap C_B$.

II. The set R_A of all $d_Ae_A \in C_A$ (i.e. of those elements of C_A with trivial U-ending) forms a complete double-coset representative system for G modulo (H,U) (and similarly for the subset R_B of all $d_Be_B \in C_B$).

III. The set of all $d_A \in C_A$ (i.e. of those elements of C_A with trivial A-ending) forms a complete double-coset representative system for G modulo (H,A) (and similarly for C_B).

In terms of any cress (C_A,C_B) for H in G, the structure of H is, according to the Karrass-Solitar subgroup theorem, as follows. For each $d_Ae_A \in C_A$, denote (somewhat as in [9]) by $A_H^{d_A}$ the intersection $d_AAd_A^{-1} \cap H$, and by $U_H^{d_Ae_A}$ the intersection $d_Ae_AU(d_Ae_A)^{-1} \cap H$, with similar conventions for B in place of A throughout. Let $\theta: C_A \to C_B$ be the map defined by $H(g\theta) = Hg$, and write

$$t_{d_Ae_A} = d_Ae_A[(d_Ae_A)\theta]^{-1}$$

for each $d_Ae_A \in C_A - C_B$.

THE SUBGROUP THEOREM. <u>Let</u> (C_A,C_B) <u>be a cress for</u> H <u>in</u> G, <u>and let the</u> $A_H^{d_A}$, $B_H^{d_B}$, $t_{d_Ae_A}$ <u>be as above</u>. <u>Then</u>:

1. <u>All the subgroups</u> $A_H^{d_A}$, $B_H^{d_B}$ <u>generate their tree product</u> P <u>say, in which neighbouring vertices are just those</u> $A_H^{d_A}$, $B_H^{d_B}$ <u>where either</u> $d_A = d_B = 1$, <u>or</u> d_A <u>and</u> d_B <u>differ by a single syllable, i.e.</u> d_B <u>is obtained from</u> d_A <u>by deleting its last syllable (its</u> B<u>-ending), or vice versa, and where if</u> d <u>is the longer of</u> d_A,d_B <u>then</u> U_H^d <u>is the subgroup amalgamated between them; thus if for instance</u> $d_B = d_Ae_A$ <u>then</u>

$$\langle A_H^{d_A}, B_H^{d_B}\rangle = (A_H^{d_A} * B_H^{d_B};\ A_H^{d_A} \cap (d_Ae_A)U(d_Ae_A)^{-1} = B_H^{d_B} \cap d_BUd_B^{-1} (= U_H^{d_B}))\ . \tag{1}$$

2. <u>The subgroup</u> H <u>is the</u> HNN <u>extension</u>

$$H = \langle t_{d_Ae_A}, P \mid \text{Rel } P,\ t_{d_Ae_A} U_H^{d_Be_B} t_{d_Ae_A}^{-1} = U_H^{d_Ae_A}\rangle\ , \tag{2}$$

<u>where</u> $(d_Ae_A)\theta = d_Be_Be_V$.

We shall now formulate our converse of this theorem; that is to say we shall define an "abstract cress" to be a pair (C_A, C_B) of subsets of $G = (A*B;\ U = V)$, supplemented by certain other natural entities, and satisfying certain natural conditions ensuring the existence of a subgroup H of G for which (C_A, C_B) is indeed a cress in the above sense.

DEFINITION. An <u>abstract cress</u> (C_A, C_B) in $G = (A*B;\ U = V)$ is a pair of non-empty subsets of G satisfying condition I above, and having the following further properties:

(i) corresponding to each $d_A \in C_A$ there is prescribed a subgroup of $d_A A d_A^{-1}$, denoted by $A_*^{d_A}$, with the property that the set $E(d_A)$ consisting of those $e_A e_U$ such that $d_A e_A e_U \in C_A$ (i.e. of those A-endings of elements of C_A which are attached to d_A) forms a right transversal for $d_A^{-1} A_*^{d_A} d_A$ in A (and similarly with B in place of A throughout). (Cf. II, III.) Moreover each pair $A_*^{d_A}$, $B_*^{d_B}$ such that $d_B = d_A e_A$ is to satisfy the condition (cf. (1))

$$(d_A e_A)^{-1} A_*^{d_A} (d_A e_A) \cap U = d_B^{-1} B_*^{d_B} d_B \cap U .$$

(The analogous conditions with the roles of A and B interchanged are also required to hold.) Note that this includes as a special case the condition $A_*^1 \cap U = B_*^1 \cap U$.

(ii) there is a bijection $\theta\colon C_A \to C_B$, with the following three properties:

(a) θ is the identity map on $C_A \cap C_B$;

(b) if $d_A e_A \in C_A - C_B$, and $(d_A e_A)\theta = d_B e_B e_V$ say, then (cf. (2))

$$(d_A e_A)^{-1} A_*^{d_A} (d_A e_A) \cap U = (d_B e_B e_V)^{-1} B_*^{d_B} (d_B e_B e_V) \cap U .$$

(c) for each (fixed) $d_A e_A \in C_A$, if $(d_A e_A)\theta = d_B e_B e_V$, then for all e_U such that $d_A e_A e_U \in C_A$, it is required that $(d_A e_A e_U)\theta = d_B e_B e_V e_U v$ $(= (d_A e_A)\theta e_U v)$ for some element

$$v \in [(d_A e_A e_U)^{-1} A_*^{d_A} (d_A e_A e_U)] \cap U$$

$$= \text{(by (ii)(b))} \ [(d_B e_B e_V e_U)^{-1} B_*^{d_B} (d_B e_B e_V e_U)] \cap U .$$

It is furthermore required that for each fixed $d_A e_A$ this defines a bijection from the set of all $d_A e_A e_U \in C_A$ to the set of all $d_B e_B \bar{e}_V \in C_B$ (with $d_B e_B$ fixed as above).

REMARKS. 1. The counterpart of II is implicit in this definition; for if $d_A e_A e_U \in C_A$, then by I we have $d_A, d_A e_A \in C_A$, whence, the set $E(d_A)$ being by (i) a right transversal for $d_A^{-1} A_*^{d_A} d_A$ in A, it follows that for any fixed d_A, the e_A such that $d_A e_A \in E(d_A)$ form a double coset representative system containing 1 for A modulo $(d_A^{-1} A_*^{d_A} d_A, U)$, and then the U-endings e_U which attach to a given $d_A e_A$ will form a right transversal for $(d_A e_A)^{-1} A_*^{d_A} (d_A e_A) \cap U$ in U.

2. It is easy to see that an arbitrary cress (defined by I, II, III) for a subgroup H of G must satisfy (ii)(c).

We are now in a position to state our converse.

THEOREM 2. Let (C_A, C_B), $\{A_*^{d_A}\}$, $\{B_*^{d_B}\}$, θ be an abstract cress in $G = (A*B); U = V)$ as in the above definition. Then there exists a unique subgroup H of G for which (C_A, C_B) is a cress (i.e. a pair of right transversals satisfying I, II, III), and for which $A_H^{d_A} = A_*^{d_A}$, $B_H^{d_B} = B_*^{d_B}$, and for all $g \in C_A$, $Hg = H(g\theta)$.

PROOF. The bulk of the proof consists in defining a (right) action of G on the set C_A. Once this is done we take H to be the stablizer of $1 \in C_A$, and it will then be clear that our action is equivalent to right multiplication of the right cosets of H by the elements of G, and that H has the properties claimed of it.

We define our action of G on C_A by defining actions of A and B separately, and then showing that these actions define the same action of $U = V$. Thus we define the action $\cdot$ of each $a \in A$ on each $g = d_A e_A e_U \in C_A$ by

$g \cdot a = d_A \alpha$, where α is determined by:

$$\alpha \in E(d_A),\ e_A e_U a \in (d_A^{-1} A_*^{d_A} d_A)\alpha . \tag{3}$$

(That this defines $g \cdot a$ uniquely follows from the requirement (in condition (i) of the definition) that $E(d_A)$ be a right transversal for $d_A^{-1} A_*^{d_A} d_A$ in A.)

For each $b \in B$ and $h = d_B e_B e_V \in C_B$, define an analogous action (of B on C_B), denoted by $\circ$, by

$h \circ b = d_B \beta$, where β is determined by:

$$\beta \in E(d_B),\ e_B e_V b \in (d_B^{-1} B_*^{d_B} d_B)\beta .$$

We then use this to define an action of B on C_A (also denoted by $\circ$) as follows: for each $g \in C_A$, $b \in B$, set

$$g \circ b = [(g\theta) \circ b]\theta^{-1} . \tag{4}$$

We now check that (3) and (4) do indeed define actions of A and B respectively on C_A. To see that (3) defines an action of A, let a_1, a_2 be arbitrary elements of A. Then on the one hand

$g \cdot (a_1 a_2) = d_A \alpha$, where α is determined by:

$$\alpha \in E(d_A),\ e_A e_U a_1 a_2 \in (d_A^{-1} A_*^{d_A} d_A)\alpha . \tag{5}$$

On the other hand,

$g \cdot a_1 = d_A \alpha_1$, where α_1 is determined by:

$$\alpha_1 \in E(d_A),\ e_A e_U a_1 \in (d_A^{-1} A_*^{d_A} d_A)\alpha_1 , \tag{6}$$

and then

$(g \cdot a_1) \cdot a_2 = (d_A \alpha_1) \cdot a_2 = d_A \bar{\alpha}$ where $\bar{\alpha}$ is determined by:

$$\bar{\alpha} \in E(d_A),\ \alpha_1 a_2 \in (d_A^{-1} A_*^{d_A} d_A)\bar{\alpha} . \tag{7}$$

From (6) and (7) it follows that

$$e_A e_U a_1 a_2 \in (d_A^{-1} A_*^{d_A} d_A)\alpha_1 a_2 = (d_A^{-1} A_*^{d_A} d_A)\bar{\alpha} ,$$

whence $\alpha = \bar{\alpha}$ since $\alpha, \bar{\alpha}$ both belong to $E(d_A)$, which is by Part (i) of the definition of an abstract cress, a right transversal for $d_A^{-1} A_*^{d_A} d_A$ in A. It being immediate that 1 acts trivially on C_A, it follows that (3) does indeed define an action of A on C_A (and, similarly, that $\circ$ is an action of B on C_B). That (4) defines an action of B on C_A is then immediate from the fact that θ is a bijection.

It only remains to check (and this is where Theorem 2 breaks new ground) that the established actions of A and B on C_A define the same action of $U = V$ on C_A. Thus let $u \in U$ (= V), and let $g = d_A e_A e_U \in C_A$. The verification breaks naturally into two cases.

<u>Case</u> 1. Suppose that we also have $g \in C_B$. Assume initially also that $e_A \neq 1$. Write $d_B = d_A e_A$. (Note that by I we have $d_B = d_A e_A \in C_A \cap C_B$.) From (3) we obtain

$g \cdot u = d_A \alpha$, where α is determined by:

$$\alpha \in E(d_A),\ e_A e_U u \in (d_A^{-1} A_*^{d_A} d_A)\alpha . \tag{8}$$

On the other hand from (4) and the fact that θ is the identity map on $C_A \cap C_B$ (see (ii)(a))

$g \cdot u = d_B \beta$, where β is determined by:

$$\beta \in E(d_B),\ e_U u \in (d_B^{-1} B_*^{d_B} d_B)\beta . \tag{9}$$

It follows, again from Remark 1 (which applies also with B replacing A), that $\beta \in U$. From (8) and (9) we deduce that $e_U u$ belongs to both of the cosets $[(d_A e_A)^{-1} A_*^{d_A}(d_A e_A)]e_A^{-1}\alpha$ and $(d_B^{-1} B_*^{d_B} d_B)\beta$, and therefore, since $e_A^{-1}\alpha = \bar{u}$ and β both lie in U, to both of the cosets

$$[(d_A e_A)^{-1} A_*^{d_A}(d_A e_A) \cap U]\, e_A^{-1}\alpha\ ,\ [(d_B^{-1} B_*^{d_B} d_B) \cap U]\,\beta . \tag{10}$$

Now since by part (i) of the definition of an abstract cress, the bracketed intersections in (10) are equal, it follows that these two cosets are equal. Since by (8), (9) and I, the elements $d_B\overline{u}$ $(= d_A\alpha)$ and $d_B\beta$ are both in C_A, it follows from Remark 1 that $\overline{u} = \beta$, whence $g \cdot u = g \circ u$, as required.

If the assumption $e_A \neq 1$ does not hold, then either $d_A = d_B e_B$ with $e_B \neq 1$, in which case the argument analogous to the above, with A and B interchanged, can be used, or else $d_A = 1$, in which case a similar, but easier, argument, invoking the condition $A_*^1 \cap U = B_*^1 \cap U$ (see Part (i) of the definition of an abstract cress); yields the desired conclusion.

Case 2. Suppose that $g \in C_A - C_B$. Then by I, in $g = d_A e_A e_U$ we must have $e_A \neq 1$. The element $g \cdot u$, the result of the action of u on g, with u regarded as an element of A, is given, as before, by (8). On the other hand, regarding u as an element of B we obtain from (4) that

$$g \circ u = [(g\theta) \circ u]\theta^{-1} .$$

Now by (ii)(c) if $(d_A e_A)\theta = d_B e_B e_V$, then

$$g\theta = (d_A e_A e_U)\theta = d_B e_B e_V e_U v ,$$

$$\text{where } v \in [(d_B e_B e_V e_U)^{-1} B_*^{d_B} (d_B e_B e_V e_U)] \cap U .$$

Hence by definition of the action $\circ$ of B on C_B, we have

$$(g\theta) \circ u = d_B\beta, \text{ where } \beta \text{ is determined by:}$$

$$\beta \in E(d_B),\ e_B e_V e_U v u \in (d_B^{-1} B_*^{d_B} d_B)\beta . \tag{11}$$

It follows from Remark 1 (with B in place of A) that $\beta = e_B w$ for some $w \in U$. Now by (ii)(c) there is an element of the form $d_A e_A u_1 \in C_A$ $(u_1 \in U)$ such that $(d_A e_A u_1)\theta = d_B e_B w$. Since, again by (ii)(c), $(d_A e_A u_1)\theta = d_B e_B e_V u_1 v_1$ for some $v_1 \in (d_B e_B e_V u_1)^{-1} B_*^{d_B} (d_B e_B e_V u_1) \cap U$, it follows that $u_1 = e_V^{-1} w v_1^{-1}$, so that

$$[(g\theta) \circ u]\theta^{-1} = (d_B\beta)\theta^{-1} = (d_B e_B w)\theta^{-1} = d_A e_A u_1 = d_A e_A e_V^{-1} w v_1^{-1} . \tag{12}$$

Now from (11) we see that vu belongs to the coset

$$[(d_Be_Be_Ve_U)^{-1}B_*^{d_B}(d_Be_Be_Ve_U) \cap U](e_Be_Ve_U)^{-1}e_Bw$$

$$= [(d_Be_Be_Ve_U)^{-1}B_*^{d_B}(d_Be_Be_Ve_U) \cap U](e_Ve_U)^{-1}w\,. \tag{13}$$

Since v belongs to the bracketed intersection in (13), it follows that u belongs to the coset in (13). On the other hand by (8), u belongs also to the coset

$$[(d_Ae_Ae_U)^{-1}A_*^{d_A}(d_Ae_Ae_U) \cap U](e_Ae_U)^{-1}\alpha\,. \tag{14}$$

Since by (ii)(b) (conjugated appropriately) the bracketed intersections in (13) and (14) are equal, it follows that the cosets are equal. Therefore $e_V^{-1}w$ and $e_A^{-1}\alpha\ (=\bar{u})$ lie in the same right coset of $[(d_Ae_A)^{-1}A_*^{d_A}(d_Ae_A) \cap U]$ in U.

Now putting $u_1 = e_V^{-1}wv_1^{-1}$ we obtain

$$v_1^{-1} \in (d_Be_Be_Vu_1)^{-1}B_*^{d_B}(d_Be_Be_Vu_1) \cap U = (d_Be_Bw)^{-1}B_*^{d_B}(d_Be_Bw) \cap U\,.$$

Hence

$$e_V^{-1}wv_1^{-1} \in e_V^{-1}w(d_Be_Bw)^{-1}B_*^{d_B}(d_Be_Be_V)e_V^{-1}w \cap U$$

$$= [(d_Be_Be_V)^{-1}B_*^{d_B}(d_Be_Be_V) \cap U]e_V^{-1}w\,.$$

Hence $e_V^{-1}wv_1^{-1}$ and $e_A^{-1}\alpha\ (=\bar{u})$ lie in the same right coset of $(d_Ae_A)^{-1}A_*^{d_A}(d_Ae_A) \cap U$ in U. Since $d_Ae_A\bar{u}$ (see Case 1) and $d_Ae_A(e_V^{-1}wv_1^{-1})$ (see (12)) both belong to C_A, it follows from Remark 1 that $e_A^{-1}\alpha = e_V^{-1}wv_1^{-1}$. From (12) we then obtain $[(g\theta)\circ u]\theta^{-1} = d_A\alpha$, which is by (8) just $g\cdot u$, completing the proof.

§2. A lemma on free groups

Before introducing the lemma we prove two propositions.

PROPOSITION 1. *Let F be a free group, let x be any nontrivial element of F, and let K be any finitely generated subgroup of F intersecting ⟨x⟩ trivially. Then there exists a positive integer r such that*

<u>for every positive integer</u> s <u>there is a finite-index normal subgroup</u> Q_s <u>of</u> F <u>with the property that</u>

$$KQ_s \cap \langle x \rangle = Q_s \cap \langle x \rangle = \langle x^{rs} \rangle \,. \tag{15}$$

REMARK. The conclusion (15) may be rephrased as follows: Corresponding to each multiple rs of r there is a finite homomorphic image of F in which the image of $\langle x \rangle$ has order rs, and intersects the image of K trivially. In the case when $K = \{1\}$, it can be shown (see [16, Lemma 1]) that r may be taken to be 1, i.e. that x takes on all possible orders modulo finite index normal subgroups of F. However rather than this condition, we shall need the property (15), which evolved from it.

PROOF OF PROPOSITION 1. It is not difficult to see that since x and K are contained in a finitely generated free factor of F, we may assume without loss of generality that F is finitely generated. By a result implicit in M. Hall's paper [7], and made explicit in [3] and [8], every finitely generated subgroup of a free group is a free factor of a subgroup of finite index. Hence there is a subgroup H of F (a "free complement" for K in F) such that the subgroup generated by H and K has finite index in F, and is the free product of H and K, i.e. $\langle H,K \rangle = H*K$. Let x^ρ be the smallest positive power of x in $\langle H,K \rangle$; it is easy to see that since $K \cap \langle x \rangle = \{1\}$, we must have $\langle K,x^\rho \rangle = K*\langle x^\rho \rangle$. Let H_1 be a free complement in the above sense for $\langle K,x^\rho \rangle$ in F; thus $\langle K,x^\rho,H_1 \rangle = K*\langle x^\rho \rangle *H_1$, and has finite index in F. Write $P = \langle K,x^\rho,H_1 \rangle$. Denote by Q the core of P in F, i.e. the intersection of all conjugates of P in F; the subgroup Q is normal in F and again has finite index. Let x^t be the smallest positive power of x contained in Q. (Clearly t is a multiple of ρ.) Finally, for each positive integer s, let Q_s denote the verbal subgroup of Q defined by the words $[y,z]$ and y^{2s}; thus Q/Q_s is free abelian of exponent $2s$, and is moreover finite since F, and therefore its finite-index subgroup Q, are finitely generated. Hence Q_s has finite index in F. Being a characteristic subgroup of the normal subgroup Q, it is also normal in F. Set $r = 2t$.

Having defined r and, for each s, Q_s, it remains to prove (15). Since x^t occurs among a set of free generators of Q, the equality $Q_s \cap \langle x \rangle = \langle x^{2ts} \rangle = \langle x^{rs} \rangle$ is immediate from the definition of Q_s. We shall now show, finally, that $KQ_s \cap \langle x \rangle \leqq \langle x^{rs} \rangle$, the reverse inclusion being trivial. Suppose that $x^m \in KQ_s$, where $0 < m < rs$. Then we cannot have $x^m \in (K \cap Q)Q_s$ for the following reasons. By the Kurosh subgroup theorem applied to the free product $P = K*\langle x^\rho \rangle *H_1$, the subgroup Q of P is a free product of the form $(K \cap Q)*\langle x^t \rangle *L$. It follows, by definition of Q_s, that $\frac{(K \cap Q)Q_s}{Q_s}$ and $\frac{\langle x^t \rangle Q_s}{Q_s} = \langle (xQ_s)^t \rangle$ of Q/Q_s, generate their direct product. Now if $x^m \in (K \cap Q)Q_s$, then $x^m \in Q$, so that m is a multiple of t, and $(xQ_s)^m$ is in both of the above subgroups of Q/Q_s. Hence $x^m \in Q_s$, contradicting the assumption that $0 < m < rs$. Thus if $x^m \in KQ_s$, then Q_s contains an element of the form ax^m where $a \in K - (K \cap Q)$. Since $a \notin Q$, we have also $x^m \notin Q$. It is not difficult to infer from the details of the Kurosh subgroup theorem as it applies to $P = K*\langle x^\rho \rangle *H_1$, that there is a "regular extended Schreier system" for Q in P, consisting of a pair of right transversals for Q in P, one corresponding to each of the two free factors K and $\langle x^\rho \rangle *H_1$, containing respectively, as words of length 1, the element a and the element x^{-m}, as representatives of the same coset, and that therefore ax^m is a free generator of Q (see [11, pp. 239-246]). It follows by definition of Q_s, that $ax^m \notin Q_s$, giving a final contradiction.

Our second proposition states that in a free group a finitely generated subgroup is finitely separable not just from any element outside it, but from any non-trivial cyclic coset (i.e. coset of a cyclic group) outside it.

PROPOSITION 2. *Let* F *be a free group, let* x *be any non-trivial element of* F, *and let* K *be any finitely generated subgroup of* F. *Then if*

$g \in F$ *is such that* $K \cap \langle x \rangle g$ *is empty, there exists a finite-index subgroup* L *of* F, *containing* K, *such that* $L \cap \langle x \rangle g$ *is also empty*.

PROOF. *Case* 1. Assume that $K \cap \langle x \rangle = \{1\}$. Let ρ, H_1, P be as in the proof of Proposition 1; thus the subgroup $P = \langle K, x^\rho, H_1 \rangle$ has finite index in F and is the free product of K, $\langle x^\rho \rangle$, and H_1, i.e. $P = K * \langle x^\rho \rangle * H_1$.

If $P \cap \langle x \rangle g$ is empty then we may take P as the subgroup L we are seeking. We may therefore assume that $x^s g \in P$ for some integer s; but then since $\langle x \rangle (x^s g) = \langle x \rangle g$, we may suppose (without loss of generality) that $g \in P$ (by taking $x^s g$ in the role of g). Then since an element of the form $x^k g$ is in P precisely if ρ divides k, it suffices to find a subgroup $L > K$ of finite index in P (and therefore also in F) such that $\langle x^\rho \rangle g \cap L$ is empty. Write $y = x^\rho$. We have thus simplified our problem to that of finding in the free group $P = K * \langle y \rangle * H_1$, a finite-index subgroup $L > K$ such that $L \cap \langle y \rangle g$ is empty (given $g \in P$ such that $K \cap \langle y \rangle g$ is empty).

We may clearly also assume of g that as an element of the free product $K * \langle y \rangle * H_1$, its initial syllable does not come from the factor $\langle y \rangle$. Let p be the largest (non-negative) integer such that $y^{\pm p}$ is a syllable of g; it then follows that the elements $g, yg, \ldots, y^p g$ do not belong to $\langle K, y^{p+1} \rangle$. Since P is free, and therefore (by [7]) subgroup separable, there is a subgroup L of finite index in P, which contains $\langle K, y^{p+1} \rangle$ and avoids $g, yg, \ldots, y^p g$; but then L avoids also all elements in $\langle y^{p+1} \rangle g$ (since L contains y^{p+1} and avoids g). Hence L avoids $\langle y \rangle g$, as required.

Case 2. If $K \cap \langle x \rangle \neq \{1\}$, we proceed as follows. Let x^t be the smallest positive power of x in K. By the subgroup separability of F, there is a finite-index subgroup L of F, which contains K and avoids the elements $g, xg, \ldots, x^{t-1} g$. It is then easy to see that $L \cap \langle x \rangle g$ is empty, completing the proof.

LEMMA. *Let* F *be a free group. Let* $x, g_1, \ldots, g_m$ *be any* $m + 1$ *non-trivial elements of* F, *and let* $K_1, \ldots, K_n$ *be any* n *finitely generated subgroups of* F $(m,n$ *finite*). *Then there exists a positive integer* k *such*

<u>that for every positive integer</u> ℓ <u>there is a normal subgroup</u> N <u>of</u> F <u>of finite index with the following five properties</u>:

(α) $N \cap \langle x\rangle = \langle x^{k\ell}\rangle$;

(β) <u>for each</u> K_j <u>such that</u> $K_j \cap \langle x\rangle = \{1\}$, <u>we have</u> $K_jN \cap \langle x\rangle = \langle x^{k\ell}\rangle$;

(γ) <u>for each</u> K_j <u>such that</u> $K_j \cap \langle x\rangle \neq \{1\}$, <u>we have</u> $K_jN \cap \langle x\rangle = K_j \cap \langle x\rangle$;

(δ) <u>for each</u> $g_i \notin \langle x\rangle$, <u>we have</u> $g_i \notin N\langle x\rangle$;

(ϵ) <u>for each pair</u> i,j <u>such that</u> $\langle x\rangle g_i \cap K_j$ <u>is empty, we have that</u> $\langle x\rangle g_i \cap K_jN$ <u>is also empty</u>.

PROOF. By the subgroup separability of F (proved in [7]) there exists a subgroup L_0 of finite index in F, containing x and avoiding those g_i outside $\langle x\rangle$. Similarly, for each K_j such that $K_j \cap \langle x\rangle \neq \{1\}$, there is a finite index subgroup L_j containing K_j and avoiding the elements $x, x^2, \ldots, x^{tj-1}$, where x^{tj} is the smallest positive power of x belonging to K_j. Denote by N_0 the intersection of all conjugates in F of all of the subgroups L_0, L_j. Then N_0 has finite index in F, and clearly it (and therefore any smaller normal subgroup) has properties (γ) and (δ).

We next turn to condition (ϵ). For each pair i,j such that $K_j \cap \langle x\rangle g_i$ is empty, there exists, by Proposition 2, a subgroup L_{ij} of finite index in F, containing K_j and avoiding $\langle x\rangle g_i$. Denote by N_1 the intersection with N_0 of all conjugates in F of all the L_{ij} (over all i,j such that $K_j \cap \langle x\rangle g_i$ is empty). Then N_1 has finite index in F, and clearly it (and therefore any smaller normal subgroup) has properties (γ), (δ) and (ϵ).

We next seek a finite-index, normal subgroup N_2 contained in N_1 (and so satisfying (γ), (δ) and (ϵ)), and also satisfying (α) and (β). Let k_0 be the smallest positive integer such that $x^{k_0} \in N_1$. For each K_j such that $K_j \cap \langle x\rangle = \{1\}$ let r_j and, for each positive integer s, $Q_{j,s}$, be as in Proposition 1. Let ℓ be any positive integer. For each K_j (with $K_j \cap \langle x\rangle = \{1\}$) define s_j to be the product of the integers k_0, ℓ and those r_i with $i \neq j$. Then by Proposition 1 (or more specifically the definition of $Q_{j,s}$)

$$Q_{j,s_j} \cap \langle x \rangle = \langle x^{k_0 \ell \pi} \rangle , \tag{16}$$

where π is the product of all of the r_j (such that $K_j \cap \langle x \rangle = \{1\}$). If we set $k = k_0\pi$, and define N_2 to be the intersection of N_1 with all of the Q_{j,s_j}, then this k and this N_2 (which has finite index, being the intersection of finitely many finite-index subgroups) fulfill conditions (α) and (β) (as well as (γ), (δ) and (Θ)). That condition (α) is satisfied follows immediately from (16) and the fact that $x^{k_0 \ell \pi} \in N_1$. Condition (β) is satisfied by virtue of the fact that since by (15) we have $K_j Q_{j,s_j} \cap \langle x \rangle = \langle x^{k\ell} \rangle$, this will continue to hold if in it Q_{j,s_j} is replaced by any smaller subgroup which also contains $x^{k\ell}$. Since the finite-index normal subgroup N_2 has all five properties, the proof is complete.

§3. Proof of Theorem 1

Step 1. Construction of an abstract cress $(\hat{C}_A, \hat{C}_B)$. We wish to show that $G = (A*B;\ U = V = \langle x \rangle \neq \{1\})$ is subgroup separable, given that A and B are free. Thus let H be any finitely generated subgroup of G, and let $g \in G - H$. Let (C_A, C_B) be a cress for H in G relative to left transversals T_A, T_B (both containing 1) for U in A, B respectively (see the definition in §1). We may assume $g \in C_A$ since clearly any subgroup of G separating H from g, will in fact separate H from Hg.

As in §1 we write a typical element of C_A in the form $d_A e_A e_U$ where $e_A \in T_A$, $e_U \in U$, and $e_A e_U$ is the A-ending of the element (with the analogous convention for C_B). By [9, Lemma 3] the base group of the finitely generated subgroup H, regarded as an HNN extension in accordance with the Subgroup Theorem (see §1), is a tree product of finitely many of its vertex groups. Denote by S_1 a finite set of $d_A \in C_A$ and $d_B \in C_B$ such that H is the tree product of the vertex groups $A_H^{d_A}, B_H^{d_B}, d_A, d_B \in S_1$.

Define S to be S_1 together with all elements $d_A e_A \in C_A - C_B$, all $d_B e_B \in C_B - C_A$, the element gu^{-1} where u is the U-ending of g, and, finally, all initial segments of these elements (i.e. ensure that the (finite) set S is closed under taking initial segments). Write $S_A = S \cap C_A$, $S_B = S \cap C_B$. For each $d_A \in S_A$ (i.e. for each element of S_A with trivial A-ending) write $K(d_A)$ for the subgroup $d_A^{-1} A_H^{d_A} d_A$ of A. By [9, Lemma 3, Theorems 4, 5], since U is cyclic and H is finitely generated, the $K(d_A)$ are finitely generated. Define $K(d_B)$, $d_B \in S_B$, analogously.

We now apply the lemma of §2, with A in the role of F, the set of all a, $a^{-1}\alpha \neq 1$, where a,α range over all A-syllables of elements of $S = S_A \cup S_B$, in the role of $\{g_1,\ldots,g_m\}$, and finally the $K(d_A)$ and their conjugates $a^{-1}K(d_A)a$ (where a ranges as above) in the roles of $K_1,\ldots,K_n$. Let k_A be the positive integer guaranteed by that lemma, and let k_B be the positive integer it guarantees when applied to the analogous set-up in B. It then follows further from that lemma with $\ell = k_B$ that there is a finite-index, normal subgroup N_A of A with the following five properties:

(α) $N_A \cap \langle x\rangle = \langle x^{k_A k_B}\rangle$; (17)

(β) for each K (equal to a $K(d_A)$ or a conjugate $a^{-1}K(d_A)a$ with a as above) such that $K \cap \langle x\rangle = \{1\}$, we have
$KN_A \cap \langle x\rangle = \langle x^{k_A k_B}\rangle$; (18)

(γ) for each K (as above) such that $K \cap \langle x\rangle \neq \{1\}$, we have
$KN_A \cap \langle x\rangle = K \cap \langle x\rangle$; (19)

(δ) each a or $a^{-1}\alpha$ outside $\langle x\rangle$ is also outside $N_A\langle x\rangle$; (20)

(ϵ) for each K (as above) and y (from the set of a and $(a^{-1}\alpha)$) such that $\langle x\rangle y \cap K$ is empty, the intersection $\langle x\rangle y \cap KN_A$ is also empty. (21)

Since in the lemma in place of k any integer multiple of k will also serve, we may by taking k_A large enough, also arrange (in view of (18) and (19)) that

(ζ) if $g \in \langle x\rangle$, then for every K such that $g \notin K$, we also have $g \notin KN_A$. (22)

Let N_B be the analogous finite-index, normal subgroup of B. Then in particular

$$N_A \cap \langle x\rangle = \langle x^{k_A k_B}\rangle = N_B \cap \langle x\rangle .$$

It follows that the natural maps $A \to A/N_A$ and $B \to B/N_B$ extend (uniquely) to an epimorphism

$$\varphi\colon (A*B;\ U = V = \langle x\rangle) \to (\frac{A}{N_A} * \frac{B}{N_B};\ xN_A = xN_B) . \tag{23}$$

We shall exploit this epimorphism in the final step of the proof.

Let $\hat{T}_A \subseteq T_A$ be a left transversal for N_AU in A, containing 1 and all A-syllables of elements of $S = S_A \cup S_B$; this is possible in view of condition (20) which ensures in particular that no such A-syllable is in the trivial coset N_AU, and that distinct A-syllables are in distinct left cosets of N_AU in A. Define a left transversal $\hat{T}_B \subseteq T_B$ for N_BU in B analogously.

We next construct a (possibly new) cress (C_A', C_B') for H in G. For any cress (C_A, C_B) for H in G it is always the case that for each $d_A \in C_A$, the set of $e_A \in T_A$ such that $d_Ae_A \in C_A$ forms a complete set of double-coset representatives for A modulo $(d_A^{-1}A_H^{d_A}d_A, U)$, and analogously for C_B. (This is the import of II in §1.) Our new cress is constructed in the usual way (as in [9, proof of Lemma 6]) by induction on double-coset length, but with the following conditions imposed at each stage of the induction: for each $d_A \in C_A'$ already defined, the e_A to be attached to that d_A are to be chosen whenever possible from $\hat{T}_A$ (in other words whenever a double coset $(d_A^{-1}A_H^{d_A}d_A)yU$, $y \in A$, contains an element of $\hat{T}_A$, then its representative e_A is to be chosen from $\hat{T}_A$), and furthermore this is to be done in such a way as to ensure that $S_A \subseteq C_A'$ (and similarly with B replacing A throughout). This is justified by the fact that in the usual construction of a cress, the e_A to be attached to a given d_A (already constructed) are chosen arbitrarily subject to belonging to T_A and forming a complete double-coset representative system for A modulo $(d_A^{-1}A_H^{d_A}d_A, U)$ (see [9, bottom of p. 241]), and secondly by the fact that the A-syllables and B-syllables of the elements of $S = S_A \cup S_B$ belong respectively to $\hat{T}_A$ and $\hat{T}_B$.

We next define, for each $d_A \in C_A'$ a subgroup $A_*^{d_A}$, and for each $d_B \in C_B'$ a subgroup $B_*^{d_B}$. (As the notation suggests, these (or rather some of them) will play the role of the subgroups figuring in condition (i) of the definition of an abstract cress.) We begin with the d_A in S_A and the d_B in S_B.

Thus for each $d_A \in S_A$, define

$$A_*^{d_A} = A_H^{d_A}(d_A N_A d_A^{-1}) \ , \tag{24}$$

and for each $d_B \in S_B$, define

$$B_*^{d_B} = B_H^{d_B}(d_B N_B d_B^{-1}) \ . \tag{25}$$

Next suppose $d_A \in C_A' - S_A$; for such d_A (and for the $d_B \in C_B' - S_B$) the definition of $A_*^{d_A}$ (and the analogous $B_*^{d_B}$) is more complicated. Let z be the longest initial segment of d_A belonging to $S = S_A \cup S_B$. (Since z is an initial segment of d_A and since by the definition of a cress $d_A \in C_A \cap C_B$, it follows that $z \in S_A \cap S_B$.) We shall define $A_*^{d_A}$ by induction on the difference in the lengths of d_A and z. If this difference is 1, then $d_A = d_B e_B$ where $z = d_B \in S_B$, and $e_B \in T_B - \{1\}$, and we define $A_*^{d_A}$, in terms of the already defined (by (25)) $B_*^{d_B}$, by

$$d_A^{-1} A_*^{d_A} d_A = [(d_B e_B)^{-1} B_*^{d_B} (d_B e_B) \cap U] N_A \ . \tag{26}$$

The point of this definition of $A_*^{d_A}$ is that it ensures firstly that $d_A^{-1} A_*^{d_A} d_A \geqq N_A$ (which we shall need later), and secondly that the equality in (i) (with A and B interchanged) is satisfied; for by (25) we have $(d_B e_B)^{-1} B_*^{d_B} (d_B e_B) \geqq N_B$, whence in view of (17)

$$\{[(d_B e_B)^{-1} B_*^{d_B} (d_B e_B) \cap U] N_A\} \cap U = (d_B e_B)^{-1} B_*^{d_B} (d_B e_B) \cap U \ ,$$

which, by (26) is equivalent to

$$d_A^{-1} A_*^{d_A} d_A \cap U = (d_B e_B)^{-1} B_*^{d_B} (d_B e_B) \cap U \ , \tag{27}$$

as required by (i). Similarly if $d_B \in C_B' - S_B$, and $d_B = d_A e_A$ where $d_A \in S_A$, then we define $B_*^{d_B}$ in terms of the group $A_*^{d_A}$ already defined by (24) as follows:

$$d_B^{-1} B_*^{d_B} d_B = [(d_A e_A)^{-1} A_*^{d_A} (d_A e_A) \cap U] N_B \ . \tag{28}$$

This definition ensures that $d_B^{-1} B_*^{d_B} d_B \geqq N_B$, and by an argument analogous to that yielding (27), that

$$(d_A e_A)^{-1} A_*^{d_A} (d_A e_A) \cap U = d_B^{-1} B_*^{d_B} d_B \cap U \ , \tag{29}$$

which is just the equation in part (i) of the definition of an abstract cress. For the inductive step suppose the difference in lengths between $d_A \in C_A' - S_A$ and z exceeds 1, and that $d_A = d_B e_B$, where it is assumed inductively that $B_*^{d_B}$ has already been defined. Define $A_*^{d_A}$ as in (26); the inclusion $d_A^{-1} A_*^{d_A} a_A \geqq N_A$ is then obvious, while (27) follows as before. The analogous inductive step for $d_B \in C_B' - S_B$ secures in the same way the inclusion $d_B^{-1} B_*^{d_B} d_B \geqq N_B$, and equation (29). This completes the definition of the $A_*^{d_A}$, $d_A \in C_A'$, and the $B_*^{d_B}$, $d_B \in C_B'$.

In connection with this definition observe that for all $d_A \in C_A'$ we have $A_*^{d_A} \geqq A_H^{d_A}$ (and similarly for the $d_B \in C_B'$); this is clear from (24) in the case $d_A \in S_A$, while if $d_A \in C_A' - S_A$ it follows from (26) in light of the fact that for such d_A we have $A_H^{d_A} \leqq d_A U d_A^{-1}$ so that by (1) (with the roles of A and B interchanged) we have $A_H^{d_A} \leqq B_H^{d_B} \cap (d_B e_B U (d_B e_B)^{-1})$.

Again with an eye to (i), for each $d_A \in C_A'$, we now denote by $\hat{E}(d_A) \subseteq \hat{T}_A$ a complete double-coset representative system for $A \bmod (d_A^{-1} A_*^{d_A} d_A, U)$, with the following two properties: firstly, for every $e_A \in \hat{E}(d_A)$, we demand that $d_A e_A \in C_A'$, and, secondly, whenever $d_A \in S_A$, then $\hat{E}(d_A)$ should contain all e_A such that also $d_A e_A \in S_A$. That the first of these requirements is satisfiable follows from the facts that for each fixed $d_A \in C_A'$ the e_A such that $d_Ae_A \in C'_A$ form, by II, a complete double-coset representative system for A mod$(d_A^{-1}A_H^{d_A}d_A, U)$, and that, as noted above, $A_*^{d_A} \geqq A_H^{d_A}$. The second requirement can of course be satisfied if and only if the e_A such that $d_Ae_A \in S_A$ belong to distinct double cosets of $d_A^{-1}A_*^{d_A}d_A$ and U in A, and this is secured for us by condition (21). To see this note first that by (24), $d_A^{-1}A_*^{d_A}d_A = (d_A^{-1}A_H^{d_A}d_A)N_A = K(d_A)N_A$, since $d_A \in S_A$. Suppose that $a,\alpha \in T_A$ are such that d_Aa, $d_A\alpha$ both belong to S_A, and yet $K(d_A)N_AaU = K(d_A)N_A\alpha U$. Then αUa^{-1} intersects $K(d_A)N_A$, or, equivalently, $\alpha^{-1}K(d_A)N_A\alpha = \alpha^{-1}K(d_A)\alpha N_A$ intersects $Ua^{-1}\alpha$. Now by (21) this implies that $\alpha^{-1}K(d_A)\alpha$ intersects $Ua^{-1}\alpha$, whence $K(d_A)aU = K(d_A)\alpha U$. However since $a,\alpha \in T_A$, and both $d_Aa, d_A\alpha \in C_A$, it follows from II in §1, that then $a = \alpha$. For each $d_B \in C'_B$, define $\hat{E}(d_B) \subseteq \hat{T}_B$ analogously.

We next define a subset $\hat{R}$ of $C'_A \cup C'_B$ inductively as follows. The elements of length $\leqq 1$ of $\hat{R}$ are defined to be just the elements of $\hat{E}(d_A) \cup \hat{E}(d_B)$ where $d_A = d_B = 1$, i.e. they are just the chosen double-coset representatives for A modulo $((H \cap A)N_A, U)$ and for B modulo $((H \cap B)N_B, U)$. Assuming inductively that $d_A \in C'_A$ has already been included in $\hat{R}$, include also all d_Ae_A with $e_A \in \hat{E}(d_A)$; the analogous procedure with B replacing A completes the inductive step of the definition. Thus $\hat{R}$ consists of those $d_Ae_A \in C'_A$ and those $d_Be_B \in C'_B$, with $e_A \in \hat{E}(d_A)$, $e_B \in \hat{E}(d_B)$, and whose initial segments share this property. Write $\hat{R}_A = \hat{R} \cap C'_A$, $\hat{R}_B = \hat{R} \cap C'_B$. It follows almost immediately from the facts that each $\hat{E}(d_A)$ with $d_A \in S_A$ contains all e_A with $d_Ae_A \in S_A$ (and similarly for those $\hat{E}(d_B)$ where $d_B \in S_B$), and that $S_A \subseteq C'_A$, $S_B \subseteq C'_B$, that $S_A \subseteq \hat{R}_A$, $S_B \subseteq \hat{R}_B$. By [9, first corollary on p. 247] the rank of the free part of H is independent of the cress chosen for H; since this rank is just the number of $d_Ae_A \in C_A - C_B$ (or of $d_Be_B \in C_B - C_A$), and since S_A contains all such d_Ae_A (and S_B all such d_Be_B), it follows that $\hat{R}_A - \hat{R}_B = S_A - S_B$ and $\hat{R}_B - \hat{R}_A = S_B - S_A$.

To complete the construction of the desired abstract cress $(\hat{C}_A, \hat{C}_B)$ all that remains to do is to attach suitable U-endings to the elements of $\hat{R}_A$ and $\hat{R}_B$ (and to define a bijection $\hat{\theta}: \hat{C}_A \to \hat{C}_B$). For each $d_A e_A \in \hat{R}_A$, if

$$[(d_A e_A)^{-1} A_*^{d_A}(d_A e_A)] \cap U = \langle x^t \rangle , \qquad (30)$$

then we simply adjoin to $\hat{R}_A$ the elements $d_A e_A x, \ldots, d_A e_A x^{t-1}$. Having done this for all $d_A e_A \in \hat{R}_A$, denote the resulting set by $\hat{C}_A$. Now let $d_B e_B \in \hat{R}_B$. If $d_B e_B \in \hat{R}_A$ also, then we make the same adjunction of elements $d_B e_B x$, $d_B e_B x^2$, etc., as already made to $\hat{R}_A$. If on the other hand $d_B e_B \notin \hat{R}_A$ (so that incidentally $e_B \neq 1$) then proceed as follows. Let in the usual way $\theta': C_A' \to C_B'$ associate representatives of the same right cosets of H in G. Since $d_B e_B \in \hat{R}_B - \hat{R}_A$, it follows that $d_B e_B \in C_B' - C_A'$. If $d_A e_A e_U \in C_A'$ is such that $(d_A e_A e_U)\theta' = d_B e_B$, then it follows readily from property II in the definition of a cress that $(d_A e_A)\theta' = d_B e_B e_V$ for some $e_V \in V = U = \langle x \rangle$. Suppose $e_V = x^\tau$. Then if $1, x, \ldots, x^{t-1}$ are the U-endings already attached to $d_A e_A$ (which, being in $C_A' - C_B'$, is also in $\hat{R}_A - \hat{R}_B$), we adjoin to $\hat{R}_B$ the elements $d_B e_B x^{(\tau+i) \bmod t}$, $i = 0, 1, \ldots, t-1$, where $(\tau+i) \bmod t$ denotes the least non-negative residue of $(\tau+i)$ modulo t. (One of these, namely $d_B e_B$ itself, is already in $\hat{R}_B$.) Having made the appropriate adjunctions to $\hat{R}_B$ for all $d_B e_B \in \hat{R}_B$, we denote the resulting set by $\hat{C}_B$.

We now define $\hat{\theta}: (\hat{C}_A - \hat{C}_B) \to (\hat{C}_B - \hat{C}_A)$ by

$$(d_A e_A x^i)\hat{\theta} = d_B e_B x^{(\tau+i) \bmod t} , \qquad (31)$$

where $i = 0, 1, \ldots, t-1$, and for each $d_A e_A \in \hat{C}_A - \hat{C}_B$, the entities t, $d_B e_B$, τ (all depending on $d_A e_A$) are as above. It is easy to see that $\hat{\theta}$ is a bijection. Extend $\hat{\theta}$ to a bijection $\hat{\theta}: \hat{C}_A \to \hat{C}_B$, by defining it to be the identity map on $\hat{C}_A \cap \hat{C}_B$.

We have thus defined $(\hat{C}_A, \hat{C}_B)$, $\hat{\theta}: \hat{C}_A \to \hat{C}_B$, and subgroups $A_*^{d_A}$, $d_A \in \hat{C}_A$; $B_*^{d_B}$, $d_B \in \hat{C}_B$. We now show that these define an abstract cress in G.

Step 2. Verification that $(\hat{C}_A, \hat{C}_B)$ is an abstract cress in G. That condition I ((a),(b),(c)) is satisfied is immediate from the definition of $(\hat{C}_A, \hat{C}_B)$. Thus it only remains to check that conditions (i) and (ii) of the definition of an abstract cress in §1 are satisfied.

We begin with condition (i). We first verify the second part of that condition, namely that for each pair $d_A \in \hat{C}_A$, $d_B \in \hat{C}_B$ with $d_B = d_A e_A$, we have

$$(d_A e_A)^{-1} A_*^{d_A} (d_A e_A) \cap U = d_B^{-1} B_*^{d_B} d_B \cap U, \tag{32}$$

(and its analogue with A, B interchanged). Now if in (32) we have $d_B \notin S_B$, then by the definition of $B_*^{d_B}$ for such d_B, (29) holds, whence (32). Thus we may assume $d_B \in S_B$, which implies that $d_A \in S_A$, so that $A_*^{d_A}$ and $B_*^{d_B}$ are defined by (24) and (25) respectively. Hence in this case (32) becomes

$$[(d_A e_A)^{-1} A_H^{d_A} (d_A e_A)] N_A \cap U = [d_B^{-1} B_H^{d_B} d_B] N_B \cap U . \tag{33}$$

Now since $d_A \in C_A$, $d_B \in C_B$ and (C_A, C_B) is a cress for H in G, we know from (1) that

$$(d_A e_A)^{-1} A_H^{d_A} d_A e_A \cap U = d_B^{-1} B_H^{d_B} d_B \cap U ,$$

or, in other notation,

$$e_A^{-1} K(d_A) e_A \cap U = K(d_B) \cap U .$$

By choice of N_A and N_B, if both sides of this equation are trivial, then (33) is immediate from (18) (and its analogue with B replacing A), while if both sides are non-trivial, then (33) is immediate from (19) (and its analogue). The analogue of (32) with A and B interchanged is established similarly.

To complete the verification of condition (i) we need to show that for each $d_A \in \hat{C}_A$, the set of $e_A e_U$ such that $d_A e_A e_U \in \hat{C}_A$ is a right transversal for $d_A^{-1} A_*^{d_A} d_A$ in A (and analogously for each $d_B \in \hat{C}_B$). Now for each $d_A \in \hat{C}_A$, the set of $e_A \in \hat{T}_A$ such that $d_A e_A \in \hat{C}_A$, is precisely the set $\hat{E}(d_A)$; this is immediate from the definition of $\hat{E}(d_A)$ (more specifically, from the first of the two requirements imposed on $\hat{E}(d_A))$ and of $\hat{C}_A$. The desired conclusion then follows from the facts that $\hat{E}(d_A)$ by definition is a complete double-coset representative system for A mod $(d_A^{-1}A_*^{d_A}d_A, U)$, and that, by (30) et seqq., for each $d_Ae_A \in \hat{C}_A$, the U-endings e_U such that $d_Ae_Ae_U \in \hat{C}_A$ form a transversal for the subgroup $(d_Ae_A)^{-1}A_*^{d_A}(d_Ae_A) \cap U$ in U. The analogous conclusion for each $d_B \in \hat{C}_B$ follows similarly, except that in the last step (32) needs to be used (for those $d_Be_B \in \hat{C}_A \cap \hat{C}_B$), and the yet-to-be-established (34) below (for those $d_Be_B \in \hat{C}_B - \hat{C}_A$).

We now turn to condition (ii) of the definition of an abstract cress. Part (a) of condition (ii) was explicitly built in to the definition of $\hat{\theta}$. We now verify that for all $d_Ae_A \in \hat{C}_A - \hat{C}_B$,

$$(d_Ae_A)^{-1}A_*^{d_A}(d_Ae_A) \cap U = (d_Be_Bx^r)^{-1}B_*^{d_B}(d_Be_Bx^r) \cap U , \tag{34}$$

where $(d_Be_Bx^r) = (d_Ae_A)\hat{\theta}$. Since $d_Ae_A \in \hat{C}_A - \hat{C}_B$, we have by an earlier remark that $d_Ae_A \in S_A$, and, similarly, $d_Be_B \in S_B$. Hence by (24) and (25) (and since the abelianness of U allows us to drop the x^r) we may rewrite (34) equivalently as

$$[(e_A^{-1}K(d_A)e_A)N_A] \cap U = [(e_B^{-1}K(d_B)e_B)N_B] \cap U . \tag{35}$$

Now by definition of $\hat{\theta}$ (see (31)) we know that $(d_Ae_A)\theta' = d_Be_Bx^\tau$ (where $r \equiv \tau$ modt, t being determined by (30) -- but this is not relevant at present). Since (C_A', C_B') is a cress for H in G, and θ' the bijection that goes with it, we have from (2) that

$$(d_Ae_A)^{-1}A_H^{d_A}(d_Ae_A) \cap U = (d_Be_Bx^\tau)^{-1}B_H^{d_B}(d_Be_Bx^\tau) \cap U ,$$

which, as before, may be rewritten as

$$e_A^{-1}K(d_A)e_A \cap U = e_B^{-1}K(d_B)e_B \cap U . \tag{36}$$

If both sides of (36) are trivial then (35) follows from (18), while if both sides are non-trivial then (36) implies (35) by virtue of (19).

There remains only property (ii)(c) to verify; this however is almost immediate from (31) and (34).

This completes the verification that the pair $(\hat{C}_A, \hat{C}_B)$ (with its associated $\hat{\theta}$, $A_*^{d_A}$, $B_*^{d_B}$) is an abstract cress. Denote by $\hat{H}$ the subgroup determined by this abstract cress in accordance with Theorem 2.

Step 3. Proof that $\hat{H} \geqq H$. Since the base group of H is generated by the $A_H^{d_A}$, $B_H^{d_B}$ with $d_A, d_B \in S = S_A \cup S_B$, and since $S_A \subseteq \hat{C}_A$, $S_B \subseteq \hat{C}_B$, and, as we have observed, for all $d_A \in \hat{C}_A$, $d_B \in \hat{C}_B$ (in fact for all $d_A \in C'_A$, $d_B \in C'_B$), we have $A_*^{d_A} \geqq A_H^{d_A}$, $B_H^{d_B} \geqq B_H^{d_B}$, it is immediate that $\hat{H}$ (in fact the base group of $\hat{H}$) contains the base group of H. Thus it only remains to show that the elements $t_{d_A e_A} = d_A e_A[(d_A e_A)\theta']^{-1}$, $d_A e_A \in C'_A - C'_B$, belong to $\hat{H}$. Now for any fixed $d_A e_A \in C'_A - C'_B$ write as before $(d_A e_A)\theta' = d_B e_B x^{\tau}$, and $(d_A e_A)\hat{\theta} = d_B e_B x^r$. By (31) we have $r \equiv \tau$ mod t where t is determined by (30); write $r = \tau + st$. Then

$$t_{d_A e_A} = d_A e_A (d_B e_B x^{\tau})^{-1} = (d_A e_A) x^{st} (d_A e_A)^{-1} \cdot d_A e_A (d_B e_B x^r)^{-1} .$$

Since $(d_A e_A) x^{st} (d_A e_A)^{-1} \in \hat{H}$ by (30), and $d_A e_A (d_B e_B x^r)^{-1} = d_A e_A (d_A e_A \hat{\theta})^{-1}$ is a generator of the free part of $\hat{H}$, we have the desired conclusion.

Step 4. Completion of proof. Write $\hat{A} = A/N_A$, $\hat{B} = B/N_B$. We return now to the epimorphism (defined in (23))

$$\varphi\colon (A*B;\ U = V = \langle x \rangle) \to (\hat{A}*\hat{B};\ \langle xN_A \rangle = \langle xN_B \rangle = \hat{U}) ,$$

extending the natural maps $A \to \hat{A}$, $B \to \hat{B}$. Since $\hat{T}_A$ and $\hat{T}_B$ are left transversals respectively for $N_A\langle x \rangle$ in A and $N_B\langle x \rangle$ in B, and since, essentially by (30) et seqq., the x^s which occur as U-endings of elements of $\hat{C} \cup \hat{C}_B$ are all such that $0 \leqq s < k_A k_B$, where $\langle x^{k_A k_B} \rangle = N_A \cap \langle x \rangle = N_B \cap \langle x \rangle$, it follows that the restriction of φ to $\hat{C}_A \cup \hat{C}_B$ is one-to-one. If we define for each $d_A\varphi \in \hat{C}_A\varphi$, and for each $d_B\varphi \in \hat{C}_B\varphi$,

$$\hat{A}_*^{d_A\varphi} = A_{\hat{H}}^{d_A}\varphi\ ,\quad \hat{B}_*^{d_B\varphi} = B_{\hat{H}}^{d_B}\varphi\ , \tag{37}$$

and define $\theta_\varphi: \hat{C}_A\varphi \to \hat{C}_B\varphi$ to be $\varphi^{-1}\hat{\theta}\varphi$ (where φ is restricted to $\hat{C}_A \cup \hat{C}_B$), then it is not difficult to see that by virtue of the facts that φ is one-to-one on $\hat{C}_A \cup \hat{C}_B$, and that $(\hat{C}_A,\hat{C}_B)$ is an abstract cress in G whose associated subgroups $A_*^{d_A}$, $B_*^{d_B}$ contains $d_A N_A d_A^{-1}$, $d_B N_B d_B^{-1}$ respectively, the pair $(\hat{C}_A\varphi,\hat{C}_B\varphi)$ with associated subgroups as in (37) and bijection θ_φ, is an abstract cress in $G\varphi$. It is moreover clear that the subgroup corresponding to this abstract cress is just $\hat{H}\varphi$.

We shall now show that $g\varphi \notin H\varphi$. Suppose first that $g \notin \langle x \rangle$. Then if u is the U-ending of g, we have $gu^{-1} \neq 1$. Recall that we chose gu^{-1} to be in $S_A \cup S_B$, so that $gu^{-1} \in \hat{C}_A \cup \hat{C}_B$. Hence $(gu^{-1})\varphi$ is a nontrivial element of $\hat{C}_A\varphi \cup \hat{C}_B\varphi$, with trivial $U\varphi$-ending. Now by definition of a cress (we are thinking of the cress $(\hat{C}_A\varphi,\hat{C}_B\varphi)$ here) a nontrivial element of $\hat{C}_A\varphi \cup \hat{C}_B\varphi$ with trivial $U\varphi$-ending must lie in a double coset $(\hat{H}\varphi)y(U\varphi)$ different from $(\hat{H}\varphi)(U\varphi)$. Hence $(gu^{-1})\varphi \notin (\hat{H}\varphi)(U\varphi)$, whence $g\varphi \notin \hat{H}\varphi$, and, finally, $g\varphi \notin H\varphi$. If on the other hand $g \in \langle x \rangle$, then by (22) and the definition of $(\hat{C}_A,\hat{C}_B)$, we have $g \notin (\hat{H} \cap U)N_A$. Hence $g\varphi \notin \hat{H}\varphi \cap U\varphi$, whence $g\varphi \notin \hat{H}\varphi$, and again $g\varphi \notin H\varphi$.

Now $G\varphi$, being an amalgamated product of two finite groups is free-by-finite ([10, Theorem 1]). Since by M. Hall's original result in [7] free groups are subgroup separable, and since a finite extension of a subgroup separable group is again subgroup separable (see Romanovskiĭ [14] or Scott [15]), it follows that $G\varphi$ is subgroup separable. Thus $G\varphi$ is a subgroup separable image of G in which $g\varphi \notin H\varphi$; hence G has a finite homomorphic image in which the image of g is outside the image of H.

REFERENCES

1. G. Baumslag, "On the residual finiteness of generalized free products of nilpotent groups," Trans. Amer. Math. Soc. 106 (1963), 192-209.

2. G. Baumslag, "Some problems on one-relator groups," In: Proceedings of the second international conference on the theory of groups, pp. 75-81. Lecture notes in mathematics No. 372. Springer-Verlag, Berlin-Heidelberg-New York.

3. R. G. Burns, "A note on free groups," Proc. Amer. Math. Soc. 23 (1969), 14-17.

4. R. G. Burns, "On finitely generated subgroups of free products," J. Austral. Math. Soc. 12 (1971), 358-364.

5. D. E. Cohen, "Subgroups of HNN groups," J. Austral. Math. Soc. 16 (1973), 394-405.

6. I. M. S. Dey, "Schreier systems in free products," Proc. Glasgow Math. Assoc. 7 (1965-66), 61-79.

7. M. Hall, Jr., "Coset representations in free groups," Trans. Amer. Math. Soc. 67 (1949), 421-432.

8. A. Karrass and D. Solitar, "On finitely generated subgroups of a free group," Proc. Amer. Math. Soc. 22 (1969), 209-213.

9. A. Karrass and D. Solitar, "The subgroups of a free product of two groups with an amalgamated subgroup," Trans. Amer. Math. Soc. 150 (1970), 227-255.

10. A. Karrass, A. Pietrowski and D. Solitar, "Finite and infinite cyclic extensions of free groups," J. Austral. Math. Soc. 16 (1973), 458-466.

11. W. Magnus, A. Karrass and D. Solitar, Combinatorial group theory: Presentations of groups in terms of generators and relations, Pure and Appl. Math. vol. 13 (Interscience, 1966).

12. A. I. Mal'cev, "On homomorphisms onto finite groups," Ivanov. Gos. Ped. Inst. Učen. Zap. 18 (1958), 49-60.

13. J. Nielsen, "Om Regning med ikke kommutative Factorer og dens Anvendelse i Gruppeteorien, Mat. Tiddskrift B (1921), 77-94.

14. N. S. Romanovskiĭ, "On the finite residuality of free products, relative to an embedding," Izv. Akad. Nauk SSSR, Ser. Matem. 33 (1969), 1324-1329.

15. Peter Scott, "Subgroups of surface groups are almost geometric," J. London Math. Soc. (2), 17 (1978), 555-565.

16. Peter F. Stebe, "Conjugacy separability of certain free products with amalgamation," Trans. Amer. Math. Soc. 156 (1971), 119-129.

A. M. BRUNNER
DIVISION OF SCIENCE
UNIVERSITY OF WISCONSIN-PARKSIDE
KENOSHA, WISCONSIN 53141

Received November 24, 1981

R. G. BURNS and D. SOLITAR
DEPARTMENT OF MATHEMATICS
YORK UNIVERSITY
DOWNSVIEW, TORONTO
ONTARIO M3J1P3, CANADA

Contemporary Mathematics
Volume 33, 1984

ANOTHER PROOF OF THE GRUSHKO-NEUMANN-WAGNER THEOREM FOR FREE PRODUCTS

R. G. Burns and T. C. Chau

0. Introduction.

The strong form, due to Wagner [5], of the theorem of the title is as follows. (See §2 for the definition of a Nielsen transformation.)

Wagner's Strong Form of the Grushko-Neumann Theorem. Any generating set for an arbitrary free product $\prod_{i\in I} {}^*A_i$ of groups A_i, can be Nielsen-transformed to a generating set contained in $\bigcup_{i\in I} A_i$.

Grushko [2] and B. H. Neumann [4] proved this (independently) for finitely generated free products only, thereby showing that the least number of generators (i.e. the rank) of a free product is the sum of the ranks of the free factors. Different proofs (of the original theorem or its strong form) have since been published by Chiswell, Higgins, Lyndon, and Stallings. (Some of the references may be found in [3].)

The idea of the present proof is to show that any subset S of a free product $A*B$ can be Nielsen-transformed to a "Kurosh generating set" for $\langle S\rangle$ (the subgroup generated by S), that is, to a generating set arising from an application of the Kurosh subgroup theorem to the subgroup $\langle S\rangle$ of $A*B$. The above theorem is then immediate in the case of a free product $A*B$ of two groups, and a simple device allows one to quickly deduce also the full result.

The lemma in §4, the centrepiece of the proof, may have independent interest. In that section we define, in terms of an arbitrary well-ordering of a free group F, a natural partial order of the set of free bases of F;

© 1984 American Mathematical Society
0271-4132/84 $1.00 + $.25 per page

the lemma states that this p.o. set of bases has minimal elements. We then show that the Kurosh-reduced subsets are just the images (under suitable homomorphisms from F to $A*B$) of the minimal free bases arising from one particular well-ordering of F. (We note in passing that the lemma is a triviality in the case of finitely generated F, which corresponds to the situation of a finitely generated free product dealt with by Grushko and Neumann.)

Our proof also depends on the converse (due to Dey [1]) of the Kurosh subgroup theorem. In §1 we state the Kurosh subgroup theorem and its converse (proving the latter), in §2 we state the main theorems, and deduce the strong form of the Grushko-Neumann theorem, and in §§3, 4 these main theorems are proved.

1. Preliminaries: The Kurosh subgroup theorem and its converse

Although we shall not be needing the Kurosh subgroup theorem, we shall state it in full, the better to motivate its converse, on which, as mentioned above, our proof of the Grushko-Neumann theorem is based. Thus let A and B be arbitrary groups and write $G = A*B$, their free product. It is well-known (see e.g. [3, p. 175]) that G is characterized by the property that each of its elements g has a unique normal form (as an "alternating product")

$$g = t_1 \dots t_n ,$$

where $n \geqq 0$, $t_1, \dots, t_n \in (A \cup B) - \{1\}$, and for $i = 1, \dots, n-1$, the elements t_i, t_{i+1} come from different factors, i.e. are not both in A or both in B. We call the t_i the <u>syllables</u> of g. If $t_n \in A$ then we shall say of g that its A-<u>ending</u> is t_n and its B-<u>ending</u> is 1, while if on the other hand $t_n \in B$, then t_n is the B-<u>ending</u> of g, and its A-<u>ending</u> is trivial. Finally, we call the elements $t_1 \dots t_i$, $i = 0, 1, \dots, n$, the <u>initial segments</u> of g, and we write $|g| = n$, the <u>length</u> of g.

Now let H be a subgroup of G. The detailed structure of H is most simply described in terms of a special pair (C_A, C_B) of right transversals for H in G. We now define what we mean by "special pair (C_A, C_B)". (Such a pair is sometimes called a "regular extended Schreier system"; however in the present context the term "Kurosh system" seems slightly better.) Following Karrass and Solitar we shall adopt the convenient notation $d_A e_A$ for a typical element $g = t_1 \ldots t_n \in C_A$, where e_A is the A-ending of G, and (consequently) $d_A = t_1 \ldots t_{n-1}$ if $t_n \in A$ (i.e. if $e_A = t_n$), while $d_A = g$ if $t_n \in B$ (i.e. if $e_A = 1$).

DEFINITION 1. A pair (C_A, C_B) of right transversals for a subgroup H of a free product $G = A*B$, is called a <u>Kurosh system</u> for H in G, if it satisfies the following three conditions:

I. Each of C_A, C_B is closed under taking initial segments;

II. Any element of $C_A \cup C_B$ whose last syllable is from A is in C_A (and analogously for C_B).

III. The set of all $d_A \in C_A$ (i.e. of those elements of C_A with trivial A-ending) forms a complete double-coset representative system for G modulo (H,A) (and analogously for C_B).

THE KUROSH SUBGROUP THEOREM. *If H is any subgroup of $G = A*B$, then there exists a Kurosh system for H in G. If (C_A, C_B) is any Kurosh system for H in G, and if for each $d_A \in C_A$, $d_B \in C_B$, we write*

$$A_H^{d_A} = H \cap d_A A d_A^{-1}, \quad B_H^{d_B} = H \cap d_B B d_B^{-1},$$

and for each $d_A e_A \in C_A$, we write

$$t_{d_A e_A} = d_A e_A [(d_A e_A)\theta]^{-1},$$

where $\theta: C_A \to C_B$ is the map defined by $H(g\theta) = Hg$, then the nontrivial elements among the $t_{d_A e_A}$ freely generate a free subgroup of G, and H is the free product of this free group and the subgroups $A_H^{d_A}$, $B_H^{d_B}$, $d_A \in C_A$, $d_B \in C_B$.

This theorem prompts the following definition.

DEFINITION 2. A set of generators S for a subgroup H of $A*B$ is called a <u>Kurosh generating set</u> for H if there is a Kurosh system (C_A,C_B) for H in G such that S is contained in the (set-theoretic) union of the $A_H^{d_A}$, $B_H^{d_B}$ and the set of non-trivial $t_{d_A e_A}$.

Thus a Kurosh generating set is a "simple" generating set for H in the sense that it is clear (up to a point) how the generators combine to give H.

We now turn to the converse of the Kurosh subgroup theorem. Before we can state the converse, we need the concept of an "abstract Kurosh system"; this is a pair (C_A,C_B) of subsets of $G = A*B$ supplemented by certain other entities and satisfying certain natural conditions which (and this forms the content of the converse) ensure the existence of a subgroup H of G for which (C_A,C_B) is indeed a Kurosh system in the sense of Definition 1. Thus this converse supplies the answer, in this special context, to the following general problem: Given a subset of a group, find a subgroup (if any) for which that subset is a transversal.

DEFINITION 3. An <u>abstract Kurosh system</u> in $A*B$ is a pair (C_A,C_B) of nonempty subsets of G satisfying conditions I, II above and having the following two further properties:

(i) corresponding to each $d_A \in C_A$ there is prescribed a subgroup of $d_A A d_A^{-1}$, denoted by $A_*^{d_A}$, with the property that the set $E(d_A)$ consisting of those $e_A \in A$ such that $d_A e_A \in C_A$ (i.e. of those A-endings of elements of C_A which are attached to d_A), forms a right transversal for $d_A^{-1} A_*^{d_A} d_A$ in A (and similarly with B in place of A throughout).

(ii) there exists a bijection $\theta\colon C_A \to C_B$, whose restriction to $C_A \cap C_B$ is the identity map.

(We note that condition (i) is natural in the sense that it is satisfied by a "real" Kurosh system (C_A,C_B) for a subgroup H, but of course with respect to the $A_H^{d_A}$ rather than their "abstract" counterparts the $A_*^{d_A}$. This follows easily from III and the basic assumption that any "real" C_A and C_B are right transversals for H in $A*B$.)

CONVERSE OF THE KUROSH SUBGROUP THEOREM (Dey [1]). *Let* (C_A, C_B), $\{A_*^{d_A}, B_*^{d_B}\}$, θ, *form an abstract Kurosh system in* $G = A*B$ *in accordance with Definition* 3. *Then there is a unique subgroup* H *of* G *for which* (C_A, C_B) *is a Kurosh system (as in Definition* 1), *and for which* $A_H^{d_A} = A_*^{d_A}$, $B_H^{d_B} = B_*^{d_B}$ *for all* $d_A \in C_A$, $d_B \in C_B$, *and* $Hg = H(g\theta)$ *for all* $g \in C_A$.

Sketch of the proof. The bulk of the proof consists in defining an action of G on the set C_A. This achieved, one shows that the action of C_A on 1 is given by $1.c = c$ for c in C_A, and then, taking H to be the stabilizer of $1 \in C_A$, it follows easily that our action is equivalent to right multiplication of the right cosets of H by the elements of G, and that H has the properties claimed of it.

We define the action of G on C_A by defining actions of A and B separately; by virtue of the "freeness" of the free product $A*B$ these actions then extend uniquely to an action of the whole of $G = A*B$ on C_A.

The action $\cdot$ of each $a \in A$ on each $g = d_A e_A \in C_A$ is defined by

$g \cdot a = d_A\alpha$, where α is determined by:

$$\alpha \in E(d_A),\ e_A a \in (d_A^{-1} A_*^{d_A} d_A)\alpha\ . \tag{1}$$

That this defines $g \cdot a$ uniquely follows from (i) above. To see that (1) does define an action of A, let a_1, a_2 be arbitrary elements of A. Then on the one hand

$g \cdot (a_1 a_2) = d_A\alpha$, where α is determined by:

$$\alpha \in E(d_A),\ e_A a_1 a_2 \in (d_A^{-1} A_*^{d_A} d_A)\alpha\ . \tag{2}$$

On the other hand

$g \cdot a_1 = d_A\alpha_1$, where α_1 is determined by:

$$\alpha_1 \in E(d_A),\ e_A a_1 \in (d_A^{-1} A_*^{d_A} d_A)\alpha_1\ , \tag{3}$$

and then

$(g \cdot a_1) \cdot a_2 = (d_A\alpha_1) \cdot a_2 = d_A\bar{\alpha}$, where $\bar{\alpha}$ is determined by:

$$\bar{\alpha} \in E(d_A),\ \alpha_1 a_2 \in (d_A^{-1} A_*^{d_A} d_A)\bar{\alpha}\ . \tag{4}$$

From (3) and (4) it follows that

$$e_A a_1 a_2 \in (d_A^{-1} A_*^{d_A} d_A)\alpha_1 a_2 = (d_A^{-1} A_*^{d_A} d_A)\bar{\alpha} .$$

However from (2) we have that $e_A a_1 a_2 \in (d_A^{-1} A_*^{d_A} d_A)\alpha$. Since $\alpha, \bar{\alpha}$ both belong to $E(d_A)$, which by (i) is a right transversal for $d_A^{-1} A_*^{d_A} d_A$ in A, it follows that $\alpha = \bar{\alpha}$. It being immediate that 1 acts trivially on C_A, it follows that (1) does indeed define an action of A on C_A.

To obtain the appropriate action of B on C_A, first define an action of B on C_B analogously to (1), and then use the bijection θ to transfer this to an action of B on C_A. Thus the action of each $b \in B$ on each $g \in C_A$ is given by $g \cdot b = [(g\theta) \cdot b]\theta^{-1}$.

The fact that $1.c = c$ for $c \in C_A$ now follows by an easy induction on the length of c, using the fact that, by I and II above, for all proper initial segments d of such c we have $d\theta = d$.

2. Statement of the main theorem and deduction of the Grushko-Neumann theorem

In this section, after defining a "Kurosh-reduced subset" of $A*B$, we state our two results concerning such sets, and then show how the strong form of the Grushko-Neumann theorem (stated in the introduction) follows from them.

We begin with some preliminary terminology. We shall say that a product rs of elements $r,s \in A*B$ is reduced if no cancellation or consolidation occurs across the juncture of r and s, i.e. if the final syllable of r and the initial syllable of s come one from A and the other from B. We shall often need to write a general element g of $A*B$ in the reduced form $g = sct^{-1}$ where the lengths of s and t are the same (in symbols $|s| = |t|$), and c is from A or B (i.e. $|c| \leq 1$). We shall call s and t^{-1} respectively the first and second halves of g, and c the central factor of g. If $s \neq t$ we shall say that g is a non-conjugate, and, in this case only, talk of sc as the augmented first half of g.
(If $s = t$ we shall take the augmented first half to be the same as the first

half, namely s.) Finally we define the meet $g \wedge h$ of elements $g,h \in A*B$, to be their longest common initial segment.

DEFINITION 4. A subset S of a free product $A*B$ is said to be Kurosh-reduced if it has the following three properties.

(a) for every pair u,v of distinct, non-trivial elements of S, we have for all $\delta, \epsilon = \pm 1$

$$|u^{\delta} \wedge v^{\epsilon}| < \min (\tfrac{1}{2}|u|, \tfrac{1}{2}|v|) . \tag{5}$$

(b) (cf. (i)) for each $d \in A*B$ with trivial A-ending (so that either d has its last syllable in B, or else $d = 1$), the set of $a \in A$ such that da is an initial segment of the augmented first half of some element of S, or $(da)^{-1}$ is a terminal segment of the second half of some element of S, forms part of a right transversal $T(d,A)$ for $\langle d^{-1}Sd \cap A \rangle$ in A (and analogously with A and B interchanged).

(c) given any element g of S with first and second halves s,t^{-1}, where $s \neq t$ (i.e. given any non-conjugate in S), the segment t is "isolated" in S, i.e. g^{-1} is the only element of $S \cup S^{-1}$ having t as an initial segment.

We define in the usual way a Nielsen transformation of a subset S of an arbitrary group to be a map φ of S induced by a free automorphism, i.e. if X is a free basis for a free group F and $f\colon F \to \langle S \rangle$ the homomorphism defined by a bijection from X to S, then there should exist an automorphism $\hat{\varphi}$ of F such that $\varphi = f^{-1}\hat{\varphi}f$.

We can now state our two theorems (to be proved in §§3, 4).

THEOREM 1. *Any Kurosh-reduced subset S of a free product $A*B$ is a Kurosh generating set for $\langle S \rangle$ (see Definition 2); i.e. there is a Kurosh system (C_A, C_B) for the subgroup $\langle S \rangle$ in $A*B$, such that for each $d_A \in C_A$, $d_B \in C_B$, we have*

$$A^{d_A}_{\langle S \rangle} = \langle S \cap d_A A d_A^{-1} \rangle , \quad B^{d_B}_{\langle S \rangle} = \langle S \cap d_B B d_B^{-1} \rangle ,$$

<u>and the set of non-trivial elements of the form</u> $t(t\theta)^{-1}$, $t \in C_A$, <u>is, up to inverses, precisely the set of non-conjugates in</u> S.

THEOREM 2. <u>Any subset of a free product</u> $A*B$ <u>can be Nielsen-transformed to a Kurosh-reduced subset of</u> $A*B$.

We now show how the strong form of the Grushko-Neumann theorem follows from these two theorems. The case of a free product $A*B$ of two groups is immediate since if S generates $A*B$, i.e. $\langle S \rangle = A*B$, then of course the only Kurosh system for $\langle S \rangle$ in $A*B$ is the trivial one: $C_A = \{1\}$, $C_B = \{1\}$, so that any Kurosh generating set for $\langle S \rangle$ is contained in $A \cup B$.

For the general case of a free product $\prod_{i\in I} * A_i$ of an arbitrary family of groups we resort to the following device. (Alternatively the lemma in §4 below may be used.) Let A be any group containing the A_i as subgroups (e.g. take $A = \Pi * A_i$). Let B be any group of cardinal equal to that of I, and index the elements of B in a one-to-one manner with the elements of I; i.e. denote them by b_i, $i \in I$. Consider the subgroup H of $A*B$ generated by the subgroups $b_i A_i B_i^{-1}$, $i \in I$; it is easy to see that H is the free product of these subgroups. If S is any generating set for H, then by Theorem 2 there is a Nielsen transform $\hat{S}$ of S which is Kurosh-reduced, and then Theorem 1 yields a Kurosh system (C_A, C_B) corresponding (as in that theorem) to the set $\hat{S}$. Among the free factors in the Kurosh decomposition corresponding to (C_A, C_B) are the groups $H \cap d_i A d_i^{-1}$ where the $d_i = h_i b_i a_i$, $h_i \in H$, $a_i \in A$, are the elements of C_A with trivial A-ending in the double cosets $Hb_i A$. Now since H is the free product of the $b_i A_i b_i^{-1}$, it follows that $d_i = h_i b_i a_i$ has the form

$$h_i b_i a_i = b_{i_1} a_{i_1} b_{i_1}^{-1} \cdot b_{i_2} a_{i_2} b_{i_2}^{-1} \cdots b_{i_n} a_{i_n} b_{i_n}^{-1} b_i a_i ,$$

where $n \geq 0$, and if $n \geq 1$ then $a_{i_k} \in A_{i_k} - \{1\}$, and $i_{k-1} \neq i_k$. Since (C_A, C_B) is a Kurosh system in $A*B$ and $d_i \in C_A$, we know that all initial segments of $d_i = h_i b_i a_i$ are also in C_A. It follows that $n \leq 1$ since otherwise we should have the two initial segments b_{i_1} and $b_{i_1} a_{i_1}$ both representing the coset Hb_{i_1}. If $n = 1$ then for the same reason we shall have to have $b_{i_1} = b_i$, whence $d_i = b_{i_1}a_{i_1}a_i$, and then since d_i has trivial A-ending it will follow that $a_{i_1}a_i = 1$, whence $d_i = b_i$. If $n = 0$, then $d_i = b_i a_i$, whence (for the same reason) $a_i = 1$ and again $d_i = b_i$. Thus $H \cap d_i A d_i^{-1} = H \cap b_i A b_i^{-1} = b_i A_i b_i^{-1}$. Hence any set S of generators of $\Pi^* b_i A_i b_i^{-1}$ can be Nielsen-transformed to a set contained in $\bigcup_{i \in I} b_i A_i b_i^{-1}$. The desired conclusion now follows by identifying A_i with $b_i A_i b_i^{-1}$.

§3. Proof of Theorem 1

The proof consists in the construction of an appropriate abstract Kurosh system (C_A, C_A) (see Definition 3) from the given Kurosh-reduced subset $S \subseteq A*B$; the desired conclusion is then immediate from the Converse of the Kurosh Subgroup Theorem.

To begin with we define sets S_A and S_B (eventually to form part of C_A and C_B respectively) as follows. Initially we distribute between S_A and S_B the augmented first halves and inverses of second halves of all elements of S in the following way: for every (reduced) element sct^{-1} belonging to S, where as usual $|s| = |t|$, $|c| \leqq 1$, if $s = t$ we put s in both S_A and S_B, while if $s \neq t$, we put sc in S_A and t in S_B in case sc ends in an element of A (so that t necessarily ends in an element of B), and the other way around if sc has its last syllable in B. (Note that we are admitting the possibility that $c = 1$, in which case the last syllable of sc is just the last syllable of s (provided s is not also trivial).) If P denotes the set of all proper initial segments of all elements so far placed in either S_A or S_B, then include P in both S_A and S_B; this in fact then completes the definition of S_A and S_B. (If S is empty, set $S_A = S_B = \{1\}$.)

We next define a pair of sequences of subsets of $A*B$:

$$S_A = C_A^0 \subseteq C_A^1 \subseteq \dots ,$$

$$S_B = C_B^0 \subseteq C_B^1 \subseteq \dots ,$$

with the following three properties:

(α) (cf. I, II.) For all n, if an element g in $C_A^n \cup C_B^n$ has its last syllable in A, then all initial segments of g (including g itself) belong to C_A^n (and analogously for B).

(β) (cf. (i).) For each d_A in C_A^n of length $< n$ which does not end in an A-syllable (so that either $d_A = 1$ or d_A ends in a B-syllable), the set $E(d_A)$ consisting of all $a \in A$ such that $d_A a \in C_A^n$ is a right transversal for $\langle d_A^{-1} S d_A \cap A \rangle$ in A (and similarly with A and B interchanged throughout).

(γ) For all n, $C_A^n - C_B^n = S_A - S_B$ and $C_B^n - C_A^n = S_B - S_A$.

Clearly the pair $C_A^0 = S_A$, $C_B^0 = S_B$ have these properties (the first property by definition of S_A, S_B, the second vacuously, and the third trivially). Assuming inductively that we have defined C_A^n, C_B^n with these three properties, we define C_A^{n+1}, C_B^{n+1} to contain C_A^n, C_B^n (respectively) together with elements obtained by adjoining syllables to the ends of certain of the length-n elements in C_A^n, C_B^n, as follows. For every element d_A of length n in C_A^n whose last syllable is not in A, include in C_A^{n+1} the set $d_A T(d_A, A)$ where $T(d_A, A)$ is the right transversal (containing 1, we note) for $\langle d_A^{-1} S d_A \cap A \rangle$ in A mentioned in the definition of the Kurosh-reduced set S (see Definition 4, Part (b)), and include in C_B^{n+1} the set $d_A T(d_A, A) - S_A$. (This ensures that $C_A^{n+1} - C_B^{n+1} = S_A$.) Make the analogous adjunctions for all elements d_B of length n in C_B^n, not ending in a B-syllable, to obtain C_B^{n+1}.

We now verify that properties (α), (β), (γ) (with $n + 1$ replacing n) hold for C_A^{n+1}, C_B^{n+1}. First consider the property (α). Suppose $g \in C_A^{n+1} \cup C_B^{n+1}$ has for instance non-trivial A-ending; we wish to show that then all initial segments of g are in C_A^{n+1}. If $|g| \leq n$, then $g \in C_A^n$, and so by inductive hypothesis all initial segments of g are in C_A^n and therefore certainly in C_A^{n+1}. Suppose on the other hand $|g| = n + 1$. If $g \notin S_A$ (the case $g \in S_A$ being easy) then by construction of C_A^{n+1} it must have been obtained by adjoining an A-syllable a to an element d of C_A^n with trivial A-ending. Hence its longest proper initial segment d lies in C_A^{n+1} (since $C_A^n \subseteq C_A^{n+1}$), and then all proper initial segments of d lie in C_A^{n+1} (in fact in C_A^n) since the longest of them (if non-trivial) has non-trivial A-ending. Properties (β) and (γ) hold by virtue of being built in explicitly in defining C_A^{n+1}, C_B^{n+1}.

Define $C_A = \bigcup_n C_A^n$, $C_B = \bigcup_n C_B^n$. Since the C_A^n, C_B^n all have property (α) above, it is clear that the pair (C_A, C_B) has properties I and II (see §1). Similarly, if we define $A_*^{d_A} = \langle S \cap d_A A d_A^{-1} \rangle$ for each $d_A \in C_A$ not ending in an A-syllable, and analogously $B_*^{d_B} = \langle S \cap d_B B d_B^{-1} \rangle$, then property ($\beta$) guarantees that the pair (C_A, C_B) has property (i) of the definition of an abstract Kurosh system (Definition 3). To complete the specification of our abstract Kurosh system (C_A, C_B) we need a bijection $\theta: C_A \to C_B$ satisfying condition (ii) of Definition 3. We first define θ as a map from $S_A - S_B$ to $S_B - S_A$ as follows. In defining S_A and S_B at the beginning of the present proof, for each non-conjugate in S with reduced from sct^{-1}, $|s| = |t|$, $|c| \leq 1$, we placed sc in one of S_A, S_B (and only one) and t in the other; let θ be the bijection associating these segments, i.e. define $(sc)\theta = t$ if $sc \in S_A$, and $t\theta = sc$ otherwise. That θ is well-defined and indeed a bijection bollows from the fact that S has properties (a) and (c) of Definition 4. From the property (γ) it follows by means of a little elementary set theory that $C_A - C_B = S_A - S_B$ and $C_B - C_A = S_B - S_A$; hence if we extend θ by defining it to be the identity map on $C_A \cap C_B$, we obtain a bijection $\theta: C_A \to C_B$ as required. That the subgroup which arises from this abstract Kurosh system (C_A, C_B) is just $\langle S \rangle$ is immediate from the definition of θ and of the $A_*^{d_A}$, $B_*^{d_B}$.

§4. Proof of Theorem 2

We begin with a lemma, which as noted in the introduction, may be of interest in its own right.

Let F be a free group and let $\leq$ be any well-ordering of the set F. We define a relation $\preceq$ on the set $\mathcal{B}$ of free bases of F as follows:

If X,Y are any two free bases, we set $X \preceq Y$ if there is a bijection $\varphi: Y \to X$, with the following two properties:

(A) For all $y \in Y$, $y\varphi \leq y$;

(B) For all $y \in Y$ for which $y\varphi \neq y$, we have $y \in \langle x \mid x \in X, x < y \rangle$, i.e. y belongs to the subgroup of F generated by the elements of X which precede y in the given well-ordering of F.

It is easy to verify that this relation is a partial order of the set $\mathcal{B}$. It is less evident, but for us crucial, that this p.o. set $\mathcal{B}$ has minimal elements.

LEMMA. The partially ordered set $(\mathcal{B}, \preceq)$ has minimal elements.

Proof. We shall show that any chain $\mathcal{C}$ in the p.o. set $\mathcal{B}$ has a lower bound in $\mathcal{B}$. The desired conclusion will then follow by Zorn's lemma.

To show that $\mathcal{C}$ has a lower bound in $\mathcal{B}$, we first construct (inductively) a certain subset of $\mathcal{B}$, well-ordered with respect to the relation inverse to $\preceq$. We shall denote the elements of this subset by X_β where β ranges over the ordinals up to some particular ordinal α (to be defined), and where $X_{\beta_1} \succ X_{\beta_2}$ precisely if $\beta_1 < \beta_2$. At the same time we shall define bijections $f_\beta: X_1 \to X_\beta$, such that for all β_1, β_2 with $\beta_1 < \beta_2$ the bijection $f_{\beta_1}^{-1} f_{\beta_2}: X_{\beta_1} \to X_{\beta_2}$ has properties (A) and (B) above.

To begin the induction, define X_1 to be any member of $\mathcal{C}$, and define $f_1: X_1 \to X_1$ to be the identity map. (If $\mathcal{C}$ is empty then a lower bound exists trivially.) Suppose inductively that we have defined $X_\gamma \in \mathcal{C}$ and $f_\gamma: X_1 \to X_\gamma$, for all $\gamma < \beta$. If β is not a limit ordinal, say $\beta = \hat{\beta} + 1$, define X_β to be any member of $\mathcal{C}$ such that $X_{\hat{\beta}} \succ X_\beta$, and take $f_\beta = f_{\hat{\beta}}\varphi_{\hat{\beta}}$ where $\varphi_{\hat{\beta}}: X_{\hat{\beta}} \to X_\beta$ is any bijection satisfying (A) and (B). That $f_{\beta_1}^{-1} f_\beta$ has properties (A) and (B) for all $\beta_1 < \beta$ then follows easily. (Note that if no such X_β exists then $X_{\hat{\beta}}$ is a lower bound for $\mathcal{C}$.) If on the other hand β is a limit ordinal then we define X_β and f_β as follows (and this is where properties (A) and (B) come into play). Let x be any particular element of X_1 and let $xf_{\hat{\gamma}}$ be the least (in the given well-ordering of F)

among all of its images xf_γ, $\gamma < \beta$. Since $f_{\hat\gamma}^{-1} f_\gamma$ has property (A) for all $\hat\gamma \leq \gamma < \beta$, and $xf_\gamma = (xf_{\hat\gamma})(f_{\hat\gamma}^{-1} f_\gamma)$, it follows that xf_γ is constant for $\hat\gamma \leq \gamma < \beta$. We define X_β to be the set of all such $xf_{\hat\gamma}$ where x ranges over X_1 (and $\hat\gamma$ depends on x), and we define $f_\beta\colon X_1 \to X_\beta$ by $xf_\beta = xf_{\hat\gamma}$, $x \in X_1$.

To complete the induction we show in turn that X_β is a free basis for the subgroup it generates, that f_β is one-to-one, that in fact $\langle X_\beta \rangle = F$, and finally that for all $\gamma < \beta$, $f_\gamma^{-1} f_\beta$ has properties (A) and (B).

If either X_β does not freely generate $\langle X_\beta \rangle$ or f_β is not one-to-one, then there exist distinct elements $x_1, \ldots, x_n$ of X_1 and a non-trivial word $w(x_1, \ldots, x_n)$ such that $w(x_1 f_\beta, \ldots, x_n f_\beta) = 1$. By definition of f_β and by the finiteness of n, there exists in this case an ordinal $\gamma_0 < \beta$ such that for all $i = 1, \ldots, n$, $x_i f_\gamma = x_i f_\beta$ for $\gamma_0 \leq \gamma < \beta$. This means that for each such γ, the elements $x_i f_\gamma$, $i = 1, \ldots, n$, satisfy the non-trivial relation $w(x_1 f_\gamma, \ldots, x_n f_\gamma) = 1$, which contradicts the inductive hypothesis that the f_γ define automorphisms of F (i.e. are bijections between free bases).

We next prove that X_β generates F. For this it suffices to show that each $x \in X_1$ belongs to $\langle X_\beta \rangle$. In terms of an arbitrary such element x of X_1, we define a sequence $\beta_1 < \beta_2 < \ldots < \beta$ of ordinals, together with elements $x_{\beta_i} \in X_{\beta_i}$, as follows. Define β_1 to be the first ordinal $< \beta$ such that $xf_{\beta_1} \neq x$ (if there are any such ordinals). Since by inductive hypothesis the map $f_\gamma^{-1} f_{\beta_1}$ has property (B) for all $\gamma < \beta_1$ and since for those γ we have $xf_\gamma = x$, it follows that $x \in \langle x_{\beta_1}^1, \ldots, x_{\beta_1}^{n_1} \rangle$ where $x > x_{\beta_1}^i \in X_{\beta_1}$. Define x_{β_1} to be any of $x_{\beta_1}^1, \ldots, x_{\beta_1}^{n_1}$, and let β_2 be the first ordinal (if there are any) such that $\beta_1 < \beta_2 < \beta$ and $x_{\beta_1} \neq x_{\beta_1}(f_{\beta_1}^{-1} f_{\beta_2})$. Since by inductive hypothesis $f_{\beta_1}^{-1} f_{\beta_2}$ has property (B), this implies that $x_{\beta_1} \in \langle x_{\beta_2}^1, \ldots, x_{\beta_2}^{n_2} \rangle$ where $x_{\beta_1} > x_{\beta_2}^i \in X_{\beta_2}$. Since every descending sequence

$$x > x_{\beta_1} > x_{\beta_2} > \ldots ,$$

defined by continuing (inductively) in this manner, must terminate (by the well-known property of a well-ordered set that any subset well-ordered with respect to the inverse order is finite), it follow that for some $\gamma_0 < \beta$ there exist elements $x^1_{\gamma_0},\ldots,x^n_{\gamma_0} \in X_{\gamma_0}$ such that $x \in \langle x^1_{\gamma_0},\ldots,x^n_{\gamma_0}\rangle$ and $x^i_{\gamma_0}(f^{-1}_{\gamma_0}f_\gamma) = x^i_{\gamma_0}$ for all γ with $\gamma_0 < \gamma < \beta$. However by definition of X_β, the elements $x^i_{\gamma_0}$ belong to X_β, so that $x \in \langle X_\beta\rangle$, as required.

Note that we have here incidentally also established that all $f^{-1}_\gamma f_\beta$, $\gamma < \beta$, have property (B), since if for any $\hat\beta < \beta$, we have an element $x_{\hat\beta} \in X_{\hat\beta}$ such that $x_{\hat\beta}(f^{-1}_{\hat\beta}f_\beta) \neq x_{\hat\beta}$, then by definition of f_β this implies that there is an ordinal γ_1 with $\hat\beta < \gamma_1 < \beta$ such that $x_{\hat\beta}(f^{-1}_{\hat\beta}f_{\gamma_1}) \neq x_{\hat\beta}$, and then the same argument as was applied above to x to yield $x \in \langle x^1_{\gamma_0},\ldots,x^n_{\gamma_0}\rangle$ where $x > x^i_{\gamma_0} \in X_\beta$, can now be applied to $x_{\hat\beta}$.

Leaving to the reader the easy verification that for all $\gamma < \beta$ the map $f^{-1}_\gamma f_\beta$ has property (A), we have thus completed the inductive step in the definition of the maps $f_\beta\colon X_1 \to X_\beta$. Now let α be the first ordinal for which X_α cannot be defined. (This will certainly occur for ordinals whose corresponding cardinals exceed the cardinal of $\mathcal{C}$.) The ordinal α cannot be a limit ordinal since for limit ordinals β the above construction always yields X_β (and f_β), given the X_γ, f_γ for all $\gamma < \beta$. Hence $\alpha = \hat\alpha + 1$ where $X_{\hat\alpha}$, $f_{\hat\alpha}$ are defined, and then $X_{\hat\alpha}$ will serve as the desired lower bound for $\mathcal{C}$.

<u>Proof of Theorem</u> 2. We wish to show that an arbitrary subset S of a free product $A*B$ can be Nielsen-transformed to a Kurosh-reduced set. In order to apply the above lemma we need to have at hand a free group F well-ordered in a way suited to our purpose. We take F to be a free group with a free basis X in one-to-one correspondence with S, and denote by $\theta\colon F \to \langle S\rangle$ the epimorphism extending this correspondence.

We now well-order F in terms of a particular well-ordering $\leqq$ of $A*B$, chosen with our object in mind. We first well-order $A*B$ lexicographically in the usual way. Thus we choose arbitrary (but henceforth fixed) well-orderings for $A - \{1\}$ and $B - \{1\}$, set every element of $A - \{1\}$ less than every element of $B - \{1\}$, and then for each pair of elements $g_1, g_2 \in A*B$ with normal forms $g_1 = t_1 \ldots t_m$, $g_2 = \tau_1 \ldots \tau_n$, define $g_1 L g_2$ (L for "lexicographically less than") if $|g_1| < |g_2|$ (i.e. $m < n$), or if $m = n$ and for some i, $1 \leqq i \leqq n$, we have that $t_1 = \tau_1, \ldots, t_{i-1} = \tau_{i-1}$, but t_i precedes τ_i in the initially chosen well-ordering of $(A - \{1\}) \cup (B - \{1\})$.

The well-ordering $\leqq$ of $A*B$ is now defined as follows. Again let g_1, g_2 be any elements of $A*B$. Write, in reduced form,

$$g_1 = s_1 c_1 t_1^{-1} \ , \ g_2 = s_2 c_2 t_2^{-1} \ , \ |s_i| = |t_i|, \ |c_i| \leqq 1 \ .$$

(Thus s_i is the first half, t_i^{-1} the second half, and c_i the central factor of g_i.) Define $g_1 < g_2$ if $|g_1| < |g_2|$, or if $|g_1| = |g_2|$ and $s_1 \neq s_2$, $s_1 L s_2$, or if $|g_1| = |g_2|$, $s_1 = s_2$, $t_1 \neq t_2$, $t_1 L t_2$, or finally if $|g_1| = |g_2|$, $s_1 = s_2$, $t_1 = t_2$, but c_1 precedes c_2 in the initially chosen well-order of $(A - \{1\}) \cup (B - \{1\})$. (In other words in deciding the order of two elements, one first compares lengths, then (if their lengths are equal) one compares their first halves lexicographically, thirdly the first halves of their inverses, and lastly (all else failing to distinguish them) their central factors.)

The appropriate well-order of F (for which we shall use the same symbol $\leqq$) is now defined in the following way. Denote by N the kernel of $\theta: F \to \langle S \rangle$, and well-order each coset of N in F arbitrarily. Then for any $w_1, w_2 \in F$, set $w_1 < w_2$ if $w_1\theta < w_2\theta$ in the well-order $\leqq$ of $A*B$ just defined, or if $w_1\theta = w_2\theta$ but $w_1 < w_2$ in the chosen well-order of the coset $Nw_1 = Nw_2$.

By the lemma the p.o. set $\mathcal{B}$ of free bases of F under the partial order $\preceq$ defined (as in the preamble to the lemma) in terms of the above well-order $\leqq$ of F, has a minimal element, Y say. We shall now show that $Y\theta$ is a Kurosh-reduced subset of $A*B$.

(a) We first verify that $Y\theta$ has property (a) of the definition (Definition 4) of a Kurosh-reduced subset of $A*B$. Thus let $u,v \in Y\theta - \{1\}$ and suppose $u < v$, and that for instance $|u \wedge v| \geq \frac{1}{2}|u|$. This implies that $|u^{-1}v| < |v|$ (since even if exactly half of u^{-1} cancels into v, there will be subsequent consolidation), whence $u^{-1}v < v$ (in the well-ordering $\leq$ of $A*B$). Let $\hat{u} \in Y$ be any particular inverse image of u under θ, and define a map φ of Y as follows: for all inverse images $\hat{v} \in Y$ of $v \in Y\theta$, set $\hat{v}\varphi = \hat{u}^{-1}\hat{v}$, and for all other $y \in Y$, set $y\varphi = y$. Clearly $Y\varphi$ is a free basis of F, and φ is a bijection. Since $u^{-1}v < v$, we have $\hat{v}\varphi < \hat{v}$ for all inverse images $\hat{v}$ of v, whence φ has property (A). Since for these $\hat{v}$ we have $\hat{v} = \hat{u}(\hat{v}\varphi)$ where $\hat{u} < \hat{v}$ and $\hat{v}\varphi < \hat{v}$, we see that φ also has property (B). Hence $Y\varphi \prec Y$, contradicting the minimality of Y.

(b) Suppose next that $Y\theta$ violates condition (b) of the definition of a Kurosh-reduced subset. Then there is an element d of $A*B$, with trivial A-ending, and there are distinct elements a_1, a_2 of A, such that $d(a_1a_2^{-1})d^{-1} \in \langle Y\theta \cap dAd^{-1}\rangle$, and da_1, da_2 are initial segments of, for instance, the augmented first halves of elements $g_1 = s_1c_1t_1^{-1}$ and $g_2 = s_2c_2t_2^{-1}$ respectively, from $Y\theta$. Suppose that $a_1 < a_2$ in the well-ordering $\leq$ of $A*B$. The map $\hat{\varphi}$ of $Y\theta$ which sends g_2 to $d(a_1a_2^{-1})d^{-1}g_2$ and fixes all other elements of $Y\theta$ is clearly a Nielsen transformation of $Y\theta$. (Note here that g_2 is not in dAd^{-1}, since if it were its augmented first half would be, by definition, just d.) Since the effect of $\hat{\varphi}$ on g_2 is to replace the initial segment da_2 by da_1, we have $g_2\hat{\varphi} < g_2$, so that the corresponding map φ of Y (defined much as in (a) above) has property (A).

To see that φ has property (B) we show first that any element in $Y\theta$ of the form dad^{-1}, $a \in A$, precedes or is equal to g_2. This is clear if $|dad^{-1}| < |g_2|$. If on the other hand $|dad^{-1}| = |g_2|$ then we must have $s_2 = d$, $c_2 = a_2$ since da_2 is an initial segment of the augmented first half of $g_2 = s_2c_2t_2^{-1}$. Now we cannot have $t_2 = s_2$ $(= d)$ since augmented first halves of elements of the form dad^{-1} are by definition just their first halves (d in this case); and we cannot have $t_2 L s_2$ since in this case $g_2^{-1} < g_2$, and then the map of Y which inverts all preimages (under θ) of $g_2 \in Y\theta$, and fixes all other elements of Y, would yield a free basis $X \prec Y$. Hence $s_2 L t_2$, whence $g_2 > dad^{-1}$ by lexicographic comparison of the inverses of their second halves. Hence $d(a_1 a_2^{-1})d^{-1}$ is in a subgroup generated by elements of $Y\theta$ which precede g_2. Since $g_2 = [d(a_1 a_2^{-1})d^{-1}]^{-1} g_2\hat{\varphi}$ and also $g_2\hat{\varphi} < g_2$, it follows that φ has property (B). Hence $Y\varphi \prec Y$, again contradicting the minimality of Y.

(c) Finally, suppose that condition (c) of Definition 4 fails for $Y\theta$, and let $g = sct^{-1}$ be, in reduced form, a non-conjugate in $Y\theta$ such that t is not isolated in $Y\theta$, i.e. there is an element h in $Y\theta - \{g\}$ with t as an initial segment or t^{-1} as a terminal segment. Suppose for instance t^{-1} is a terminal segment of h. As noted in (b) above we must have sLt (and $s \neq t$ since g is a non-conjugate). It follows that since less than half of h and g^{-1} cancels in the product hg^{-1} (by property (a) already established for $Y\theta$), the reduced form of the product hg^{-1} will have the same first half as h, but the inverse of its second half will be lexicographically smaller than that of h. Hence $hg^{-1} < h$. Thus if $g < h$ then the Nielsen transformation of $Y\theta$ which sends h to hg^{-1} and fixes all other elements has properties (A) and (b), contradicting, as before, the minimality of Y. If on the other hand $h < g$ then certainly $hg^{-1} < g$, and to obtain a contradiction we take instead the Nielsen transformation which sends g to hg^{-1} and fixes the other elements of $Y\theta$.

REFERENCES

[1] I. M. S. Dey, "Schreier systems in free products," Proc. Glasgow Math. Soc., 67 (1949), 421-432.

[2] I. A. Grushko, "On the bases of a free product of groups," Mat. Sbornik, 3 (1938), 543-551.

[3] R. C. Lyndon and P. E. Schupp, Combinatorial Group Theory. Ergebnisse der Mathmetik und ihrer Grenzgebiete, Bd. 89. Springer-Verlag, Berlin-Heidelberg-New York (1977).

[4] B. H. Neumann, "On the number of generators of a free product," J. London Math. Soc. 18 (1943), 12-20.

[5] D. H. Wagner, "On free products of groups," Trans. Amer. Math. Soc. 84 (1957), 352-378.

YORK UNIVERSITY
DOWNSVIEW, ONTARIO,
CANADA

LAURENTIAN UNIVERSITY
SUDBURY, ONTARIO,
CANADA

Received November 24, 1981

Contemporary Mathematics
Volume 33, 1984

ON VARIETAL ANALOGS OF HIGMAN'S EMBEDDING THEOREM

Frank B. Cannonito

(To Roger C. Lyndon on his 65th birthday)

The purpose of this article is to record and discuss some recent results in combinatorial group theory which suggest that analogs of Higman's embedding theorem may still be found for the varieties $\underline{\underline{C}}$ (center-by-metabelian) and $\underline{\underline{N}}_c\underline{\underline{A}}$ (class c-by-abelian, $c \geq 2$) despite the belief held by some that this is not possible.

The idea of varietal analogs of Higman's theorem stems from the discovery in 1972 by Gilbert Baumslag [2] that each finitely generated metabelian group can be embedded in a finitely presented metabelian group (see also [3] and [12]). Baumslag suggested that analogs of Higman's theorem may be found for other varieties. Nevertheless no new analogs have been found and some authors have asserted that such analogs do not exist for, for example, the variety $\underline{\underline{C}}$ (cf. [7]). I believe this view is mistaken, especially if one understands an analog of Higman's theorem to mean a <u>characterization of the finitely generated subgroups of the finitely presented groups of the variety</u>.

Until now, it seems to me, very little attention has been paid to finding those conditions that are necessarily imposed on the finitely generated subgroups of finitely presented groups in a variety, and perhaps this is to be expected. For, in the case of the variety of all groups, the necessary condition - recursive presentation, the "easy" part of Higman's proof - is a trivial consequence of the fact that the intersection of two recursively

© 1984 American Mathematical Society
0271-4132/84 $1.00 + $.25 per page

enumerable sets is recursively enumerable. Furthermore, in the case of Baumslag's proof for the metabelian variety, P. Hall's discovery [8] that finitely generated metabelian groups satisfy Max-n automatically confers recursive presentability on all finitely generated metabelian groups, which, as it turned out, was the only necessary condition required. (Another fact which may bear is this: Baumslag was not attempting to obtain a characterization for the metabelian variety at the time he was proving his theorem [personal communication] and so this approach seems not to have occurred to him or others.) But when we move from these varieties to an arbitrary variety $\underline{\underline{V}}$ very little may be known about the locally hereditary properties of the finitely presented groups in $\underline{\underline{V}}$. Indeed it may be just as difficult to determine these "necessary conditions" as it is to discover "sufficient" conditions which enable the "embedding" to succeed. Let me now take up these ideas in connection with the varieties $\underline{\underline{C}}$ and $\underline{\underline{N}}_c\underline{\underline{A}}$.

Concerning the variety $\underline{\underline{C}}$ the most basic result is P. Hall's discovery [8] of a finitely generated center-by-metabelian group whose center is free abelian of countably infinite rank. This result seems to have suggested to some authors, on mere cardinality grounds, that there were too many finitely generated center-by-metabelian groups and so no way to embed all of them. Apparently such notions derived from Baumslag's success at embedding all finitely generated metabelians, which is a result peculiar only to the metabelian variety. Obviously any attempt to embed a continuum of finitely generated groups was doomed to failure since only countably many of these groups could be recursively presentable and this condition is a necessary consequence of Higman's theorem. Thus it is foolish to conclude that because something, which is inherently impossible at the outset, turns out on (further) analysis to be still impossible, implies no analog of Higman's theorem exists for $\underline{\underline{C}}$. The sensible thing to have done was to have gone immediately to the class of finitely generated recursively presented center-by-metabelian groups to see which of them can be embedded, for as we will see it is not possible to embed all such.

The first non-obvious locally hereditary property possessed by the finitely presented groups in $\underline{\underline{C}}$ was discovered by J. R. J. Groves, who proved a finitely presented center-by-metabelian group is abelian-by-polycyclic [7]. Since finitely generated abelian-by-polycyclic groups have Max-n (P. Hall [8]) and are residually finite (Roseblade [11], Jategoankar [10]) we see that finitely presented center-by-metabelian groups have solvable word problem. Now, having solvable word problem is a locally hereditary property of groups and so any finitely generated center-by-metabelian group which stands a chance of being embedded must, accordingly, have solvable word problem. But it is easy to construct from P. Hall's example, mentioned above, a finitely generated recursively presented center-by-metabelian group with unsolvable word problem. One kills off a recursively enumerable non-recursive set of the generators of the center. (There is some delicacy involved here, but this is essentially correct.) Thus we see that it makes no sense to seek to embed all finitely generated recursively presented center-by-metabelian groups. It likewise makes no sense to conclude at this point that no analog of Higman's theorem exists for $\underline{\underline{C}}$ because not all the recursively presented groups are embeddable! Since we could only hope to embed those with solvable word problem we had no hope of embedding all recurisvely presented ones once we knew not all of these had solvable word problem. We will see it is also not reasonable to hope to embed all those with solvable word problem.

As Groves observed, it is not possible to embed a finitely generated (non-cyclic) free center-by-metabelian group in a finitely presented center-by-metabelian group for the following reasons. First, as mentioned above Groves discovered that finitely presented center-by-metabelian groups are abelian-by-polycyclic. This is a hereditary property of groups. To see this let $1 \to A \to G \to P \to 1$ be an extension where A is abelian and P is polycyclic and $H \leq G$. Then $HA/A \cong H/H \cap A$ is polycyclic and $H \cap A$ is abelian and so

$1 \to H \cap A \to H \to H/H \cap A \to 1$ is an extension of the required kind. Now a finitely generated free center-by-metabelian group can have a monstrous center and, hence, need not have Max-n. This means it cannot be abelian-by-polycyclic and hence, it is not embeddable in a finitely presented center-by-metabelian group, even though it has solvable word problem (e.g., as a linear group).

Since this is the extent of the necessary properties presently known by me, I wish to propose the following problem which, if solved affirmatively, would lead to the first new example of a Higman embedding theorem for a variety.

Problem 1. Prove or disprove. A necessary and sufficient condition for a finitely generated center-by-metabelian group to be embeddable in a finitely presented center-by-metabelian group is that it be abelian-by-polycyclic.

Notice that finitely generated abelian-by-polycyclic groups have solvable word problem because of Max-n (e.g., see Theorem 1) and so this condition is omitted from Problem 1.

Now I turn to the varieties $\underline{\underline{N}}_c\underline{\underline{A}}$ where as we will see the situation is again quite interesting. The first result I will mention is needed for the sequel.

Theorem 1 (Baumslag, Cannonito, Miller [4]). A nilpotent-by-polycyclic group with Max-n has solvable word problem. ("Finitely generated" has been omitted from the statement of the theorem since P. Hall has shown [8] solvable groups with Max-n are finitely generated.)

Since Bieri and Strebel have shown [5] finitely presented $\underline{\underline{N}}_2\underline{\underline{A}}$ groups have Max-n, Theorem 1 implies all finitely presented $\underline{\underline{N}}_2\underline{\underline{A}}$ groups have solvable word problem. (Bieri and Strebel also showed finitely presented $\underline{\underline{N}}_2\underline{\underline{A}}$ groups are residually finite and thus have solvable word problem.) But Yepanchintesev and Kukin [13] have proved any variety $\underline{\underline{V}}$ containing the variety $\underline{\underline{N}}_2\underline{\underline{A}}$ has a group finitely presented in $\underline{\underline{V}}$ (and, hence, recursively presented in the variety of all groups) with unsolvable word problem. The existence of finitely generated recursively presented $\underline{\underline{N}}_2\underline{\underline{A}}$ groups with unsolvable word problem also follows from

Theorem 2 (Cannonito and N. D. Gupta [6]). For each $k \geq 2$ there exists a group which is finitely generated recursively presented center-by-(free solvable of step k and rank 3) with unsolvable word problem.

It follows, just as discussed above for the variety $\underline{\underline{C}}$, that there is no hope of embedding an arbitrary finitely generated $\underline{\underline{N}}_2\underline{\underline{A}}$ group, which is merely recursively presented, in a finitely presented $\underline{\underline{N}}_2\underline{\underline{A}}$ group. Since this together with residual finiteness is the complete state of our knowledge at present for the variety $\underline{\underline{N}}_2\underline{\underline{A}}$ regarding locally hereditary properties of finitely presented $\underline{\underline{N}}_2\underline{\underline{A}}$ groups, I am encouraged to frame.

Problem 2. Prove or disprove. A necessary and sufficient condition that a finitely generated $\underline{\underline{N}}_2\underline{\underline{A}}$ group be embeddable in a finitely presented $\underline{\underline{N}}_2\underline{\underline{A}}$ group is that it have solvable word problem and be residually finite.

As before, should Problem 2 turn out to have an affirmative answer a Higman type characterization for the variety $\underline{\underline{N}}_2\underline{\underline{A}}$ would result. In connection with this problem we have the partial information provided by the next

Theorem 3 (Cannonito, C. K. Gupta, N. D. Gupta). A finitely generated free center-by-metabelian group can be embedded in a finitely presented $\underline{\underline{N}}_2\underline{\underline{A}}$ group.*

The point of view I have been developing throughout the discussion above is intended to lead to the conclusion: in general, there is very little likelihood of being in position to declare there exists no Higman embedding theorem for a particular variety $\underline{\underline{V}}$. For to make this assertion we must be in a position of knowing all locally hereditary properties of the finitely presented groups in the variety and of having ruled (the union of) them out,

*Added in proof: G. Baumslag has pointed out that this result was already present implicitly in the Ph.D. thesis of M. Thompson.

in that finitely generated groups in the variety with these properties turn out not to be embeddable. To paraphrase Quine, this bites off more outology than I knew how to chew. Rather, the best we can do, it seems to me, is pursue a program for testing whether a given variety has a Higman analog. That is, first discover if there <u>are</u> any finitely presented groups in the variety (that are non-trivial) and second, discover what their locally hereditary properties are, one by one. Having done this, seek to embed finitely generated groups of the variety with these properties in finitely presented groups of the variety. If the attempt is successful a Higman characterization results. If not continue the search for further properties and repeat the process.

I find it quite interesting that the necessity of solvability of the word problem has appeared in all the varieties, other than the variety of all groups, for which either a Higman analog has been found, or is promising. This is even the case for the varieties $\underline{\underline{AN}}_c$: abelian-by-nilpotent of class c for which only recently has it been shown finitely presented groups which are not nilpotent-by-abelian-by finite exist (D. J. S. Robinson and R. Strebel, unpublished). These groups being residually finite all have solvable word problem. It would indeed be interesting if it turns out that those solvable varieties which have Higman analogs all turn out to have finitely presented groups with solvable word problem. This leads me to mention extremely interesting result of Olga G. Harlampovich [9] who has shown there exists a finitely presented 3-step solvable group in the variety $\underline{\underline{N}}_4\underline{\underline{A}}$ with unsolvable word problem! Her result together with Theorem 1 shows this group is a second example of a finitely presented solvable group without Max-n. The first example of such a group was found by Abels [1] in the variety $\underline{\underline{N}}_3\underline{\underline{A}}$ but Abels' group, being linear, has solvable word problem. It would be extremely interesting therefore to have the answer to

<u>Problem 3</u>. Do all finitely presented $\underline{\underline{N}}_3\underline{\underline{A}}$ groups have solvable word problem?

Most likely the answer to Problem 3 will be "no," but I do not expect to see this resolved in the near future.

In conclusion, it seems to me the question of the existence of analogs of Higman's theorem for the varieties $\underline{\underline{C}}$, $\underline{\underline{N}}_c\underline{\underline{A}}$ $(c \geq 2)$ and $\underline{\underline{A}}\underline{\underline{N}}_c$ $(c \geq 2)$ is still very much alive. It remains to be seen what bearing the bouquet of intriging results I have discussed above will have on the eventual resolution of what, for the moment, remains a tantalizing enigma.

REFERENCES

1. H. Abels, "An example of a finitely presented soluble group," Homological Group Theory, London Math. Soc. Lecture Notes, Series 36, Cambridge Univ. Press, 1979, 205-211.

2. G. Baumslag, "On finitely presented metabelian groups," Bull. AMS 78 (1972), 279.

3. G. Baumslag, "Subgroups of finitely presented metabelian groups," Jour. Austral. Math. Soc. 16 (1973), 98-110.

4. G. Baumslag, F. B. Cannonito and C. F. Miller, III, "Some recognizable properties of solvable groups," M. Zeit 178 (1981), 289-295.

5. R. Bieri and R. Strebel, "Valuations and finitely presented metabelian groups," Proc. London Math. Soc. (III), Vol. XLI (1980), 439-464.

6. F. B. Cannonito and N. D. Gupta, "On centre-by-free solvable groups," Archiv d. Math. 41 (1983), Fasc. 6, 493-497.

7. J. R. J. Groves, "Finitely presented centre-by-metabelian groups," J. London Math. Soc. (2) 18 (1978), 65-69.

8. P. Hall, "Finiteness conditions for soluble groups," Proc. London Math. Soc. (III) 4 (1954), 419-436.

9. O. G. Harlampovich, "Konyechno opredelennaya razreshimaya gruppa s nerazreshimoi problemoi ravensta," Izvestija Akad. Nauk USSR, ser. Mat. Tom 45, No. 4 (1981), 852-873 (Russian).

10. A. V. Jategoankar, "Integral group rings of polycyclic-by-finite groups," J. Pure Appl. Algebra 4 (1974), 337-343.

11. J. E. Roseblade, "Applications of the Artin-Rees lemma to group rings," Symp. Math. 17 (1976), 471-478.

12. V. N. Remeslennikov, "On finitely presented groups," 4th Vsyesoyuzni simpozium po terorii grupp, Novosibirsk (1973), 164-169 (Russian).

13. V. I. Yepanchintsev and G. P. Kukin, "The word problem in varieties of groups containing $\underline{\underline{N}}_2\underline{\underline{A}}$," Algebra i Logika 18, No. 3 (1979), 259-285 (Russian).

DEPARTMENT OF MATHEMATICS
UNIVERSITY OF CALIFORNIA-IRVINE
IRVINE, CALIFORNIA 92717

Received February 5, 1982

Contemporary Mathematics
Volume 33, 1984

ON THE WHITEHEAD METHOD IN FREE PRODUCTS

Donald J. Collins* and Heiner Zieschang

For Roger Lyndon on his 65th birthday

Introduction

Let $F = F(S)$ be the free group on a finite set S. J. H. C. Whitehead [6,7] proved, by topological means, a theorem that yields an algorithm that enables one to decide of any two elements of F whether or not there exists an automorphism of F that carries one element to the other. E. S. Rapaport [5] gave a purely algebraic proof of this result and a substantial simplification of Rapaport's argument was obtained by P. J. Higgins and R. C. Lyndon [3].

Here we consider the problem of generalising Whitehead's theorem to the case of a free product $G = \mathop{*}_{i \in I} G_i$ of finitely many indecomposable factors. We shall show that the natural analogue of Whitehead's theorem is valid if G has only two factors or if G has three factors, none of which are infinite cyclic, but is false in general. Tantalisingly, the argument goes through in full generality save for one particular situation.** We also observe that even when the analogue of Whitehead's theorem is true, the construction of an algorithm to decide if two elements lie in the same automorphism class does not seem to be immediate unless those factors which are not infinite cyclic are finite.**

---

* The first author wishes to acknowledge financial support from the Alexander-von-Humboldt-Stiftung and hospitality from the Ruhr-Universität, Bochum during the preparation of this paper.

** (Added May 1984) A generalisation of Whitehead's theorem, valid without restriction, is established in [0].

© 1984 American Mathematical Society
0271-4132/84 $1.00 + $.25 per page

We begin by describing Whitehead's theorem, following Higgins-Lyndon [3] and the similar account in Lyndon-Schupp [4] closely. For this we consider two types of automorphism of F.

(i) We say α is a <u>permutation automorphism</u> of F if α permutes the elements of the set $S \cup S^{-1}$.

(ii) We say α is a <u>Whitehead automorphism</u> of F if there is a fixed $x \in S \cup S^{-1}$ such that under α each $s \in S$ is carried to one of $s, sx, x^{-1}s, x^{-1}sx$.

A <u>cyclic word</u> w of F is a cyclically ordered, reduced string of elements of $S \cup S^{-1}$. Clearly automorphisms of F act on cyclic words.

<u>Whitehead's Theorem</u>

<u>Let</u> u <u>and</u> v <u>be cyclic words of</u> F <u>and</u> α <u>an automorphism such that</u> $u\alpha = v$. <u>Then</u> α <u>can be expressed as a product</u>

$$\alpha = \rho_1 \rho_2 \cdots \rho_n$$

<u>of permutation and Whitehead automorphisms such that for some</u> $p, q, 1 \leq p \leq q \leq n$,

(a) $|u\rho_1 \cdots \rho_{i-1}| > |u\rho_1 \cdots \rho_i|, \quad 1 \leq i \leq p$

(b) $|u\rho_1 \cdots \rho_{j-1}| = |u\rho_1 \cdots \rho_j|, \quad p+1 \leq j \leq q$

(c) $|u\rho_1 \cdots \rho_{k-1}| < |u\rho_1 \cdots \rho_k|, \quad q+1 \leq k \leq n.$

The inequalities (a), (b) and (c) are illustrated by the following diagram in which length is plotted upwards.

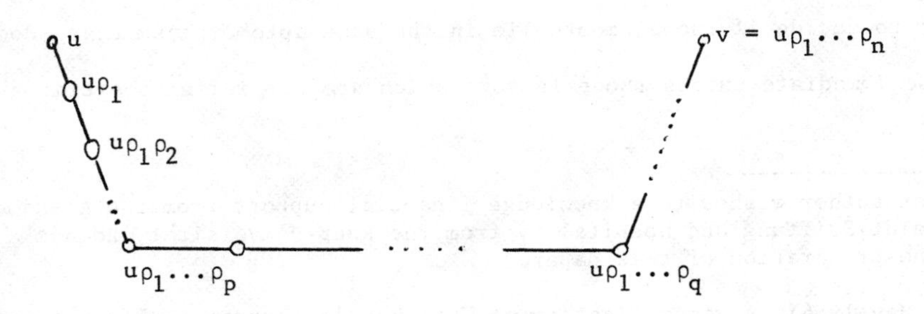

The algorithm to determine when two elements of F lie in the same automorphism class is obtained from the theorem by the following observations.

(1) There are only finitely many Whitehead automorphisms.

(2) There are only finitely many cyclic words of a given length.

(3) Two elements of F are conjugate if and only if they define the same cyclic word.

The theorem is a simple inductive consequence of the fact that permutation and Whitehead automorphisms generate the automorphism group of F, and the following.

Peak-Reduction Lemma

Let σ and τ be permutations or Whitehead automorphisms, and let u, w and v be cyclic words such that

(a) $u\sigma^{-1} = w$, $w\tau = v$

(b) $|u| \leq |w| \geq |v|$

(c) $|u| < |w|$ or $|w| > |v|$.

Then $\sigma^{-1}\tau$ can be expressed as a product

$$\sigma^{-1}\tau = \rho_1\rho_2 \cdots \rho_n$$

of permutation or Whitehead automormphisms such that $|u\rho_1 \cdots \rho_i| < |w|$ for $1 \leq i \leq n-1$.

(The use of σ^{-1} rather than σ follows Higgins-Lyndon and is later advantageous.)

The lemma is illustrated by the following diagram, with length again plotted vertically,

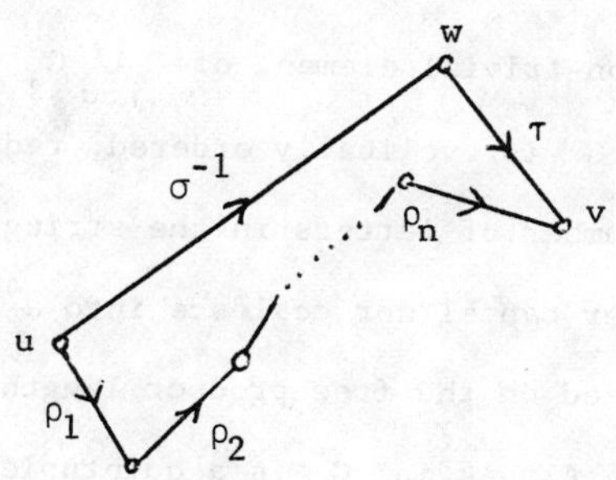

and the analogous diagrams for the cases $|u| = |w|$ or $|w| = |v|$. It is upon this "peak-reduction" lemma that our attention will be focused.

Now let $G = \ast_{i \in I} G_i$ be a free product of finitely many factors. Unless specifically noted, it is not assumed that the factors G_i are indecomposable. We can write, for some $J \subset I$,

$$G = (\ast_{j \in J} G_j) * F(S)$$

where the groups G_j, $j \in J$, are not infinite cyclic and $F(S)$ is the free group on the (possibly empty) set S. We consider three types of automorphism of G.

(i) It may happen that various G_i are isomorphic. We assume some fixed isomorphism to be given in all such cases and call α a <u>permutation automorphism</u> if α permutes some isomorphic G_i via the fixed isomorphisms.

(ii) For each $i \in I$, let φ_i be an automorphism of G_i. We say α is a <u>factor automorphism</u> if α restricted to G_i is φ_i.

(iii) We say α is a <u>Whitehead automorphism</u> if there is some fixed x in $(\bigcup_{j \in J} G_j) \cup (S \cup S^{-1})$ such that under α each G_j, $j \in J$ is conjugated by x or left fixed (in both cases pointwise), and each $s \in S$ is carried to one of s, sx, $x^{-1}s$, $x^{-1}sx$. We also require, as a matter of convention, that if $x \in G_k$, $k \in J$, then G_k is left fixed by α. (If $x \in S \cup S^{-1}$, then, necessarily, $x\alpha = x$.)

If all G_i are actually indecomposable then Fouxe-Rabinovitch [1,2] proved that the elements (i)-(iii) generate the automorphism group of G and, in addition, gave a complete set of relations for these generators.

A <u>letter</u> of G is a non-trivial element of $\bigcup_{j \in J} G_j$ or an element of $S \cup S^{-1}$. A <u>cyclic word</u> of G is cyclically ordered, reduced string of letters whose length $|w|$ is the number of letters in the string. Here <u>reduced</u> means that adjacent letters neither cancel nor coalesce into a new letter, and $|w|$ is thus a "mixed" length based on the free product length in $\ast_{j \in J} G_j$ and the free group length in $F(S)$. A <u>peak</u> in G is a quintuple $(u, w, v, \sigma^{-1}, \tau)$, where u, v and w are cyclic words and σ^{-1} and τ are automorphisms of types (i)-(iii) such that

(a) $u\sigma^{-1} = w$, $w\tau = v$

(b) $|u| \leq |w| \geq |v|$

(c) $|u| < |w|$ or $|w| > |v|$.

We say the peak admits reduction if $\sigma^{-1}\tau$ can be expressed as a product

$$\sigma^{-1}\tau = \rho_1\rho_2 \ldots \rho_n$$

where ρ_i, $1 \leq i \leq n$ are of types (i)-(iii) and $|u\rho_1 \ldots \rho_i| < |w|$, $1 \leq i \leq n-1$.

Our main result is the following.

Peak-Reduction Lemma

Let $(u, w, v, \sigma^{-1}, \tau)$ be a peak in G. Then this peak admits reduction unless σ^{-1} and τ are Whitehead automorphisms whose associated fixed elements lie in the same factor G_k, where G_k is not infinite cyclic.

The proof of the Peak-Reduction Lemma will be finished at the end of §1. The necessity of the condition in the lemma is demonstrated by the examples given in §2 below.

§1

We follow the method of Higgins-Lyndon as described in [3] or [4].

Let $G = \ast_{i\in I} G_i = (\ast_{j\in J} G_j) * F(S)$, where $J \subset I$ and the G_j are not infinite cyclic. Let $L = J \cup S \cup S^{-1}$, let $A \subset L$ and let $x \in G_k$, $k \in J$. We shall use and abuse notation variously, writing X for G_k, $x \in A$ in place of $x \in G_k$ and $k \in A$, A-X for $A - \{k\}$, $A + X$ for $A \cup \{k\}$ and $X \in A$ instead of $k \in A$. We write $A' = L - A$.

If α is a Whitehead automorphism with fixed element x, we put

$$A = \{j \in J \mid G_j\alpha = x^{-1}G_jx\} \cup \{s \in S \cup S^{-1} \mid s\alpha = sx \text{ or } s\alpha = x^{-1}sx\} \cup A_0$$

where A_0 is the singleton $\{k\}$ if $x \in G_k$, $k \in J$ and A_0 is $\{x\}$ if $x \in S \cup S^{-1}$. We often write (A,x) for α since α is completely specified by A and x. We note that if $x \in J \cap A$, then $x^{-1} \in A$, while if $x \in S \cup S^{-1}$, then $x^{-1} \notin A$.

Let w be any cyclic word and let A and B be subsets of L. We write $A.B$ for the number of two-letter subwords of w of the form $(ab^{-1})^{\pm 1}$, where $a \in A$ and $b \in B$. If x is a letter we write $A.x$ for the number of two-letter subwords $(ax^{-1})^{\pm 1}$, $a \in A$, of w and $A.x.B$ for the number of three-letter subwords $(xb^{-1})^{\pm 1}$ of w with $a \in A$ and $b \in B$. Note that $A.x = A.x^{-1}.B + A.x^{-1}.B'$ and that if $x \in J$ then $A.x.B = (A-X).x.B = (A-X).x.(B-X)$. Also $A.B = B.A$, $A.x = x.A$, $A.x.B = B.x^{-1}.A$ and $L.x = L.x^{-1}$.

Lemma 1.1

Let w *be a cyclic word and* (A,x) *a Whitehead automorphism.*

(i) *If* $x \in X \cup S^{-1}$, *then* $|w(A,x)| - |w| = A.A' - L.x$.

(ii) *If* $x \in J$, *then* $|w(A,x)| - |w| = (A-X).A' - A.x^{-1}.A'$.

Proof. (i) Under (A,x) an increase in length of 1 occurs for each two-letter subword $(ab^{-1})^{\pm 1}$ with $a \in A-x$, $b \in A'$, and a decrease in length of 1 occurs for each two-letter subword $(ax^{-1})^{\pm 1}$, with $a \in A-x$. No other two-letter subword gives rise to a change in length and hence

$$\begin{aligned}|w(A,x)| - |w| &= (A-x).A' - (A-x).x\\ &= A.A' - x.A' - A.x\\ &= A.A' - L.x.\end{aligned}$$

(The analogous argument is given in detail in [3] and [4].)

(ii) In this case, (A,x) causes an increase in length of 1 for each two-letter subword $(ab^{-1})^{\pm 1}$, where $a \in A-X$, $b \in A'$ and a decrease in length of 1 for each three letter subword $(ax^{-1}b^{-1})^{\pm 1}$, where $a \in A-X$ and $b \in A'$. Consideration of two-letter subwords not involving a letter of X and of three-letter subwords involving an element of X shows that these are the only length changes caused by (A,x). Hence

$$\begin{aligned}|w(A,x)| - |w| &= (A-X).A' - (A-X).x^{-1}.A'\\ &= (A-X).A' - A.x^{-1}.A' . \quad \square\end{aligned}$$

Given $x \in G_k$, $k \in J$, write $\gamma(x)$ for the factor automorphism defined by the inner automorphism of G_k corresponding to x and the identity automorphism for all G_i, $i \neq k$.

Lemma 1.2

Let (A,x) be a Whitehead automorphism and w a cyclic word.

(i) If $x \in S \cup S^{-1}$, then $(A,x)^{-1} = (A-x+x^{-1}, x^{-1})$.

(ii) If $x \in J$, then $(A,x)^{-1} = (A,x^{-1})$.

(iii) If $x \in S \cup S^{-1}$, then $w(A,x) = w(A', x^{-1})$.

(iv) If $x \in J$, then $w(A,x) = w\gamma(x^{-1})(A'+X, x^{-1})$.

Proof. This is routine to verify. The $\gamma(x^{-1})$ in (iv) is forced on us by our convention regarding the effect of (A,x) on X. □

We record one more lemma for later use.

Lemma 1.3

Let (A,x) and (B,y) be Whitehead automorphisms, and let $P_{11} = A \cap B$, $P_{12} = A \cap B'$, $P_{21} = A' \cap B$ and $P_{22} = A' \cap B'$.

(i) Then $P_{11}.P'_{11} + P_{22}.P'_{22} \leq A.A' + B.B'$.

(ii) Then $P_{12}.P'_{12} + P_{21}.P'_{21} \leq A.A' + B.B'$.

(iii) If $x \in P_{11}$, then

$(A,x) = (P_{11},x)(P_{12}+x,x) = (P_{12}+x,x)(P_{11},x)$, when $x \in S \cup S^{-1}$ and

$(A,x) = (P_{11},x)(P_{12}+X,x) = (P_{12}+X,x)(P_{11},x)$, when $x \in J$.

(iv) If $x \in P_{12}$, then

$(A,x) = (P_{11}+x,x)(P_{12},x) = (P_{12},x)(P_{11}+x,x)$, when $x \in S \cup S^{-1}$ and

$(A,x) = (P_{11}+X,x)(P_{12},x) = (P_{12},x)(P_{11}+X,x)$, when $x \in J$.

Proof. Parts (i) and (ii) follow from the equalities

$$A.A + B.B' = P_{11}.P'_{11} + P_{22}.P'_{22} + 2P_{12}.P_{21}$$

and $$A.A + B.B' = P_{12}.P'_{12} + P_{21}.P'_{21} + 2P_{11}.P_{22}.$$

Routine verification gives (iii) and (iv). □

In what follows we shall maintain the notation P_{ij}, $i,j = 1,2$ introduced above.

We now turn to the proof of the Peak-Reduction Lemma. We note firstly that if either σ^{-1} or τ is a permutation or factor automorphism, then the lemma is easy to establish. For example suppose $\sigma^{-1} = (A,x^{-1})$ and τ is a factor automorphism. Then $(A,x^{-1})\tau = \tau(A,(x\tau)^{-1})$ and since $|w| = |v|$ and $|u\tau| = |u|$, then $|u\tau| < |w|$ and the required rewriting is simply $\sigma^{-1}\tau = \tau(A,(x\tau)^{-1})$. If τ is a permutation automorphism, then $(A,x^{-1})\tau = \tau(A\tau,(x\tau)^{-1})$, where $A\tau$ has the obvious meaning, is the required rewriting. Henceforth, then, we may suppose that both σ^{-1} and τ are Whitehead automorphisms, say $\sigma^{-1} = (A,x^{-1})$ and $\tau = (B,y)$.

There are three principal cases to consider:

I. Both x and y lie in $S \cup S^{-1}$.

II. Both x and y lie in J (but not in the same factor).

III. One of x and y lies in $S \cup S^{-1}$ and the other in J.

Case I.

This case is easily dealt with since it can be derived from the Peak-Reduction Lemma for free groups. Specifically, replace u, w and v by words $\tilde{u}$, $\tilde{w}$ and $\tilde{v}$ obtained by replacing each letter of G_k, $k \in J$ by the letter $\bar{k}$. Then $\tilde{u}$, $\tilde{w}$ and $\tilde{v}$ lie in the free group $F(\bar{J} \cup S)$ and $|\tilde{u}| = |u|$, $|\tilde{w}| = |w|$ and $|\tilde{v}| = |v|$. Moreover σ^{-1} and τ induce Whitehead automorphisms $\tilde{\sigma}^{-1}$ and $\tilde{\tau}$ such that $\tilde{u}\tilde{\sigma}^{-1} = \tilde{w}$ and $\tilde{w}\tilde{\tau} = \tilde{v}$. Hence $\tilde{\sigma}^{-1}\tilde{\tau}$ can be written as a product $\tilde{\rho}_1\tilde{\rho}_2 \ldots \tilde{\rho}_n$ of permutation or Whitehead automorphisms with the successive images of $\tilde{u}$ satisfying $|\tilde{u}\tilde{\rho}_1 \ldots \tilde{\rho}_i| < |\tilde{w}|$, $i \leq i \leq n-1$. Examination of the proof of the Peak-Reduction Lemma for free groups shows that any permutation automorphisms used in the rewriting operate non-trivially on x and y only. Hence $\tilde{\rho}_1\tilde{\rho}_2 \ldots \tilde{\rho}_n$ can be "lifted" back to a sequence $\rho_1\rho_2 \ldots \rho_n$ that provides the necessary rewriting of $\sigma^{-1}\tau$.

Case II.

We suppose $x \in X$, $y \in Y$, $X \neq Y$.

II.1 $A \cap B = \emptyset$. We shall show that $\sigma^{-1}\tau = \tau\sigma^{-1}$, $|v\sigma| - |v| = |w\sigma| - |w|$ and $|u\tau| - |u| = |w\tau| - |w|$. Then an easy exercise in inequalities gives

$|u\tau| < |w|$ and thus $\sigma^{-1}\tau = \tau\sigma^{-1}$ is the required rewriting.

It is routine to check that $\sigma^{-1}\tau = \tau\sigma^{-1}$, bearing in mind that since, conventionally, $x^{-1} \in A$ and $y \in B$ then $A \cap B = \emptyset$ gives $x^{-1} \notin B$ and $y \notin A$.

To show that $|u\tau| - |u| = |w\tau| - |w|$ we argue that length changes that occur when τ is applied to $u = w\sigma$ take place at the same "spots" as when τ is applied to w. Formally a positive spot for τ in w is a subword $(bc^{-1})^{\pm 1}$ with $b \in B-Y$, $c \in B'$. If $c \notin X$, then, since $b \notin A$, we obtain either $(bc^{-1})^{\pm 1}$ or $(bx^{-1}c^{-1})^{\pm 1}$ when σ is applied to w, according as $c \in A'$ or $c \in A-X$. Then $(bc^{-1})^{\pm 1}$ or $(bx^{-1})^{\pm 1}$ constitutes a positive spot for τ in u. If $c \in X$, we have to consider $(bc^{-1})^{\pm 1}$ as part of a three-letter subword $(bc^{-1}d^{-1})^{\pm 1}$. When σ is applied, three possibilities occur:

(a) $d \in A'$ and $bc^{-1}d^{-1} \overset{\sigma}{\mapsto} bc^{-1}d^{-1}$

(b) $d \in A-X$, $c^{-1} \neq x$ and $bc^{-1}d^{-1} \overset{\sigma}{\mapsto} b(c^{-1}x^{-1})d^{-1}$

(c) $d \in A-X$, $c^{-1} = x$ and $b^{-1}c^{-1}d^{-1} \overset{\sigma}{\mapsto} bd^{-1}$.

In all three cases, there arises a positive spot for τ in u.

We conclude from the above that the number of positive spots for τ in u is at least the number of positive spots for τ in w. However we can repeat the argument with the roles of u and w exchanged and hence the two numbers coincide.

Now we consider a negative spot for τ in w, that is, a three-letter subword $(by^{-1}c^{-1})^{\pm 1}$, where $b \in B-Y$, $c \in B'$, of w. If $c \notin X$ then under σ we obtain $(by^{-1}c^{-1})^{\pm 1}$ or $(by^{-1}x^{-1}c^{-1})^{\pm 1}$ according as $c \in A'$ or $c \in A-X$. In both cases a negative spot for τ in u results. If $c \in X$, we have to consider the four-letter subword $(by^{-1}c^{-1}d^{-1})^{\pm 1}$. (If $|w| = 3$ then b and d^{-1} coincide but our argument remains valid). Under σ, there are three possibilities, namely:

(a) $d \in A'$ and $by^{-1}c^{-1}d^{-1} \overset{\sigma}{\mapsto} by^{-1}c^{-1}d^{-1}$

(b) $d \in A-X$, $c^{-1} \neq x$ and $by^{-1}c^{-1}d^{-1} \overset{\sigma}{\mapsto} by^{-1}(c^{-1}x^{-1})d^{-1}$

(c) $d \in A-X$, $c^{-1} = x$ and $by^{-1}c^{-1}d^{-1} \overset{\sigma}{\mapsto} by^{-1}d^{-1}$.

In all three cases a negative spot for τ in u results and so, applying the same argument with roles of u and w exchanged, we conclude that u and w have the same number of negative spots for τ.

This proves that $|u\tau| - |u| = |w\tau| - |w|$ and, by symmetry, $|v\sigma| - |v| = |w\sigma| - |w|$.

II.2 $A \cap B \neq \emptyset$. There are four separate subcases:

II.2.1 $x \in B', y \in A'$; II.2.2 $x \in B', y \in A$;

II.2.3 $x \in B, y \in A'$; II.2.4 $x \in B, y \in A$.

By symmetry II.2.3 reduces to II.2.2. We can reduce II.2.2 to II.2.1 by <u>complementation</u>, that is, an application of Lemma 1.2. For in II.2.2, $w = u(A,x^{-1}) = u\gamma(x)(A' + X,x)$ and $y \notin A' + X$ since $X \neq Y$. Since $|u\gamma(x)| = |u|$, $(u\gamma(x),w,v,(A' + X,x), (B,y))$ is a peak which, by II.2.1, admits reduction. If $|u| < |w|$, then we see, immediately, that (u,w,v,σ^{-1},τ) admits reduction while if $|u| = |w|$ then after reducing $(u\gamma(x),w,v,(A' + X,x),(B,y))$ there will still be a peak involving $\gamma(x)$ that needs to be reduced. However, this is trivial to do since $\gamma(x)$ is a factor automorphism. We further observe that complementation also reduces II.2.4 to II.2.2 and, hence, to II.2.1. We are therefore left to examine II.2.1 in detail.

From the assumption that (u,w,v,σ^{-1},τ) is a peak, we have

$$(A-X).A' - A.x^{-1}.A' + (B-Y).B' - B.y^{-1}.B' < 0 .$$

By our case hypothesis, $x \in P_{12} = A \cap B'$ and $y \in P_{21} = A' \cap B$. We consider

$$\begin{aligned}
&|w(P_{12},x)| - |w| + |w(P_{21},y)| - |w| \\
&= (P_{12}-X).P'_{12} - P_{12}.x^{-1}.P'_{12} + (P_{21}-Y).P'_{21} - P_{21}.y^{-1}.P'_{21} \\
&= P_{12}.P'_{12} - X.P'_{12} - P_{12}.x^{-1}.P'_{12} + P_{21}.P'_{21} - Y.P'_{21} - P_{21}.y^{-1}.P'_{21} \\
&\leq A.A' - X.P'_{12} - P_{12}.x^{-1}.P'_{12} + B.B' - Y.P'_{21} - P_{21}.y^{-1}.P'_{21} \\
&\leq A.A' - X.A' - X.P_{11} - P_{12}.x^{-1}.A' + B.B' - Y.B' - Y.P_{11}-P_{21}.y^{-1}.B' \\
&\leq (A-X).A'-P_{11}.x^{-1}.A' - P_{12}.x^{-1}.A' + (B-Y).B'-P_{11}.y^{-1}.B'-P_{21}.y^{-1}.B' \\
&= (A-X).A' - A.x^{-1}.A' + (B-Y).B' - B.y^{-1}.B' \\
&< 0 .
\end{aligned}$$

Hence $|w(P_{12},x)| - |w| < 0$ or $|w(P_{21},y)| - |w| < 0$. Suppose, without loss of generality, the former occurs. Then we write $\sigma^{-1} = (P_{11} + X,x^{-1})(P_{12},x^{-1})$ and observe that

$$|u(P_{11} + X,x^{-1})| = |w(P_{12},x)| < w .$$

Furthermore $(u(P_{11} + X,x^{-1}),w,v,(P_{12},x^{-1}),\tau)$ is a peak and $P_{12} \cap B = \emptyset$. By case II.1, this peak is reducible and hence (u,w,v,σ^{-1},τ) is reducible.

<u>Case III</u>.

By an appeal to symmetry, we may suppose $x \in J$ and $y \in S \cup S^{-1}$.

III.1 $A \cap B = \emptyset$. Here $(A,x^{-1})(B,y) = (B,y)(A \cup B-y,x^{-1})$ or $(A,x^{-1})(B,y) = (B,y)(A,x^{-1})$ according as $y^{-1} \in A$ or $y^{-1} \in A'$. These equalities are routinely checked, noting that since $A \cap B = \emptyset$, $x,x^{-1} \notin B$ and $y \notin A$. In both cases $|u(B,y)| < |w|$ and this is proved by an argument similar to that given for case II.1 above (and virtually identical to that given for the corresponding situation in [3] and [4]).

III.2 $A \cap B \neq \emptyset$. If $A \subset B$, then we write $(A,x^{-1})(B,y) = (A,x^{-1})(B',y^{-1})$ and reduce to case III.1 since $A \cap B' = \emptyset$. Henceforth, then, we also assume $A \cap B' \neq \emptyset$. There are four subcases to be considered:

III.2.1 $x \in B$, $y^{-1} \in A'$; III.2.2 $x \in B$, $y^{-1} \in A$;

III.2.3 $x \in B'$, $y \in A'$; III.2.4 $x \in B'$, $y \in A$.

However complementation reduces III.2.3 to III.2.1, III.2.4 to III.2.2 and, further, III.2.2 to III.2.1. Thus it remains to examine III.2.1 in detail.

Since (u,w,v,σ^{-1},τ) is a peak, we have

$$(A-X).A' - A.x^{-1}.A' + B.B' - L.y < 0 .$$

(Recall $L = J \cup S \cup S^{-1}$.) Now $x \in P_{11} = A \cap B$ and $y^{-1} \in P_{22} = A' \cap B'$. So we consider

$$
\begin{aligned}
&|w(P_{11},x)| - |x| + |w(P_{22},y^{-1})| - |w| \\
&\quad = (P_{11}-X).P'_{11} - P_{11}.x^{-1}.P'_{11} + P_{22}.P'_{22} - L.y^{-1} \\
&\quad = P_{11}.P'_{11} - X.A' - X.P_{12} - P_{11}.x^{-1}.P'_{11} + P_{22}.P'_{22} - L.y \\
&\quad \leq A.A' - X.A' - P_{12}.x^{-1}.A' - P_{11}.x^{-1}.A' + B.B' - L.y \\
&\quad = (A-X).A' - A.x^{-1}.A' + B.B' - L.y \\
&\quad < 0 .
\end{aligned}
$$

Hence $|w(P_{11},x)| - |w| < 0$ or $|w(P_{22},y^{-1})| - |w| < 0$. If the former occurs, we write $(A,x^{-1})(B,y) = (P_{12} + X,x^{-1})(P_{11},x^{-1})(B,y)$. Now $|u(P_{12} + X,x^{-1})| = |w(P_{11},x)| < |w|$ and we can deal with the peak involving $(P_{11},x^{-1})(B,y)$ since $P_{11} \subset B$. If $|w(P_{22},x^{-1})| - |w| < 0$, then we write $(A,x^{-1})(B,y) = (A,x^{-1})(B',y^{-1}) = (A,x^{-1})(P_{22},y^{-1})(P_{12} + y^{-1},y^{-1})$ and observe that $A \cap P_{22} = \emptyset$ whence we may appeal to case III.1.

We have now dealt with all the required cases and the proof of the Peak-Reduction Lemma is complete.

§2

In this section we show that in certain circumstances, the restriction in the Peak-Reduction Lemma can be avoided. We also give examples to show that in general the restriction is necessary.

Proposition 2.1

Let $G = G_1 * G_2$ be the free product of two factors. Then every peak in G admits reduction.

Proof. By the Peak-Reduction Lemma, we have to examine a peak involving Whitehead automorphisms (A,x^{-1}), (B,y) where x^{-1} and y lie in the same factor X, X not infinite cyclic.

If $A = L$, then (A,x^{-1}) is equivalent to a factor automorphism and the result is immediate. So, using symmetry we may suppose $A \neq L$ and $B \neq L$. Then one factor must be infinite cyclic and we may take

$$S = \{s\},\ A = \{s,X\} \text{ and } B = \{s,X\} \text{ or } B = \{s^{-1},X\} .$$

Then $A = B$ or $A = B' + X$ and therefore $(A,x^{-1})(B,y) = (A,yx^{-1})$ or $(A,x^{-1})(B,y) = \gamma(x)B(yx)$ and the peak has been reduced. □

Proposition 2.2

Let $G = G_1 * G_2 * G_3$, *where none of* G_i, $1 \leq i \leq 3$, *is infinite cyclic. Then every peak in* G *admits reduction.*

Proof. Suppose we have a peak involving (A,x^{-1}) and (B,y) with x, y in X. We may assume $A \neq L$, $B \neq L$. Then $A = B$ or $A = B' + X$ and the result follows as in Proposition 2.1. □

Proposition 2.3

Let $G = \ast_{i \in I} G_i$ *where each* G_i *is either infinite cyclic or cyclic of order two. Then every peak in* G *admits reduction.*

Proof. Here the awkward case involves (A,x) and (B,x) where $x^2 = 1$. However we may write

$$\begin{aligned}(A,x)(B,x) &= (A-B+X,x)(A \cap B,x)(A \cap B,x)(B-A+X,x)\\ &= (A-B+X,x)(B-A+X,x)\\ &= ((A-B) + (B-A) + X,x)\end{aligned}$$

and the peak is reduced. □

We give two examples to show that the conditions in our Peak-Reduction Lemma cannot be avoided.

Example 1

Let $G = X * Z * F(s)$ where X is cyclic, on x, of order three and Z is not isomorphic to X or $F(s)$.

Let z be any non-trivial element of Z. We put

$$u = szx^{-1}s^{-1}x^{-1}zsx^{-1}szx^{-1}s^{-1}x^{-1}zs$$

and $\sigma^{-1} = (s+X,x^{-1})$, $\tau = (s^{-1}+X,x)$. Then
$w = u\sigma^{-1} = sx^{-1}zs^{-1}x^{-1}zsxsx^{-1}zs^{-1}x^{-1}zsx^{-1}$ and
$v = w\tau = sx^{-1}zs^{-1}zx^{-1}ssx^{-1}zs^{-1}zx^{-1}sx$. Then
$|u| = 15 = |v|$ and $|w| = 16$ so that (u,w,v,σ^{-1},τ) is a peak.

There are no permutation automorphisms and clearly it is impossible to transform u into v using only factor automorphisms. Thus if the peak were reducible, Whitehead automorphisms would be employed in the rewriting of $\sigma^{-1}\tau$. Now a Whitehead automorphism $\rho = (C,y)$ is equivalent to a factor automorphism if $y \in \{X,Z\}$ and $|C| = 4$ or if $y \in \{s,s^{-1}\}$ and $|C| = 3$. Thus the rewriting must involve some $\rho = (C,y)$ where

(*) $y \in \{X,Z\}$ implies $|C| = 2$ or 3 and $y \in \{s,s^{-1}\}$ implies $|C| = 2$.

Moreover we may assume that such a ρ is applied to u. Hence, to show the peak is not reducible, it suffices to show that $|u\rho| > |u|$ for every ρ satisfying (*). This is routine to check and we simply enumerate the results of the necessary computations.

(a) $y = x$ or $y = x^{-1}$. By complementation, it suffices to consider the six cases tabulated below, with $y = x^{-1}$.

C	$u\rho$	$\|u\rho\|$
$s + X$	w	16
$s^{-1} + X$	$xszx^{-1}s^{-1}xzxsszx^{-1}s^{-1}xzxs$	17
$Z + X$	$sxzxs^{-1}zx^{-1}sx^{-1}sxzxs^{-1}zx^{-1}s$	17
$s + s^{-1} + X$	$sx^{-1}zs^{-1}xzxsx^{-1}sx^{-1}zs^{-1}xzxs$	17
$s + Z + X$	$szx^{-1}s^{-1}zx^{-1}sxszx^{-1}s^{-1}zx^{-1}sx^{-1}$	16
$s^{-1} + Z + X$	$xsxzxs^{-1}x^{-1}zssxzxs^{-1}x^{-1}zs$	17

(b) $y = s$ or $y = s^{-1}$. Here we may assume $|C| = 2$ and, by complementation, it suffices to consider the two cases below - with $y = s$.

C	$u\rho$	$\|u\rho\|$
$X + s$	$szs^{-1}x^{-1}s^{-1}x^{-1}szx^{-1}sszs^{-1}x^{-1}s^{-1}x^{-1}szs$	19
$Z + s$	$zsx^{-1}s^{-1}x^{-1}s^{-1}zssx^{-1}zsx^{-1}s^{-1}x^{-1}s^{-1}zss$	19

(c) $y \in Z$. By complementation, it suffices to consider the three cases below.

C	$u\rho$	$\|u\rho\|$
$s + Z$	$s(yz)x^{-1}y^{-1}s^{-1}x^{-1}zsyx^{-1}s(yz)x^{-1}y^{-1}s^{-1}x^{-1}zsy$	≥ 18
$s^{-1} + Z$	$y^{-1}szx^{-1}s^{-1}yx^{-1}(zy^{-1})sx^{-1}y^{-1}szx^{-1}s^{-1}yx^{-1}(zy^{-1})s$	≥ 17
$X + Z$	$s(zy^{-1})x^{-1}ys^{-1}y^{-1}x^{-1}(yz)sy^{-1}x^{-1}ys(zy^{-1})x^{-1}ys^{-1}y^{-1}x^{-1}(yz)s$	≥ 17

It may be instructive to say something of the construction of this example. We have $\sigma^{-1}\tau = (s+X,x^{-1})(s^{-1}+X,x)$. It is easy to check that the following are possible ways to rewrite $\sigma^{-1}\tau$:

(1) $(s^{-1}+X,x)(s+X,x^{-1}) = \tau\sigma^{-1}$

(2) $(s+X,x)\,\gamma(x^{-1})\,(Z+X,x^{-1})$

(3) $(Z+X,x)\,\gamma(x)\,(s^{-1}+X,x^{-1})$.

These correspond to the diagrams:

(1)

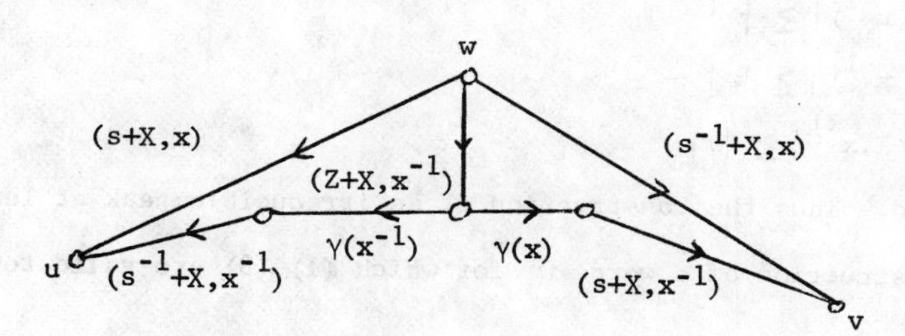

(2)

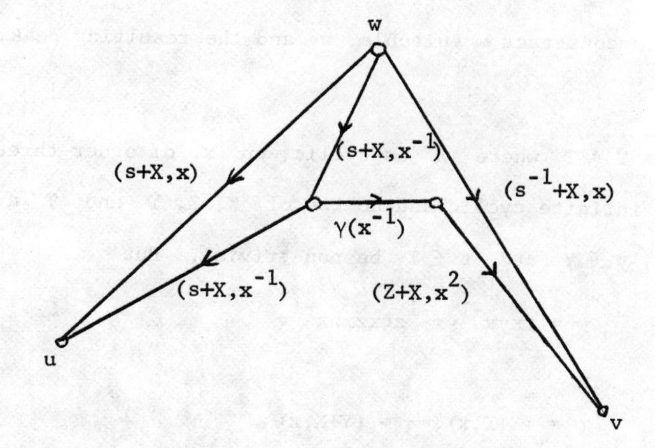

(3)

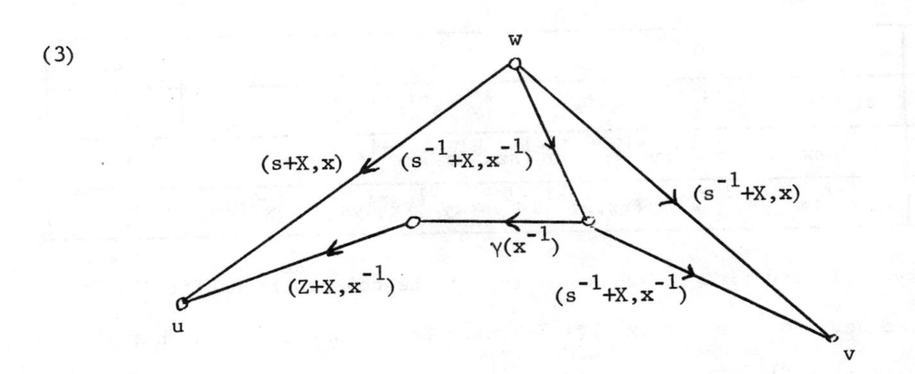

In (2) we have "complemented" $(s^{-1}+X,x)$ and in (3) $(s+X,x)$. The diagrams are all drawn as if the rewriting involved has reduced the peak - if we take length as plotted upwards. So to construct a peak that does not admit reduction, such diagrams must be impossible. In particular the conditions

(1) $|w(Z+X,x^{-1})| \geq |w|$

(2) $|w(s+X,x^{-1})| \geq |w|$

(3) $|w(s^{-1}+X,x^{-1}| \geq |w|$

must be satisfied. Thus the construction of an irreducible peak at least requires the construction of a word w for which (1)-(3) are valid together with

(4) $|w(s+X,x)| - |w| + |w(s^{-1}+X,x)| - |w| < 0.$

However conditions (1)-(4) can be spelled out in great detail using Lemma 1.1 and this enabled us to construct a suitable w and the resulting peak.

<u>Example 2</u>

Let $G = X * Z * Y * T$ where X is cyclic, on x, of order three, none of Z, Y and T is infinite cyclic and no two of X, Z, Y and T are isomorphic. Let $z \in Z$, $y \in Y$ and $t \in T$ be non-trivial. Put

$$w = txyx^{-1}yx^{-1}ztxzxzx^{-1}y$$

and

$$\sigma = (Z+X,x), \ \tau = (Y+X,x) \ .$$

Then it is easy to check, that with $u = w\sigma$, $v = w\tau$, (u,w,v,σ^{-1},τ) is a peak - indeed $|u| = |v| = 13$, $|w| = 14$. Moreover the conditions

(1) $|w(T+X,x^{-1})| \geq |w|$

(2) $|w(Z+X,x^{-1})| \geq |w|$

(3) $|w(Y+X,x^{-1})| \geq |w|$,

which are the analogues of (1)-(3) in Example 1 are all valid, and hence none of the corresponding rewritings

(1) $(Y+X,x)(Z+X,x^{-1}) = \tau\sigma^{-1}$

(2) $(Z+X,x)\gamma(x^{-1})(T+X,x^{-1})$

(3) $(T+X,x)\gamma(x)(Y+X,x^{-1})$

reduces the peak. We leave it to the reader to check that $|u\rho| > |u|$ for every Whitehead automorphism ρ and hence that the peak does not admit reduction.

We conclude with a few remarks concerning the derivation of the analogue of Whitehead's theorem in the cases where the Peak-Reduction Lemma is valid and the construction of an algorithm for deciding when two elements lie in the same automorphism class. In the first instance we must assume the factors G_i are indecomposable, for only in these circumstances do the permutation, factor and Whitehead automorphisms together generate the whole automorphism group. Thus Whitehead's theorem is valid for the case of two indecomposable factors, for the case of three indecomposable factors, none of which are infinite cyclic and for the case when the factors that are not infinite cyclic are of order two.

In these cases the first step in the algorithm does go through. Although there may now be infinitely many Whitehead automorphisms, it is clear that $|u(A,x)| < |u|$ only if x appears as a letter in u and hence only finitely many automorphisms need be tested to see if u is of minimal length in its automorphism class. Thus two arbitrary words can always be replaced by two words of minimal length. At present, however, we see no way to deal with this situation other than by assuming that all factors, which are not infinite cyclic, are finite so that the same procedure as in the free group case can be employed.

REFERENCES

[0] Collins, D. J., Zieschang, H.: Rescuing the Whitehead method for free products I: Peak Reduction, Math. Z. 185 (1984), 487-504, II: The Algorithm, Math. Z. (to appear).

[1] Fouxe-Rabinovitch, D. I.: Ober die Automorphismengruppen der freien Produkte, I. Mat. Sbornik 8 (1940), 265-276.

[2] Fouxe-Rabinovitch, D. I.: Ober die Automorphismengruppen der freien Produkte, II. Mat. Sbornik 9 (1941), 183-220.

[3] Higgins, P. J., Lyndon, R. C.: Equivalence of elements under automorphisms of a free group, J. London Math. Soc. 8 (1974), 254-258.

[4] Lyndon, R. C., Schupp, P. E.: Combinatorial Group Theory. Springer Verlag, Berlin 1977.

[5] Rapaport, E. S.: On free groups and their automorphisms. Acta Math. 99 (1958), 139-163.

[6] Whitehead, J. H. C.: On certain elements in a free group. Proc. London Math. Soc. 41 (1936), 48-56.

[7] Whitehead, J. H. C.: On equivalent sets of elements in a free group. Ann. of Math. 37 (1936), 782-800.

DEPARTMENT OF PURE MATHEMATICS
QUEEN MARY COLLEGE
MILE END ROAD
LONDON E1 4NS
U.K.

Received September 9, 1982

INSTITUT FÜR MATHEMATIK
RUHR-UNIVERSITÄT, BOCHUM
POSTFACH 10 21 48
D-4630 BOCHUM 1
W. GERMANY

Contemporary Mathematics
Volume 33, 1984

QUADRATIC PARAMETRIC EQUATIONS OVER FREE GROUPS

Leo P. Comerford, Jr. and Charles C. Edmunds*

For Roger Lyndon on his 65th birthday

§1. INTRODUCTION

An (ordinary) equation over a group $G = \langle a_1, a_2, \ldots ; R_1, R_2, \ldots \rangle$ is an expression $W(a_1, a_2, \ldots, x_1, x_2, \ldots) = 1$, where W is a member of the free product of the free group H on $a_1, a_2, \ldots$ (constants) and the free group F on $x_1, x_2, \ldots$ (variables). A solution to this equation is a map ϕ from $F*H$ to H which fixes each element of H such that $W\phi$ defines the identity in G. Many important decision problems for G can be formulated as asking if a class of equations over G is solvable, that is, asking if there is an algorithm which tells whether or not an equation in the class has a solution. For example, the word and conjugacy problems for G are the problems of solving equations $\{U = 1 : U \in H\}$ and $\{x_1^{-1} U x_1 = V : U, V \in H\}$, respectively, over G.

Apart from the word and conjugacy problems, most of the positive results about solving classes of equations have dealt with equations over free groups. Over free groups, Lyndon [13] showed that one-variable equations are solvable, and Schupp [16] proved solvability of the class $\{W = U : W \in \langle x_1, x_2 \rangle, U \in H\}$. Comerford and Edmunds [5] showed that quadratic equations (i.e., those in which each variable which occurs appears exactly twice, each time with exponent $+1$ or -1) over free groups are solvable.

*Research support from the Natural Sciences and Engineering Research Council of Canada.

© 1984 American Mathematical Society
0271-4132/84 $1.00 + $.25 per page

A more general sort of equation over G is one which involves not only (ordinary) variables ranging over H, but also integer-valued <u>parameters</u> $\lambda_1, \lambda_2, \ldots$ appearing as exponents on elements of H. Formally, a <u>parametric equation</u> over G is an expression

$$U_0 V_1^{\lambda_1} U_1 V_2^{\lambda_2} \ldots U_{k-1} V_k^{\lambda_k} U_k = 1$$

with $U_0, U_1, \ldots, U_k$ in $F*H$ and $V_1, \ldots, V_k$ in H, together with a finite (possibly empty) set $\mathcal{L}$ of side conditions or constraints which are linear equations or inequalities in $\lambda_1, \lambda_2, \ldots$. A <u>solution</u> to such an equation is a map ϕ assigning an element of H to each variable in the equation and a map π sending parameters $\lambda_1, \lambda_2, \ldots$ to integers such that π is a solution to $\mathcal{L}$ and the image of $U_0 V_1^{\lambda_1} U_1 V_2^{\lambda_2} \ldots U_{k-1} V_k^{\lambda_k} U_k$ under ϕ and π defines the identity in G.

An important example of a decision problem which can be stated using parametric equations is the <u>power-conjugacy problem</u> for G, which asks whether or not equations $x^{-1} U^{\lambda_1} x V^{\lambda_2} = 1$, $U, V \in H$, have solutions subject to $\lambda_1 \neq 0$, $\lambda_2 \neq 0$. Some other decision problems of this sort are the <u>order problem</u> ($U^{\lambda} = 1$, $U \in H$, $\lambda \neq 0$), the <u>power problem</u> ($U^{\lambda} = V$, $U, V \in H$), and the <u>generalized power problem</u> ($U^{\lambda_1} = V^{\lambda_2}$, $U, V \in H$, $\lambda_1 \neq 0$, $\lambda_2 \neq 0$). These problems are discussed in some detail in [12]. Each of the equations above is <u>quadratic</u> in that its ordinary variables are quadratic.

All of the decision problems that we have mentioned are solvable over free groups. The power-conjugacy problem has received considerable attention, especially for free products with cyclic amalgamation and HNN extensions with cyclic conjugated subgroups; see, for example, [2], [3], [4], [10], [11].

In addition to their intrinsic interest, parametric equations sometimes arise in the process of solving ordinary equations over groups. For instance, to solve the conjugacy problem for a free product with cyclic amalgamation $\langle G_1 * G_2 ; A_1 = A_2 \rangle$ ($A_1 \in G_1$, $A_2 \in G_2$), one must be able to solve equations of the type $U A_1^{\lambda_1} V A_1^{\lambda_2} = 1$ in G_1 and $U A_2^{\lambda_1} V A_2^{\lambda_2} = 1$ in G_2. In another context, Lyndon [13] used parametric words to describe solution sets of one-variable equations over free groups. In fact, Lyndon [13] showed that the class of all "parametric equations" is solvable over a free group; Lyndon's parametric equations differ from ours in that they do not contain ordinary variables, but do admit multiple layers of exponents which involve parameters.

The main result of this paper is:

Theorem 1.1. The class of quadratic parametric equations over a free group is solvable.

For this theorem, the restriction that the constraints on parameters be linear is essential. The negative solution to Hilbert's Tenth Problem (see [6]) shows that one cannot, in general, tell if a system of diophantine equations of degree two or higher has a solution.

We begin in Section 2 by defining "parametric word," which is a pair $(W,\mathcal{S})$ where W is a representation for the left-hand side of a parametric equation and $\mathcal{S}$ is a finite system of linear constraints on the parameters. We formalize the notion of image of a parametric word under maps ϕ of variables to elements of H and π of parameters to integers. The problem of solving $W = 1$ subject to $\mathcal{S}$ then becomes the problem of deciding whether or not 1 is an image of $(W,\mathcal{S})$. We define several notions of reduction for parametric words and several transformations which may be applied to parametric words to simplify them or to reduce the amount of cancellation in their images.

In Section 3, we associate with a quadratic parametric word a set $D_{(W,\mathcal{S})}$ of "normal forms" with the property that a word U in $F*H$ is an image of $(W,\mathcal{S})$ if and only if it is a "cancellation-free" image of some word $(W',\mathcal{S}')$ in $D_{(W,\mathcal{S})}$ (Theorem 3.1). This set is analogous to the set D_W used to study nonparametric quadratic equations in [5]. As is explained in Section 3, the problem of determining if $W = 1$ has a solution subject to $\mathcal{S}$ then becomes the problem of determining if $D_{(W,\mathcal{S})}$ has an element of the form $(1,\mathcal{T})$. This last problem is settled in Section 4, where it is shown that for a quadratic parametric word $(W,\mathcal{S})$, the set $D_{(W,\mathcal{S})}$ is finite and effectively calculable (Theorem 4.9).

Before proceeding, we wish to acknowledge explicitly that in this work we are traveling along a path opened by Roger Lyndon. Lyndon was the first to study systematically cancellation in images of parametric words [14]. Much of what we do here uses variations of techniques developed by Lyndon. The problem that we examine in this paper is more general than Lyndon's in that we allow variables over group elements, and less general in that we permit only one level of parameters in exponents ("height one" in Lyndon's terminology).

§2. PARAMETRIC WORDS

Our terminology and notation will be consistent, as far as is practical, with that of [5]. One major difference is that we shall make no use of "cyclic words."

We call $x_1, x_1^{-1}, x_2, x_2^{-1}, \ldots$ F-letters and $a_1, a_1^{-1}, a_2, a_2^{-1}, \ldots$ H-letters, and let F and H be the free groups they generate. Our usual convention will be to let upper case letters denote words and lower case letters denote single letters.

Let M be the free $\mathbb{Z}$-module $\mathbb{Z}[1,\lambda_1,\lambda_2,\ldots]$. We shall think of the elements of M as polynomials of degree at most one over $\mathbb{Z}$. We call $\lambda_1,\lambda_2,\ldots$ parameters.

If B is a cyclically reduced, nonempty element of H which is not a proper power (i.e., $B = C^n$ for C in H implies that $n = \pm 1$), and if $f \in M$, then the expression

$$(B \uparrow f)$$

is called a parametric expression (over H). Technically, a parametric expression is an ordered pair (B,f), but the "exponential" notation is more suggestive of the intended application. We denote the set of all parametric expressions over H by P and refer to the elements of P as P-letters. It may be of interest to note the parallel between our "parametric expressions" and the "periodic words" of Adian [1].

We let B^{∞} and $B^{-\infty}$ represent the infinite sequences $BBB\ldots$ and $B^{-1}B^{-1}B^{-1}\ldots$ respectively. For $k \in \mathbb{Z}$, we define $(B \uparrow k)$ to be the initial segment of B^{∞} of length k if $k \geq 0$, and the initial segment of $B^{-\infty}$ of length $-k$ if $k < 0$. Note that our notation is very different from standard exponential notation; $(B \uparrow k)$ is an initial subword of a power of B or B^{-1} of length $|k|$, not the product of k factors of B or B^{-1}. Also, note that $(B \uparrow k)$ and $(B \uparrow -k)$ are not inverses of one another unless k is divisible by the length of B. For example, $(abc \uparrow 5) = (c^{-1}b^{-1}a^{-1} \uparrow -5) = abcab$, while $(abc \uparrow -5) = (c^{-1}b^{-1}a^{-1} \uparrow 5) = c^{-1}b^{-1}a^{-1}c^{-1}b^{-1}$.

We make a distinction, which is harmlessly suppressed by the notation, between $(B \uparrow k)$ where k is a nonzero integer and $(B \uparrow k)$ where k is a constant polynomial in M. The former is an element of H, while the latter is a P-letter.

An <u>unconstrained parametric word</u> (or more simply a <u>word</u>) W is a finite sequence $y_1 y_2 \ldots y_k$ of F-letters, H-letters, and P-letters. We use the symbols "$\equiv$" to denote identical equality of words and "1" to denote the empty word. If $W \equiv ABC$, then B is called a <u>subword</u> of W. If $W \equiv AB$ then A and B are called, respectively, <u>initial</u> and <u>terminal</u> subwords of W. A word B <u>occurs</u> in W if there are (possibly empty) words A and C such that $W \equiv ABC$. Following Adian [1], we distinguish among different occurrences of the same subword in a word W as follows. If $W \equiv ABC$ we call the triple (A,B,C) an <u>occurrence of</u> B <u>in</u> W.

The <u>F-length</u>, <u>H-length</u>, and <u>P-length</u> of W, denoted $|W|_F$, $|W|_H$, and $|W|_P$, are the numbers of terms in the sequence $y_1 y_2 \ldots y_k$ which are, respectively, F-letters, H-letters, and P-letters. The <u>length</u> of W, denoted $|W|$, is defined by

$$|W| = |W|_F + |W|_H + |W|_P .$$

If $|W|_P = 0$ we call W an <u>ordinary word</u>; this may be expressed by writing $W \in F*H$. Equality of words in $F*H$ (modulo free reduction) will be denoted by "=".

Consider a finite system, $\mathcal{A}$, of linear "side conditions" partitioned as follows:

$$\mathcal{S} = \mathcal{C} \dot{\cup} \mathcal{E} \dot{\cup} \mathcal{I}$$

where

(i) $\mathcal{C}$ is a finite system of linear congruences of the form $f \equiv 0 \pmod{m}$ where $f \in M$ and m is a positive integer,

(ii) $\mathcal{E}$ is a finite system of linear diophantine equations of the form $g = 0$ for $g \in M$, and

(iii) $\mathcal{I}$ is a finite system of linear diophantine inequalities of the form $h \geq 0$ for $h \in M$.

We could, of course, replace each linear congruence $f \equiv 0 \pmod{m}$ by an equation $f - \lambda m = 0$ with λ a new parameter, but we find it convenient not to do so. For any such system we shall assume henceforth without further mention that

(i) if c is a coefficient or constant term of f in any congruence $f \equiv 0 \pmod{m}$, then $0 \leq c < m$, and

(ii) each congruence in $\mathcal{C}$ and each equation in $\mathcal{E}$ involves at least one parameter.

We say that a system $\mathcal{S}$ with these properties is in <u>standard form</u>.

A <u>solution</u> of $\mathcal{S}$ is a retraction π of M onto $\mathbb{Z}$ (i.e., a $\mathbb{Z}$-module homomorphism of M onto $\mathbb{Z}$ fixing $\mathbb{Z}$ elementwise) such that the resulting congruences, inequalities, and equations

$$\ldots, \pi(f) \equiv 0 \pmod{m}, \ldots$$

$$\ldots, \pi(g) = 0, \ldots$$

$$\ldots, \pi(h) \geq 0, \ldots$$

hold in $\mathbb{Z}$. The system is said to be <u>consistent</u> if it has a solution.

If a P-letter $(B \uparrow f)$ occurs in W, we call f an <u>exponential polynomial</u> of the pair $(W, \mathcal{S})$. If $\mathcal{E}$ contains an equation $g = 0$, we call g a <u>constraint polynomial</u> of $(W, \mathcal{S})$. Together, all such f and g are called the <u>polynomials</u> of $(W, \mathcal{S})$. A parameter, λ_i, <u>occurs</u> in a polynomial if it has a nonzero coefficient in the polynomial.

We call a pair $(W,\mathscr{A})$, as above, a <u>constrained parametric word</u> if:

(i) $\mathscr{A}$ is consistent, and

(ii) if $(A \uparrow f)$ is a P-letter occurring in W, then $\mathscr{A}$ implies a congruence

$$f \equiv k \pmod{|A|} .$$

Note that (i) implies the uniqueness of the k in (ii). We will use the term "word" for both constrained and unconstrained words and will abuse the language somewhat by saying, for example, that if $W \equiv AB$, then B is a terminal subword of $(W,\mathscr{A})$.

If $(A \uparrow f)$ is a P-letter occurring in a word $(W,\mathscr{A})$ and $A \equiv A_1A_2$ where $\mathscr{A}$ implies that

$$f \equiv |A_1| \pmod{|A|},$$

then the <u>inverse of</u> $(A \uparrow f)$ <u>relative to</u> $\mathscr{A}$, denoted

$$(A \uparrow f)_{\mathscr{A}}^{-1} ,$$

is the P-letter

$$(A_1^{-1}A_2^{-1} \uparrow f) .$$

As noted previously, this is not necessarily the same as $(A^{-1} \uparrow f)$ or $(A \uparrow -f)$. If $(W,\mathscr{A})$ is a word with $W \equiv y_1y_2\ldots y_k$, then the <u>inverse of</u> $(W,\mathscr{A})$, denoted $(W,\mathscr{A})^{-1}$, is $(W',\mathscr{A})$ where

$$W' \equiv y_k^{-1} \ldots y_1^{-1}$$

with the inverse of each P-letter taken relative to $\mathscr{A}$. We shall usually suppress the reference to $\mathscr{A}$ in the inverse notation.

IMAGES AND EQUIVALENCE

We now discuss images in $F*H$ of a parametric word $(W,\mathscr{A})$. These are called "values" by Lyndon [13]. The idea is simply to produce a word in $F*H$ by substituting words in $F*H$ for the F-letters of W and integers, satisfying $\mathscr{A}$, for the parameters in W. Note that if $(A \uparrow f)$ is a P-letter and π is a retraction of M onto $\mathbb{Z}$, then $\pi(f) \in \mathbb{Z}$; thus $(B \uparrow \pi(f))$ has been defined as a word in H.

An endomorphism of $F*H$ fixing H elementwise is called an H-map, exactly as in [5]. If ϕ is an H-map and π is a retraction of M onto $\mathbb{Z}$, then the pair (π,ϕ) is called a retractive mapping. The domain of (π,ϕ) is the set of all words $(W,\mathcal{S})$ for which π is a solution of $\mathcal{S}$. The range is a subset of $F*H$. The mapping sends $(W,\mathcal{S})$ to the word in $F*M$ obtained by replacing each F- and H-letter by its image under ϕ and replacing each P-letter, $(A \uparrow f)$, by $(A \uparrow \pi(f))$. Note that this yields a (possibly unreduced) word in $F*H$. We denote the word thus produced $(W,\mathcal{S})\ (\pi,\phi)$ and refer to this as the formal image of $(W,\mathcal{S})$ under (π,ϕ). If $U = (W,\mathcal{S})$ (π,ϕ), we call U an image of $(W,\mathcal{S})$ under (π,ϕ).

We let $Im(W,\mathcal{S})$ denote the set of all images of $(W,\mathcal{S})$. If $Im(W_1,\mathcal{S}_1) = Im(W_2,\mathcal{S}_2)$ we say that $(W_1,\mathcal{S}_1)$ and $(W_2,\mathcal{S}_2)$ are equivalent. This is clearly an equivalence relation.

If $(W,\mathcal{S})$ is a word in the domain of (π,ϕ), we call the formal image $(W,\mathcal{S})\ (\pi,\phi)$ the cancellation free (or c-free) image of $(W,\mathcal{S})$ under (π,ϕ) if

(i) $x\phi \neq 1$ for each F-letter x occurring in W,

(ii) $\pi(f) > 0$ for each exponential polynomial f of $(W,\mathcal{S})$ and

(iii) $(W,\mathcal{S})\ (\pi,\phi)$ is a freely reduced word in $F*H$.

REDUCTION

In what follows we will have occasion to transform one word into another. This will involve insertions and deletions of side conditions from $\mathcal{S}$ and modification of certain occurrences in W. In general terms if $U \equiv (L,S,R)$ and $V \equiv (L,T,R)$ are, respectively, occurrences of S in LSR and T in LTR, we say LTR is the word resulting from LSR by replacement of S by T. Our goal will be to apply a sequence of transformations to replace a word by an equivalent word which is "fully reduced" in the sense defined below.

Prereduction

Clearly we wish to eliminate inverse pairs of F-letters and H-letters and to "consolidate" pairs of P-letters where possible. A word $(W,\mathcal{S})$ is

<u>prereduced</u> if it contains no occurrences of subwords yy^{-1} where y is an F-letter or an H-letter and no occurrences of subwords

$$(A \uparrow f)(B \uparrow g)$$

for which there is an $\epsilon = \pm 1$ such that $A \equiv A_1A_2$, $B^\epsilon \equiv A_2A_1$, and $\mathscr{A}$ implies that $f \equiv |A_1| \pmod{|A|}$.

If $(W,\mathscr{A})$ is not prereduced, we may delete an inverse pair of F-letters or H-letters or replace a subword $(A \uparrow f)(B \uparrow g)$ as described above by $(A \uparrow f + \epsilon g)$. Since these operations reduce length without changing the set of images of $(W,\mathscr{A})$, we have an effective procedure to replace any word with an equivalent prereduced word.

<u>Reduction</u>

Next we wish to remove F-letters and parameters which are redundant from the point of view of producing images of a word. We do not always simplify as much as may be possible; rather, we perform those simplifications which will facilitate our constructions at a later stage.

If x and y are F-letters, a subword xy of $(W,\mathscr{A})$ is called <u>F-reducible</u> if $x^{\pm 1}$ and $y^{\pm 1}$ occur only in subwords xy and $y^{-1}x^{-1}$ of W. If $(A \uparrow f)$ occurs in $(W,\mathscr{A})$ and $\mathscr{A}$ implies that $f = 0$, we call $(A \uparrow f)$ a <u>P-reducible</u> subword of $(W,\mathscr{A})$.

A parameter λ_i occurring in $(W,\mathscr{A})$ is <u>reducible</u> if it does not occur in $\mathscr{I}$ and either:

(i) λ_i occurs only in $\mathscr{C}$, or

(ii) λ_i occurs in an equation $a\lambda_i + b = 0$ of $\mathscr{E}$ with a and b integers and $a \neq 0$, and λ_i occurs in no other polynomial of $(W,\mathscr{A})$, or

(iii) λ_i occurs in an equation of $\mathscr{E}$, $a\lambda_i + g = 0$, with $a \neq 0$, λ_i occurs in at most one other polynomial of $(W,\mathscr{A})$ and occurs there with coefficient a or $-a$, and for every congruence $e\lambda_i + h \equiv 0 \pmod{m}$ in which λ_i appears, $ax \equiv e \pmod{m}$ has a solution, that is, $\gcd(a,m)$ divides e, or

(iv) there is a λ_j, $j \neq i$, which does not occur in $\mathscr{I}$ and there are nonzero integers a and b such that if either λ_i or λ_j occurs in a polynomial of $(W,\mathcal{S})$, $d(a\lambda_1 + b\lambda_j)$ occurs in that polynomial for some integer d, and for every congruence $e\lambda_j + h \equiv 0 \pmod{m}$ in which λ_j appears, $bx \equiv e \pmod{m}$ has a solution, that is, $\gcd(b,m)$ divides e.

A prereduced word $(W,\mathcal{S})$ is <u>reduced</u> if it contains no reducible subwords or parameters.

We now introduce several transformations which may be used to eliminate reducible subwords and parameters from a prereduced word $(W,\mathcal{S})$ which is not reduced. If xy is an F-reducible subword of $(W,\mathcal{S})$, we transform $(W,\mathcal{S})$ into $(W',\mathcal{S})$, where W' results from W by deletion of all occurrences of y and y^{-1}. If $(A \uparrow f)$ is a P-reducible subword of $(W,\mathcal{S})$, we transform $(W,\mathcal{S})$ into $(W',\mathcal{S})$ where W' results by deletion of every occurrence of $(A \uparrow f)$ from W. If λ_i is a reducible parameter of $(W,\mathcal{S})$, there are three cases to consider:

(i) If λ_i occurs only in $\mathcal{C}$, we transform $(W,\mathcal{S})$ into $(W,\mathcal{S}')$, where $\mathcal{S}'$ results by replacing all occurrences of λ_i in $\mathcal{S}$ by an integer such that the system $\mathcal{S}'$ is consistent.

(ii) If λ_i is reducible by condition (ii) of the definition, we replace λ_i by $-b/a$ throughout $\mathcal{S}$. (Note that a divides b since $\mathcal{S}$ is consistent.)

(iii) If λ_i is reducible by condition (iii) of the definition, we form $(W',\mathcal{S}')$ by replacing $a\lambda_i$ by $-g$ in all polynomials of $(W,\mathcal{S})$ and, for each congruence $e\lambda_i + h \equiv 0 \pmod{m}$ in which λ_i appears, replacing $e\lambda_i$ by $-xg$ where $ax \equiv e \pmod{m}$.

(iv) If λ_i is reducible by condition (iv) of the definition, we form $(W',\mathcal{S}')$ by replacing $b\lambda_j$ by $c\lambda_j - a\lambda_i$ in all polynomials of $(W,\mathcal{S})$, where $c = \gcd(a,b)$, and, for each congruence $e\lambda_j + h \equiv 0 \pmod{m}$ in which λ_j appears, replacing $e\lambda_j$ by $x(c\lambda_j - a\lambda_i)$ where $bx \equiv e \pmod{m}$. (Note that if λ_i remains in $(W',\mathcal{S}')$, it is reducible there by condition (i).)

Following any of these moves, we return $\mathcal{S}'$ to standard form, if necessary.

We note that each of these transformations reduces $|W|$ or the number of parameters occurring in $(W,\mathscr{A})$ and, except in case (i) of removal of a reducible parameters, results in a parametric word equivalent to $(W,\mathscr{A})$. In the exceptional case, we still have $\mathrm{Im}(W,\mathscr{A}') \subseteq \mathrm{Im}(W,\mathscr{A})$.

Dualization

In forming images, we shall want the exponential polynomials of our word to take on positive values. With this in mind, we say that a word $(W,\mathscr{A})$ is consistent if

$$\mathscr{A} \cup \left(\bigcup_{i=1}^{k} \{f_i \geq 0\}\right)$$

is consistent, where the f_i's are the exponential polynomials of $(W,\mathscr{A})$. If $(L,(A \uparrow f),R)$ is an occurrence of a P-letter in $(W,\mathscr{A})$, we dualize this occurrence by replacing $(W,\mathscr{A})$ by $(W',\mathscr{A})$, where $W' \equiv L(A^{-1} \uparrow -f)R$. Note that the resulting word is equivalent to $(W,\mathscr{A})$. If $\mathscr{A}$ is consistent, there is clearly at least one sequence of dualizations, starting with $(W,\mathscr{A})$, which produces a consistent word $(W',\mathscr{A})$. Any such sequence of dualizations is called an admissible dualization for $(W,\mathscr{A})$.

Full Reduction

We define one final type of reduction. A consistent, reduced word $(W,\mathscr{A})$ is called fully reduced if it has no occurrence of a subword of the form:

(i) $(A_1A_2 \uparrow f)(B \uparrow g)$ where $\mathscr{A}$ implies that $f \equiv |A_1|(\mathrm{mod}\,|A_1A_2|)$ and the last letter of A_2A_1 cancels the first letter of B, or

(ii) $a(a^{-1}A \uparrow f)$, or

(iii) $(A_1A_2 \uparrow f)a$ where $\mathscr{A}$ implies that $f \equiv |A_1|(\mathrm{mod}\,|A_1A_2|)$ and a^{-1} is the last letter of A_2A_1.

To eliminate such subwords from $(W,\mathscr{A})$, we use the following transformations:

(i) For a subword of type (i), if $B \equiv aB_0$, we replace

$$(L,(A_1A_2 \uparrow f)(B \uparrow g),R)$$

by $(L,(A_1A_2 \uparrow f\text{-}1)(B_0a \uparrow g\text{-}1),R)$ in $(W,\mathscr{A})$. We call this a PP-cancellation.

(ii) For subwords of type (ii), we replace

$$(L,a(a^{-1}A \uparrow f),R)$$

by $(L,(Aa^{-1} \uparrow f-1),R)$ in $(W,\mathscr{A})$. We call this an HP-cancellation.

(iii) For subwords of type (iii), we replace

$$(L,(A_1A_2 \uparrow f)a,R)$$

by $(L,(A_1A_2 \uparrow f-1),R)$ in $(W,\mathscr{A})$. We call this a PH-cancellation.

Note that PP-, HP-, and PH-cancellations preserve equivalence of constrained words.

CANCELLATION

In the various notions of reduction that we have defined, our concern was to put a word $(W,\mathscr{A})$ into a reasonably simplified form. We next describe transformations that will be used to eliminate cancellations in the image of a word. The first three of these, τ-maps, ρ-maps, and μ-maps, are essentially as they were defined in [5]. These maps will be denoted by

$$\tau_x,\ \rho_{x,y,z},\ \text{and}\ \mu_{x,a}$$

where x,y,z are F-letters and a is an H-letter.

A τ-map τ_x is admissible for a reduced word $(W,\mathscr{A})$ if $x^{\pm 1}$ occurs in W. The effect of τ_x is to transform $(W,\mathscr{A})$ into $(W',\mathscr{A})$ where W' results by deleting all occurrences of x and x^{-1} from W. We use this map to eliminate x from W when its image under an H-map is empty.

A ρ-map $\rho_{x,y,z}$ is admissible for a reduced word $(W,\mathscr{A})$ if xy^{-1} or yx^{-1} occurs in W but $z^{\pm 1}$ does not. This transformation produces a word $(W',\mathscr{A})$ where W' results from W by replacing each x^{ϵ_1} and y^{ϵ_2}, $\epsilon_1 = \pm 1$, $\epsilon_2 = \pm 1$, which does not occur in a subword $(xy^{-1})^{\pm 1}$ by $(xz)^{\epsilon_1}$ and $(yz)^{\epsilon_2}$, respectively. This map is used to eliminate cancellation between the images of a pair of adjacent F-letters in W.

A μ-map $\mu_{x,a}$ is admissible for a reduced word $(W,\mathscr{A})$ if xa^{-1} or ax^{-1} occurs in W. The map $\mu_{x,a}$ sends $(W,\mathscr{A})$ to $(W',\mathscr{A})$ where W'

results from W by replacing each occurrence of x^{ϵ}, $\epsilon = \pm 1$, which does not occur in a subword $(xa^{-1})^{\pm 1}$ by $(xa)^{\epsilon}$, and by replacing occurrences of xa^{-1} and ax^{-1} by x and x^{-1}, respectively. This map is used when cancellation occurs between the image of an F-letter and an adjacent H-letter of W.

We need a map, called a <u>ν-map</u>, to handle cancellation which occurs between the images of an F-letter and an adjacent P-letter of W. We denote a ν-map by

$$\nu_{x,(A \uparrow \lambda_i)}$$

where x is an F-letter and $(A \uparrow \lambda_i)$ is a P-letter, with λ_i a parameter. This map is <u>admissible</u> for a reduced word $(W,\mathscr{A})$ if W has a subword $(x(B \uparrow f))^{\pm 1}$ where $B \equiv a^{-1}A_0^{-1}$ and $A \equiv aA_0$ and if λ_i does not occur in $(W,\mathscr{A})$. The map transforms $(W,\mathscr{A})$ into $(W',\mathscr{A}')$ where $\mathscr{A}' = \mathscr{A} \cup \{\lambda_i \equiv 1(\text{mod}\,|A|)\}$ and W' results from W by replacing each subword $(xy)^{\epsilon}$, $\epsilon = \pm 1$, of W by a subword which depends on the nature of the letter y.

<u>Case 1</u>. If y is $(a^{-1}A_0^{-1} \uparrow g)$ for some $g \in M$, we replace $(xy)^{\epsilon}$ by $(x(A_0^{-1}a^{-1} \uparrow g-\lambda_i))^{\epsilon}$.

<u>Case 2</u>. If y is $(A_0 a \uparrow g)$ for some $g \in M$, we replace $(xy)^{\epsilon}$ by $(x(aA_0 \uparrow g+\lambda_i))^{\epsilon}$.

<u>Case 3</u>. Otherwise, we replace $(xy)^{\epsilon}$ by $(x(aA_0 \uparrow \lambda_i)y)^{\epsilon}$.

Finally, we introduce a second type of <u>τ-map</u>, denoted

$$\tau_{(L,(A \uparrow f),R)}$$

where $(L,(A \uparrow f),R)$ is an occurrence of a P-letter. This map, which is used to eliminate P-letters whose images are empty, is <u>admissible</u> for a reduced word $(W,\mathscr{A})$ if $(L,(A \uparrow f),R)$ is an occurrence of $(A \uparrow f)$ in W and $\mathscr{A} \cup \{f = 0\}$ is consistent. The map sends $(W,\mathscr{A})$ to $(W',\mathscr{A}')$ where $W' \equiv LR$ and $\mathscr{A}' = \mathscr{A} \cup \{f = 0\}$.

STAR GRAPHS

We now describe a graphical representation of a parametric word $(W,\mathscr{A})$ in terms of a star graph or coinitial graph $\Gamma(W,\mathscr{A})$. This is an adaptation for parametric words of a well-known construction; see, for example, [5], [8], and [17].

With each P-letter $(A \uparrow f)$ such that $(A \uparrow f)$ or its inverse occurs in $(W, \mathscr{A})$, we associate a simple closed edge-path ∂A of length $|A|$, oriented counterclockwise, called a parametric cycle. The edges of ∂A are labeled by H-letters so that the label on ∂A, beginning at a particular vertex $\alpha(A \uparrow f)$, is A. Since A is not a proper power, the vertex $\alpha(A \uparrow f)$ is unique. If k is an integer such that $0 \leq k < |A|$ and $\mathscr{A}$ implies that $f \equiv k \pmod{|A|}$, we distinguish a second vertex $\omega(A \uparrow f)$ on ∂A which is located by traversing k edges in a counterclockwise direction from $\alpha(A \uparrow f)$. If B is a cyclic permutation of A such that $(B \uparrow g)$ or its inverse occurs in $(W, \mathscr{A})$, we use the same parametric cycle for $(A \uparrow f)$ and $(B \uparrow g)$ although, of course, $\alpha(B \uparrow g)$ and $\omega(B \uparrow g)$ may differ from $\alpha(A \uparrow f)$ and $\omega(A \uparrow f)$. A P-letter and its inverse, however, are represented by different parametric cycles.

We now define the directed graph $\Gamma(W, \mathscr{A})$. The vertices of $\Gamma(W, \mathscr{A})$ are:

(i) the vertices of ∂A if $(A \uparrow f)$ is a P-letter such that $(A \uparrow f)$ or its inverse occurs in W,

(ii) x and x^{-1} for each F-letter x occurring in W, and

(iii) a distinguished vertex $*$ which will be associated with H-letters.

The edges of $\Gamma(W, \mathscr{A})$ are as follows.

(i) Each edge of a parametric cycle is an edge of $\Gamma(W, \mathscr{A})$, oriented to coincide with the counterclockwise orientation of the cycle.

(ii) The graph $\Gamma(W, \mathscr{A})$ has an edge $e(s, t^{-1})$ for each subword st of W and also for s the terminal letter and t the initial letter of W. The initial vertex of $e(s, t^{-1})$ is s if s is an F-letter, $*$ if s is an H-letter, and $\omega(s)$ if s is a P-letter. The terminal vertex of $e(s, t^{-1})$ is t^{-1} if t is an F-letter, $*$ if t is an H-letter, and $\alpha(\bar{t})$ if t is a P-letter and $\bar{t}$ denotes the dual of t.

Note that $\Gamma(W,\mathscr{A})$ may have multiple edges joining vertices and loops at a vertex.

We use $\Gamma(W,\mathscr{A})$ to associate with $(W,\mathscr{A})$ two inter-valued quantities, $\Delta(W,\mathscr{A})$ and $\pi(W,\mathscr{A})$, which will measure the complexity of $(W,\mathscr{A})$ in ways which will be applied later. We let

$$\Delta(W,\mathscr{A}) = f(W,\mathscr{A}) - c(W,\mathscr{A})$$

where $f(W,\mathscr{A})$ is the number of distinct generators x_i of F such that x_i or x_i^{-1} occurs in W and $c(W,\mathscr{A})$ is the number of F-components of $\Gamma(W,\mathscr{A})$, that is, components whose only vertices are F-letters. For W in $F{*}H$, $\Delta(W,\mathscr{A})$ is one minus the Euler characteric of W.

The definition of $\Pi(W,\mathscr{A})$ is a bit more involved. If Q_1 and Q_2 are vertices on parametric cycles ∂A_1 and ∂A_2 of $\Gamma(W,\mathscr{A})$, we say that Q_1 and Q_2 are <u>congruent</u> if the labels around ∂A_1 and ∂A_2 read counterclockwise from Q_1 to Q_1 on ∂A_1 and from Q_2 to Q_2 on ∂A_2 are either identical or inverse words in H. By an <u>F-path</u> in $\Gamma(W,\mathscr{A})$, we mean a simple edge path γ containing no edges from parametric cycles and with each vertex along γ, except possibly its initial and terminal vertices, an F-letter. We say that an F-path γ <u>links</u> vertices Q_1 and Q_2 if Q_1 is the initial vertex of γ and Q_2 is its terminal vertex.

For each edge e of $\Gamma(W,\mathscr{A})$, we define an integer $\pi(e)$ as follows:

(i) $\pi(e) = 0$ if e is part of a parametric cycle, if the endpoints of e are both $*$, or if the endpoints of e are a pair of congruent vertices on parametric cycles.

(ii) $\pi(e) = 1$ if one endpoint of e is $*$ and the other is not, or if one endpoint of e, say Q_1, is on a parametric cycle and the other is an F-letter with e or e^{-1} forming part of an F-path linking Q_1 to a congruent vertex.

(iii) $\pi(e) = 2$ otherwise.

We then define $\Pi(W,\mathscr{A})$ by

$$\Pi(W,\mathscr{A}) = \Sigma\, \pi(e) - 4c(W,\mathscr{A})$$

where the sum is taken over all edges of $\Gamma(W,\mathscr{A})$.

§3. THE SET $D_{(W,\mathscr{A})}$

The "parametric equations" mentioned in the introduction were of the form

$$(1) \qquad U_0 V_1^{\lambda_1} U_1 V_2^{\lambda_2} \cdots U_{k-1} V_k^{\lambda_k} U_k = 1 ,$$

where $U_i \in F{*}H$, $V_i \in H$, and the λ_i are integer-valued parameters. This equation had an associated (possibly empty) finite set of $\mathscr{L}$ of "side conditions," which were linear equations and inequalities in the λ_i's. (We could have allowed arbitrary linear polynomials in $\lambda_1, \lambda_2, \ldots$ as exponents in (1), but this would not have been a real generalization.) We may associate with equation (1) and its side conditions $\mathscr{L}$ a parametric word $(W,\mathscr{A})$. We form W from the left-hand side of (1) by deleting V_i if $V_i = 1$ in H and otherwise by replacing V_i by $C_i^{-1}(A_i \uparrow n_i\lambda_i)C_i$ where $V_i = C_i^{-1}A_i^{n_i}C_i$ in H with A_i cyclically reduced and not a proper power and n_i a positive integer. The set $\mathscr{A}$ consists of $\mathscr{L}$ together with the congruences

$$\lambda_i \equiv 0 \pmod{|A_i|}$$

for each i such that $V_i \neq 1$ in H. The problem of deciding whether or not (1) has a solution subject to $\mathscr{L}$ can then be restated as deciding whether or not 1 is an image of $(W,\mathscr{A})$.

From now on, we shall assume that our words $(W,\mathscr{A})$ are <u>quadratic</u>, that is, that each F-letter which occurs in W occurs exactly twice, each time with exponent 1 or -1. In this section we shall show how to associate with a quadratic word $(W,\mathscr{A})$ a set of words $D_{(W,\mathscr{A})}$ such that a reduced word V of $F{*}H$ is an image of $(W,\mathscr{A})$ if and only if it is a c-free image of some word in $D_{(W,\mathscr{A})}$. Since the empty word is a c-free image only of itself, the question of deciding whether or not (1) has a solution subject to $\mathscr{L}$ reduces to the question of whether or not $D_{(W,\mathscr{A})}$ contains a word $(1,\mathscr{T})$. In the next section, we show that for $(W,\mathscr{A})$ quadratic, $D_{(W,\mathscr{A})}$ is a finite, effectively calculable set; thus in this case the question of whether or not (1) has a solution subject to $\mathscr{L}$ can be answered effectively.

The set $D_{(W,\mathscr{A})}$ is defined to be the set of words which are output when $(W,\mathscr{A})$ is used as input for the nondeterministic algorithm defined below. We consider two words produced as output to define the same element of $D_{(W,\mathscr{A})}$ if they differ by a relabeling of F-letters and parameters, that is, if one word can be changed to the other by permutations of the collections of all F-letters and of all parameters. Thus, we may assume that the words in $D_{(W,\mathscr{A})}$ are written on initial segments of the sequences $x_1,x_2,\ldots$ and $\lambda_1,\lambda_2,\ldots$. We call our nondeterministic algorithm the $D_{(W,\mathscr{A})}$-process. We first give a flowchart (see Figure 1) and then more fully describe the process.

The $D_{(W,\mathscr{A})}$-process begins with the replacement of $(W,\mathscr{A})$ by an equivalent prereduced word. We next apply any admissible dualization to the resulting word. To this we apply an "admissible step" as follows:

(i) If the word is not reduced, we apply any one of the transformations needed to remove a reducible subword or parameter.

(ii) If the word is reduced, we apply either a PP-, PH-, or HP-cancellation (if applicable), an admissible ρ-, μ-, ν-, or τ-map, or the identity transformation.

The flowchart then defines the rest of the process.

The usefulness of $D_{(W,\mathscr{A})}$ for describing images of $(W,\mathscr{A})$ is shown in the following analogue of Theorem 2.3 [5].

Theorem 3.1. *If $(W,\mathscr{A})$ is a quadratic word, then a reduced word U in $F*H$ is an image of $(W,\mathscr{A})$ if and only if U is a c-free image of some $(W',\mathscr{A}')$ in $D_{(W,\mathscr{A})}$.*

A major portion of the proof of this theorem is contained in the following lemma.

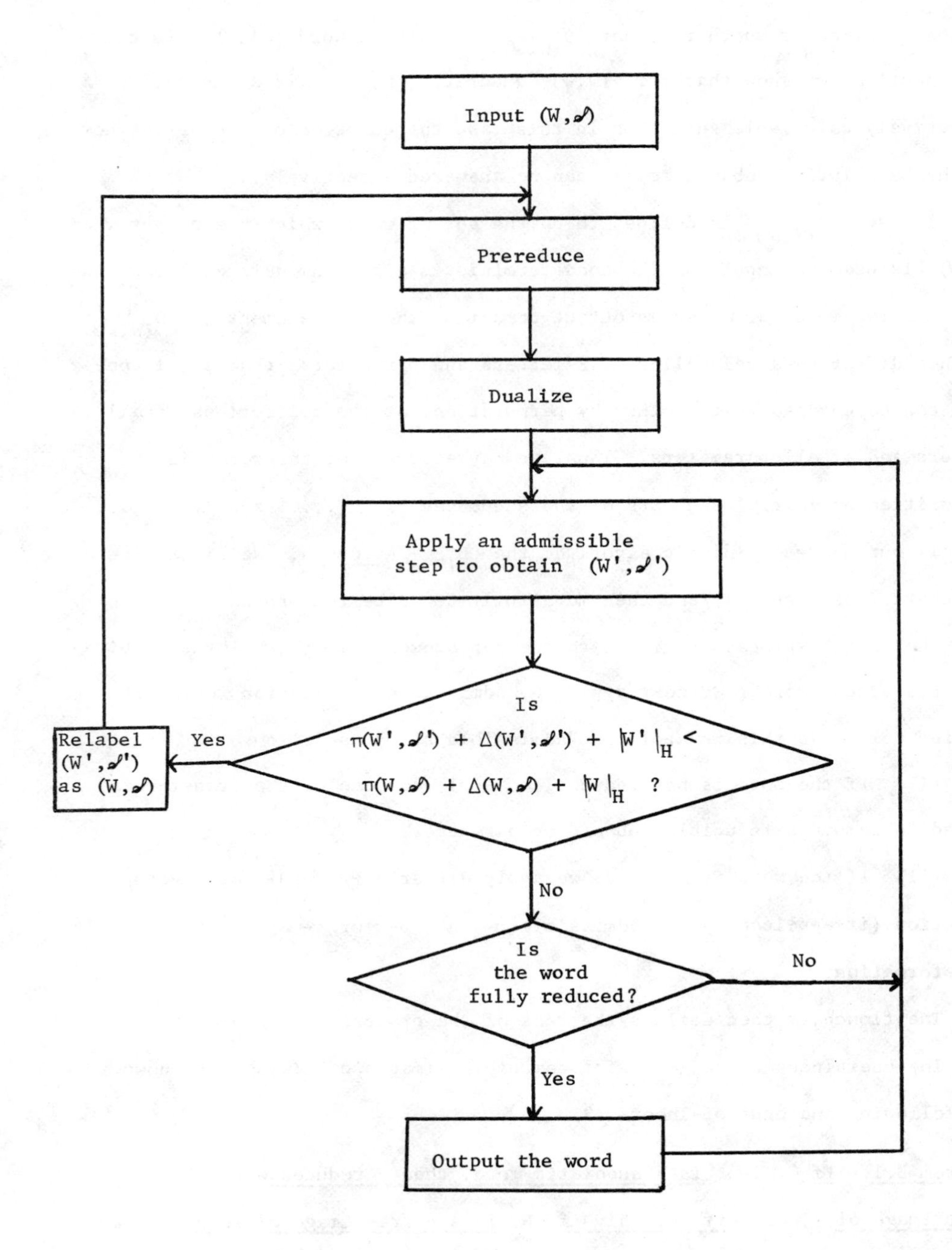
Input (W,𝒜)
Prereduce
Dualize
Apply an admissible step to obtain (W',𝒜')
Is
π(W',𝒜') + Δ(W',𝒜') + |W'|_H <
π(W,𝒜) + Δ(W,𝒜) + |W|_H ?
Yes
Relabel (W',𝒜') as (W,𝒜)
No
Is the word fully reduced?
No
Yes
Output the word

Figure 1

Lemma 3.2. If $(W,\mathscr{A})$ is a quadratic word and $(W',\mathscr{A}')$ is a word produced at some stage of the $D_{(W,\mathscr{A})}$-process with W' not the empty word, then

a) $\Pi(W',\mathscr{A}') \leq \Pi(W,\mathscr{A})$, $\Delta(W',\mathscr{A}') \leq \Delta(W,\mathscr{A})$, and $|W'|_H \leq |W|_H$, and

b) if $(W,\mathscr{A})$ is prereduced and

$\Pi(W',\mathscr{A}') + \Delta(W',\mathscr{A}') + |W'|_H = \Pi(W,\mathscr{A}) + \Delta(W,\mathscr{A}) + |W|_H$,

then $(W',\mathscr{A}')$ is prereduced.

Proof of Lemma 3.2. Note that if $(W,\mathscr{A})$ is quadratic, every word produced in the $D_{(W,\mathscr{A})}$-process is also quadratic. Thus, using induction on the number of steps in the $D_{(W,\mathscr{A})}$-process used to produce $(W',\mathscr{A}')$ from $(W,\mathscr{A})$, we may assume that $(W',\mathscr{A}')$ was produced from $(W,\mathscr{A})$ in one step.

It is clear from the definition of the $D_{(W,\mathscr{A})}$-process that $|W'|_H \leq |W|_H$. In verifying b), we observe that by a), any of $\pi(W',\mathscr{A}') < \pi(W,\mathscr{A})$, $\Delta(W',\mathscr{A}') < \Delta(W,\mathscr{A})$, $|W'|_H < |W|_H$ imply that $\pi(W',\mathscr{A}') + \Delta(W',\mathscr{A}') + |W'|_H < \pi(W,\mathscr{A}) + \Delta(W,\mathscr{A}) + |W|_H$.

The proof of the lemma proceeds by considering, in turn, the steps of the $D_{(W,\mathscr{A})}$-process by which $(W',\mathscr{A}')$ might have been produced from $(W,\mathscr{A})$. Much of this is similar to the analyses carried out in Section 3 of [7] and in the proof of Theorem 2.4 of [5]. We present here a few sample cases, leaving the others to the reader.

Suppose first that W' results from deleting a P-reducible subword $(A \uparrow f)$ from W, with $\mathscr{A}' = \mathscr{A}$. We assume that $|W| \geq 3$; other cases are done by similar arguments. The word W has a subword, cyclically, $s(A \uparrow f)t^{-1}$, which is replaced by st^{-1} in W'. The resulting change in $\Gamma(W,\mathscr{A})$ is shown in Figure 2.

Since W is prereduced, $\pi(e_1)$ and $\pi(e_2)$ are each at least one, so $\pi(e_1) + \pi(e_2) \geq \pi(f_1)$. If there is an edge e_3 of $\Gamma(W,\mathscr{A})$ different from e_1 and e_2 with $\pi(e_3)$ larger in $\Gamma(W',\mathscr{A}')$ than in $\Gamma(W,\mathscr{A})$, it must be that e_3 is part of an F-path linking either P or Q, say P, to a congruent vertex in $\Gamma(W,\mathscr{A})$ but that e_3 is not part of a linking F-path in $\Gamma(W',\mathscr{A}')$. Vertices P and Q are congruent, so e_2 is not part of a linking F-path in

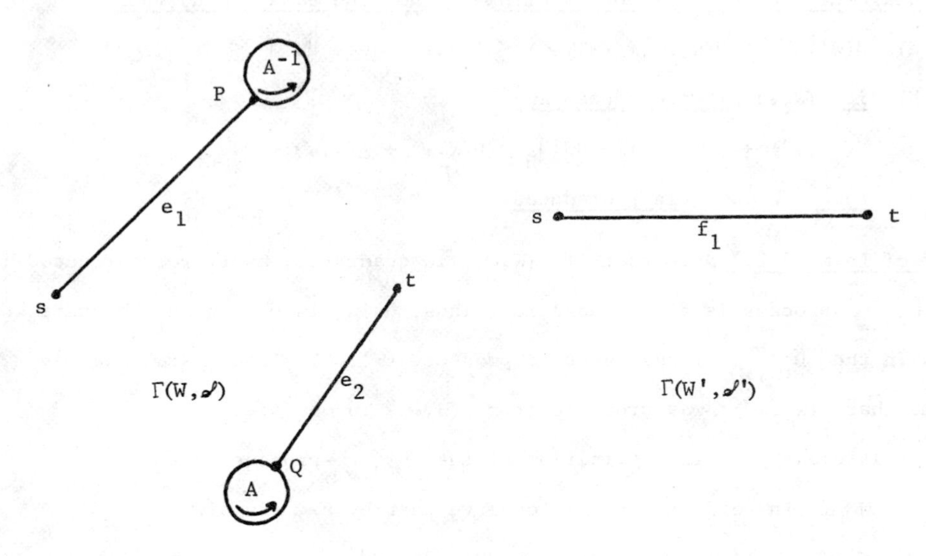

Figure 2

$(W,\mathscr{A})$. It follows, then, that there can be at most one such edge e_3 and that $\pi(e_1) = 1$ and $\pi(f_1) = \pi(e_2)$, since s must be an F-letter. The increase in $\pi(e_3)$ in passing from $\Gamma(W,\mathscr{A})$ to $\Gamma(W',\mathscr{A}')$ is offset by the loss of $\pi(e_1)$, and so in all cases $\Sigma\, \pi(e)$ over the edges of $\Gamma(W,\mathscr{A})$ is at least as large as the corresponding sum for $\Gamma(W',\mathscr{A}')$. Since $f(W',\mathscr{A}') = f(W,\mathscr{A})$ and $c(W',\mathscr{A}') \geq c(W,\mathscr{A})$, we thus have $\pi(W',\mathscr{A}') \leq \pi(W,\mathscr{A})$ and $\Delta(W',\mathscr{A}') \leq \Delta(W,\mathscr{A})$.

If W' is not prereduced, either s and t^{-1} are both F-letters or both H-letters with $s = t$ or s and t^{-1} are both P-letters and st^{-1} is a subword of W' to be "consolidated" in the prereduction process. We observe that if s and t^{-1} are F-letters, $\Gamma(W',\mathscr{A}')$ contains a new F-component whose only vertex is s. If s and t^{-1} are H-letters, $\pi(e_1) = \pi(e_2) = 1$ and $\pi(f_1) = 0$. If s and t^{-1} are P-letters, $\pi(e_1) = \pi(e_2) = 2$ while $\pi(f_1) = 0$. In any event, $\pi(W',\mathscr{A}') < \pi(W,\mathscr{A})$.

Suppose next that $(W',\mathscr{A}')$ is produced by applying an admissible ν-map $\nu_{x,(A \uparrow \lambda_i)}$ to $(W,\mathscr{A})$. The word W, then, has a subword $(x(B \uparrow f))^{\pm 1}$, where $B \equiv a^{-1}A_0^{-1}$ and $A \equiv aA_0$, which is replaced by $(x(A_0^{-1}a^{-1} \uparrow f-\lambda_i))^{\pm 1}$ in W'.

If the other occurrence of x in W is in a subword $(xy)^{\pm 1}$, the effect of the ν-map depends on what y is. If y is $(a^{-1}A_0^{-1} \uparrow g)$, $(xy)^{\pm 1}$ is replaced by $(x(A_0^{-1}a^{-1} \uparrow g-\lambda_i))^{\pm 1}$ as shown in Figure 3(a). If y is $(A_0 a \uparrow g)$, $(xy)^{\pm 1}$ is replaced by $(x(aA_0 \uparrow g+\lambda_i))^{\pm 1}$ as shown in Figure 3(b). Otherwise, $(xy)^{\pm 1}$ is replaced by $(x(aA_0 \uparrow \lambda_i)y)^{\pm 1}$ as shown in Figure 3(c).

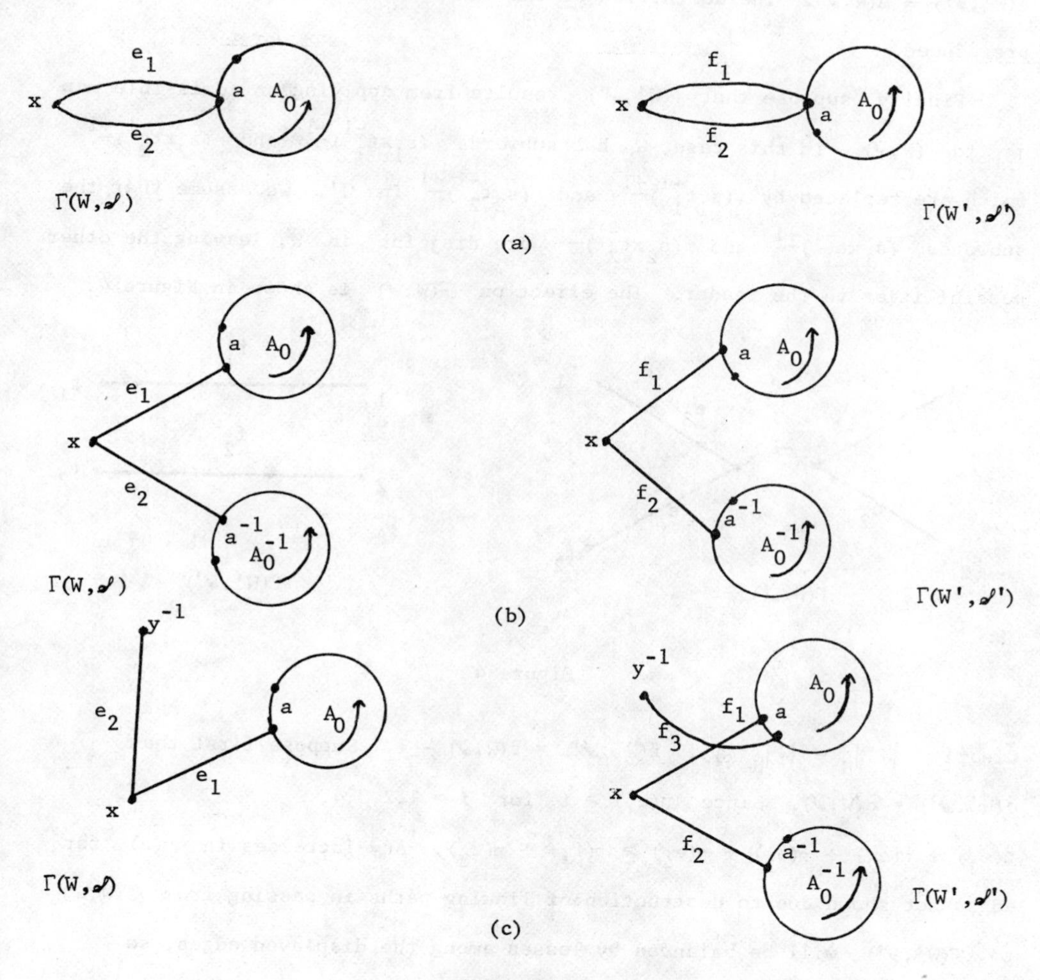

Figure 3

In the first two cases, $\pi(e_1) = \pi(e_2) = \pi(f_1) = \pi(f_2) = 1$. In the third case, we observe that $\pi(f_1) = \pi(f_2) = 1$, and consider the possibilities for y. If y is an H-letter, $\pi(e_1) = 2$ and $\pi(e_2) = \pi(f_3) = 1$. If y is an

F-letter or a P-letter, edges e_1 and e_2 are part of a linking F-path in $\Gamma(W,\mathscr{A})$ if and only if f_3 is part of a linking F-path in $\Gamma(W',\mathscr{A}')$. No matter what y is, we find that $\pi(e_1) + \pi(e_2) = \pi(f_1) + \pi(f_2) + \pi(f_3)$. Since $\pi(e)$ is unchanged for all edges e which are not shown and since $f(W',\mathscr{A}') = f(W,\mathscr{A})$ and $c(W',\mathscr{A}') = c(W,\mathscr{A})$, we have $\pi(W',\mathscr{A}') = \pi(W,\mathscr{A})$ and $\Delta(W',\mathscr{A}') = \Delta(W,\mathscr{A})$. The definition of the ν-map ensures that $(W',\mathscr{A}')$ is prereduced.

Finally, suppose that $(W',\mathscr{A}')$ results from applying an admissible map τ_x to $(W,\mathscr{A})$. In this case, W has subwords $(s_1xt_1^{-1})^{\pm 1}$ and $(s_2xt_2^{-1})^{\pm 1}$ which are replaced by $(s_1t_1^{-1})^{\pm 1}$ and $(s_2t_2^{-1})^{\pm 1}$ in W'. We assume that the subwords $(s_1xt_1^{-1})^{\pm 1}$ and $(s_2xt_2^{-1})^{\pm 1}$ are disjoint in W, leaving the other possibilities to the reader. The effect on $\Gamma(W,\mathscr{A})$ is shown in Figure 4.

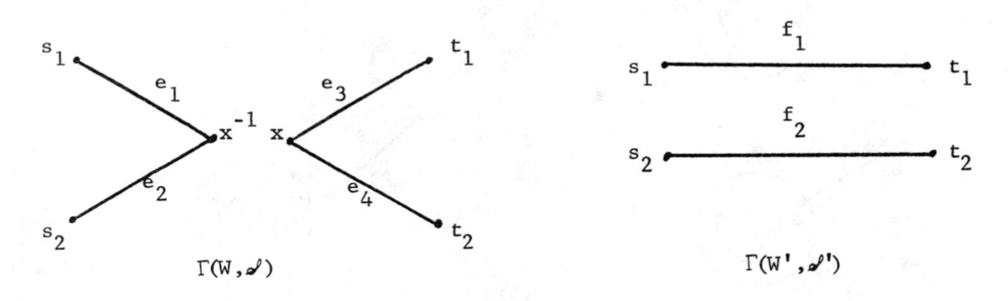

Figure 4

Clearly $|W'|_H = |W|_H$ and $f(W',\mathscr{A}') = f(W,\mathscr{A}) - 1$. Suppose first that $c(W',\mathscr{A}') = c(W,\mathscr{A})$. Since $\pi(e_i) \geq 1$ for $i = 1,2,3,4$, $\pi(e_1) + \pi(e_2) + \pi(e_3) + \pi(e_4) \geq \pi(f_1) + \pi(f_2)$. Any increases in $\pi(e)$ for edges not shown due to destruction of linking paths in passing from $\Gamma(W,\mathscr{A})$ to $\Gamma(W',\mathscr{A}')$ will be balanced by losses among the displayed edges, so $\pi(W',\mathscr{A}') \leq \pi(W,\mathscr{A})$ and $\Delta(W',\mathscr{A}') = \Delta(W,\mathscr{A}) - 1$.

It may also be that $c(W',\mathscr{A}') = c(W,\mathscr{A}) - 1$. This occurs if x and x^{-1} are in different components of $\Gamma(W,\mathscr{A})$ and at least one of these components, say that of x, is an F-component. In this case, we see that $\pi(e_3) = \pi(e_4) = 2$, that $\pi(e_1) = \pi(f_1)$ and $\pi(e_2) = \pi(f_2)$, and that no linking F-paths are created or destroyed in passing from $\Gamma(W,\mathscr{A})$ to $\Gamma(W',\mathscr{A}')$, although a linking F-path through s_1 and s_2 would be lengthened. We find, then, that $\pi(W',\mathscr{A}') = \pi(W,\mathscr{A})$, $\Delta(W',\mathscr{A}') = \Delta(W,\mathscr{A})$, and that the word W' is prereduced.

This concludes our proof of Lemma 3.2.

Proof of Theorem 3.1. To verify the "if" assertion of Theorem 3.1, we examine the steps of the $D_{(W,\mathscr{A})}$-process. Each of these steps yields a word equivalent to the word on which it acted, with the exception of case (i) of removal of a reducible parameter and application of τ-maps. Even in these exceptional cases, the image set of the resulting word is contained in the image set of the original word. We thus find that if $(W',\mathscr{A}') \in D_{(W,\mathscr{A})}$, $\mathrm{Im}(W',\mathscr{A}') \subseteq \mathrm{Im}(W,\mathscr{A})$, verifying this assertion of Theorem 3.1.

The proof of the "only if" portion of Theorem 3.1 requires somewhat more work. We shall show, given that U is an image of $(W,\mathscr{A})$ under (π,ϕ), how to construct $(W',\mathscr{A}')$ in $D_{(W,\mathscr{A})}$ and (π',ϕ') such that U is the c-free image of $(W',\mathscr{A}')$ under (π',ϕ'). We first observe that if $(V,\mathscr{T})$ is a word that is produced by applying steps of the $D_{(W,\mathscr{A})}$-process, $(V,\mathscr{T})$ is quadratic, $D_{(V,\mathscr{T})} \subseteq D_{(W,\mathscr{A})}$, and if $(V,\mathscr{T})$ is fully reduced, $(V,\mathscr{T}) \in D_{(W,\mathscr{A})}$.

Our proof proceeds by induction on the following five parameters of induction, listed in order of priority: (1) $L((W,\mathscr{A}),(\pi,\phi))$, the length of the formal image of $(W,\mathscr{A})$ under (π,ϕ), (2) $|W|_H$, (3) $|W|_P$, (4) $|W|_F$, and (5) the number of distinct parameters, λ_i, involved in $(W,\mathscr{A})$. If $(W,\mathscr{A})$ is not fully reduced or if U is not a c-free image of $(W,\mathscr{A})$ under (π,ϕ), we shall show that we can use steps of the $D_{(W,\mathscr{A})}$-process to produce a word $(W',\mathscr{A}')$ such that U is an image of $(W',\mathscr{A}')$ under a retractive mapping (π',ϕ') and such that replacing $(W,\mathscr{A})$ and (π,ϕ) by $(W',\mathscr{A}')$ and (π',ϕ') reduces at least one parameter of induction without increasing parameters of higher priority. Since lexicographic ordering is a well-ordering on $\mathbb{N}^5$, this will prove the theorem.

First suppose that $(W,\mathscr{A})$ is not prereduced. If W has a subword yy^{-1} with y an F-letter or H-letter, we form W' by deleting that subword from W. If W has a subword $(A \uparrow f)(B \uparrow g)$ in which the two parametric occurrences can be "consolidated" into $(A \uparrow f + \epsilon g)$, we form W' by performing this consolidation. In all cases, $\mathscr{A}' = \mathscr{A}$ and $(\pi',\phi') = (\pi,\phi)$. None of these moves increases any parameter of induction, but we find that either $|W'|_H < |W|_H$, $|W'|_P < |W|_P$, or $|W'|_F < |W|_F$. We may assume, then, that $(W,\mathscr{A})$ is prereduced.

Next we see whether or not $(W,\mathscr{A})$ has an occurrence $(L,(A \uparrow f),R)$ with $\pi(f) < 0$. If such occurrences exist, we dualize them. This plainly has no effect on the parameters of induction. We may thus assume henceforth that $(W,\mathscr{A})$ is consistent, in fact, that $\pi(f) \geq 0$ for each exponential polynomial of $(W,\mathscr{A})$. We shall refer to this last property as having "nonnegative exponents" (under π).

Note that Lemma 3.2(b) tells us that if $(W,\mathscr{A})$ is prereduced and $(W',\mathscr{A}')$ arises from a step in the $D_{(W,\mathscr{A})}$-process, then $(W',\mathscr{A}')$ is prereduced unless $\Pi(W',\mathscr{A}') + \Delta(W',\mathscr{A}') + |W'|_H < \Pi(W,\mathscr{A}) + \Delta(W,\mathscr{A}) + |W|_H$. In this last case, the $D_{(W,\mathscr{A})}$-process allows us to return to the prereduction and dualization steps. We shall take care in this proof that if $(W,\mathscr{A})$ is prereduced and has nonnegative exponents under π, the word $(W',\mathscr{A}')$ that we produce will have nonnegative exponents under π'.

Suppose now that $(W,\mathscr{A})$ is prereduced and has nonnegative exponents but is not reduced. Here, and throughout the proof of this theorem, we let $(W',\mathscr{A}')$ and (π',ϕ') agree with $(W,\mathscr{A})$ and (π,ϕ) except as specified. If W has an F-reducible subword xy, we delete all occurrences of $y^{\pm 1}$ from W and let $x\phi' = (x\phi)(y\phi)$. If W has a P-reducible subword $(A \uparrow f)$, we simply delete all occurrences of this subword from W. If $(W,\mathscr{A})$ has a reducible parameter λ_i, we eliminate λ_i by the means described in Section 2 (in case (i), replacing λ_i by $\pi(\lambda_i)$), we let $\phi' = \phi$, and we define π' as follows: in cases (i) and (ii), $\pi' = \pi$; in case (iii), $\pi'(\lambda_i) = -\pi(g)/a$; in case (iv),

$\pi'(\lambda_j) = (a\pi(\lambda_i) + b\pi(\lambda_j))/c$. After elimination of a reducible parameter, we return the set $\mathcal{S}'$ to standard form. We note that in each of the above situations, no parameter of induction increases, while either F-length, P-length, or the number of parameters drops. Further, $(W',\mathcal{S}')$ has nonnegative exponents under π'.

Given that $(W,\mathcal{S})$ is now reduced and has nonnegative exponents under π, we ask if $(W,\mathcal{S})$ has "positive exponents" under π, that is, if $\pi(f) > 0$ for each exponential polynomial f of $(W,\mathcal{S})$. If $(L,(A \uparrow f),R)$ is an occurrence in W and if $\pi(f) = 0$, we form $(W',\mathcal{S}')$ by applying $\tau_{(L,(A\uparrow f),R)}$ to $(W,\mathcal{S})$. Clearly $\mathcal{S}' = \mathcal{S} \cup \{f = 0\}$ is consistent since it has π as a solution, and $(W',\mathcal{S}')$ has nonnegative exponents under π. The first two parameters of induction are unchanged by this transformation, while the third parameter, P-length, drops.

Now suppose that $(W,\mathcal{S})$ is reduced and has positive exponents, but has $x\phi = 1$ for some F-letters x occurring in W. In this event, we form $(W',\mathcal{S}')$ by applying τ_x to $(W,\mathcal{S})$. This yields a word with positive exponents under π and does not change the first three parameters of induction, while decreasing the fourth parameter, F-length.

We may now assume that $(W,\mathcal{S})$ is reduced, has positive exponents, and that $x\phi \neq 1$ for each F-letter x occurring in W. If U is not a c-free image of $(W,\mathcal{S})$ under (π,ϕ), then it must be that W has a subword st such that there is cancellation between the images of s and of t under (π,ϕ). We consider each of the possibilities for s and t.

Not both s and t could be H-letters, since $(W,\mathcal{S})$ is prereduced. If one is an H-letter and the other is a P-letter, or if both are P-letters, we may apply an HP-, PH-, or PP-cancellation to st, yielding a word $(W',\mathcal{S})$ which has nonnegative exponents and with $L((W',\mathcal{S}),(\pi,\phi)) = L((W,\mathcal{S}),(\pi,\phi)) - 2$.

We suppose then that at least one of s and t, say s, is an F-letter x. If t is also an F-letter y, y cannot be x^{-1} since $(W,\mathcal{S})$ is prereduced. We let $(W',\mathcal{S}')$ be the result of applying $\rho_{x,y^{-1},z}$ to $(W,\mathcal{S})$, where $z^{\pm 1}$

does not occur in $(W,\mathscr{A})$, let $\pi' = \pi$, and let ϕ' agree with ϕ except that $x\phi' = A$, $y\phi' = B$, and $z\phi' = a$, where $x\phi \equiv Aa$ and $y\phi \equiv a^{-1}B$. If t is an H-letter a, we let $(W',\mathscr{A}')$ be the image of $(W,\mathscr{A})$ under $\mu_{x,a^{-1}}$ and define $x\phi'$ to be A, where $x\phi \equiv Aa^{-1}$. If t is a P-letter $(B \uparrow f)$, we let $(W',\mathscr{A}')$ result from applying $\nu_{x,(A \uparrow \lambda_i)}$ to $(W,\mathscr{A})$, where $B \equiv a^{-1}A_0^{-1}$, $A \equiv aA_0$, and λ_i does not occur in $(W,\mathscr{A})$, and let $\pi'(\lambda_i) = 1$ and $x\phi' = C$ where $x\phi \equiv Ca$. The cases that t is an F-letter are similar. In all cases, we find that $(W',\mathscr{A}')$ has nonnegative exponents under π' and that $L((W',\mathscr{A}'),(\pi',\phi')) < L((W,\mathscr{A}),(\pi,\phi))$.

We have shown that we can produce a reduced word $(W',\mathscr{A}')$ by the $D_{(W,\mathscr{A})}$-process and a retractive mapping (π',ϕ') such that U is the c-free image of $(W',\mathscr{A}')$ under (π',ϕ'). An inspection of the definition of "fully reduced" shows that such a word $(W',\mathscr{A}')$ is fully reduced. Finally, since the prereduction process does not change a word which is already prereduced, and since for a reduced, consistent word the identity transformation is both an admissible dualization and an "admissible step," any fully reduced word which arises in the $D_{(W,\mathscr{A})}$-process is output by the process. Thus $(W',\mathscr{A}')$ is in $D_{(W,\mathscr{A})}$ and our proof of Theorem 3.1 is complete.

§4. FINITENESS OF $D_{(W,\mathscr{A})}$

Our principal goal in this final section is to show that for a quadratic word $(W,\mathscr{A})$, the set $D_{(W,\mathscr{A})}$ is finite and effectively calculable. To do this, we show that the set $\mathcal{D}_{(W,\mathscr{A})}$ of all words produced by the $D_{(W,\mathscr{A})}$-process is finite. We compute bounds depending on $(W,\mathscr{A})$ on the F-length, H-length, and P-length of words $(W',\mathscr{A}')$ in $\mathcal{D}_{(W,\mathscr{A})}$ (Lemma 4.1), a bound on the sum of the absolute values of the coefficients of parameters in the polynomials of $(W',\mathscr{A}')$ (Lemma 4.4), and a bound on the sum of the absolute values of the constant terms in the polynomials of $(W',\mathscr{A}')$ (Lemma 4.8). We are then in a position to show that $\mathcal{D}_{(W,\mathscr{A})}$ is finite and that $D_{(W,\mathscr{A})}$ is finite and effectively calculable (Theorem 4.9). As we observed at the beginning of Section 3, this and Theorem 3.1 show that one can effectively determine whether or not a quadratic parametric equation over a free group has a solution.

Lemma 4.1. If $(W,\mathscr{A})$ is quadratic and $(W',\mathscr{A}')$ is in $\mathscr{D}_{(W,\mathscr{A})}$, then $|W'|_F \leq \max\{|W|_F, 6\Delta(W,\mathscr{A}) + 6\}$, $|W'|_H \leq |W|_H$, and $|W'|_P \leq \max\{|W|_P, \frac{1}{2}\Pi(W,\mathscr{A}) + 2\}$.

Before proving Lemma 4.2, we establish the following result.

Lemma 4.2. If $(W,\mathscr{A})$ is a reduced quadratic word, then $|W|_F = 2f(W,\mathscr{A}) \leq 6\Delta(W,\mathscr{A}) + 4$ and $|W|_P \leq \frac{1}{2}\Pi(W,\mathscr{A}) + 1$.

Proof. If $W \equiv 1$ or $W \equiv x^2$, we find that $\Delta(W,\mathscr{A}) = \Pi(W,\mathscr{A}) = 0$. Clearly the inequalities of Lemma 4.2 hold in these cases. We suppose, then, that W is neither 1 nor x^2.

If $\Gamma(W,\mathscr{A})$ has an F-component with only one vertex, an F-letter x, then xx^{-1} is a subword (cyclically) of W. Since W is reduced, it follows that $W \equiv x^{-1}W_0x$. If $\Gamma(W,\mathscr{A})$ has an F-component with two vertices, F-letters x and y, these F-letters occur only in subwords (cyclically) $(xy^{-1})^{\pm 1}$ of W. Since $W \not\equiv x^2$, either $W \equiv yW_0(xy^{-1})^{\pm 1}W_1x^{-1}$ or $W \equiv x^{-1}W_0(xy)^{\pm 1}W_1y$. In any event, we see that $\Gamma(W,\mathscr{A})$ can have at most one F-component with fewer than three vertices. Thus,

$$c(W,\mathscr{A}) \leq (2f(W,\mathscr{A}) - 1)/3 + 1 = (2f(W,\mathscr{A}) + 2)/3$$

and so

$$\Delta(W,\mathscr{A}) = f(W,\mathscr{A}) - c(W,\mathscr{A}) \geq f(W,\mathscr{A}) - (2f(W,\mathscr{A}) + 2)/3 = (f(W,\mathscr{A}) - 2)/3 ,$$

from which the first assertion of Lemma 4.2 follows.

To establish the second assertion of the lemma, recall that by definition,

$$\Pi(W,\mathscr{A}) = \Sigma\, \pi(e) - 4c(W,\mathscr{A}) ,$$

where the indicated sum is over all edges e of $\Gamma(W,\mathscr{A})$. If C is an F-component of $\Gamma(W,\mathscr{A})$, C is a simple closed edge-path and $\pi(e) = 2$ for each edge e of C. It follows that C can make a net negative contribution to $\Pi(W,\mathscr{A})$ only if it has just one vertex. Since there can be at most one such component in $\Gamma(W,\mathcal{A})$, we have

$$\Pi(W,\mathcal{A}) \geq \sum_E \pi(e) - 2 ,$$

where E is the collection of edges of $\Gamma(W,\mathcal{A})$ which do not lie in F-components. If e is an edge of $\Gamma(W,\mathcal{A})$ which has an endpoint on a parametric cycle and e is not $e(s,t^{-1})$ where $W \equiv tW_0 s$, $\pi(e) \geq 1$ and, since $(W,\mathcal{A})$ is reduced, $\pi(e) = 2$ if both endpoints of e lie on parametric cycles. Since each occurrence of a P-letter $(A \uparrow f)$ in W is responsible for one edge with terminal vertex on $\partial(A^{-1})$ and another with initial vertex on ∂A in $\Gamma(W,\mathcal{A})$, we find that $\sum_E \pi(e) \geq 2|W|_P - 2$, with $\sum_E \pi(e) \geq 2|W|_P$ unless W begins and ends with P-letters. In this last case, $\Gamma(W,\mathcal{A})$ has no F-components with fewer than three vertices, so that in any event

$$\Pi(W,\mathcal{A}) \geq 2|W|_P - 2 .$$

This concludes the proof of Lemma 4.2.

Proof of Lemma 4.1.

First note that no step in the $D_{(W,\mathcal{A})}$-process increases H-length, so $|W'|_H \leq |W|_H$. To establish the other two assertions of the lemma, consider a sequence of words

$$(W,\mathcal{A}) \equiv (W_0,\mathcal{A}_0), (W_1,\mathcal{A}_1), \ldots, (W_n,\mathcal{A}_n) \equiv (W',\mathcal{A}')$$

where $(W_i,\mathcal{A}_i)$ results from applying one step of the $D_{(W,\mathcal{A})}$-process to $(W_{i-1},\mathcal{A}_{i-1})$. Observe that the only step in the $D_{(W,\mathcal{A})}$-process that increases F-length is application of a ρ-map, and that the only step that may increase P-length is application of a ν-map. Since ν-maps and ρ-maps can be applied only to reduced words, we find that if none of $(W_0,\mathcal{A}_0),\ldots,(W_{n-1},\mathcal{A}_{n-1})$ is reduced, $|W'|_F \leq |W|_F$ and $|W'|_P \leq |W|_P$.

Suppose, then, that at least one of $(W_0,\mathcal{A}_0),\ldots,(W_{n-1},\mathcal{A}_{n-1})$ is reduced and let k be the largest integer less than n such that $(W_k,\mathcal{A}_k)$ is reduced. Now $|W_{k+1}|_F \leq |W_k|_F + 2$ and $|W_{k+1}|_P \leq |W_k|_P + 1$ and, as above, $|W'|_F \leq |W_{k+1}|_F$ and $|W'|_P \leq |W_{k+1}|_P$, so $|W'|_F \leq |W_k|_F + 2$ and

$|W'|_P \leq |W_k|_P + 1$. By Lemma 4.2, $|W_k|_F \leq 6\Delta(W_k,\mathscr{A}_k) + 4$ and $|W_k|_P \leq \frac{1}{2}\Pi(W_k,\mathscr{A}_k) + 1$, while $\Delta(W_k,\mathscr{A}_k) \leq \Delta(W,\mathscr{A})$ and $\Pi(W_k,\mathscr{A}_k) \leq \Pi(W,\mathscr{A})$ by Lemma 3.2a). (If W_k is empty, W' is also empty and the conclusion of the lemma clearly holds.) Combining the above inequalities, we find that

$$|W'|_F \leq 6\Delta(W,\mathscr{A}) + 6 \quad \text{and} \quad |W'|_P \leq \tfrac{1}{2}\Pi(W,\mathscr{A}) + 2 .$$

This concludes the proof of Lemma 4.1.

Our next step is to bound the coefficients of the parameters in the polynomials of a word $(W',\mathscr{A}')$ in $\mathscr{D}_{(W,\mathscr{A})}$. To do this, we must define some new terms. Let $(W,\mathscr{A})$ be a quadratic word with exponential polynomials $f_1,\ldots,f_m$ and constraint polynomials $g_1,\ldots,g_n$, that is, $\mathscr{E} = \{g_1 = 0,\ldots,g_n = 0\}$ where $\mathscr{E}$ is the set of equations of $\mathscr{A}$. We define $|\mathscr{E}|$ to be n and let $c_j(f)$ be the sum of the coefficients of parameter λ_j in polynomial f. We say that λ_j <u>occurs</u> in f if $c_j(f) \neq 0$. We define

$$\Lambda_j(W,\mathscr{A}) = \sum_{i=1}^{m} |c_j(f_i)| + \sum_{i=1}^{n} |c_j(g_i)|$$

and
$$\Lambda(W,\mathscr{A}) = \Sigma\, \Lambda_j(W,\mathscr{A}) ,$$

the last sum taken over all j such that λ_j occurs in one of the polynomials of $(W,\mathscr{A})$. We say that a parameter is <u>quadratic</u> in $(W,\mathscr{A})$ if λ_j does not occur in $\mathscr{J}$, the set of inequalities of $\mathscr{A}$, and either λ_j occurs in exactly two polynomials p_1 and p_2 of $(W,\mathscr{A})$ with $|c_j(p_1)| = |c_j(p_2)|$ and, for every congruence $e\lambda_j + h \equiv 0 \pmod{m}$ in which λ_j appears, $c_j(p_i)x \equiv e \pmod{m}$ has a solution (λ_j <u>split quadratic</u>) or λ_j occurs in exactly one polynomial p of $(W,\mathscr{A})$ and, for every congruence $e\lambda_j + h \equiv 0 \pmod{m}$ in which λ_j appears, either $c_j(p)x \equiv e \pmod{m}$ has a solution or $c_j(p)$ is even and $(c_j(p)/2)x \equiv e \pmod{m}$ has a solution (λ_j <u>nonsplit quadratic</u>). We use $Q(W,\mathscr{A})$ for the number of quadratic parameters in $(W,\mathscr{A})$.

<u>Lemma 4.3</u>. <u>If</u> $(W,\mathscr{A})$ <u>is a reduced quadratic word</u>, $Q(W,\mathscr{A}) \leq |W|_P{}^2 + |\mathscr{E}|$.

Proof: We begin by observing that no split quadratic parameter can occur in a constraint polynomial, for such a parameter would be reducible by clause (iii) of the definition.

We first bound the number of split quadratic parameters of $(W,\mathscr{A})$. If that number exceeded $2\binom{|W|_P}{2} = |W|_P(|W|_P - 1)$, there would be a pair of exponential polynomials f_r and f_s and split quadratic parameters $\lambda_1,\lambda_2,\lambda_3$ such that λ_1,λ_2, and λ_3 occur in both f_r and f_s. Since $c_i(f_r) = \varepsilon_i c_i(f_s)$ with $\varepsilon_i = \pm 1$ for $i = 1,2,3$, at least two of $\varepsilon_1,\varepsilon_2,\varepsilon_3$, say ε_1 and ε_2, are the same. We find that λ_1, taken with λ_2, is a redundant parameter of type (iv), violating our hypothesis that $(W,\mathscr{A})$ is reduced.

It is clear that no pair of nonsplit quadratic parameters could occur in the same polynomial of $(W,\mathscr{A})$, for this would give a reducible parameter of type (iv) in $(W,\mathscr{A})$. Since the number of polynomials of $(W,\mathscr{A})$ is $|W|_P + |\mathscr{E}|$, we combine this result with that of the previous paragraph to obtain the conclusion of the lemma.

Lemma 4.4. If $(W,\mathscr{A})$ is quadratic and $(W',\mathscr{A}')$ is in $\mathscr{D}_{(W,\mathscr{A})}$, then $\Lambda(W',\mathscr{A}') \leq \Lambda(W,\mathscr{A}) + \frac{1}{2}(\Pi(W,\mathscr{A}))^2 + 2\Pi(W,\mathscr{A}) + 2|W|_P + 2|\mathscr{E}| + 4.$

Proof: Consider a sequence $(W,\mathscr{A}) \equiv (W_0,\mathscr{A}_0),\ldots,(W_n,\mathscr{A}_n) \equiv (W',\mathscr{A}')$ where $(W_i,\mathscr{A}_i)$ results from $(W_{i-1},\mathscr{A}_{i-1})$ by one step of the $D_{(W,\mathscr{A})}$-process. Examining the rules for the $D_{(W,\mathscr{A})}$-process, we find that we can have $\Lambda_j(W_i,\mathscr{A}_i) > \Lambda_j(W_{i-1},\mathscr{A}_{i-1})$ only if $(W_i,\mathscr{A}_i)$ is the result of applying a map $\nu_{x,(A\uparrow\lambda_j)}$ to $(W_{i-1},\mathscr{A}_{i-1})$. Thus if no ν-maps are used in producing $(W',\mathscr{A}')$, then we have $\Lambda(W',\mathscr{A}') \leq \Lambda(W,\mathscr{A})$. Suppose then that k is the largest integer less than n such that $(W_{k+1},\mathscr{A}_{k+1})$ is the result of applying a map $\nu_{x,(A\uparrow\lambda_j)}$ to $(W_k,\mathscr{A}_k)$. It follows, then, that

$$\Lambda(W',\mathscr{A}') \leq \Lambda(W_{k+1},\mathscr{A}_{k+1}) \leq \Lambda(W_k,\mathscr{A}_k) + 2 .$$

Suppose, now, that $\Lambda_i(W_k,\mathscr{A}_k) > 0$. If $\Lambda_i(W,\mathscr{A}) > 0$ as well, $\Lambda_i(W_k,\mathscr{A}_k) \leq \Lambda_i(W,\mathscr{A})$. If $\Lambda_i(W,\mathscr{A}) = 0$, λ_i was introduced at some stage by a ν-map and so $\Lambda_i(W_k,\mathscr{A}_k) \leq 2$ and λ_i does not occur in $\mathscr{I}$, the set of inequalities of $\mathcal{A}_0, \mathcal{A}_1, \ldots, \mathcal{A}_n$. It follows that λ_i is a quadratic parameter. Using Lemma 4.3, then, we have

$$\begin{aligned} \Lambda(W', \mathcal{A}') &\leq \Lambda(W_k, \mathcal{A}_k) + 2 \leq \\ \Lambda(W, \mathcal{A}) &+ 2Q(W_k, \mathcal{A}_k) + 2 \leq \\ \Lambda(W, \mathcal{A}) &+ 2\,|W_k|_P^2 + 2\,|\mathcal{E}_k| + 2 . \end{aligned}$$

Since $(W_k, \mathcal{A}_k)$ is reduced, each equation in $\mathcal{E}_k$ must contain at least one nonquadratic parameter. We saw in the proof of Lemma 4.3 that no split quadratic parameter can occur in a polynomial of $\mathcal{E}_k$. If there were an equation in $\mathcal{E}_k$ whose only parameters were nonsplit quadratic parameters, at least one of those parameters would be reducible by either clause (ii) or clause (iv) of the definition. The $D_{(W,\mathcal{A})}$-process never increases the number of polynomials of a word which contain nonquadratic parameters, so $|\mathcal{E}_k| \leq |W|_P + |\mathcal{E}|$. Using this fact, the bound $|W_k|_P \leq \frac{1}{2}\Pi(W_k, \mathcal{A}_k) + 1$ provided by Lemma 4.2, and the fact that $\Pi(W_k, \mathcal{A}_k) \leq \Pi(W, \mathcal{A})$, we find that

$$\begin{aligned} \Lambda(W', \mathcal{A}') &\leq \Lambda(W, \mathcal{A}) + 2\,|W_k|_P^2 + 2\,|\mathcal{E}_k| + 2 \\ &\leq \Lambda(W, \mathcal{A}) + 2(\tfrac{1}{2}\Pi(W, \mathcal{A}) + 1)^2 + 2\,|W|_P + 2\,|\mathcal{E}| + 2 \\ &\leq \Lambda(W, \mathcal{A}) + \tfrac{1}{2}(\Pi(W, \mathcal{A}))^2 + 2\Pi(W, \mathcal{A}) + 2\,|W|_P + 2\,|\mathcal{E}| + 4 . \end{aligned}$$

Our next task is to establish a bound on the sum of the absolute values of the constant terms of the polynomials of a word $(W', \mathcal{A}')$ in $\mathcal{D}_{(W,\mathcal{A})}$; we denote this sum by $K(W', \mathcal{A}')$. To do this we require a new integer-valued function LS on quadratic words, which we call the *Lyndon-Schützenberger number*. Our application of this function will be based on the following lemma.

Lemma 4.5 (Lyndon, Schützenberger [15]). *If U and V are cyclically reduced words in a free group H and if some power of U and some power of V have a common initial segment of length $|U| + |V|$, then U and V are powers of a common element, that is, there are a Z in H and integers m and n such that $U \equiv Z^m$ and $V \equiv Z^n$.*

Now suppose that $(W, \mathcal{A})$ is a quadratic word and that γ is an F-path whose initial vertex Q_1 is on a parametric cycle ∂A and whose terminal vertex Q_2 is on a parametric cycle ∂B. We define $LS(\gamma)$ to be zero if Q_1 and Q_2 are congruent vertices, and otherwise define $LS(\gamma)$ to be the length of the longest word produced in common by reading ∂A and ∂B in the clockwise direction from Q_1 and Q_2, respectively. Note that by Lemma 4.5, $LS(\gamma) < |A| + |B|$. We define $LS(W,\mathcal{A})$ by

$$LS(W,\mathcal{A}) = \Sigma\, LS(\gamma) \ ,$$

where the sum is taken over all unoriented F-paths of $\Gamma(W,\mathcal{A})$ which have both endpoints on parametric cycles. For example, if

$$W \equiv (abcdbc \uparrow f)x^{-1}zxy^{-1}zy(c^{-1}b^{-1}a^{-1} \uparrow g)$$

where $\mathcal{A}$ implies that $f \equiv 3 \pmod 6$ and $g \equiv 0 \pmod 3$, $\Gamma(W,\mathcal{A})$ is shown in Figure 5.

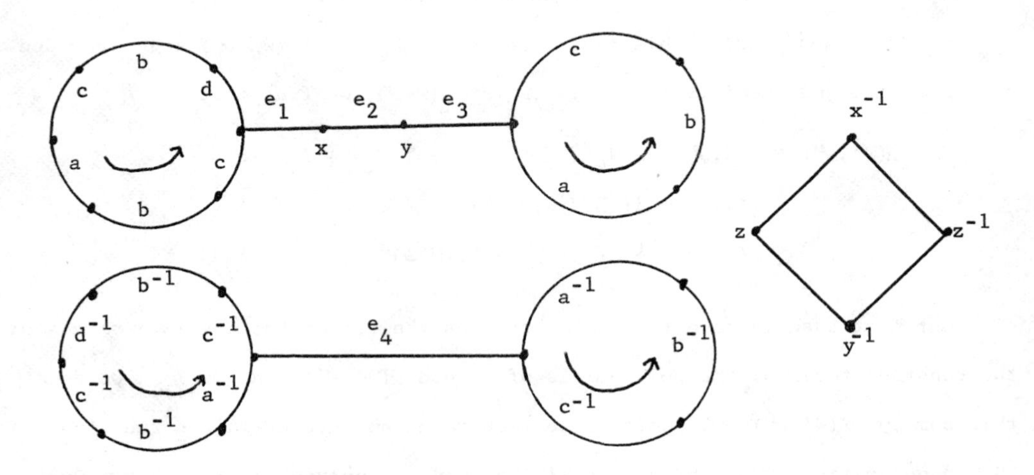

Figure 5

Here $LS(e_1e_2e_3) = 5$ and $LS(e_4) = 3$, so $LS(W,\mathcal{A}) = 8$.

A critical fact about the function LS is the following.

Lemma 4.6. If $(W,\mathcal{A})$ *is a quadratic word,* $(W',\mathcal{A}')$ *is in* $\mathcal{D}_{(W,\mathcal{A})}$, *and* $(W'',\mathcal{A}'')$ *results from* $(W',\mathcal{A}')$ *by one step of the* $D_{(W,\mathcal{A})}$*-process,* $LS(W'',\mathcal{A}'') \leq LS(W',\mathcal{A}')$ *unless the step was a dualization or*

$$\Pi(W'',\mathcal{A}'') + \Delta(W'',\mathcal{A}'') + |W''|_H < \Pi(W',\mathcal{A}') + \Delta(W',\mathcal{A}') + |W'|_H \ .$$

Proof: Our proof is similar to that of Lemma 3.2 in that we consider the possible steps of the $D_{(W,\mathscr{A})}$-process by which $(W'',\mathscr{A}'')$ is produced from $(W',\mathscr{A}')$. As we did before, we provide the details for a few representative steps in the process, leaving verification in the remaining cases to the reader. In referring to the proof of Lemma 3.2, note that $(W',\mathscr{A}')$ and $(W'',\mathscr{A}'')$ are now playing the roles played by $(W,\mathscr{A})$ and $(W',\mathscr{A}')$ in that proof. Also, recall that by Lemma 3.2a), a decrease in the value of any one of Π, Δ, or H-length implies a decrease in their sum.

Suppose that we have deleted a P-reducible subword $(A \uparrow f)$ from $(W',\mathscr{A}')$, replacing a subword (cyclically) $s(A \uparrow f)t^{-1}$ by st^{-1} in $(W'',\mathscr{A}'')$. This situation is depicted in Figure 2. Resuming the notation of Figure 2, let γ_1, γ_2, and η be maximal F-paths containing e_1, e_2, and f_1, respectively. As that portion of the proof of Lemma 3.2 shows, we can have $\Pi(W'',\mathscr{A}'') + \Delta(W'',\mathscr{A}'') + |W''|_H = \Pi(W',\mathscr{A}') + \Delta(W',\mathscr{A}') + |W'|_H$ only if γ_1, γ_2 and η all link pairs of congruent vertices or if exactly one of γ_1 and γ_2, say γ_1, links a pair of congruent vertices and γ_2 and η do not. In the former case, $LS(\gamma_1) = LS(\gamma_2) = LS(\eta) = 0$, while in the latter case, since vertices P and Q are congruent, $LS(\gamma_1) = 0$ and $LS(\gamma_2) = LS(\eta)$. In any event, $LS(W'',\mathscr{A}'') = LS(W',\mathscr{A}')$.

Next suppose that $(W'',\mathscr{A}'')$ results from the application of an admissible map $\nu_{x,(A \uparrow \lambda_i)}$ as shown in Figure 3. In cases (a) and (b), all displayed F-paths in $\Gamma(W',\mathscr{A}')$ and $\Gamma(W'',\mathscr{A}'')$ have Lyndon-Schützenberger number zero. In case (c), the F-path through y^{-1} is shortened, while an F-path γ through x with $LS(\gamma) = 0$ is created. In each of these cases, $LS(W'',\mathscr{A}'') = LS(W',\mathscr{A}')$.

Finally, suppose that $(W'',\mathscr{A}'')$ is obtained from $(W',\mathscr{A}')$ by applying an admissible map τ_x, replacing subwords $(s_1xt_1^{-1})^{\pm 1}$ and $(s_2xt_2^{-1})^{\pm 1}$ in W' by $(s_1t_1^{-1})^{\pm 1}$ and $(s_2t_2^{-1})^{\pm 1}$ in W''. As indicated in the proof of Lemma 3.2, $f(W'',\mathscr{A}'') = f(W',\mathscr{A}') - 1$, so $\Delta(W'',\mathscr{A}'') < \Delta(W',\mathscr{A}')$ unless $c(W'',\mathscr{A}'') = c(W',\mathscr{A}') - 1$. In this case, one of the vertices x and x^{-1}, say x, lies in an F-component of $\Gamma(W',\mathscr{A}')$; see Figure 4. The only possible effect on F-paths is that an F-path which passed through x^{-1} in $\Gamma(W',\mathscr{A}')$ is replaced by a longer F-path in $\Gamma(W'',\mathscr{A}'')$. Thus $LS(W'',\mathscr{A}'') = LS(W',\mathscr{A}')$, and our proof of Lemma 4.6 is finished.

Lemma 4.7. *If* $(W,\mathscr{A})$ *is a quadratic word and* $(W',\mathscr{A}')$ *is in* $\mathcal{D}_{(W,\mathscr{A})}$, *then* $LS(W',\mathscr{A}') \leq M_{LS}$ *where*

$$M_{LS} = 2\max\{|W|_P, \tfrac{1}{2}\Pi(W,\mathscr{A}) + 2\}\max\{|A| : (A \uparrow f) \text{ occurs in } W\}.$$

Proof: First note that if γ is an F-path joining vertices on ∂A and ∂B in $\Gamma(W',\mathscr{A}')$, then either γ is a linking F-path, in which case $LS(\gamma) = 0$, or $LS(\gamma) < |A| + |B|$ by Lemma 4.5. Since the "base" B of a P-letter $(B \uparrow g)$ occurring in W' must be a cyclic permutation of A or A^{-1} for some P-letter $(A \uparrow f)$ of W, it follows that

$$LS(\gamma) \leq 2\max\{|A| : (A \uparrow f) \text{ occurs in } W\}$$

for all F-paths γ of $\Gamma(W',\mathscr{A}')$. The number of F-paths γ joining vertices on parametric cycles in $\Gamma(W',\mathscr{A}')$ is bounded by $|W'|_P$, which is in turn bounded by $\max\{|W|_P, \tfrac{1}{2}\Pi(W,\mathscr{A}) + 2\}$ by Lemma 4.1. This completes the proof of Lemma 4.7.

Lemma 4.8. *If* $(W,\mathscr{A})$ *is a quadratic word and* $(W',\mathscr{A}')$ *is in* $\mathcal{D}_{(W,\mathscr{A})}$, *then*

$$K(W',\mathscr{A}') \leq K(W,\mathscr{A}) + (2M_{LS} + 1)(\Pi(W,\mathscr{A}) + \Delta(W,\mathscr{A}) + |W|_H + 1).$$

Proof: Consider a sequence

$$(W,\mathscr{A}) \equiv (W_0,\mathscr{A}_0), (W_1,\mathscr{A}_1), \ldots, (W_n,\mathscr{A}_n) \equiv (W',\mathscr{A}')$$

where each $(W_i,\mathscr{A}_i)$ is produced from $(W_{i-1},\mathscr{A}_{i-1})$ by one step of the $D_{(W,\mathscr{A})}$-process. We call a maximal subsequence on which the value of the "Π plus Δ plus H-length" function is constant a plateau. A review of the $D_{(W,\mathscr{A})}$-process shows that the only steps which can give $K(W_i,\mathscr{A}_i) > K(W_{i-1},\mathscr{A}_{i-1})$ are HP-, PH-, and PP-cancellations. Since for quadratic words $(V,\mathcal{T})$, $\Pi(V,\mathcal{T}) + \Delta(V,\mathcal{T}) + |V|_H < 0$ only if V freely reduces to the empty word, we may assume that the number of plateaus in our derivation of $(W',\mathscr{A}')$ is at most $\Pi(W,\mathscr{A}) + \Delta(W,\mathscr{A}) + |W|_H + 1$.

An HP- or PH-cancellation gives $K(W_i,\mathscr{A}_i) = K(W_{i-1},\mathscr{A}_{i-1}) + 1$ and results in a change of plateau. The total increase in the value of K from such moves is at most $\Pi(W,\mathscr{A}) + \Delta(W,\mathscr{A}) + |W|_H$. A PP-cancellation gives $K(W_i,\mathscr{A}_i) \leq K(W_{i-1},\mathscr{A}_{i-1}) + 2$ and also $LS(W_i,\mathscr{A}_i) = LS(W_{i-1},\mathscr{A}_{i-1}) - 1$. Since by Lemma 4.6, the value of LS can be increased within a plateau only by dualizations, and since the structure of the $D_{(W,\mathscr{A})}$-process requires that within a plateau any dualizations precede any PP-cancellations, Lemma 4.7 tells us that the increase in K within a plateau is bounded by $2M_{LS}$. The total increase in the value of K in passing from $(W,\mathscr{A})$ to $(W',\mathscr{A}')$, then, is bounded by

$$(2M_{LS} + 1)(\Pi(W,\mathscr{A}) + \Delta(W,\mathscr{A}) + |W|_H + 1) ,$$

and our proof of Lemma 4.8 is complete.

<u>Theorem 4.9</u>. <u>If</u> $(W,\mathscr{A})$ <u>is a quadratic word, the sets</u> $\mathscr{D}_{(W,\mathscr{A})}$ <u>and</u> $D_{(W,\mathscr{A})}$ <u>are finite and effectively calculable</u>.

<u>Proof</u>. We first show that $\mathscr{D}_{(W,\mathscr{A})}$ is finite. Since $D_{(W,\mathscr{A})}$ is a subset of $\mathscr{D}_{(W,\mathscr{A})}$, it follows that it is finite as well.

Since we consider words which differ only by a relabeling of F-letters to define the same element of $\mathscr{D}_{(W,\mathscr{A})}$, we may assume, using Lemma 4.1, that all F-letters occurring in words $(W',\mathscr{A}')$ of $\mathscr{D}_{(W,\mathscr{A})}$ come from the set

$$\{x_1^{\pm 1}, x_2^{\pm 1}, \ldots, x_n^{\pm 1}\},\ n = \max\{|W|_F,\ 6\Delta(W,\mathscr{A}) + 6\} .$$

The H-letters occurring in W' form a subset of the H-letters occurring in W and their inverses. If $(B \uparrow g)$ is a P-letter occurring in W', there is a P-letter $(A \uparrow f)$ in W with B a cyclic permutation of A or A^{-1}. The bounds of Lemma 4.1 then tells us that, except possibly for the exponential polynomials, the set of words W' for which $(W',\mathscr{A}')$ is in $\mathscr{D}_{(W,\mathscr{A})}$ is finite.

We now work on bounding the occurrences of parameters in words $(W',\mathscr{A}')$ of $\mathscr{D}_{(W,\mathscr{A})}$. Lemma 4.4 bounds the sum of the absolute values of the coefficients of the parameters in the exponential and constraint polynomials of $(W',\mathscr{A}')$, and hence also the number of these parameters. Lemma 4.8 puts a bound on the sum of the absolute values of the constant terms of these same polynomials. Since we have been careful never to change the set of inequalities of $\mathscr{A}$ in the $D_{(W,\mathscr{A})}$-process, we need only look at the set of congruences $\mathscr{C}'$ of a word $(W',\mathscr{A}')$ in $\mathscr{D}_{(W,\mathscr{A})}$.

New parameters and new congruences are introduced only by ν-maps. Thus, the only possible moduli of congruences of $\mathscr{C}'$ are those of congruences in $\mathscr{A}$ and lengths of bases A of P-letters $(A \uparrow f)$ occurring in $(W,\mathscr{A})$. A reduced word cannot have a parameter which occurs only in its set of congruences, and a ν-map is admissible only to reduced words. It follows that the number of parameters occurring in $\mathscr{C}'$ is bounded by the bound of Lemma 4.4 and the number of parameters of $(W,\mathscr{A})$. Since we again consider words obtained from one another by permuting labels of parameters to be the same in $\mathscr{D}_{(W,\mathscr{A})}$, we have bounded the number of possibilities for sets of exponential polynomials and constraint sets $\mathscr{A}'$ for words $(W',\mathscr{A}')$ of $\mathscr{D}_{(W,\mathscr{A})}$. This, then, completes the proof that $\mathscr{D}_{(W,\mathscr{A})}$ is finite.

To show that $\mathscr{D}_{(W,\mathscr{A})}$ is effectively calculable, we shall show that given any $(W',\mathscr{A}')$ which has been produced at any stage of the $D_{(W,\mathscr{A})}$-process, we can tell effectively which steps of the $D_{(W,\mathscr{A})}$-process are then applicable to $(W',\mathscr{A}')$. We may then produce $\mathscr{D}_{(W,\mathscr{A})}$ by starting with $(W,\mathscr{A})$ and enumerating all possible sequences of steps of the $D_{(W,\mathscr{A})}$-process, terminating a sequence whenever it produces a word that duplicates one previously produced.

Observe first that we can certainly tell whether or not W' contains an inverse pair of F-letters or H-letters or a subword $(A \uparrow f)(B \uparrow g)$ with B or B^{-1} a cyclic permutation of A. To find an integer k such that $\mathscr{A}$ implies that $f \equiv k \pmod{|A|}$, we check each of $\mathscr{A} \cup \{f \equiv k \pmod{|A|}\}$ for consistency for $k = 0,1,\ldots |A| - 1$. Checking consistency of a finite set of linear equations and inequalities is an integer linear programming problem, which is well-known to be an $\mathcal{NP}$-complete problem, and hence solvable; see, for example, [9]. (A congruence $h \equiv 0 \pmod m$ is equivalent to an equation $h - \lambda m = 0$, with λ a new parameter.) We can therefore effectively apply the prereduction procedure and tell whether or not $(W', \mathscr{S}')$ is prereduced.

To tell whether or not a sequence of dualizations is admissible for $(W', \mathscr{S}')$ we must again solve an integer linear programming problem.

One can clearly recognize F-reducible subwords and reducible parameters in $(W', \mathscr{S}')$. A subword $(A \uparrow f)$ of $(W', \mathscr{S}')$ is reducible if and only if $\mathscr{S}' \cup \{f = 0\}$ is consistent but neither $\mathscr{S}' \cup \{f > 0\}$ nor $\mathscr{S}' \cup \{f < 0\}$ is; these are integer linear programming problems. We can therefore effectively tell whether or not $(W', \mathscr{S}')$ is reduced. For a reduced word, we can readily tell whether or not HP-, PH-, or PP-cancellations, or ρ-, μ-, ν-, or τ-maps are admissible. Finally, since the functions $\Pi(W', \mathscr{S}')$, $\Delta(W', \mathscr{S}')$, and $|W'|_H$ are all effectively calculable, we can tell when the $D_{(W,\mathscr{S})}$-process allows us to return to the prereduction and dualization stages.

We have now shown that $\mathcal{D}_{(W,\mathscr{S})}$ is effectively calculable. To show the same for $D_{(W,\mathscr{S})}$, we need only notice that one can tell whether or not a reduced word is fully reduced. This completes the proof of Theorem 4.9.

REFERENCES

1. S. I. Adian, The Burnside Problem and Identities in Groups, Springer-Verlag, Berlin, Heidelberg, New York, 1979.
2. M. Anshel and P. Stebe, "The solvability of the conjugacy problem for certain HNN groups," Bull. Amer. Math. Soc. 80 (1974), 266-270.
3. M. Anshel and P. Stebe, "Conjugated powers in free products with amalgamation," Houston J. Math. 2 (1976), 139-147.
4. L. P. Comerford, Jr., "A note on power-conjugacy," Houston J. Math. 3 (1977), 337-341.
5. L. P. Comerford, Jr. and C. C. Edmunds, "Quadratic equations over free groups and free products," J. Algebra, 68 (1981), 276-297.
6. M. Davis, "Hilbert's tenth problem is unsolvable," Amer. Math. Monthly 80 (1973), 233-269.
7. C. C. Edmunds, "On the endomorphism problem for free groups II," Proc. London Math. Soc. (3) 38 (1979), 153-168.

8. A. H. M. Hoare, A. Karrass, and D. Solitar, "Subgroups of finite index of Fuchsian groups," Math. Z. 120 (1971), 289-298.

9. J. E. Hopcroft and J. D. Ullman, Introduction to Automata Theory, Languages, and Computation, Addison-Wesley, Reading, Mass., 1979.

10. L. Larsen, "The conjugacy problem and cyclic HNN constructions," J. Austral. Math. Soc. 23 (Series A) (1977), 385-401.

11. L. Larsen, "On the computability of conjugate powers in finitely generated Fuchisan groups," Acta Math. 139 (1977), 267-291.

12. S. Lipschutz and C. F. Miller III, "Groups with certain solvable and unsolvable decision problems," Comm. Pure Appl. Math. 24 (1971), 7-15.

13. R. C. Lyndon, "Equations in free groups," Trans. Amer. Math. Soc. 96 (1960), 445-457.

14. R. C. Lyndon, "Groups with parametric exponents," Trans. Amer. Math. Soc. 96 (1960), 518-533.

15. R. C. Lyndon and M. P. Schützenberger, "The equation $a^M = b^N c^P$ in a free group," Michigan Math. J. 9 (1962), 289-298.

16. P. E. Schupp, "On the substitution problem for free groups," Proc. Amer. Math. Soc. 23 (1969), 421-423.

17. J. H. C. Whitehead, "On certain sets of elements in a free group," Proc. London Math. Soc. (2) 41 (1936), 48-56.

DIVISION OF SCIENCE
UNIVERSITY OF WISCONSIN-PARKSIDE
KENOSHA, WISCONSIN 53141

DEPARTMENT OF MATHEMATICS
MOUNT SAINT VINCENT UNIVERSITY
HALIFAX, NOVA SCOTIA B3M 2J6
CANADA

Received April 10, 1982

Contemporary Mathematics
Volume 33, 1984

FINITE GROUPS OF OUTER AUTOMORPHISMS OF A FREE GROUP

Marc Culler

for Roger Lyndon

0. Introduction

The known results on periodicity properties of automorphisms of free groups appear to come from two different directions. On one hand, the work of Nielsen [N], Jaco-Shalen [JS], and Thurston [T] on the structure of surface homeomorphisms gives considerable information about those automorphisms which are induced by homeomorphisms of bounded surfaces. Using Kerckhoff's realization theorem [K], one can use a geometric approach to study certain finite subgroups of outer automorphisms. However, these geometric automorphisms form a restricted class of automorphisms. On the other hand, Dyer and Scott [DS] use the structure theorem for free-by-finite groups to study arbitrary automorphisms of finite order.

In this note we use a geometric approach to study arbitrary finite subgroups of the outer automorphism group of a free group. Our main observation is that the structure theorem for free-by-finite groups actually implies a 1-dimensional realization theorem. We show that any finite group of outer automorphisms of a free group is realized by a group of homeomorphisms of a graph. We use this to study the problem of lifting groups of outer automorphisms to groups of automorphisms, and to understand the Dyer-Scott results from a geometric perspective.

© 1984 American Mathematical Society
0271-4132/84 $1.00 + $.25 per page

I thank Bill Dunbar, Philip Harrington, John Hempel, and Ken Brown for helpful discussions regarding this work.

1. Preliminaries

By a graph we shall mean a connected 1-dimensional CW-complex. An automorphism of a graph will be a cellular homeomorphism which is linear on the edges and has no inversions -- that is, no edge is mapped to itself with orientation reversed. Of course, any cellular homeomorphism which is linear on edges can be made into an automorphism by introducing a vertex at the midpoint of each edge that is inverted.

A simply-connected graph will be called a tree. Any graph which is not a tree contains a unique subgraph, which we will call its core, that is the smallest subgraph for which the inclusion map is a homotopy equivalence. The core of X can be characterized, for example, as the union of the edges e of X such that $X\text{-Int}(e)$ is either connected or has no simply-connected component. The core of X is compact if and only if $\pi_1(X)$ is finitely generated.

If X is any arc-connected space then any homeomorphism of X induces a unique outer automorphism of $\pi_1(X)$ -- the indeterminacy up to inner automorphisms arising because the homeomorphism may move the base point. Thus there is a homomorphism $I_*: \mathcal{H}(X) \to \mathrm{Out}(\pi_1(X))$, where $\mathcal{H}(X)$ denotes the group of homeomorphisms of X. If π is a group and Γ a subgroup of $\mathrm{Out}(\pi)$ then we will say that Γ is realized by a group of homeomorphisms of X if there exists a subgroup G of $\mathcal{H}(X)$ and an isomorphism $\phi: \pi_1(X) \to \pi$ such that $\phi_* \circ I_*$ restricts to an isomorphism from G to Γ, where $\phi_*: \mathrm{Out}(\pi_1(X)) \to \mathrm{Out}(\pi)$ is induced by ϕ.

2. A realization theorem.

The following theorem is a 1-dimensional analogue of Kerckhoff's realization theorem [K].

2.1 Theorem. Let F be a free group. Any finite subgroup of $\mathrm{Out}(F)$ is realized by a group of automorphisms of a graph. If F is finitely generated then the graph can be taken to be compact.

Proof. Let Γ be a finite subgroup of $\mathrm{Out}(F)$, and let $\tilde{\Gamma}$ be the inverse image in $\mathrm{Aut}(F)$ of Γ. If each element x of F is identified with the inner automorphism $y \to xyx^{-1}$, $\tilde{\Gamma}$ is an extension

$$1 \to F \to \tilde{\Gamma} \to \Gamma \to 1 .$$

The conjugation action of $\tilde{\Gamma}$ on F agrees with the action of $\tilde{\Gamma}$ as a group of automorphisms.

Now, since $\tilde{\Gamma}$ is a free-by-finite group, the structure theorem states that $\tilde{\Gamma}$ is a graph product with finite edge and vertex groups. (This theorem was proved for finite rank by Karrass-Pietrowski-Solitar [KPS], using results of Stallings [St], & extended by Cohen [C] & Scott [Sc].) Following Bass and Serre [BS], this means that $\tilde{\Gamma}$ acts by automorphisms on a tree T so that the vertex and edge groups are stabilizers, respectively, of vertices and edges of T. The free group F must act as a group of covering translations on T since all stabilizers have finite order. Thus F can be identified with the fundamental group of the graph T/F, which inherits an action by Γ.

We claim that, with this identification of F with $\pi_1(T/F)$, Γ is realized by its own action as a group of automorphisms of T/F. Specifically, the identification of F with $\pi_1(T/F)$ is this: If $\sigma\colon [0,1] \to T/F$ is a loop based at the base point in T/F, then the homotopy class of σ is identified with the element f of F such that $f(\tilde{\sigma}(0)) = \tilde{\sigma}(1)$, where $\tilde{\sigma}$ is the lift of σ such that $\tilde{\sigma}(0)$ is the base point in T. It follows that if $\sigma\colon [0,1] \to T/F$ is an arbitrary loop and if $f \in F$ is such that $f(\tilde{\sigma}(0)) = \tilde{\sigma}(1)$ for some lift $\tilde{\sigma}$ of σ, then f is identified with an element of the conjugacy class, $[\sigma]$, in $\pi_1(T/F)$ that is determined by σ. Thus the conjugacy class, $[f]$, of f in F is identified with the conjugacy class $[\sigma]$. Suppose that f and σ are so related, that γ is an element of Γ, and that $\tilde{\gamma}$ is an element of $\tilde{\Gamma}$ in the preimage of γ. To verify our claim we must show that $\gamma\cdot[f]$ is identified with $\gamma\cdot[\sigma]$. Now,

$$\gamma\cdot[f] = [\tilde{\gamma}f\tilde{\gamma}^{-1}],$$

and

$$\gamma\cdot[\sigma] = [\gamma\circ\sigma] .$$

Also, $\tilde{\gamma}\circ\tilde{\sigma}$ is a lift of $\gamma\circ\sigma$. Finally, observe that

$$\tilde{\gamma}f\tilde{\gamma}^{-1}(\tilde{\gamma}\circ\tilde{\sigma}(0)) = \tilde{\gamma}f(\tilde{\sigma}(0)) = \tilde{\gamma}\circ\tilde{\sigma}(1) .$$

Thus $[\tilde{\gamma}f\tilde{\gamma}^{-1}] \subset F$ is identified with $[\gamma\circ\sigma] \subset \pi_1(T/F)$, as desired.

If F is finitely generated, then we may replace T/F by its core, which will be compact and invariant under the action of Γ. □

3. Fixed subgroups.

Let F be a free group and $\alpha \in \mathrm{Aut}(F)$ be an automorphism whose image in $\mathrm{Out}(F)$ has finite order. Here we prove a theorem which gives a geometric description of the fixed subgroup of α. As a corollary we show that the fixed subgroup is either cyclic or is a free factor. This statement also follows from a result of Dyer and Scott [DS].

If g is an automorphism of a finite graph X then $\mathfrak{L}(g)$ will denote the Lefschetz number of g and $\mathfrak{Fix}(G)$ will denote the subgraph of X fixed by g. (This is a subgraph because g acts without inversions.) Computing the Lefschetz number on the chain level, it is immediate that $\mathfrak{L}(g) = \chi(\mathfrak{Fix}(g))$.

Let Y be a subspace of a topological space X. Any based map from Y to X which is homotopic to the inclusion determines a homomorphism $\pi_1(Y) \to \pi_1(X)$ -- the images of these homomorphisms form a conjugacy class of subgroups of $\pi_1(X)$. If Y is a subgraph of a graph X then the inclusion homomorphisms are injective and their images are free factors of $\pi_1(X)$. In this situation we will say that a subgroup of $\pi_1(X)$ is conjugate to $\pi_1(Y)$ if it is in the conjugacy class of images of inclusion monomorphisms.

3.1 Theorem. Let $\alpha \in \mathrm{Out}(F)$ be an element of finite order, and let $\tilde{\alpha} \in \mathrm{Aut}(F)$ be an automorphism in the coset α. Let g be any realization of

α by an automorphism of a graph. Then the fixed subgroup of $\tilde{\alpha}$ is either cyclic or is conjugate to the fundamental group of a component of $\mathfrak{Fix}(G)$.

Proof. Let A be any finitely generated non-cyclic subgroup of $\mathfrak{Fix}(\tilde{\alpha})$. Let $\tilde{X}$ be the covering space of X corresponding to A, and let $\tilde{X}_0$ be the core of $\tilde{X}$. Since A is finitely generated, $\tilde{X}_0$ is compact. The automorphism g lifts to an automorphism $\tilde{g}$ of $\tilde{X}$, and $\tilde{X}_0$ is invariant under $\tilde{g}$. Since $\tilde{g}$ fixes every conjugacy class in A, it must act trivially on the homology of $\tilde{X}_0$. Thus

$$\chi(\tilde{X}_0) = \mathcal{L}(\tilde{g}) = \chi(\mathrm{Fix}(\tilde{g}|_{\tilde{X}_0})) .$$

If a finite graph has the same Euler characteristic as a subgraph, then the inclusion is a homotopy equivalence. Thus, by the minimality of $\tilde{X}_0$, $\mathrm{Fix}(\tilde{g}|_{\tilde{X}_0}) = \tilde{X}_0$. Let X_0 be the image of $\tilde{X}_0$ in X; then X_0 is a subgraph contained in $\mathfrak{Fix}(G)$, and A is contained in a conjugate of the free factor $\pi_1(X_0)$. This free factor is, in turn, contained in a free factor G determined by the component of $\mathfrak{Fix}(g)$ that contains X_0.

We claim that G is equal to the fixed subgroup of $\tilde{\alpha}$. Since G corresponds to a subgraph of $\mathfrak{Fix}(g)$, and since g is homotopic to a map which induces $\tilde{\alpha}$, there exists $y \in F$ such that $\tilde{\alpha}(x) = yxy^{-1}$ for all $x \in G$. But $\tilde{\alpha}(a) = a$ for all $a \in A \subset G$. Thus A is contained in the centralizer of y. Hence, since $\mathrm{rank}(A) > 1$, $y = 1$. Thus G is contained in the fixed subgroup of $\tilde{\alpha}$. Conversely, suppose that $\tilde{\alpha}(a_1) = a_1$. Let A_1 be the subgroup generated by A and a_1. By the above argument, A_1 is contained in a free factor G_1 determined by a component of $\mathrm{Fix}(g)$. Since distinct conjugates of a free factor meet trivially, as do free factors corresponding to distinct components of $\mathrm{Fix}(g)$, it follows that $G = G_1$. Thus $a_1 \in G$, and we conclude that G is the fixed subgroup of $\tilde{\alpha}$. □

3.2 Corollary (Dyer and Scott). Let $\alpha \in \mathrm{Out}(F)$ be an element of finite order, and let $\tilde{\alpha} \in \mathrm{Aut}(F)$ be an automorphism in the coset α. Then the fixed subgroup of $\tilde{\alpha}$ is either cyclic or is a free factor of F.

4. A lifting criterion.

The next theorem characterizes the groups of outer automorphisms which lift to $\mathrm{Aut}(F)$.

4.1 Theorem. Let Γ be a finite subgroup of $\mathrm{Out}(F)$. The following are equivalent.

1) Γ is the isomorphic image of a subgroup of $\mathrm{Aut}(F)$.

2) Any realization of Γ has a common fixed point.

3) Some realization of Γ has a common fixed point.

Proof. The theorem is clear in the case where F is cyclic, so we assume $\mathrm{rank}(F) > 1$. Obviously (2) implies (3). It is also clear that (3) implies (1), since the realization of Γ will induce the desired subgroup of $\mathrm{Aut}(F)$. Suppose now that Γ lifts to $\mathrm{Aut}(F)$, and let G be a realization of Γ as a group of automorphisms of a graph X.

Consider the inverse image $\tilde{\Gamma}$ of Γ in $\mathrm{Aut}(F)$. By hypothesis this is a split extension of F by Γ, where F is identified with $\mathrm{Inn}(F)$. Now consider the group $\tilde{G}$ of all lifts of elements of G to the universal cover $\tilde{X}$ of X. Identifying F with the group of covering translations and G with Γ, $\tilde{G}$ is also an extension of F by Γ. Moreover, it was shown in the proof of our realization theorem 2.1 that these extensions are in the same automorphism class. Since F has trivial center, these two extensions are equivalent. (This can be seen by a direct covering space argument as well.) In particular $\tilde{G}$ is a split extension.

Thus there is a subgroup H of $\tilde{G}$ which maps isomorphically to G. Since H acts on the tree $\tilde{X}$ and cannot be a nontrivial graph product, it must be contained in the stabilizer of some vertex of $\tilde{X}$ (see [BS], p. 90). It follows that G has a common fixed point. □

4.2 Corollary. Let $\alpha \in \mathrm{Out}(F)$ have finite order and let α_* be the automorphism of $H_1(F)$ induced by α. If $\mathrm{tr}(\alpha_*) \neq 1$, then the coset α contains an automorphism of finite order.

Proof. By the Lefschetz fixed point theorem any realization of α has a fixed point. □

4.3 Remark. The proof of Theorem 4.1 implies the analogous statement for automorphisms of surface groups; one uses Kerckhoff's realization theorem.

5. Normal forms.

In [DS], Dyer and Scott give a normal form for automorphisms of F of prime order. We will explain how to interpret this normal form geometrically. Also, we will derive a normal form for automorphisms whose image in $\mathrm{Out}(F)$ has prime order but which are not expressible as an automorphism of finite order composed with an inner automorphism.

Let α be an outer automorphism of a free group F which is realized by an automorphism g' of a graph X'. Suppose that X' contains a subcomplex Y which is invariant under the action of g' and such that each component of Y is a tree. We may form a new graph X by collapsing each component of Y to a point. Then X is homotopy equivalent to X' and inherits an automorphism g which realizes α. By performing this operation with a maximal subcomplex Y with the above properties, we obtain a realization g of α with the following property. For any edge e of X', $\bigcup_{n\in Z} g^n(e)$ is not a disjoint union of trees.

Suppose now that α is the image of an automorphism $\tilde{\alpha}$ of prime order p. Then g has order p and must fix a vertex of X. We claim that, in fact, g fixes every vertex. For otherwise there is an edge e of X with one fixed vertex and one vertex whose orbit under g consists of p points. Clearly $\bigcup_{n=1}^{p} g^n(e)$ is a tree, contrary to our assumption. Similarly, an edge with distinct endpoints cannot be fixed by g.

Let T be a maximal tree in X, and v be a vertex. We may identify F with $\pi_1(X,v)$ so that $\alpha = g_*\colon \pi_1(X,v) \to \pi_1(X,v)$. Let E be a collection of edges which contains every edge of T, and which contains exactly one edge from each g-orbit. Then $E = A \cup B \cup C \cup D$ where

$A = \{e \in E \mid e$ is fixed by $g\}$,

$B = \{e \in E \mid e$ has one endpoint but is moved by $g\}$,

$C = \{e \in E \mid e \in T\}$, and

$D = \{e \in E \mid e$ has two endpoints but is not in $T\}$.

For each $e \in E$ let i_e and t_e be minimal edge paths in T joining v to the endpoints of e. We may assume that i_e is not longer than t_e. Let x_e denote the edge path

$$\begin{cases} i_e \cdot e \cdot t_e^{-1} & \text{if } e \in A \cup B \cup D \\ t_e \cdot g(t_e^{-1}) & \text{if } e \in C \end{cases}.$$

It is not difficult to see that the following edge paths represent a basis for $\pi_1(X)$:

$$\bigcup_{e \in A} \{x_e\} \cup \bigcup_{e \in B \cup D} \{x_e, g(x_e), \ldots, g^{p-1}(x_e)\} \cup \bigcup_{e \in C} \{x_e, g(x_e), \ldots, g^{p-2}(x_e)\}.$$

Observe that, for $e \in C$,

$$\begin{aligned} g^{p-1}(x_e) &= g^{p-1}(t_e) \cdot t_e^{-1} \\ &= g^{p-1}(t_e) g^{p-2}(t_e^{-1}) g^{p-2}(t_e) \ldots g(t_e^{-1}) g(t_e) t_e^{-1} \\ &= g^{p-2}(x_e)^{-1} \ldots g(x_e)^{-1} x_e^{-1}. \end{aligned}$$

Also, if $e \in A$ and the endpoint of e is the common endpoint of f and t_f for $f \in C$, then

$$\begin{aligned} g(x_e) = g(t_e) e g(t_e^{-1}) = g(t_f) e g(t_f)^{-1} &= g(t_f) t_f^{-1} t_f e t_f^{-1} t_f g(g_f^{-1}) \\ &= x_f^{-1} x_e x_f. \end{aligned}$$

We have proven the following.

5.1 <u>Theorem</u> (Dyer and Scott). Let $\tilde{\alpha}$ be an automorphism of a free group F with $\tilde{\alpha}^p = 1$ for some prime p. Then $F = G_0 * (\underset{i \in I}{*} G_i) * (\underset{j \in J}{*} G_j)$ where each factor is invariant and

(1) G_0 is the fixed subgroup of α,

(2) For $i \in I$, G_i has a basis $\{x_1, \ldots, x_p\}$ with $\tilde{\alpha}(x_s) = x_{s+1} \pmod{p}$,

(3) For $j \in J$, G_j has a basis $\{y_1,\ldots,y_{p-1}\} \cup \{z_k \mid k \in K\}$ with

$\tilde{\alpha}(y_s) = y_{s+1}$, $s = 1,\ldots,p-2$

$\tilde{\alpha}(y_{p-1}) = y_{p-2}^{-1}\cdots y_1^{-1}$

$\tilde{\alpha}(z_k) = y_1^{-1}z_k y_1$.

The figure below illustrates this choice of basis for an automorphism of order 3.

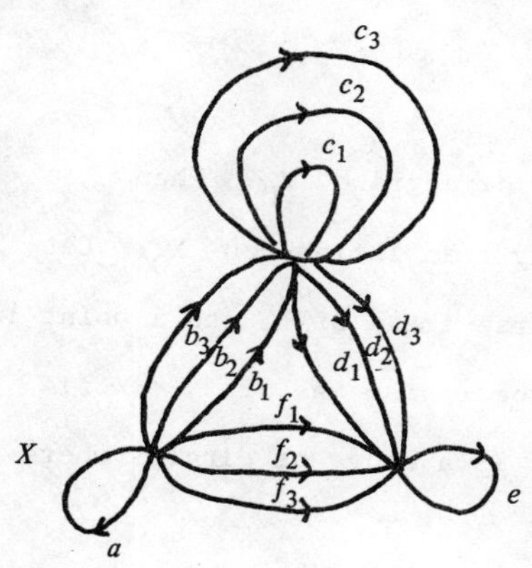

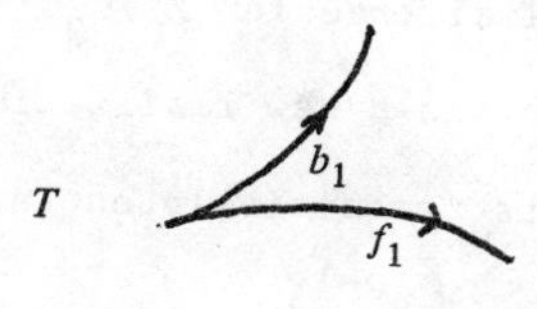

Action of g:

$g(a) = a$,

$g(b_i) = b_{i+1}$ (mod 3),

$g(c_i) = c_{i+1}$ (mod 3),

$g(d_i) = d_{i+1}$ (mod 3),

$g(e) = e$,

$g(f_i) = f_{i+1}$ (mod 3).

Basis of $\pi_1(X)$:

$x = a$,

$y_1 = b_1b_2^{-1}$, $y_2 = b_2b_3^{-1}$,

$z_1 = b_ic_ib_i^{-1}$, $i = 1,2,3$,

$w_i = b_id_if_i^{-1}$, $i = 1,2,3$,

$v = f_1ef_1^{-1}$,

$u_1 = f_1f_2^{-1}$, $u_2 = f_2f_3^{-1}$.

Action of α:

$\tilde{\alpha}(x) = x$,

$\tilde{\alpha}(y_1) = y_2$, $\tilde{\alpha}(y_2) = y_2^{-1}y_1^{-1}$,

$\tilde{\alpha}(z_i) = z_{i+1}$ (mod 3),

$\tilde{\alpha}(w_i) = w_{i+1}$ (mod 3),

$\tilde{\alpha}(u_1) = u_2$, $\tilde{\alpha}(u_2) = u_2^{-1}u_1^{-1}$,

$\tilde{\alpha}(v) = u_1^{-1}vu_1$.

5.2 Theorem. Let α be an outer automorphism of F of order p, p a prime, and suppose that α is not the image of any automorphism of finite order. Then there exists an automorphism $\tilde{\alpha}$, with image α, so that $F = C*(\underset{i\in I}{*} G_i)$ where each factor is invariant under $\tilde{\alpha}$ and

(1) C is infinite cyclic with generator C and $\alpha(C) = C$

(2) G_i has a basis $x_1,\ldots,x_p$ such that

$$\alpha(x_s) = x_{s+1},\ s = 1,\ldots,p-1$$
$$\alpha(x_p) = cx_1c^{-1}.$$

Proof. Realize α by an automorphism g_0 of a graph X_0. Then $G_0 = \{1,g_0,\ldots,g_0^{p-1}\}$ acts as a group of covering translations on X_0. Let T be a maximal tree in X_0/G_0. Contract each inverse image of T to a point in X_0, to obtain a new realization of α by an automorphism g of $X = X_0/T$ which acts transitively on vertices. Thus X/G is a wedge of circles where $G = \{1,g,\ldots,g^{p-1}\}$.

Choose an edge e of X. Since the conclusion remains unchanged if α is replaced by a power of α, we may assume that one endpoint, v, of e is mapped to the other. Identify $\pi_1(X,v)$ with $\pi_1(X,g(v))$ via the path e from v to $g(v)$. Then $\alpha = g_*: \pi_1(X,v) \to \pi_1(X,g(v))$.

Let c be the class of the path $e\cdot g(e)\ldots g^{p-1}(e)$.

Let $\{f_i \mid i \in I\}$ be a set of paths incident to v, consisting of one edge from each orbit other than that of e.

Set $x_{i,1} = f_i\cdot g^{k-1}(e^{-1})g^{k-2}(e^{-1})\ldots e^{-1}$, where v and $g_k(v)$ are the endpoints of f_i.

Set $x_{i,s} = e\cdot g(e)\ldots g^{s-1}(e)g^{s-1}(x_{i,1})g^{s-1}(e^{-1})\ldots e^{-1}$. Let T be the maximal tree $e \cup g(e) \cup \ldots \cup g^{p-2}(e)$. Note that $\{[c],[x_{i,s}] \mid i \in I,\ s = 1,\ldots,p\}$ form a free basis for $\pi_1(X)$ since each $[x_{i,s}]$ equals, to within a factor of $[c]$, an element of the standard set of generators associated with edges of $X-T$. One checks easily that the action of α is as specified in the statement. $\square$

REFERENCES

[BS] J. P. Serre, "arbres, amalgames, SL_2," asterisque 46 (1977).

[C] D. E. Cohen, "Groups with Free Subgroups of Finite Index," Conference on Group Theory, Springer Lecture Notes in Mathematics, 319 (1973), pp. 26-44.

[DS] J. L. Dyer and G. P. Scott, "Periodic Automorphisms of Free Groups," Communications in Algebra, 3 (1975), pp. 195-201.

[JS] W. Jaco and P. Shalen, "Surface Homeomorphisms and Periodicity," Topology 16 (1977), pp. 347-367.

[K] S. Kerckhoff, "The Nielsen Realization Problem," preprint.

[KPS] A. Karass, A. Pietrowski and D. Solitar, "Finitely Generated Groups with a Free Subgroup of Finite Index," J. Austral. Math. Soc. 16 (1973), pp. 458-466.

[N] J. Nielsen, "Abildungsklassen endlichen Ordnung," Acta Math. 75 (1943), pp. 23-115.

[Sc] G. P. Scott, "An Embedding Theorem for Groups with a Free Subgroup of Finite Index," Bull. London Math. Soc. 6 (1974), pp. 304-306.

[St] J. R. Stallings, "Group Theory and 3-dimensional Manifolds," Yale Monographs 4 (1971).

[T] W. Thurston, "On the Geometry and Dynamics of Surface Homeomorphisms I," preprint.

DEPARTMENT OF MATHEMATICS
RUTGERS UNIVERSITY
NEW BRUNSWICK, N.J. 08903

Received August 24, 1982

Contemporary Mathematics
Volume 33, 1984

A REMARK ON AUTOMORPHISM GROUPS

Joan L. Dyer

To Roger Lyndon on his 65th birthday

Which groups occur as automorphism groups of groups? This and related questions are treated in [1], [2], [4], [5], and [6]; and see also the references therein. It is the point of this brief remark to provide two examples of non-automorphism groups. The arguments used here are elementary, stopping just shy of the point at which cohomological techniques appear to be of use.

<u>Example 1</u>. Let F be a nontrivial free group. There exists no group G whose automorphism group, $Aut(G)$, is isomorphic to F.

<u>Example 2</u>. There is no finitely generated group G for which $Aut(G)$ is isomorphic to $SL(2,Z)$.

Recall that $Aut(G)$ contains the normal subgroup $Inn(G)$, and the sequence

$$1 \to C(G) \to G \to Inn(G) \to 1 \tag{1}$$

is exact, where $C(G)$ is the center of G.

In the first example, $Inn(G)$ is free, and so the extension (1) splits whence G is isomorphic to the direct product of $C(G)$ with a free group. Consequencely $Aut(G)$ contains subgroups isomorphic to $Aut(C(G))$ and $Aut(Inn(G))$; these either have elements of order two or are trivial. Assuming that $Aut(G)$ is free therefore forces $Inn(G)$ to be trivial and G to be of order at most 2, a clear contradiction.

© 1984 American Mathematical Society
0271-4132/84 $1.00 + $.25 per page

Let C be any characteristic central subgroup of G. Then restriction and projection induce a homomorphism from $\mathrm{Aut}(G)$ to the direct product of $\mathrm{Aut}(C)$ with $\mathrm{Aut}(G/C)$, and the sequence

$$(2) \qquad 1 \to \mathrm{Hom}(G/C,C) \to \mathrm{Aut}(G) \to \mathrm{Aut}(C) \times \mathrm{Aut}(G/C)$$

is exact.

For the second example, note first that G has non-trivial center because $C(SL(2,Z))$ is non-trivial. Secondly, the exact sequence (2) shows that $\mathrm{Hom}(G/C,C)$ is a normal abelian subgroup of $SL(2,Z)$ for any characteristic central subgroup C of G whence its order is at most 2.

As is well known, $SL(2,Z)$ is isomorphic to the free product of finite cyclic groups of orders 4 and 6, amalgamating the subgroups of order 2:

$$SL(2,Z) \sim \langle C_4 * C_6 : C_2 \rangle .$$

It follows from the subgroup theorem for amalgamated free products (or from the Kurosh subgroup theorem applied to the central quotient $PSL(2,Z)$; see [3]) that the non-cyclic normal subgroup $\mathrm{Inn}(G)$ is isomorphic to one of the following:

(1) Trivial

(2) Free of rank at least 2

(3) The direct product of a group of order 2 with a free group of rank at least 2

(4) The free product of three cyclic groups of order 4 with amalgamated subgroup of order 2

(5) The free product of two cyclic groups of order 6 with amalgamated subgroup of order 2

(6) The free product of cyclic groups of orders 4 and 6 with amalgamated subgroup of order 2

(7) The free product of two cyclic groups, both of order 3

We shall eliminate each of these possibilities in turn.

Case (1): G is finitely generated and abelian, with (characteristic) torsion subgroup T and free abelian quotient G/T. The restriction on the order of $\mathrm{Hom}(G/T,T)$ noted above forces the rank of G/T to be at most one, whence G has finite automorphism group.

Cases (2) and (3): The restriction on the order of $\mathrm{Hom}(G/C(G), C(G))$ eliminates these cases.

Cases (4), (5) and (6): In case (4) (the others are similar), using the first exact sequence above, G has a presentation over its center

$$\langle C(G), x, y, z: x^4, y^4, z^4, x^2y^{-2}, y^2z^{-2} \text{ in } C(G)\rangle$$

where explicit mention of generators and relations for $C(G)$, the relations that say $C(G)$ is central, et al., are omitted. The last two relations imply that y^2 commutes with both x and z; and so y^2 is in $C(G)$, a contradiction.

Case (7): G has a presentation over its center

$$\langle C(G), x, y: x^3, y^3 \text{ in } C(G)\rangle .$$

The map which sends x and y to their inverses and acts as inversion on $C(G)$ extends to an automorphism of G, which is of order 2 and not inner. The subgroup of $\mathrm{Aut}(G)$ generated by $\mathrm{Inn}(G)$ together with this automorphism has the presentation

$$\langle x, y, z: x^3 = y^3 = z^2 = (zx)^2 = (zy)^2 = 1\rangle$$

which, abelianized, is cyclic of order two. On the other hand, $\mathrm{Inn}(G)$ is in this case of index 4 in $\mathrm{Aut}(G)$ whence the subgroup described above is the unique subgroup of order 2 in $\mathrm{Aut}(G)$; it is the fifth group in the list above, and its abelianization yields the final contradiction.

REFERENCES

[1] H. deVries and A. B. deMiranda, Groups with a small number of automorphisms, M. Zeit. 68 (1958), pp. 450-464.

[2] H. Iyer, On solving the equation Aut(X) = G, Rocky Mountain J. Math. 9 (1979), pp. 653-670.

[3] R. J. Lyndon and P. Schupp, Combinatorial Group Theory, Erg. der Math. 89, Springer-Verlag (1977).

[4] D. J. S. Robinson, A contribution to the theory of groups with finitely many automorphisms, Proc. London Math. Soc., series 3, vol. 35 (1977), pp. 34-54.

[5] D. J. S. Robinson, Infinite torsion groups as automorphism groups, Quart. J. Math. 30 (1979), pp. 351-364.

[6] D. J. S. Robinson, Groups with prescribed automorphism group, to appear.

DEPARTMENT OF COMPUTER SCIENCE
IBM THOMAS J. WATSON RESEARCH CENTER
YORKTOWN HEIGHTS, N.Y. 10598

Received March 1, 1982

Contemporary Mathematics
Volume 33, 1984

ON THREE OF LYNDON'S RESULTS ABOUT MAPS

Leonard Greendlinger and Martin Greendlinger

The results in the title, for which we give simpler proofs, are a strengthened version of Corollary 2.5 and variants of two cases of Corollary 2.4 in [1]. See that pioneering paper and [2] for their many important applications. We shall denote vertices by v, regions by D, their degrees by d and the total number of vertices, edges and regions by V, E, and F, respectively.

Lemma. Let M be a map on the sphere with $d(D) = i$ for a_i regions $(i = 2,\ldots,p-1)$, $d(D) \geq p$ for all other regions and $d(v) \geq q$ for all vertices, where $1/p + 1/q = 1/2$. Then $\Sigma(p-i)a_i \geq 2p$.

Proof. $2E = \Sigma d(v) \geq Vq$, $2E = \Sigma d(D) \geq (F - \Sigma a_i)p + \Sigma i a_i$, yielding $2 + E = V + F \leq 2E/q + 2E/p + 1/p\ \Sigma(p-i)a_i$, whence $2p \leq \Sigma(p-i)a_i$.

Theorem. Let M be a connected, simply connected map in the plane with ≥ 2 regions, a simple closed path as bounday, $d(D) \geq p$ for all regions and $d(v) \geq q$ for all interior vertices. Let b_i be the number of M's regions with exactly i interior edges, all of them forming a consecutive part of M. If $p = 4 = q$, then $2b_1 + b_2 \geq 4$; if $p = 6$ and $q = 3$, then $3b_1 + 2b_2 + b_3 \geq 6$.

Proof. Case 1. $p = 4 = q$. Delete all vertices of degree 2 and deform two copies of the resulting map K into hemispheres whose corresponding boundary points coincide, getting a map on the sphere with $d(v) \geq 4$ for all vertices, $d(D) \geq 4$ for all interior regions of K, all regions whose interior edges do not form a consecutive part of K and all regions with ≥ 3 interior edges of K, $d(D) = 3$ for $2b_2$ regions and $d(D) = 2$ for $2b_1$ regions. It follows from the lemma that $(4 - 2)2b_1 + (4 - 3)2b_2 \geq 8$.

Case 2. $p = 6$, $q = 3$. Construct a map on the sphere as in Case 1, and then rotate one of the hemispheres by s/r radians, where $2s$ is the length of K's shortest boundary edge and r is the sphere's radius. The resulting map has $d(V) \geq 3$ for all vertices, $d(D) \geq 6$ for all interior regions of K, all regions whose interior edges do not form a consecutive part of K and all regions with ≥ 4 interior edges of K, $d(D) = 5$ for $2b_3$ regions, $d(D) = 4$ for $2b_2$ regions, $d(D) = 3$ for $2b_1$ regions and $d(D) = 2$ for no region. It follows from the lemma that

$$(6 - 2)0 + (6 - 3)2b_1 + (6 - 4)2b_2 + (6 - 5)2b_3 \geq 12.$$

REFERENCES

1. Lyndon, R. C.: On Dehn's algorithm. Math. Ann. 166, 208-228 (1966).
2. Lyndon, R. C., Schupp, P. E.; Combinatorial Group Theory. Springer-Verlag, Berlin (1977).

TULA STATE TOLSTOY PEDAGOGICAL INST.
TULA USSR

Received December 2, 1982

Contemporary Mathematics
Volume 33, 1984

ON THE INDECOMPOSABILITY OF CERTAIN RELATION CORES

K. W. Gruenberg

Introduction

A free presentation θ of a group G is a homomorphism of a free group F onto G. If R is the kernel of θ, then $\bar{R} = R/[R,R]$ is a G-module and is called the <u>relation module</u> for G determined by θ. Rather little is known about the structure of such modules in general. But for finite groups, a coherent theory of relation modules has begun to emerge. The present note deals with this fledgling theory and concerns the direct sum decomposition of relation modules. We begin with a summary account of the facts that we shall need. For details and proofs the reader may consult [2] or [7].

Let G be a finite group and let us agree to consider only free presentations in which the free group has finite rank. Any relation module $\bar{R}$ can then obviously be decomposed as $\bar{R} = A \oplus P$, where P is $\mathbb{Z}G$-projective and A has no projective summand. Of course, P could be 0, but A is never 0. Both A and P are determined to within genus by $\bar{R}$ and A is called a <u>relation core</u> for G. There always exists a decomposition in which the projective summand is free:

$$\bar{R} = B \oplus (\mathbb{Z}G)^m . \tag{1}$$

The relation cores determined by all possible relation modules from all possible finitely generated free presentations form a single complete genus class. It follows that one relation core decomposes if, and only if, they all do.

© 1984 American Mathematical Society
0271-4132/84 $1.00 + $.25 per page

If $\overline{R}$ is a minimal relation module (meaning that the rank $d(F)$ of the ambient free group equals $d(G)$) and $\overline{R}$ decomposes as in (1), then B is a relation core and m is an invariant of G, called the presentation rank of G and written $pr(G)$. Mostly this number is 0.

Suppose the relation cores decompose. This implies that the prime graph $\Pi(G)$ is not connected. This graph is defined as follows. Let $\pi(G)$ be the set of prime divisors of $|G|$, the order of G. Then $\Pi(G)$ is the graph whose vertex set is $\pi(G)$, and there is an edge joining p and q if, and only if, G contains an element of order pq. A subset π of $\pi(G)$ will be called connected (closed) in $\Pi(G)$ if the subgraph of $\Pi(G)$ on π is connected (closed).

We have some information about groups with non-connected graphs. Suppose that the non-empty, proper subset π of $\pi(G)$ is closed. Write $S_\pi = S_\pi(G)$ for the maximal π-separable normal subgroup of G and $S^\pi = S^\pi(G)$ for the smallest normal subgroup of G containing S_π and so that G/S^π is π-separable. Then the following results hold (cf. [9] for the proofs):

(i) If $S_\pi = G$, then G is a Frobenius group or a 2-Frobenius group;

(ii) if $S_\pi < G$, then $S = S^\pi/S_\pi$ is the only non-trivial minimal normal subgroup in G/S_π and it is a simple group whose order involves both π and its complement; moreover, every Sylow p-subgroup of G/S^π is cyclic for $p > 2$ and, if $p = 2$, it is cyclic, generalised quaternion or dihedral;

(iii) if $1 < S_\pi < G$, then $2 \in \pi$, G/S^π is a (possibly trivial) cyclic π-group, and S_π is a nilpotent π-group.

Groups with decomposable relation cores can therefore be found only among the above types. The exact subclass of π-separable groups (case (i)) was determined by Klaus Roggenkamp and myself: If G is π-soluble, where π is a non-empty closed subset of $\pi(G)$ (but π is here not necessarily a proper subset), then the relation cores of G are decomposable if, and only if, G is a Frobenius group with cyclic complements or G is a special 2-Frobenius group (cf. [5]).

We shall here deal with the case G/S^{π} non-cyclic. In view of (iii) above, this forces $S_{\pi} = 1$.

THEOREM A. *If* S *is a* 2-*generator simple group and* $S \triangleleft G \leq \mathrm{Aut}\, S$, *with* $\pi(G/S)$ *connected in* $\Pi(G)$ *and* G/S *not cyclic, then the relation cores of* G *are indecomposable*.

Note that, if $\Pi(G/S)$ is connected, then certainly $\pi(G/S)$ is connected in $\Pi(G)$. Julian Williams has analysed in [9] the structure of groups with non-connected prime graphs when the simple group S involved is of known type. He found that $\pi(G/S^{\pi})$ is always connected in $\Pi(G)$.

It is of interest to observe that Theorem A combines with cases (i) and (iii) of the structure theorem, the known result in the π-separable case [5], and Williams' work [9] to yield the following fact: Assuming that the classification of the finite simple groups is complete, *if* G *has a non-cyclic image* Q *with* $\pi(Q)$ *connected in* $\Pi(G)$, *then the relation cores of* G *are indecomposable*.

Theorem A is a consequence of the following result. We shall write $\mathfrak{g}$ for the augmentation ideal of G and $d_G(\mathfrak{g})$ for the minimum number of elements needed to generate $\mathfrak{g}$ as G-module. A similar notation is used for H (and for other groups that occur later).

THEOREM B. *If* H *is an image of* G *so that* $\pi(H)$ *is connected in* $\Pi(G)$ *and* $d_G(\mathfrak{g}) = d_H(\mathfrak{h})$, *then the relation cores of* G *are indecomposable*.

Some of the ideas involved in the proof of Theorem B came from my study of the interesting work of Peter Linnell (cf. [6]). Also I am grateful to him for very useful critical comments on a first draft of this note.

Proofs

We shall use the following notation. If π is a set of primes, $\mathbb{Z}_{\pi} = \bigcap_{p \in \pi} \mathbb{Z}_{(p)}$, where $\mathbb{Z}_{(p)}$ is the local ring at p. The ring of p-adic integers is $\mathbb{Z}_p$. If A is a $\mathbb{Z}G$-lattice, $A_{\pi} = A \otimes_{\mathbb{Z}} \mathbb{Z}_{\pi}$; and $d_G(A_{\pi})$ denotes the minimum number of $\mathbb{Z}_{\pi}G$-module generators of A_{π}.

Proof of Theorem B. Let us suppose the result is false and that the relation core A decomposes as $A = A_1 \oplus A_2$. If π_i is the set of primes p at which $(A_i)_{(p)}$ is not $\mathbb{Z}_{(p)}G$-projective, then π_1, π_2 are closed subsets of $\pi(G)$ and $\pi(G)$ is their disjoint union ([2], 8.10, 8.11, 8.16). Hence if X_i is the set of all non-identity π_i-elements in G, then G is the disjoint union of X_1, X_2 and $\{1\}$.

The module A must arise in an exact sequence

$$0 \to A_{\pi} \to (\mathbb{Z}_{\pi}G)^d \to \mathcal{G}_{\pi} \to 0 , \tag{2}$$

where $\pi = \pi(G)$ and $d = d_G(\mathcal{G})$ ([2], 5.11, 7.8). Consequently the character α of A satisfies $\alpha(g) = 1$ for all $g \neq 1$. If α_i is the character of A_i, then

$$\begin{aligned} \alpha_1(g) &= 1 \quad \text{if } g \in X_1 , \\ &= 0 \quad \text{if } g \in X_2 , \end{aligned}$$

because if $g \in X_2$, then $(A_1)_{\pi_2}$ is free as $\mathbb{Z}_{\pi_2}\langle g \rangle$-module. By orthogonality,

$$\alpha_1(1) = -|X_1| + \ell_1 \, |G| , \tag{3}$$

where $\ell_1 = \dim(\mathbb{Q}A_1)^G$. Similar formulae hold for α_2.

Let $d_1 = d_G((A_1)_{\pi_2})$. Then $\mathbb{Q}A_1$ is a direct summand of $(\mathbb{Q}G)^{d_1}$ and therefore $\ell_1 \leq d_1$. The following lemma will show that also $\ell_1 \geq d_1$, so that actually $\ell_1 = d_1$.

LEMMA 1. *Let* π *be a closed subset of* $\pi(G)$, π' *its complement in* $\pi(G)$, X' *the set of all non-identity* π'*-elements in* G *and* P *a finitely generated projective* $\mathbb{Z}_\pi G$*-module with character* χ. *Then* $d_G(P)$ *is the smallest integer* d *satisfying*

$$\chi(1)\phi(1) + \sum_{g\in X'} \chi(g)\phi(g^{-1}) \leq d|G|\phi(1) ,$$

for all irreducible Brauer characters ϕ *at all primes* p *in* π.

Proof. For $p \in \pi$, choose a finite extension I of $\mathbb{Z}_p$ such that the residue class field of I and the field of fractions of I are both splitting fields for G and all its subgroups ([1], §§49,50).

Let U be an indecomposable projective IG-module with associated irreducible Brauer character ϕ. Then the number of occurrences of U in $P \otimes I$ is

$$\frac{1}{|G|} \sum_{g\in(p')} \chi(g)\phi(g^{-1}) , \tag{4}$$

where (p') means the set of all p'-elements in G ([1], §60). In our case, $(p') \geq X'$ and $\chi(g) = 0$ if $g \neq 1$ and $g \in (p') - X'$, because $\pi\langle g\rangle \leq \pi$. Hence (4) becomes

$$\frac{1}{|G|} \{\chi(1)\phi(1) + \sum_{g\in X'} \chi(g)\phi(g^{-1})\} .$$

Since the number of occurrences of U in IG is $\phi(1)$, we conclude that $d_G(P \otimes I)$ is the smallest integer e_p satisfying

$$\chi(1)\phi(1) + \sum_{g\in X'} \chi(g)\phi(g^{-1}) \leq e_p|G|\phi(1) ,$$

for all irreducible Brauer characters ϕ at the prime p. But I is a finitely generated free $\mathbb{Z}_p$-module and the Krull-Schmidt theorem holds for $\mathbb{Z}_p G$-lattices. Hence

$$(P \otimes I) \oplus W \cong (IG)^{e_p} ,$$

viewed as a relation between $\mathbb{Z}_p G$-modules, yields

$$(P \otimes \mathbb{Z}_p) \oplus V \cong (\mathbb{Z}_p G)^{e_p},$$

for some suitable V. Now $e_p = d_G(P \otimes \mathbb{Z}_p) = d_G(P_{(p)})$ and $d_G(P) = \max_{p \in \pi} \{e_p\}$ ([2], 6.3). This completes the proof of the lemma.

To obtain our required inequality $\ell_1 \geq d_1$, apply the lemma with $\pi = \pi_2$, $X' = X_1$ and $P = (A_1)_{\pi_2}$. The left-hand side of the inequality in the lemma is, by (3),

$$\phi(1)\{ - |X_1| + \ell_1 |G|\} + \sum_{g \in X_1} \phi(g)$$

and this quantity is $\leq \phi(1)\ell_1|G|$ (as each $\phi(g)$ is a sum of suitable roots of unity). Thus $\ell_1 \geq d_1$, by the lemma.

Recall that our hypothesis says $\pi(H)$ is connected in $\Pi(G)$. Hence $\pi(H)$ is contained in π_1 or π_2: say $\pi(H) \leq \pi_2$. This determines that we now work with A_1 and not A_2. Since $\ell_1 = d_1$, we may write

$$(A_1)_{\pi_2} \oplus P = (\mathbb{Z}_{\pi_2} G)^{\ell_1}$$

for some projective module P. Then $P^G = 0$ and therefore $P = P\mathfrak{g}$. Now (2) yields

$$0 \to (A_2)_{\pi_2} \to (\mathbb{Z}_{\pi_2} G)^{d-\ell_1} \oplus P \to \mathfrak{g}_{\pi_2} \to 0 .$$

If N is the kernel of the given homomorphism of G onto H, then this exact sequence provides an epimorphism

$$(\mathbb{Z}_{\pi_2} H)^{d-\ell_1} \oplus (P/P\mathfrak{n}) \twoheadrightarrow \mathfrak{g}_{\pi_2}/\mathfrak{g}_{\pi_2}\mathfrak{n}.$$

The $\mathbb{Z}_{\pi_2} H$-module $P/P\mathfrak{n}$ is projective and as $\pi_2 \geq \pi(H)$, $P/P\mathfrak{n}$ is actually free. Since $P = P\mathfrak{g}$ implies $(P/P\mathfrak{n})\mathfrak{h} = P/P\mathfrak{n}$, we conclude $P/P\mathfrak{n} = 0$ and thus

$$(\mathbb{Z}_{\pi_2} H)^{d-\ell_1} \twoheadrightarrow \mathcal{f}_{\pi_2} .$$

Hence $d_H(\mathcal{f}) = d_H(\mathcal{f}_{\pi_2}) \leq d - \ell_1 < d = d_G(\mathcal{g})$, which is a contradiction. Our theorem is therefore proved.

Proof of Theorem A. If $\Pi(G)$ is connected, then we are done. We shall henceforth assume $\Pi(G)$ is not connected. By the structure theorem quoted in the introduction, G/S has its Sylow p-subgroups cyclic for $p > 2$ and, for $p = 2$, cyclic or generalized quaternion or dihedral. Such a group needs at most 2 elements to generate it. This may be a known fact, but as I have been unable to find an explicit proof in the literature, I give one here.

LEMMA 2. If every Sylow p-subgroup of the group H is cyclic when $p > 2$, and is cyclic or generalized quaternion or dihedral when $p = 2$, then $d(H) \leq 2$.

Proof. The case when H is soluble is proved in [2], 9.9. Assume therefore that H is insoluble. Then by results of Suzuki [8], H has a subgroup H_1 of index ≤ 2 so that $H_1 = Z \times L$, where Z is a metacyclic group of odd order and L is $SL(2,p)$ or $PSL(2,p)$ for $p \geq 5$. If $H = H_1$ then $d(H) = 2$ (directly or by using [3], 3.2). If $H > H_1$ then $H = Z \wr K$, where $K/\text{centre}(K) \cong PGL(2,p)$ and of course, $\text{centre}(K) = \text{centre}(L)$ has order 2 or 1. It is well known that $GL(2,p)$ is a 2 generator group and hence, as $\text{centre}(K) \leq \text{Frattini}(K)$, $d(K) = 2$. The required result, that $d(H) = 2$, can now be seen by a direct calculation or a double application of Proposition 4 of [4]: first to $A \wr K$ and then to $B \wr (A \wr K)$, where Z is the split extension of the cyclic group B by the cyclic group A.

In the situation of Theorem A, let $H = G/S$. By the above lemma, $d(H) = 2$. Thus $pr(H) = 0$ and therefore $d_H(\mathcal{f}) = 2$. We shall see in the next lemma that always $d_G(\mathcal{g}) = 2$. Thus Theorem B applies and completes the proof of

Theorem A. It is quite possible (but I cannot prove it) that actually $d(G) = 2$. A counterexample to this would be rather interesting, for it would be a group of non-zero presentation rank of a totally different sort to the known ones (which are large direct powers of perfect groups). (Cf. [3], (C).)

LEMMA 3. *If* $1 \to S \to G \to H \to 1$ *is the extension of Theorem A, then* $d_G(\mathcal{G}) = 2$.

Proof. By [2], 6.6, $pr(G) \geq m$ if, and only if, for every $p \in \pi(G)$ and every irreducible $\mathbb{F}_pG$-module M,

$$|H^1(G,M)| \leq |M|^{d(G)-\zeta_M-m}, \tag{5}$$

where ζ_M is 0 if $M = \mathbb{F}_p$ and 1 otherwise.

If $M^S = M$, then M is really an irreducible H-module and the 5 term exact cohomology sequence yields

$$H^1(G,M) \simeq H^1(H,M)$$

because $H^1(S,M) = \mathrm{Hom}(S,M) = 0$, as S is perfect. Now ([2], 2.13)

$$|H^1(H,M)| \leq |M|^{d(H)-\zeta_M},$$

where $|H^1(G,M)| \leq |M|^{2-\zeta_M}$. Suppose, on the other hand, $M^S = 0$. Then the cohomology sequence yields

$$H^1(G,M) \simeq H^1(S,M)^G .$$

Let $M = M_1 \oplus \ldots \oplus M_r$ be a complete decomposition of M as S-module. Then for all i,

$$|H^1(S,M_i)| \leq |M_i| ,$$

again by [2], 2.13, because $d(S) = 2$ and $\zeta_{M_i} = 1$. Hence

$$|H^1(G,M)| \leq \prod_i |H^1(S,M_i)| \leq \prod_i |M_i| = |M| .$$

Thus in both cases

$$|H^1(G,M)| \leq |M|^{2-\zeta_M} .$$

By (5),

$$d(G) - \zeta_M - pr(G) \leq 2 - \zeta_M$$

and consequently $d_G(\mathfrak{g}) = d(G) - pr(G) \leq 2$. But $d_G(\mathfrak{g}) \neq 1$, as G is non-cyclic, and so the lemma is proved.

REFERENCES

1. L. Dornhoff, Group representation theory, Marcel Dekker, New York, 1972.

2. K. W. Gruenberg, Relation modules of finite groups, CBMS no. 25, Amer. Math. Soc., Providence, 1976.

3. K. W. Gruenberg, "Groups of non-zero presentation rank," Symposia Math. 17 (1976), 215-224.

4. K. W. Gruenberg and K. W. Roggenkamp, "Decomposition of the relation modules of a finite group," J. London Math. Soc. 12 (1976), 262-266.

5. K. W. Gruenberg and K. W. Roggenkamp, "The decomposition of relation modules: a correction," Proc. London Math. Soc. 45 (1982), 89-96.

6. P. A. Linnell, "Relation modules and augmentation ideals of finite groups," J. Pure Appl. Algebra 22 (1981), 143-164.

7. K. W. Roggenkamp, Integral representations and presentations of finite groups, Lecture Notes Math. no. 744, Springer, 1979.

8. M. Suzuki, "On finite groups with cyclic Sylow subgroups for all odd primes," Amer. J. Math. 77 (1955), 657-691.

9. J. S. Williams, "Prime graph components of finite groups," J. Algebra 69 (1981), 487-513.

QUEEN MARY COLLEGE
LONDON, ENGLAND

Received November 1, 1981

Contemporary Mathematics
Volume 33, 1984

FOX SUBGROUPS OF FREE GROUPS II

Narain Gupta

(To Roger Lyndon on his 65th Birthday)

1. Introduction. Let R be a normal subgroup of a finitely generated free group F and let ZF be the integral group ring of F. A long standing problem due to R. H. Fox [1] is the identification of $F(n,R) = F \cap (1 + \underline{\underline{f}}^n\underline{\underline{r}})$, $n \geq 1$, where $\underline{\underline{f}} = ZF(F - 1)$ and $\underline{\underline{r}} = ZF(R - 1)$ are ideals in ZF. For an up-to-date account of this problem we refer the reader to the survey article [4]. In [3], $F(n,R)$ was identified as $\sqrt{[R \cap \gamma_n(F), R \cap \gamma_n(F)]\gamma_{n+1}(R)}$ whenever F/R is a periodic group. In this paper we go someway towards proving the general case. We present a reduction theorem (Theorem 4.2), which identifies $F(n,R)$ under a separability restriction on R. A criterion for separability (Theorem 5.1) yields identification of $F(n,R)$ for small values of n (e.g. $n \leq 7$).

2. Notation. The lower central series of a group G is defined as follows: $\gamma_1(G) = G$, $\gamma_{i+1}(G) = [\gamma_i(G),G]$ for $i \geq 1$, where for subgroups A,B of G, $[A,B] = \text{sgp}\{[a,b] \mid a \in A,\ b \in B\}$, where $[a,b] = a^{-1}b^{-1}ab$ is the commutator of a and b. For G/H nilpotent, $\sqrt{H}$ denotes the isolator $\{x \in G \mid x^m \in H \text{ for some } m > 0\}$. If R is a normal subgroup of a free group F, then $\underline{\underline{r}} = \ker(ZF \to Z(F/R)) = \text{ideal}_{ZF}\{r - 1 \mid r \in R\}$. If F is free on X and R (as a subgroup of F) is free on Y, then for any $n \geq 1$, $\underline{\underline{f}}^n\underline{\underline{r}} = \text{ideal}_{ZF}\{(x_1 - 1) \ldots (x_n - 1(y - 1) \mid x_i \in X,\ y \in Y\}$ and

Research supported by NSERC, Canada.

© 1984 American Mathematical Society
0271-4132/84 $1.00 + $.25 per page

$F(n,R) = \{w \in F \mid w - 1 \in \underline{\underline{f}}^n \underline{\underline{r}}\}$. As a left ZF-module, $\underline{\underline{r}}$ is free on $\{y - 1 \mid y \in Y\}$ (see for instance [1], Theorem 1, §3.1). Rest of the notation is explained in the text.

3. A filteration of the lower central series of F. The basis theorem of P. Hall allows us to write an arbitrary element of a free nilpotent group as a product of certain basic commutators in a canonical fashion. We refer the reader to Chapter 5 of [6] for an excellent account of the theory of basic commutators.

For a given finitely generated free group F, let

$$B_i = \{b_{i,1} < \dots < b_{i,r(i)}\}, \quad i = 1,2,\dots, \tag{3.1}$$

be the ordered set of (standard) basic commutators of weight $i(r(i) = \text{rank } \gamma_i(F)/\gamma_{i+1}(F))$. Between $\gamma_i(F)$ and $\gamma_{i+1}(F)$ we introduce normal subgroups $\lambda_{i,j}(F)$, $j = 1,\dots,r(i)$ as follows:

$$\gamma_i(F) = \lambda_{i,1}(F) \triangleright \lambda_{i,2}(F) \triangleright \dots \quad \lambda_{i,r(i)}(F) \triangleright \gamma_{i+1}(F), \tag{3.2}$$

where $\lambda_{i,j}(F)/\lambda_{i,j+1}(F)$ is the infinite cyclic group generated by $b_{i,j}\lambda_{i,j+1}(F)$ $(\lambda_{i,r(i)+1}(F) = \gamma_{i+1}(F) = \lambda_{i+1,1}(F))$. For convenience we put

$$\lambda^+_{i,j}(F) = \lambda_{i,j+1}(F).$$

Then, for any $w \in F$, $w \neq 1$, there exists a unique pair (i,j) of positive integers such that $w \in \lambda_{i,j}(F)$, $w \notin \lambda^+_{i,j}(F)$. We order (i,j)'s lexicographically. Then, for $u \in \lambda_{i,j}(F)\backslash\lambda^+_{i,j}(F)$, $v \in \lambda_{i',j'}(F)\backslash\lambda^+_{i',j'}(F)$, we define

$$u <^* v \quad \text{if} \quad (i,j) < (i',j'), \tag{3.3}$$

and

$$u \,\underline{\overline{*}}\, v \quad \text{if} \quad (i,j) = (i',j').$$

With respect to $<^*$, $\underline{\overline{*}}$, F is a linearly ordered set.

Now, let R be a non-trivial normal subgroup of F. Then, for each $k \geq 1$, the quotient $\overline{\gamma_k(R)} = \gamma_k(R)/\gamma_{k+1}(R)$ can be regarded as a right

F/R-module via conjugation in F. Let

(3.4) $$M_k = \overline{\gamma_k(R)} \underset{Z}{\otimes} Q$$

be the $Q(F/R)$-module with a basis $\overline{Y_k}$ whose elements are of the form $y_{k,(i,j)}\gamma_{k+1}(R)$, where $y_{k,(i,j)} \in \gamma_k(R) \cap \lambda_{i,j}(F)$ and $y_{k,(i,j)} \notin \gamma_k(R) \cap \lambda^+_{i,j}(F)$. We put

(3.5) $$Y_k = \{y_{k,(i,j)} \mid y_{k,(i,j)}\gamma_{k+1}(R) \in \overline{Y_k}\}.$$

For each $n \geq 1$, let

(3.6) $$Y_k^{(n)} = \{y_{k,(i,j)} \in Y_k \mid i \leq n\}.$$

Then $Y_k^{(n)}$ is a (possibly empty) finite set (in particular, $Y_k^{(n)} = \emptyset$ for $k > n$). Set

(3.7) $$U^{(n)} = \bigcup_{k=1}^{n} Y_k^{(n)}.$$

If, for some $k' > k$, $y_{k,(i,j)} \overset{*}{=} y_{k',(i,j)}$, then $y^a_{k,(i,j)}y^b_{k',(i,j)} \in \lambda^+_{i,j}(F)$ for some $a,b \in Z$. Thus, replacing $y_{k,(i,j)}\gamma_{(k+1)}(R)$ by $y^a_{k,(i,j)}y^b_{k',(i,j)}\gamma_{k+1}(R)$ in $\overline{Y_k}$, if necessary, we may assume that no two elements of $U^{(n)}$ are related by $\overset{*}{=}$. In other words:

(3.8) If $y_{k,(i,j)}, y_{k'(i',j')} \in U^{(n)}$ then $(i,j) \neq (i',j')$.

4. <u>A reduction theorem</u>. For $t \geq 1$, let $R_t = R \cap r_t(F)$. Then $R_t \leq (1 + \underline{\underline{r}} \wedge \underline{\underline{f}}^t)$, and using the identity $[a,b] - 1 = a^{-1}b^{-1}((a-1)(b-1) - (b-1)(a-1))$, it follows that $[R_{t_1}, R_{t_2}] \leq F \cap (1 + \underline{\underline{f}}^t\underline{\underline{r}})$, where $t = \min\{t_1,t_2\}$. More generally, for $n \geq 1$, let $\underset{\sim}{t}(m) = (t_1,\ldots,t_m)$, $2 \leq m \leq n+1$ be an m-tuple of positive integers satisfying

(4.1) $$t_1 + \ldots + \hat{t}_i + \ldots + t_m \geq n, \quad \text{for all } i,$$

($\hat{i}$ indicates i missing). Then, it follows that

$$[R_{t_1},\dots,R_{t_m}] \leq F \cap (1 + \underline{\underline{f}}^n\underline{\underline{r}}) = F(n,R) .$$

We define

$$(4.2) \qquad G(n,R) = \prod_{m=2}^{n+1} \prod_{\underset{\sim}{t}(m)} [R_{t_1},\dots,R_{t_m}] ,$$

where the product is taken over all m-tuples satisfying (4.1). Then $G(n,R) \leq F(n,R)$, and in turn, $\sqrt{G(n,R)} \leq F(n,R)$, since $F/F(n,R)$ is torsion-free (see [4]). As we shall prove in our main theorem, under a certain separability restriction on R, the reverse inequality also holds. In preparation, we make the following definition.

Definition 4.1. Let R be a normal subgroup of a finitely generated free group F. Set $H(0,R) = \{1\}$, $H(1,R) = R'$ and $H(n,R) = \prod_{i+j=n} [R_i,R_j]$ for $n \geq 2$. For $n \geq 0$, R is said to be n-separable if $R' \cap \gamma_m(F) = \sqrt{H(m,R)}$, for all $m \leq n$ ($\gamma_o(F) = \{1\}$).

We now state and prove our main theorem.

Theorem 4.2. Let R be a normal subgroup of a finitely generated free group F. If for some $n \geq 1$, R is $(n-1)$-separable, then $F(n,R) = \sqrt{G(n,R)}$, where $G(n,R)$ is the subgroup given by (4.2).

Proof. We shall prove by induction on $n \geq 1$, that $F(n,R) \leq \sqrt{G(n,R)}$. For $n = 1$, $F(1,R) = R' = G(1,R)$ is a well-known result due to Magnus [5]. For the inductive step, we assume that R is n-separable for some $n \geq 1$ and that $F(n,R) = \sqrt{G(n,R)}$. We proceed to conclude that $F(n+1,R) \leq \sqrt{G(n+1,R)}$.

Let $w \in F(n+1,R)$ be an arbitrary element. Since $F(n+1,R) \leq F(n,R)$, it follows by the induction hypothesis that some power of w lies in $G(n,R)$. Then, working modulo $\sqrt{G(n+1,R)}$, we can assume that w is a non-zero element of the right $Q(F/R)$-module $\overline{G(n,R)} \otimes_Z Q$, where $\overline{G(n,R)} = G(n,R)/G(n+1,R)$. For each $p = 2,\dots,n+1$, we set

$$(4.3) \qquad K_p^+(R) = \prod_{m=p+1}^{n+1} \prod_{\underset{\sim}{t}(m)} [R_{t_1},\dots,R_{t_m}] ,$$

where the product is taken over all m-tuples $(t_1,\ldots,t_m)$ of positive integers satisfying $t_1 + \ldots + \hat{t_i} + \ldots + t_m \geq n$ (here $K^+_{n+1}(R) = \{1\}$). Then, by (4.2), modulo $\sqrt{G(n+1,R)}$,

$$w \equiv w_2 \ldots w_{n+1},$$

where $w_p \in \prod_{\underset{\sim}{t}(p)} [R_{t_1},\ldots,R_{t_p}]$ and $w_p \notin K^+_p(R)$ for $p = 2,\ldots,n+1$. The proof consists in showing that each w_p is empty, contrary to the choice of w.

Let $p \in \{2,\ldots,n+1\}$ be least such that w_p is non-empty. Then, modulo $\sqrt{K^+_p(R)\ G(n+1,R)}$, w_p is a product of commutators of the form $[u_1,\ldots,u_p]$, $u_k \in R_{t_k}$. If, for some $k \in \{1,\ldots,p\}$, some power of u_k lies in $R' \cap \gamma_{t_k}(F)$, then, using the t_k-separability of R, some power of $[u_1,\ldots,u_p]$ lies in $K^+_p(R)$. Similarly, if $t_1 + \ldots + \hat{t}_i + \ldots + t_p \geq n+1$, for all i, then $[u_1,\ldots,u_p] \in G(n+1,R)$. Thus, modulo $\sqrt{K^+_p(R)\ G(n+1,R)}$ we may assume that w_p is a product of commutators of the form:

(4.4) $\quad u = [u_1,\ldots,u_p]$, satisfying

(a) $u_i \in R_{t_i}$, $u_i \notin R'$;

(b) $t_1 + \ldots + \hat{t}_i + \ldots + t_p \geq n$, for all i ;

(c) $t_1 + \ldots + \hat{t}_k + \ldots + t_p = n$, for some k .

Let $u = [u_1,\ldots,u_p]$ satisfying (4.4) (a) - (c). Then

$$u \equiv [u_k, [u_1,\ldots,u_{k-1}], u_{k+1},\ldots,u_p]^{-1}$$

$$\equiv \prod_{\sigma} [u_k, u_{1\sigma},\ldots,u_{(k-1)\sigma}, u_{k+1},\ldots,u_p]^{\varepsilon(\sigma)},$$

where $\varepsilon(\sigma) \in Z$, σ is a permutation of $1,\ldots,k-1$ (this follows by the Jacobi congruence). Thus, instead of (4.4), we may now assume that w_p is a product of commutators of the form:

(4.5) $\quad v = [v_0, v_{p-1},\ldots,v_1]$, satisfying

(a) $v_i \in R_{t_i}$, $v_i \notin R'$;

(b) $v_i <^* v_0$ or $v_i \overline{\underline{*}}\ V_0$ for $i = 1,\ldots,p-1$;

(c) $t_1 + \ldots + t_{p-1} = n$.

Now, let $v = [v_0, v_{p-1}, \ldots, v_{i+1}, v_i, \ldots, v_1]$ be a factor of w_p satisfying (4.5) (a) - (c), with $v_{i+1} <^* v_i$. Then, modulo $\sqrt{K_p^+(R)G(n+1,R)}$,

$$v \equiv [v_0, v_{p-1}, \ldots, v_i, v_{i+1}, \ldots, v_1]$$

$$\cdot [v_0, v_{p-1}, \ldots, [v_{i+1}, v_i], \ldots, v_1] ,$$

where $[v_{i+1}, v_i] \in \gamma_2(R) \cap \gamma_{t_i + t_{i+1}}(F)$. Thus by the repeated application of the above procedure, instead of (4.5), we may assume that w_p is a product of commutators of the form:

(4.6) $$v = [v_0, y_{k(\ell),(i(\ell),j(\ell))}, \ldots, y_{k(1),(i(1),j(1))}] ,$$

satisfying

(a) $y_{k,(i,j)} \in U^{(n)}$ (given by (3.7)) ;

(b) $k(1) + \ldots + k(\ell) = p - 1$;

(c) $i(1) + \ldots + i(\ell) = n$;

(d) $(i(1), j(1)) \leq \ldots \leq (i(\ell), j(\ell))$;

(e) $v_0 \in R \cap \gamma_{t_0}(F)$, $v_0 \notin R'$.

To complete the proof of the theorem we first recall that $w = w_p w_p^+$, where $w_p^+ \in K_p^+(R)$ and w_p has a non-trivial factor $v = [v_0, y_{k(\ell),(i(\ell),j(\ell))}, \ldots, y_{k(1),(i(1),j(1))}]$ satisfying (4.6) (a) - (e). Further $w \in F(n+1), R)$; so that $w-1 \in \underline{\underline{f}}^{n+1}\underline{\underline{r}}$. In the expansion of $w - 1$, modulo $\underline{\underline{f}}^{n+1}\underline{\underline{r}}$, we note that $v - 1$ contributes a unique term

$$(-1)^{\ell}(y_{k(1),(i(1),j(1))}^{-1}) \ldots (y_{k(\ell),(i(\ell),j(\ell))}^{-1})(v_0 - 1)$$

which, in turn, yields the term

(4.7) $$\alpha(b_{i(1),j(1)}^{-1}) \ldots (b_{i(\ell),j(\ell)}^{-1})(v_0 - 1) ,$$

where $y_{k,(i,j)} \equiv b_{i,j}^{\alpha_{i,j}} (\lambda_{i,j}^+(F))$ and $\alpha = (-1)^{\ell} \prod_{q=1}^{\ell} \alpha_{i(q),j(q)} \neq 0$. It follows from (3.8) and (4.6)(b), that no factor of $w_p^+ - 1$ can yield the basic product of the term (4.7) in its expansion modulo $\underline{\underline{f}}^{n+1}\underline{\underline{r}}$. Thus $w - 1 \in \underline{\underline{f}}^{n+1}\underline{\underline{r}}$ implies that

(4.8) $$(b_{i(1),j(1)}-1) \cdots (b_{i(\ell),j(\ell)}-1)(v_0^{\alpha} - 1) \in \underline{\underline{f}}^{n+1}\underline{\underline{r}} .$$

Since $(b_{i(1),j(1)}-1) \cdots (b_{i(\ell),j(\ell)}-1)$ is a part of the basis for $\underline{\underline{f}}^n$ (see, for instance, Theorem 5.8 of [6]), we conclude that $v_0^{\alpha} - 1 \in \underline{\underline{f}}\,\underline{\underline{r}}$ which, in turn, gives $v_0^{\alpha} \in R'$ or $v_0 \in R'$ (since F/R' is torsion free). This is contrary to (4.6)(e) and it follows that w_p is empty. This completes the proof of the theorem.

5. A Separability Criterion. Theorem 4.1 identifies $F(n,R)$ whenever R is $(n-1)$-separable. While we do not know any R which is not $(n-1)$-separable for all $n \geq 1$, here we give a criterion which is particularly useful for small values of n. Let $Y_1^{(n)}$ be as given by (3.6) with $k = 1$. We define

(5.1) $$S_1^{(n)} = \{b_{i,j} \in B_i \,|\, y_{1,(i,j)} \in Y_1^{(n)}\} ,$$

where B_i is given by (3.1).

Using $S_1^{(n)}$ as an ordered set of symbols (via lexicographic order on (i,j)'s), denote by $C_{1,p,q}^{(n)}$, the inductively defined and ordered set of basic commutators of weight p in $S_1^{(n)}$ and of weight q in the generators of F (e.g. if $y_{1,(1,i)}, y_{1,(2,j)}, y_{1,(3,k)} \in Y_1^{(n)}$ then $[b_{2,j}, b_{1,i}, b_{3,k}] \in C_{1,3,6}^{(n)}$).

Theorem 5.1. If for all $2 \leq p \leq q \leq n$, elements of $C_{1,p,q}^{(n)}$ are linearly independent modulo $\gamma_{q+1}(F)$, then R is $(n+1)$-separable.

Proof. We may assume R is n-separable and prove that R is $(n+1)$-separable. Let $w \in R' \cap \gamma_{n+1}(F)$ be an arbitrary element. Since $R' \cap \gamma_{n+1}(F) \leq R' \cap \gamma_n(F)$, by the induction hypothesis (some power of) w lies in $H(n,R)$ (given by Definition 4.1). Modulo $\sqrt{H(n+1,R)}$, we may assume that w lies in $\prod_{m=2}^{n} \prod_{\underset{\sim}{t}(m)} [R_{t_1},\ldots,R_{t_m}]$, where the product is taken over all m-tuples $(t_1,\ldots,t_m)$ of positive integers satisfying $t_1 + \cdots + t_m = n$. As in the proof of Theorem 4.2, we set

(5.2) $$L_p^{+}(R) = \prod_{m=p+1}^{n} \prod_{\underset{\sim}{t}(m)} [R_{t_1},\ldots,R_{t_m}] .$$

Thus $w \equiv w_2 \ldots w_n$, where $w_p \in \prod_{\underset{\sim}{t}(m)} [R_{t_1}, \ldots, R_{t_m}]$ and $w_p \notin L_p^+(R)$ (here $L_n^+(R) = \{1\}$). Let $p \in \{2,\ldots,n\}$ be least such that $w_p \notin \gamma_{n+1}(F)$. Then, modulo $L_p^+(R)$, w_p is a product of elements of $Y_p^{(n)}$ of the form $y_{p,(n,j)}$ which are clearly independent modulo $\gamma_{n+1}(F)$. By (3.8), the elements of $Y_{p+1}^{(n)}, \ldots, Y_n^{(n)}$ are disjoint from those of $Y_p^{(n)}$. Since $w \in \gamma_{n+1}(F)$, it follows that $w_p \in \gamma_{n+1}(F)$, contrary to the choice of w_p. Thus we may assume that each $w_p \in \gamma_{n+1}(F)$. Now, $w_p \in \prod_{\underset{\sim}{t}(p)} [R_{t_1}, \ldots, R_{t_p}]$ implies that w_p is a product of commutators of the form

$$u = [u_1, \ldots, u_p] ,$$

where $u_i \in R_{t_i}$, $t_1 + \ldots + t_p = n$, $u_i \notin R'$. Thus, modulo $\gamma_{n+1}(F)$, w_p is a product of basic commutators of weight p in $y_1^{(n)}$ and of total weight n in F. Since each such basic commutator contributes a unique element of $C_{1,p,n}^{(n)}$, it follows, by hypothesis, that if $w_p \in \gamma_{n+1}(F)$ then w_p is empty. This completes the proof of the theorem.

Let F/R be given by an m-generator $(m \geq 2)$ pre-abelian presentation (see [6], Theorem 3.5). So, we may assume that F is generated by $x_1, \ldots, x_m$ and $x_1^{e_1}, \ldots, x_m^{e_m} \in RF'$ with $e_m | e_{m-1} | \ldots | e_1$, $e_i \geq 0$. If $e_1 \neq 0$ (i.e. F/RF' is periodic), then $S_1^{(n)} = \{x_1, \ldots, x_m\}$ and non-trivial elements of $C_{1,p,q}^{(n)}$ occur with $p = q$ and coincide with the elements of B_q which are linearly independent modulo γ_{q+1}. Thus in this case, R is n-separable for all n and consequently we have,

<u>Theorem 5.2</u> (see [3]). If F/RF' is finite then $F(n,R) = \sqrt{G(n,R)}$ for all n.

Similarly, if $e_1 = 0$, $e_2 \neq 0$, then $S_1^{(n)} = \{x_2, \ldots, x_m, [x_i, \underbrace{x_1, \ldots, x_1}_{p}]$ for $i = 2, \ldots, m$, $p = 1, \ldots, n-1\}$; and the basic commutators with entries from $S_1^{(n)}$ occur as the basic commutators with entries from $\{x_1, \ldots, x_m\}$. Thus R is n-separable for all n and as a consequence we have,

<u>Theorem 5.3</u>. If F/R is 2-generator with $R \nleq F'$ then $F(n,R) = \sqrt{G(n,R)}$ for all n.

If $e_1 = 0 = e_2 = \dots = e_s$, $s \geq 2$, then it follows that $S_1^{(4)}$ is a subset of $\{x_{s+1},\dots,x_m; [x_i,x_j], i \in \{s+1,\dots,m\}, j \in \{1,\dots,s\};$ $[x_i,x_j,x_k], i \in \{s+1,\dots,m\}, j,k \in \{1,\dots,s\}, j < k;$ $[x_i,x_j,x_k],$ $i,j \in \{1,\dots,s\}, i > j, k \in \{s+1,\dots,m\};$ $[x_i,x_j,x_k,x_\ell], i \in \{s+1,\dots,m\},$ $j,k,\ell \in \{1,\dots,s\}, j \leq k \leq \ell;$ $[x_i,x_j; x_k,x_\ell], i \in \{s+1,\dots,m\},$ $j,k,\ell \in \{1,\dots,s\}, k > \ell;$ all basic commutators of weight 2,3,4 in $\{x_1,\dots,x_s\}\}$. If we compute $C_{1,p,q}^{(5)}$'s with $2 \leq p \leq q \leq 5$ we only need to use elements of $S_1^{(4)}$. It is not difficult to verify that the elements of $C_{1,p,q}^{(5)}$ with $2 \leq p \leq q \leq 5$ are linearly independent modulo $\gamma_{q+1}(F)$. So, by Theorem 5.1, R is 6-separable. This yields,

Theorem 5.4. If R is a normal subgroup of a finitely generated free group F then $F(n,R) = \sqrt{G(n,R)}$ for $n \leq 7$.

Concluding Remark. There are several other situations where Theorem 5.1 yields n-separability. For instance when F/R is 2-generator with $R \not\leq \gamma_4(F)$ then R is n-separable for all n and if $R \leq \gamma_4(F)$ then R can be shown to be n-separable for $n \leq 16$. We omit the details.

REFERENCES

1. R. H. Fox, Free differential calculus I - Derivations in free group rings, Ann. of Math. 57 (1953), 547-560.

2. Karl W. Gruenberg, Cohomological topics in group theory, Lecture Notes in Mathematics, No. 143, Springer-Verlag, New York (1970).

3. Narain Gupta, Fox subgroups of free groups, J. Pure and Appl. Algebra 11 (1977), 1-7.

4. Narain Gupta, A problem of R. H. Fox, Canad. Math. Bull. 24(2) (1981), 129-136.

5. W. Magnus, On a theorem of Marshall Hall, Ann. of Math. Ser. II 40 (1939), 764-768.

6. Wilhelm Magnus, Abraham Karrass and Donald Solitar, Combinatorial Group Theory, Interscience, New York (1966).

DEPARTMENT OF MATHEMATICS
UNIVERSITY OF MANITOBA
WINNIPEG, R3T 2N2, CANADA

Received December 1, 1981

Contemporary Mathematics
Volume 33, 1984

EXTENSION OF GROUPS BY TREE AUTOMORPHISMS

Narain Gupta* and Said Sidki**

(To Roger Lyndon on his 65th Birthday)

Let G be a non-trivial group. We construct a tree $\mathcal{T}(e)$ with G embedded in $\mathcal{A} = \mathrm{Aut}(\mathcal{T}(e))$. We select an extra automorphism $\gamma \in \mathcal{A}$ and consider the group $\mathcal{E} = \langle G, \gamma \rangle$ of $\mathcal{A}$. We refer to $\mathcal{E}$ as the extension of G by the tree automorphism γ. The extension $\mathcal{E}$ provides a new tool in group theory as is evidenced by some of its remarkable properties (e.g. property 2.1.1, corollary 2.2.2, Theorems 3.2.1, 4.1.1, 4.2.1)

1. Trees and automorphisms.

1.1. Definition of the tree $\mathcal{T}(e)$. Let $T(X)$ be the free monoid, with identity e, freely generated by a non-empty set X. We define a partial order "$\geq$" on $T(X)$ by:

$$u \geq v \iff v = uw \tag{1}$$

for some $w \in T(X)$. Let $\mathcal{T}(e)$ denote the graph of $(T(X), \geq)$ with $T(X)$ as its set of vertices, where two vertices u,v are connected by an edge u—v (vertical, u above v) provided $v = ux$ for some $x \in X$. Clearly $\mathcal{T}(e)$ is a one-rooted infinite descending regular tree

*Research supported by NSERC, Canada.

**Research supported by CNPq, Brazil. The second author thanks the Mathematics Department of the University of Manitoba for its warm hospitality.

© 1984 American Mathematical Society
0271-4132/84 $1.00 + $.25 per page

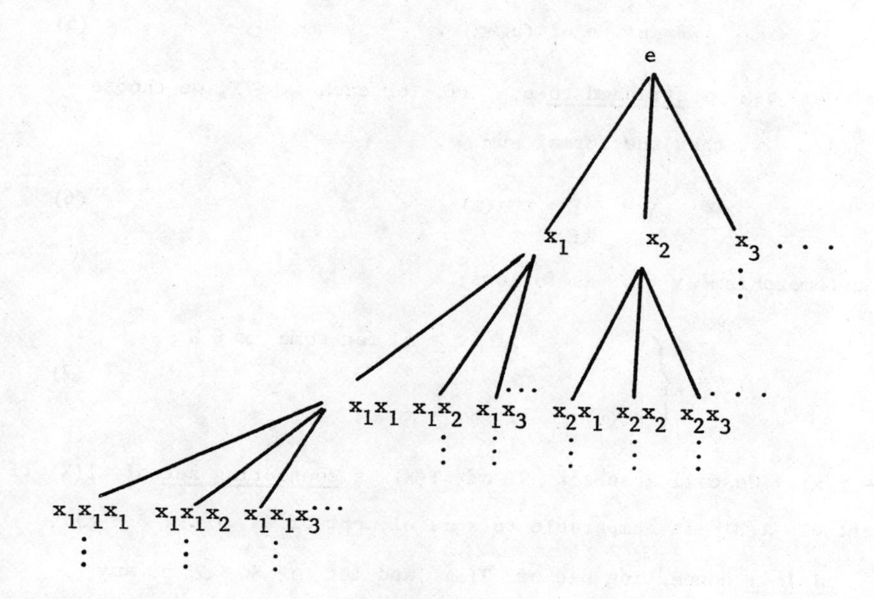

where, for any vertex $u \in T(X)$, the subtree $\mathcal{T}(u)$, rooted by u, is isomorphic to $\mathcal{T}(e)$ via the map : $uv \to v$ for all $v \in T(X)$.

1.2. Structure of $\mathrm{Aut}(\mathcal{T}(e))$. Let $\mathcal{A} = \mathcal{A}_{[e]} = \mathrm{Aut}(\mathcal{T}(e))$ be the group of automorphisms of $\mathcal{T}(e)$ and for $u \in T(X)$, let $\mathcal{A}_{[u]} = \mathrm{Aut}(\mathcal{T}(u))$. Then, for any $\alpha \in \mathcal{A}$ and $u \in T(X)$, α induces $\alpha_{[u]} \in \mathcal{A}_{[u]}$ as follows:

$$(uv)\alpha_{[u]} = u.(v)\alpha \ . \tag{2}$$

Conversely, if $\beta \in \mathcal{A}_{[u]}$ for some $u \in T(X)$, then β can be extended to an automorphism $\alpha \in \mathcal{A}_u \leq \mathcal{A}$ as follows:

$$(v)\alpha = \begin{cases} v & \text{if } v \notin \mathcal{T}(u) \\ (v)\beta & \text{if } v \in \mathcal{T}(u) \end{cases} \tag{3}$$

and $\beta = \alpha_{[u]}$. It is readily seen that $\mathcal{A} \cong \mathcal{A}_u \cong \mathcal{A}_{[u]}$ for all $u \in T(X)$. For any $u \in T(X)$ and $\alpha \in \mathcal{A}$, (2) and (3) allow us to define a unique automorphism $u*\alpha \in \mathcal{A}_u$ as follows:

$$(v)\,u*\alpha = \begin{cases} v & \text{if } v \notin \mathcal{T}(u) \\ u.(w)\alpha & \text{if } v = uw \end{cases} \ . \tag{4}$$

We note that for $u = u_1u_2 \in T(X)$,

$$u*\alpha = u_1*(u_2*\alpha) . \tag{5}$$

We refer to $u*\alpha$ as: α attached to u. If, for each $x \in X$, we choose $\alpha(x)$ attached to x, then the formal sum

$$\gamma = \sum_{x \in X} x*\alpha(x) \tag{6}$$

defines an automorphism $\gamma \in \mathcal{A}$ as follows:

$$(v)\gamma = \begin{cases} v & \text{if } v \geq x \text{ for some } x \in X \\ x.(u)\alpha_x & \text{if } v = xu \end{cases} , \tag{7}$$

where $\alpha_x = \alpha(x)$. We call a subset S of T(X) a connecting set of T(X) if every element of T(X) is comparable to some element of S (e.g. S = X). Let S be a minimal connecting set of T(X) and let $\alpha: S \to \mathcal{A}$ be any function. Then as in (6) and (7), the formal sum

$$\gamma = \sum_{s \in S} s*\alpha(s) \tag{8}$$

defines an automorphism $\gamma \in \mathcal{A}$ given by:

$$(v)\gamma = \begin{cases} v & \text{if } v \geq s \text{ for some } s \in S \\ s.(u)\alpha_s & \text{if } v = su \text{ for some } s \in S \end{cases} , \tag{9}$$

where $\alpha_s = \alpha(s)$. We refer to $\alpha: S \to \mathcal{A}$ as a decorating function for the tree $\mathcal{T}(e)$. Let $\mathcal{A}_X = \prod_{x \in X} \mathcal{A}_x$ be the cartesian product of the subgroups $\mathcal{A}_x$ of $\mathcal{A}$. Then

$$\mathcal{A}_X = \{ \sum_{x \in X} x*\alpha(x) \mid \alpha: X \to \mathcal{A} \} \tag{10}$$

is the set of all automorphisms of $\mathcal{T}(e)$ which leave elements of X point-wise fixed. Let $\mathcal{B} = \mathcal{B}(X)$ be the group of all bijections on X. Then $\mathcal{B}$ is embeddable in $\mathcal{A}$ as follows: given $\sigma \in \mathcal{B}$, $v = xu \in T(X)$ with $x \in X$, define $\sigma \in \mathcal{A}$ by

$$\sigma: xu \to (x)\sigma.u \quad . \tag{11}$$

The group $\mathcal{A}$, then, is the semi-direct product

$$\mathcal{A} = \mathcal{A}_X \mathcal{B}, \tag{12}$$

where $\mathcal{B}$ acts on $\mathcal{A}_X$ as follows: given $\gamma = \sum_{x \in X} x*\alpha(x) \in \mathcal{A}_X$, $\sigma \in \mathcal{B}$, $u \in T(X)$,

$$\begin{aligned}(xu)\sigma^{-1}\gamma\sigma &= ((x)\sigma^{-1}.u)\gamma\sigma, && \text{by (11)}\\ &= ((x)\sigma^{-1}.(u)\alpha_{(x)\sigma^{-1}})\sigma, && \text{by (7)}\\ &= x.(u)\alpha_{(x)\sigma^{-1}}, && \text{by (11)}\\ &= (xu)x*\alpha_{(x)\sigma^{-1}}, && \text{by (7)}\end{aligned}$$

where $\alpha_{(x)\sigma^{-1}} = \alpha((x)\sigma^{-1})$. Thus,

$$\begin{aligned}\sigma^{-1}\gamma\sigma &= \sum_{x \in X} x*\alpha((x)\sigma^{-1})\\ &= \sum_{x \in X} (x)\sigma*\alpha(x).\end{aligned} \tag{13}$$

Remark. For $k \geq 1$, let X^k be the set of elements of length k in the monoid $T(X)$. Let $\mathcal{A}^{(k)}$ be the subgroup of $\mathcal{A}$ consisting of all the automorphisms which fix X^k elementwise. Then for all $k \geq 1$, $\mathcal{A}^{(k)} \trianglelefteq \mathcal{A}$, $\mathcal{A}^{(k+1)} < \mathcal{A}^{(k)}$, $\bigcap_{k \geq 1} \mathcal{A}^{(k)} = \{e\}$. In particular $\mathcal{A}$ is residually finite if X is finite.

1.3. Trees on groups and decorating functions. Let G be a non-trivial group with identity e, $X = \underline{G} = \{\underline{g} \mid g \in G\}$, and $\mathcal{T}(e) = (T(X), \geq)$. The group G is embeddable in $\mathcal{B}(\underline{G})$ by its regular representation:

$$h: \underline{g} \to \underline{gh} \tag{14}$$

for all $g,h \in G$. This, in turn, yields the embedding of G in $\mathcal{A} = \mathrm{Aut}(\mathcal{T}(e))$ in accordance with (11), (12), (13).

Let $X^{\#} = X \backslash \{\underline{e}\}$ and define the following infinite subset S of $T(X)$:

$$S = X^{\#} \cup \underline{e}X^{\#} \cup \underline{e}\,\underline{e}X^{\#} \cup \ldots, \tag{15}$$

where for $Y \subset T(X)$, $\underline{e}Y = \{\underline{e}y \mid y \in Y\}$. Then S is a minimal connecting set of $T(X)$. A function $\alpha: S \to G$ is called a special decorating function provided:

$$\alpha(\underline{g}) = \alpha(\underline{e}\,\underline{g}) = \alpha(\underline{e}\,\underline{e}\,\underline{g}) = \ldots \tag{16}$$

for all $g \in G\backslash\{e\}$, and

$$G = \langle \alpha(\underline{g}) \mid g \neq e \rangle \ . \tag{17}$$

Let $\alpha: S \to G$ be a special decorating function and let $\gamma = \sum_{s \in S} s{*}\alpha(s)$ be the associated automorphism of $\mathcal{T}(e)$ as given by (8) and (9), where S is given by (15). In view of (16), γ may be written recursively as

$$\gamma = \underline{e}{*}\gamma + \sum_{g \neq e} \underline{g}{*}\alpha(\underline{g}) \ . \tag{18}$$

Further, for any $h \in G^{\#}$, by (13),

$$\begin{aligned} \gamma^{h} &= \underline{h}{*}\gamma + \sum_{g \neq e} \underline{gh}{*}\alpha(\underline{g}) \\ &= \underline{e}{*}\alpha(h^{-1}) + \underline{h}{*}\gamma + \sum_{g \neq e, h} \underline{g}{*}\alpha(\underline{gh}^{-1}) \ . \end{aligned} \tag{19}$$

For simplicity of notation we set $\alpha(\underline{e}) = \gamma$. Then we can write (18), (19) as:

$$\gamma = \sum_{g \in G} \underline{g}{*}\alpha(\underline{g}) \tag{20}$$

$$\begin{aligned} \gamma^{h} &= \sum_{g \in G} \underline{gh}{*}\alpha(\underline{g}) \\ &= \sum_{g \in G} \underline{g}{*}\alpha(\underline{gh}^{-1}) \ . \end{aligned} \tag{21}$$

2. The extension $\mathcal{G}$

Let G be a non-trivial group, $\alpha: S \to G$ a special decorating function and γ its associated tree automorphism as described in §1.3. Consider the group $\mathcal{G} = \langle G, \gamma \rangle$ (as a subgroup of $\mathcal{A} = \mathrm{Aut}(\mathcal{T}(e))$) with G acting on γ as in (21).

2.1. Some general properties of $\mathcal{G}$. Let

$$\Gamma = \langle \gamma^{h} \mid h \in G \rangle \ .$$

Then $\Gamma \lhd \mathcal{G}$ and $\mathcal{G} = \Gamma . G$ is the semi-direct product. Further, $\Gamma \leq \mathcal{A}_X$ with $X = \underline{G}$. For $g \in G$, let

$$\pi_g : \Gamma \to \mathcal{G} \tag{22}$$

be the projection on the $\underline{g}$-th co-ordinate given by

$$\pi_g : \gamma^h \to \alpha(\underline{gh}^{-1}) \tag{23}$$

for all $h \in G$ (with $\alpha(\underline{e}) = \gamma$). Then

$$w(\gamma^{h(1)^{-1}g}, \ldots, \gamma^{h(k)^{-1}g})\pi_g = w(\alpha(h(1)), \ldots, \alpha(h(k)))$$

implies that

$$\Gamma \text{ projects onto } \mathcal{G}. \tag{24}$$

Let $g(1), g(2), \ldots$ be an infinite sequence of elements of G (not necessarily distinct). Consider the following diagram:

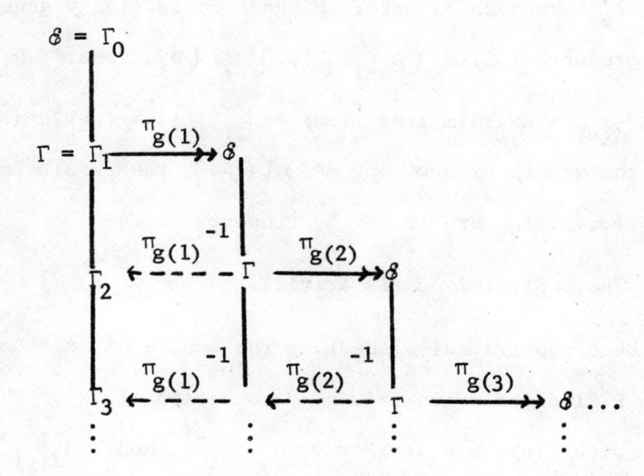

The following two properties are immediate from the diagram.

Property 2.1.1. Let $g(1), g(2), \ldots$ be an infinite sequence of elements of G. Then $\mathcal{G}$ contains an infinite subnormal series $\Gamma_0 = \mathcal{G}$, $\Gamma_1 = \Gamma$, $\Gamma_{i+1} = (\Gamma)\pi_{g(i)}^{-1} \ldots \pi_{g(1)}^{-1}$ for $i \geq 1$, with $\Gamma_i/\Gamma_{i+1} \cong G$ for all $i \geq 0$.

Property 2.1.2. The stabilizer in Γ of any vertex $u = g(1) \ldots g(k)$ projects onto $\mathcal{G}$. In other words given $\beta \in \mathcal{G}$, there exists $\hat{\beta} \in \Gamma$ such that $\hat{\beta}|_{\mathcal{T}(u)} = u*\beta$.

Since G is transitive on $X = \underline{G}$ and the stabilizer of any vertex projects onto $\mathcal{G}$, we conclude, by induction,

Property 2.1.3. $\mathcal{G}$ is transitive on X^k for all $k \geq 1$.

Let $K \lhd \mathcal{S}$ such that $\underline{e}*K \leq K$. Conjugating by $g \in G$ yields $\underline{g}*K \leq K$. Further, by (5), $\underline{gh}*K = \underline{g}*(\underline{h}*K) \leq \underline{g}*K \leq K$. Thus by iteration, $u*K \leq K$ for all $u \in T(X)$. We record this as,

Property 2.1.4. If $K \lhd \mathcal{S}$ with $\underline{e}*K \leq K$ then $u*K \leq K$ for all $u \in T(X)$.

Another easy consequence is,

Property 2.1.5. Let $G = \langle a \mid a^n = 1\rangle$ be cyclic of order n. Let $K \lhd \mathcal{S}$ with $\underline{e}*K \leq K$. Assume further that there exists $\delta \in K$ with $|\delta| = n$ and $\delta|_{\mathcal{T}(u)} = u*a$. Define $\delta_1 = \delta$, $\delta_{i+1} = (u\underline{e})^i*\delta$ for $i \geq 1$. Then for all $k \geq 2$, $\delta_1, \delta_2, \ldots, \delta_k$ are each of order n and collectively generate the iterated wreath product $(\ldots(D_k \wr D_{k-1}) \wr \cdots \wr D_2) \wr D_1$, where $D_i = \langle \delta_i \rangle$.

Proof. Since $\delta|_{\mathcal{T}(u)} = u*a$ implies $(u\underline{e})^i*\delta|_{\mathcal{T}(u\underline{e})^i u} = (u\underline{e})^i u*a$ and, $v*a$, $v\ \underline{e}*\delta$ generate the wreath product $\langle v\ \underline{e}*\delta\rangle \wr \langle v*a\rangle$, the result follows by iteration since $\delta_1, \ldots, \delta_k$ are in K by Property 2.1.4.

Property 2.1.6. The centre of $\mathcal{S}$ is trivial.

Proof. Let β be a non-trivial element in the centre of $\mathcal{S}$. Let $k \geq 1$ be least such that β induces a non-trivial permutation of X^k. Then there exists $u = v\,\underline{g}$ with $(u)\beta \neq u$, $(v)\beta = v$, $u \in X^{k-1}$, and $\beta|_{\mathcal{T}(v)} = v*\beta_1$, where $\beta_1 = wh$ for some $w \in \Gamma$ and $h \in G^{\#}$. Now, as the stabilizer in Γ of the vertex v projects on $\mathcal{S}$, it follows that β_1 is in the centre of $\mathcal{S}$. However,

$$[\beta_1, \gamma] = [wh, \gamma] = \gamma^{-wh}\gamma$$

and $(\gamma^{-wh}\gamma)\pi_e = h'\gamma \neq e$, since $wh \neq e$. This contradicts the assumption that β is non-trivial.

The elements of Γ afford a more detailed description. Let $w \in \Gamma$. Then $w = w(\gamma^{h(1)}, \ldots, \gamma^{h(k)})$ is a word in $\gamma^{h(1)}, \ldots, \gamma^{h(k)}$ for some $h(1), \ldots, h(k)$ distinct elements of G. Also

$$w = \sum_{g \in G} \underline{g}*w_g, \tag{26}$$

where by (23),

$$w_g = w(\alpha(gh(1)^{-1}),\ldots,\alpha(gh(k)^{-1})) \ . \tag{27}$$

We define the syllable length $\ell(w)$ of $w \in \Gamma$ as the syllable length of w when viewed as an element of $\hat{\Gamma}$, the free product of the set of groups $\{\langle\gamma^g\rangle | g \in G\}$, and extend the definition to the syllable length of an arbitrary word $y = wg \in \mathcal{G}$ by defining $\ell(y) = \ell(w) + 1$ for $w \in \Gamma$ and $g \in G^{\#}$. If $w \in \Gamma$ is given by (26), then clearly $\ell(w_g) \le \ell(w)$, with $\ell(w_g) = \ell(w)$ only if $w \in \langle\gamma^h\rangle$ or $w \in \langle\gamma^h\rangle\langle\gamma^{h'}\rangle$ or $w \in \langle\gamma^h\rangle\langle\gamma^{h'}\rangle\langle\gamma^h\rangle$ for some h,h' distinct in G. We record this as,

Property 2.1.7. Let $w = \sum_{g\in G} g{*}w_g$ be an arbitrary element of Γ. Then $\ell(w_g) < \ell(w)$ for all $g \in G$ except, possibly, when $\ell(w) \le 3$.

2.2. The identity decorating function. Let $\alpha\colon S \to G$ be defined by

$$\alpha((e)^i g) = g,\ i \ge 0,\ g \in G^{\#} \ . \tag{28}$$

Then the tree decorated by α represents the automorphism

$$\gamma = \underline{e}{*}\gamma + \sum_{g\ne e} \underline{g}{*}g \ . \tag{29}$$

Again, let $\mathcal{G} = \langle G,\gamma\rangle$, $\Gamma = \langle\gamma^g | g \in G\rangle$ and let Γ' be the commutator subgroup of Γ. We shall show that barring certain exceptional groups G, $\mathcal{G}$ enjoys the important property

$$\underline{e}{*}\Gamma' \le \Gamma' \ .$$

For any $h \in G$, $h^2 \ne e$, we have, in turn,

$$\gamma = \underline{e}{*}\gamma + \underline{h}{*}h + \underline{h}^2{*}h^2 + \sum_{g\ne e,h,h^2} g{*}g$$

$$\gamma^h = \underline{e}{*}h^{-1} + \underline{h}{*}\gamma + \underline{h}^2{*}h + \sum_{g\ne e,h,h^2} \underline{g}{*}gh^{-1}$$

$$[\gamma,h] = \underline{e}{*}\gamma^{-1}h^{-1} + \underline{h}{*}h^{-1}\gamma + \underline{h}^2{*}h^{-1} + \sum_{g\ne e,h,h^2} \underline{g}{*}h^{-1}$$

$$[\gamma,h]^h = \underline{e}{*}h^{-1} + \underline{h}{*}\gamma^{-1}h^{-1} + \underline{h}^2{*}h^{-1}r + \sum_{g\ne e,h,h^2} \underline{g}{*}h^{-1}$$

$$[\gamma,h,h] = \underline{e}{*}h\gamma h^{-1} + \underline{h}{*}\gamma^{-1}h\gamma^{-1}h^{-1} + \underline{h}^2{*}\gamma \ .$$

Thus if $h_1^2 \neq e \neq h_2^2$ and $\{h_1, h_1^2\} \cap \{h_2, h_2^2\} = \emptyset$, then

$$[[\gamma, h_1, h_2], [\gamma, h_2, h_2]] = \underline{e}*[h_1 \gamma h_1^{-1}, h_2 \gamma h_2^{-1}] . \tag{30}$$

Since π_e projects Γ onto $\mathcal{G}$, by suitably conjugating (30) with elements of Γ we obtain

$$\underline{e}*[\gamma^g, \gamma]^{\mathcal{G}} \leq \Gamma , \tag{31}$$

where $g = h_1 h_2^{-1}$. Given $g \in G^{\#}$, we say that g is <u>pairable</u> if there exists $h \in G^{\#}$ such that

$$(hg)^2 \neq e \neq h^2, \ \{hg, (hg)^2\} \cap \{h, h^2\} = \emptyset .$$

Thus if every element of G is pairable then $\underline{e}*[\gamma^g, \gamma] \in \Gamma'$ for all $g \in G$ and, in particular

$$\underline{e}*\Gamma' \leq \Gamma' . \tag{32}$$

We state without proof the following elementary lemma.

<u>Lemma 2.2.1</u>. Let G be a non-trivial finite group which contains a non-pairable element $g \neq e$. Then $G = \langle g \,|\, g^3 = 1 \rangle$, G has a central involution g or G is generalized dihedral: $G = A\langle g \rangle$, $g^2 = e$, $a^g = a^{-1}$ for all $a \in A$.

Together with Property 2.1.4, Lemma 2.2.1 yields the following interesting corollary:

<u>Corollary 2.2.2</u>. For n odd, $n > 3$, there exists a group K which contains a subgroup H of finite index such that $H \cong \underbrace{K \times \ldots \times K}_{n}$ and a subgroup $L \cong \bigtimes_{j \geq 1} K_j$, $K_j \cong K$.

<u>Proof</u>. Let G be cyclic of order n and $K = \Gamma'$. Since $n > 3$, every element of G is pairable and hence $\underline{e}*K \leq K$ by (32). The result now follows by Property 2.1.4 with $H = \sum_{g \in G} \underline{g}*K \leq K$ and $L = \sum_{i \geq 0} \sum_{g \neq e} (\underline{e})^i \underline{g}*K \leq K$.

3. <u>Periodic groups</u>

Let G be a non-trivial finitely generated periodic group. We recall some notation:

$$X = \underline{G},\ X^{\#} = X \backslash \{\underline{e}\}\ ,\ S = \bigcup_{i \geq 0} (\underline{e})^{i} X^{\#}\ .$$

3.1. Periodicity preserving decorating functions. A decorating function $\alpha: S \to G$ is called periodicity preserving if it satisfies the following conditions:

(i) $\alpha((\underline{e})^{i}\underline{g}) = \alpha(\underline{g})$, $i \geq 0$, $g \in G^{\#}$,

(ii) $G = \langle \alpha(\underline{g}) \mid g \neq e \rangle$,

(iii) $\{\underline{g} \mid \alpha(\underline{g}) \neq e\}$ is finite ,

(iv) $\{\alpha(\underline{h}^{i}) \mid i \neq 0\}$ is a commutative set for all $h \in G^{\#}$,

(v) $\prod_{i=1}^{|h|-1} \alpha(\underline{h}^{i}) = e$ for all $h \in G^{\#}$.

A cyclic group of order 2 does not admit a periodicity preserving decorating function, for, here conditions (ii) and (iv) are incompatible. Nevertheless, almost all finitely generated periodic groups admit periodicity preserving functions. In particular we note,

Lemma 3.1.1. Let G be a finite group of odd order and let $\alpha: S \to G$ be the identity function: $\alpha((\underline{e})^{i}\underline{g}) = g$, $g \in G^{\#}$, $i \geq 0$. Then α is periodicity preserving.

Lemma 3.1.2. Let G be a periodic group generated by a finite minimal set Y without elements of order 2. Define $\alpha: S \to G$ by

$$\alpha((\underline{e})^{i}\underline{g}) = \begin{cases} g & \text{if } \langle g \rangle = \langle y \rangle \text{ for some } y \in Y \\ e & \text{otherwise} \end{cases} .$$

Then α is periodicity preserving.

Proof. By the minimality of Y, α is well defined. The conditions (i) - (iv) are trivially satisfied. As for the condition (v), this follows from the fact that for $y \in Y$, $|y| = m$, $\prod_{\substack{1 \leq j \leq m-1 \\ (j,m)=1}} y^{j} = e$.

Remark. It is not difficult to see that the only finitely generated periodic groups which do not admit a periodicity preserving function are "generalized dihedral": $G = A\langle y\rangle$, A abelian, $y^2 = e$, $a^y = a^{-1}$ for all $a \in A$.

3.2. A theorem on periodicity.

Theorem 3.2.1. Let G be a non-trivial finitely generated periodic group which admits a periodicity preserving decorating function α. Let γ be the associated automorphism of $\mathcal{T}(e)$. Then $\mathcal{G} = \langle G, \gamma\rangle$ is a periodic group. Furthermore, the orders of elements of $\mathcal{G}$ involve the same primes as the orders of elements of G.

Proof. Let $y = wh$, $w \in \Gamma$, $h \in G$, be an arbitrary element of $\mathcal{G}$. We shall prove by induction on $\ell(y)$ that y has finite order. When $\ell(y) = 1$, $y \in G$ or $y \in \langle\gamma^h\rangle$ where $\gamma = \sum_{g\in G} \underline{g}*\alpha(\underline{g})$, $\alpha(\underline{e}) = \gamma$. Since each $\alpha(g)$ is periodic and $\{\alpha(\underline{g}) \mid g \in G^{\#}\}$ is finite, it follows that γ is periodic. We next treat the case $\ell(y) = 2$. Here $y \in \langle\gamma^h\rangle G$ or $y \in \langle\gamma^h\rangle\langle\gamma^{h'}\rangle$.

Case (a). $y \in \langle\gamma^h\rangle G$.

Here conjugating by h^{-1}, if necessary, we may assume that $y = \gamma^i h$, $|h| = m > 1$. Then

$$y^m = \gamma^i(\gamma^i)^{h^{-1}}\cdots(\gamma^i)^{h^{-(m-1)}} = \sum_{g\in G} \underline{g}*\alpha(\underline{g})^i\alpha(\underline{gh})^i\cdots\alpha(\underline{gh^{m-1}})^i$$

by (27), where $\alpha(\underline{e}) = \gamma$. If $g \notin \langle h\rangle$, then $\alpha(\underline{g})^i\alpha(\underline{gh})^i\cdots\alpha(\underline{gh^{m-1}})^i \in G$ and hence is periodic. If $g \in \langle h\rangle$, then exactly one of $\underline{g}, \underline{gh}, \ldots, \underline{gh^{m-1}}$ is $\underline{e}$ and by the periodicity properties (iv) and (v) it follows that $\alpha(\underline{g})^i\alpha(\underline{gh})^i\cdots\alpha(\underline{gh^{m-1}})^i$ is conjugate of γ^i and hence is periodic. Thus y^m is periodic and so is y.

Case (b). $y \in \langle\gamma^h\rangle\langle\gamma^{h'}\rangle$.

As before we may assume $y = \gamma^i(\gamma^j)^h$. Then $y = \sum_{g\in G} \underline{g}*\alpha(\underline{g})^i\alpha(\underline{gh^{-1}})^j$. Since at most one of $\underline{g}$, $\underline{gh^{-1}}$ is $\underline{e}$, it follows that $\alpha(\underline{g})^i\alpha(\underline{gh^{-1}})^j$ lies in G or in $\langle\gamma^{h'}\rangle G$ and hence is periodic by the Case (a). Further, by the property (iii) of periodicity, only finitely many $\alpha(\underline{g})$'s are non-trivial. Thus y is periodic.

Let $\ell(y) \geq 3$ and assume that elements of $\mathcal{S}$ of syllable length less than $\ell(y)$ are periodic. We proceed to prove that y is periodic. Let $y = wh^{-1}$ where $w \in \Gamma$, $h \in G$. There are two cases to be considered: $h = e$ or $h \neq e$.

Case (c). $y = w \in \Gamma$.

Here, $w = \sum_{g \in G} g{*}w_g$ by (26); and by Property 2.1.7, $\ell(w_g) < \ell(w)$, unless $w \in \langle\gamma^h\rangle\langle\gamma^{h'}\rangle\langle\gamma^h\rangle$, in which case a conjugate of w lies in $\langle\gamma^{h'}\rangle\langle\gamma^h\rangle$ and w is periodic by the case (b) above.

Case (d). $y = wh^{-1}$, $|h| = m > 1$, $\ell(w) = k > 1$.

Let $w = (\gamma^{t(1)})^{g(1)} \cdots (\gamma^{t(k)})^{g(k)}$ for some $g(i) \in G$ and $t(i) \in \{1,\ldots,|\gamma|-1\}$. Then

$$\hat{w} = (wh^{-1})^m = ww^h \cdots w^{h^{m-1}}$$
$$= \prod_{i=0}^{m-1} (\gamma^{t(1)})^{g(1)h^i} \cdots (\gamma^{t(k)})^{g(k)h^i}$$

so that, by (23),

$$\hat{w}_g = (\hat{w})\pi_g = \prod_{i=0}^{m-1} (\alpha(\underline{gh^{-i}g(1)^{-1}}))^{t(1)} \cdots (\alpha(gh^{-i}g(k)^{-1}))^{t(k)} \tag{33}$$
$$= c_g h_g \text{ , say ,}$$

where $c_g \in \Gamma$ and $h_g \in G$. For each $j = 1,\ldots,k$, the set

$$\{gh^{-i}g(j)^{-1},\ i = 0,\ldots,m-1\} \tag{34}$$

has at most one element which is e. Thus $\ell(c_g) \leq k$. If $\ell(c_g) < k$ then $\ell(\hat{w}_g) = \ell(c_g) + \ell(h_g) \leq k < \ell(wh^{-1}) = \ell(y)$ and by the induction hypothesis $\hat{w}_g$ is periodic. If $\ell(c_g) = k$, then for each $j = 1,\ldots,k$, the set (34) has exactly one element which is e. In other words, for each $j = 1,\ldots,k$, there exists $g(j) \in \{0,\ldots,m-1\}$, such that

$$g(j)^{-1} = h^{q(j)}g^{-1} .$$

Substituting in (33) shows that

$$\hat{w}_g = \prod_{i=0}^{m-1} \alpha(gh^{q(1)-i}g^{-1})^{t(1)} \cdots \alpha(gh^{q(k)-i}g^{-1})^{t(k)} .$$

Since h_g is the product of $(\alpha(gh^{q(j)-i}g^{-1})^{t(j)}$'s in $\hat{w}_g$ with $q(j) \neq i$, in the order of their appearance, it follows by the periodicity property (iv) that

$$h_g = \prod_{j\neq q(1)} (\alpha(gh^{q(1)-j}g^{-1}))^{t(1)} \cdots \prod_{j\neq q(k)} (\alpha(gh^{q(k)-j}g^{-1}))^{t(k)}$$

each factor of which is trivial by the periodicity property (v). Thus in this case $\hat{w}_g = c_g$ and since $\ell(c_g) = k$, by the induction hypothesis $\hat{w}_g$ is periodic and, in turn, $\hat{w}$ is periodic, wh^{-1} is periodic, as was to be shown.

4. Applications to p-groups

4.1. p-groups, p odd. Let $G = \langle a \mid a^p = 1 \rangle$ be a cyclic group of order p, p an odd prime. With $S = \bigcup_{i\geq 0} (\underline{e})^i \underline{G}^{\#}$, define the decorating function $\alpha: S \to G$ by

$$\alpha((\underline{e})^i a^j) = \begin{cases} a & \text{if } j \text{ is odd} \\ a^{-1} & \text{if } j \text{ is even}, \end{cases} \qquad (35)$$

for $1 \leq j \leq p-1$, $i \geq 0$. It is straightforward to verify that α is a periodicity preserving function. The associated automorphism γ is given by:

$$\gamma = \underline{e}*\gamma + (\underline{a} + \underline{a}^3 + \ldots + \underline{a}^{p-2})*a + (\underline{a}^2 + \underline{a}^4 + \ldots + \underline{a}^{p-1})*a^{-1}, \qquad (36)$$

and clearly $|\gamma| = p$. Let $\mathcal{G} = \langle G, \gamma\rangle = \langle a, \gamma\rangle$ be the tree extension of G by γ. Our principal result of this section is the following theorem.

Theorem 4.1.1. The group $\mathcal{G}$ has the following properties:

(a) $\mathcal{G}$ is a 2-generator infinite p-group with generators of order p;

(b) $\mathcal{G}$ is residually finite;

(c) $\underline{e}*\mathcal{G}' < \mathcal{G}'$;

(d) $\mathcal{G}$ contains, as a subgroup, an isomorphic copy of every finite p-group;

(e) Every proper quotient of $\mathcal{G}$ is finite.

Proof. Proof of (a) follows from Theorem 3.2.1. Proof of (b) follows from the Remark in §1.2. For the proof of (c), we note, in turn,

$$\gamma = \underline{e}*\gamma + \underline{a}*a + (\underline{a}^3 + \dots + \underline{a}^{p-2})*a + (\underline{a}^2 + \dots + \underline{a}^{p-1})*a^{-1} \;;$$

$$\gamma^a = \underline{e}*a^{-1} + \underline{a}*\gamma + (\underline{a}^3 + \dots + \underline{a}^{p-2})*a^{-1} + (\underline{a}^2 + \dots + \underline{a}^{p-1})*a \;;$$

$$\gamma\gamma^a = \underline{e}*\gamma a^{-1} + \underline{a}*a\gamma \;;$$

$$(\gamma\gamma^a)^{a^{-1}} = \underline{e}*a\gamma + \underline{a}^{p-1}*\gamma a^{-1} \;;$$

$$[\gamma\gamma^a, (\gamma\gamma^a)^{a^{-1}}] = \underline{e}*[\gamma a^{-1}, a\gamma]$$
$$= \underline{e}*[a^{-1}, \gamma^2] .$$

Since p is odd, $\mathcal{G}'$ is the normal closure of $[a^{-1}, \gamma^2]$ and since Γ projects onto $\mathcal{G}$, it follows that $\underline{e}*\mathcal{G}' \leq \mathcal{G}'$, as required. The proofs of (d) and (e) are more technical and will be published elsewhere.

4.2. p-groups, p even. Let $G = \langle a \mid a^4 = 1 \rangle$ be cyclic of order 4. Define the decorating function $\alpha: S \to G$ by

$$\alpha: \underline{a} \to a, \ \underline{a}^2 \to e, \ \underline{a}^3 \to a^3 ,$$
$$(\underline{e})^i \underline{g} \to \alpha(\underline{g}) , \ i \geq 0, \ g \in G^{\#} . \tag{37}$$

Once again α is periodicity preserving and the associated automorphism γ given by

$$\gamma = \underline{e}*\gamma + \underline{a}*a + \underline{a}^2*e + \underline{a}^3*a^3 \tag{38}$$

is of order 4. Let $\mathcal{G} = \langle G, \gamma \rangle = \langle a, \gamma \rangle$ be the tree extension of G by γ. Our principal result here is the following theorem.

Theorem 4.2.1. The group $\mathcal{G}$ has the following properties:

(a) $\mathcal{G}$ is a 2-generator infinite 2-group generated by elements of order 4;

(b) $\mathcal{G}$ is residually finite;

(c) $\underline{e}*\Gamma' < \Gamma'$

(d) $\mathcal{G}$ contains, as a subgroup, an isomorphic copy of every finite 2-group.

Proof. The proofs of (a), (b) are as in Theorem 4.1.1. For the proof of (c) we observe, as in Theorem 4.1.1, that

$$[\gamma\gamma^{a^2}, [\gamma,\gamma^a]] = \underline{e}*[\gamma,\gamma^{a^3}] .$$

Since Γ' is the normal closure of $[\gamma,\gamma^{a^3}]$ in $\mathcal{S}$ and Γ projects onto $\mathcal{S}$, it follows that $\underline{e}*\Gamma' < \Gamma'$. The proof of (d) will be published elsewhere.

UNIVERSITY OF MANITOBA
WINNIPEG, R3T 2N2
CANADA

UNIVERSITY OF BRASILIA
BRASILIA 70,000 - DF
BRAZIL

Received January 24, 1983

Note added April 1984

For the proofs of Theorem 4.1.1(d)-(e) and the proof of Theorem 4.2.1(d), see the authors' paper "Some infinite p-groups," Algebra i Logika, 22 (1983), 584-589 (Russian Edition).

Contemporary Mathematics
Volume 33, 1984

LENGTH FUNCTIONS AND ALGORITHMS IN FREE GROUPS

A. H. M. Hoare

To Roger Lyndon

Length functions in groups were first used by Lyndon [5] to systematize cancellation arguments in free groups and free products. Since then other applications have been found (see [4] for a brief survey). In this paper we express the algorithms of Petresco [7] and others in terms of length functions and extend them to include double cosets.

In Section 1 we prove preliminary results that apply to all length functions, in Section 2 we specialize to free groups, and in Section 3 we consider double cosets.

§1. Let G be a group, with identity e, which has a normalized integer-valued length function, that is a function $x \mapsto |x| \in \mathbb{Z}$ satisfying

A1'. $|e| = 0$,

A2. $|x| = |x^{-1}|$, and

A4. $d(x,y) > d(y,z)$ implies $d(x,z) = d(y,z)$, where

$$2d(x,y) = |x| + |y| - |xy^{-1}| .$$

We deduce immediately (see [2]) that $|x| \geq d(x,y) = d(y,x) \geq 0$, and hence

$$|x| + |y| \geq |xy^{-1}| \geq ||x| - |y|| . \tag{1}$$

The following version of Lyndon's Proposition 2.3 ([5], p. 211) is fundamental in what follows.

<u>Proposition 1</u>. <u>If</u> $|ba| + |ac| > |b| + |c|$ <u>then</u> $|bac| = |ba| + |ac| - |a|$.

© 1984 American Mathematical Society
0271-4132/84 $1.00 + $.25 per page

Proof.
$$\begin{aligned} 2d(ba,a) &= |ba| + |a| - |b| \\ &> |c| + |a| - |ac| \\ &= 2d(c^{-1},a) \end{aligned}$$

therefore by A4

$$2d(ba,c^{-1}) = 2d(c^{-1},a)$$

that is $|ba| + |c| - |bac| = |c| + |a| - |ac|$.

Let w be a word $x_0x_1\ldots x_{n+1}$, where $x_i \in G$ and $n \geq 0$. (We will use the same notation for a word and for the corresponding product in G.) For $i = 0,1,\ldots,n$ let

$$p_i = x_0x_1\ldots x_i, \text{ and}$$
$$q_i = x_{i+1}\ldots x_{n+1}\ .$$

The next Corollary follows immediately from the Proposition above.

Corollary. *If* $|p_k| + |q_{k-1}| > |p_{k-1}| + |q_k|$ *where* $1 \leq k \leq n$, *then* $|p_k| + |q_{k-1}| = |x_k| + |w|$,

Lemma 1. *Suppose* $|p_i| \geq |p_{i-1}|$ *and* $|q_{i-1}| \geq |q_i|$ *for* $i = 1,2,\ldots,n$, *then*

(i) $|w| \geq |p_n|$ *implies* $2d(p_0,q_0^{-1}) \leq \max\{|x_1|,\ldots,|x_{n+1}|\}$,

(ii) $|w| \leq |p_k|$ *implies* $|q_{k-1}| \leq \max\{|q_k|,|x_k|\}$,

(iii) $|w| < |p_n|$ *and* $|q_{i-1}| \geq |x_i|$ *for* $i = 1,\ldots,n$ *implies* $|p_i| = |p_{i-1}|$ *and* $|q_{i-1}| = |q_i|$ *for* $i = 1,\ldots,n$,

(iv) $|w| \leq \max\{|x_0|,|x_1|,\ldots,|x_{n+1}|\}$ *and* $|p_i|,|q_{i-1}| \geq |x_i|$ *for* $i = 1,\ldots,n$ *implies* $|p_i| = \max\{|p_{i-1}|,|x_i|\}$ *and* $|q_{i-1}| = \max\{|q_i|,|x_i|\}$ *for* $i = 1,\ldots,n$.

Proof. (i) Let k be the least integer, if any, for which $|p_k| + |q_{k-1}| > |p_{k-1}| + |q_k|$. Then $|p_0| = |p_1| = \ldots = |p_{k-1}|$ and $|q_0| = |q_1| = \ldots = |q_{k-1}|$ and by the Corollary $|p_k| + |q_{k-1}| = |x_k| + |w|$. Therefore

$$\begin{aligned} 2d(p_0,q_0^{-1}) &= |p_0| + |q_0| - |w| \\ &= |p_{k-1}| + |q_{k-1}| - |w| \\ &= |p_{k-1}| + |x_k| - |p_k| \leq |x_k|\ . \end{aligned}$$

If there is no such k then

$$2d(p_0,q_0^{-1}) = |p_n| + |q_n| - |w| \leq |x_{n+1}|.$$

Thus $2d(p_0,q_0^{-1}) \leq \max\{|x_1|,\ldots,|x_{n+1}|\}$.

(ii) Suppose $|q_{k-1}| > |q_k|,|x_k|$ then by the Corollary $|p_k| + |q_{k-1}| = |x_k| + |w|$, giving $|w| > |p_k|$.

(iii) Let k be the greatest integer, if any, such that $|p_k| + |q_{k-1}| > |p_{k-1}| + |q_k|$. Then $|p_k| = |p_{k+1}| = \ldots = |p_n|$ and $|q_k| = \ldots = |q_n|$, and by the Corollary $|p_k| + |q_{k-1}| = |x_k| + |w|$. Thus

$$|p_n| - |w| = |p_k| - |w| = |x_k| - |q_{k-1}|$$

contradicting the assumptions.

(iv) Suppose that either $|p_k| > |p_{k-1}|,|x_k|$ or $|q_{k-1}| > |q_k|,|x_k|$. Then $|p_{k-1}|,|x_k|,|q_k| < |p_k| + |q_{k-1}| - |x_k|$, and by the Corollary $|p_k| + |q_{k-1}| = |x_k| + |w|$. Now $|p_i| \geq |p_{i-1}|,|x_i|$ and $|q_{i-1}| \geq |q_i|,|x_i|$ by hypothesis, so by induction $|p_{k-1}| \geq |x_0|,|x_1|,\ldots,|x_{k-1}|$ and $|q_k| \geq |x_{k+1}|,\ldots,|x_{n+1}|$. Thus $|w| = |p_k| + |q_{k-1}| - |x_k| > |x_0|,\ldots,|x_k|,\ldots,|x_{n+1}|$ contradicting the hypotheses.

In [3] we introduced the following definition.

Definition. A reduced word $x_0x_1\ldots x_{n+1}$ is a *sink* if

$$|x_0x_1\ldots x_{n+1}| < |x_0x_1\ldots x_n|$$

and no proper subword or its inverse satisfies the corresponding inequality. A reduced word is *sink-free* if no subword or its inverse is a sink.

Lemma 2 [3]. *If every proper subword of the reduced word* $w = x_0x_1\ldots x_{n+1}$ *is sink-free and if*

$$|w| \leq \max\{|x_0|,|x_1|,\ldots,|x_{n+1}|\}$$

then

$$|x_ix_{i+1}\ldots x_j| = \max\{|x_i|,|x_{i+1}|,\ldots,|x_j|\}$$

for all proper subwords $x_i\ldots x_j$. *Moreover strict inequality implies*

$$|x_0|, |x_{n+1}| \geq |x_1|, \ldots, |x_n| \ .$$

Proof. By part (iv) of Lemma 1 we have $|p_i| = \max\{|p_{i-1}|, |x_i|\}$ and $|q_{i-1}| = \max\{|q_i|, |x_i|\}$ for $i = 1,\ldots,n$. Hence by induction

$$|p_j| = \max\{|x_0|, |x_1|, \ldots, |x_j|\} \text{ for } j = 1,\ldots,n \text{ and}$$
$$|q_{i-1}| = \max\{|x_i|, \ldots, |x_{n+1}|\}, \text{ for } i = 1,\ldots,n \ .$$

This proves the first part of the Lemma for the proper subwords p_j and q_{i-1}. However it also means that each of the reduced words q_{i-1} satisfies the hypothesis of this Lemma. Applying the case just proved to q_{i-1} we have

$$|x_i \ldots x_j| = \max\{|x_i|, \ldots, |x_j|\}$$

for $1 \leq i \leq j \leq n$ which, with the results for $|p_j|$ and $|q_{i-1}|$ above, proves the first part of the Lemma for all proper subwords.

Suppose strict inequality holds. We may assume by symmetry that $|x_0| \geq |x_{n+1}|$, so that

$$|w| < \max\{|x_0|, \ldots, |x_n|\} = |p_n| \text{ by the first part.}$$

Therefore the hypotheses of part (iii) of Lemma 1 hold, giving $|p_i| = |p_{i-1}|$ and $|q_{i-1}| = |q_i|$ for $i = 1,\ldots,n$. Thus

$$|x_0| = |p_0| = |p_n| = \max\{|x_0|, \ldots, |x_n|\}$$
$$|x_{n+1}| = |q_n| = |q_0| = \max\{|x_1|, \ldots, |x_{n+1}|\}$$

which proves the second part of the Lemma.

Following Lyndon [5] we define

$$N = \{x \in G: 2d(x, x^{-1}) \geq |x|\} \ .$$

Lemma 3. *Let* $w = x_0 x_1 \ldots x_{n+1}$ *satisfy the hypotheses of Lemma 2 and suppose* $|x_1|, |x_2|, \ldots, |x_n| \leq |x_{n+1}| = |x_0|$. *Then* $x_0 x_1 \ldots x_n \in N$ *whenever* $x_0 = x_{n+1}$, *and* $x_1 \ldots x_n \in N$, $n \geq 1$, *whenever* $x_0 = x_{n+1}^{-1}$.

Proof. Put $y = x_1 x_2 \ldots x_n$. By hypothesis $|w| \leq |x_{n+1}|$ and, by Lemma 2, $|x_0| \geq |y|$ so

$$2d(x_0^{-1}, (x_0 y)^{-1}) = |x_0| + |x_0 y| - |y| \geq |x_0 y|$$

and

$$2d(x_0y, x_{n+1}^{-1}) = |x_0y| + |x_{n+1}| - |w| \geq |x_0y| .$$

If $x_0 = x_{n+1}$ then by A4,

$$2d(x_0y, (x_0y)^{-1}) \geq |x_0y| ,$$

which proves (i).

Also, by Lemma 2, $|x_0| = |x_0y|$ and $|yx_{n+1}| = |x_{n+1}|$ so

$$2d(x_0, y^{-1}) = |x_0| + |y| - |x_0y| = |y| \text{ and}$$
$$2d(y, x_{n+1}^{-1}) = |y| + |x_{n+1}| - |yx_{n+1}| = |y| .$$

If $x_0 = x_{n+1}^{-1}$ then by A4,

$$2d(y, y^{-1}) \geq |y| ,$$

which proves (ii).

§2. We now assume

A0. $N = \{e\}$

in which case G is a free group. (See Lyndon [5].)

Lemma 4. *Let X be a subset of G with m elements, and let b be any element of G. If $x_1x_2 \ldots x_{n+1}$ is sink free, $x_i \in X^{\pm 1}$, and if*

$$|b| \leq |bx_1| = \ldots = |bx_1 \ldots x_n| \geq |bx_1 \ldots x_{n+1}|$$

then $n+1 < 2^m$ and the inequalities are not both strict.

Proof. Put $b = x_0$ and define p_i and q_i as in Lemma 1. Then for $i = 1, \ldots, n$, $|p_i| \geq |w|$ and $|p_i| \geq |p_{i-1}|$ by hypothesis, and $|q_{i-1}| \geq |q_i|, |x_i|$ because $x_1 \ldots x_{n+1}$ is sink-free. Thus, by part (ii) of Lemma 1, $|q_{i-1}| = \max\{|q_i|, |x_i|\}$ for $i = 1, \ldots, n$. Hence by induction

$$|x_1x_2 \ldots x_{n+1}| = |q_0| = \max\{|x_1|, \ldots, |x_{n+1}|\} .$$

Thus Lemma 2 applies to the word q_0, and so

$$|x_i \ldots x_j| = \max\{|x_i|, \ldots, |x_j|\}$$

for $1 \leq i \leq j \leq n+1$.

Suppose $n+1 \geq 2^m$. It can be shown by induction on m that, in any sequence of length 2^m with terms chosen from a partially ordered set of m elements, there are two terms which are equal and which have no term between them which is greater in the partial order. Applying this to the sequence $x_1^{\pm 1}, x_2^{\pm 1}, \ldots, x_{n+1}^{\pm 1}$ with partial order defined on the pairs $x^{\pm 1}$ by length we have a subword $x_1 \ldots x_j$, $i < j$, satisfying the conditions for Lemma 3. But no subword of w gives an element of $N = \{e\}$, thus $n+1 < 2^m$.

If the last inequality is strict then, by part (iii) of Lemma 1, $|p_0| = |p_1| = \ldots$ i.e. $|b| = |bx_1|$.

Corollary. *If* $|b| < |bx_1|$, *then* $|bx_1x_2 \ldots x_s|$ *is a non-decreasing function of* s.

Proof. By the Lemma no increase at any point can be followed by a decrease.

Definition. A subset X of G, not containing e, is *progressive* [7], or *weakly reduced* [2], or *level* [1], if every reduced word in $X^{\pm 1}$ is sink-free. In this paper we will use the term progressive. Clearly a progressive set is free.

Algorithm 1 [7]. *If* X *is a finite subset of* G *then there is an algorithm*

(i) *to decide if* X *is progressive*,

(ii) *if not to find a Nielsen transformation of* X *taking an element of* X *to a shorter element*,

(iii) *hence to find a Nielsen transformation taking* X *to a progressive set*.

Remark. All the Algorithms in this paper require an effective procedure for calculating the lengths of an element given as a product of certain elements of G. We will assume the existence of such a procedure whenever necessary.

Proof. (i) Suppose $w = x_0x_1 \ldots x_{n+1}$, $x_i \in X^{\pm 1}$, is a sink, then $x_0 \ldots x_n$ is sink-free, so $|x_0 \ldots x_n| \geq |x_0|, \ldots, |x_n|$. Moreover, by part (iii) of Lemma 1

$$|x_0| = |x_0x_1| = \ldots = |x_0x_1 \ldots x_n| ,$$

and applying Lemma 4 with $b = x_0$ we have $n+1 < 2^m$. Thus to decide if X

is progressive or not we need only consider the finite set of word $x_0x_1\ldots x_{n+1}$, $x_i \in X^{\pm 1}$, $n+1 < 2^m$.

(ii) If $w = x_0x_1\ldots x_{n+1}$, $x_i \in X^{\pm 1}$, is a sink, then $x_0 \neq x_i^{\pm 1}$, $i = 1,\ldots,n+1$ otherwise $|x_0\ldots x_i| = |x_0| = \max\{|x_0|,\ldots,|x_i|\}$ from above, and Lemma 3 gives a contradiction. Therefore $x_0 \to x_0\ldots x_{n+1}$ is a Nielsen transformation taking x_0 to a shorter element.

(iii) Since the sum of the lengths of the elements of X is finite, this process terminates in a progressive set (deleting the identity whenever it occurs in the process).

We now let H denote the subgroup generated by X.

Algorithm 2 [7]. *If X is any finite subset of G, and if a is any element of G then there is an algorithm*

(i) *to find an element of aH of minimal length*

(ii) *to find all elements of aH of any given length ℓ.*

Proof. We may assume that X has been transformed by Algorithm 1 to a progressive set. Since X is a finite set, there are only finitely many reduced words $x_1\ldots x_{n+1}$ for any given n. By Lemma 4 the sequence

$$|a|, |ax_1|, |ax_1x_2|, \ldots, |ax_1x_2\ldots x_{n+1}|$$

cannot have more than 2^m-1 successive terms equal where $m = |X|$, and cannot have an increase followed by a decrease. Thus after a finite number of steps we can find a minimal element in aH. When a is minimal in aH, the sequence above is monotonic non-decreasing and

$$|ax_1x_2\ldots x_{n+1}| \geq |a| + [(n+1)/2^m].$$

Thus we can find all elements of given length in a finite number of steps.

Lemma 5. *Suppose X is progressive and $\max\{|x|, x \in X\} = r$. If $|a|$ is minimal in aH, and if B is a subset of G with $2d(b,a) > r$ for all $b \in B$, then $|b|$ is minimal in bH, for all $b \in B$, and the cosets bH are distinct.*

Proof. For each $h \in H$, $h \neq e$, there is a reduced word $x_1 x_2 \ldots x_{n+1}, x_i \in X^{\pm 1}$, which is equal to h in G. Since $|a|$ is minimal we may apply part (i) of Lemma 1 and Lemma 4 to obtain

$$2d(a,h^{-1}) \leq r .$$

Now $2d(b,a) > r$ by hypothesis so, by A4, $2d(b,h^{-1}) = 2d(a,h^{-1})$, i.e.

$$|b| + |h| - |bh| = |a| + |h| - |ah| ,$$

which shows that $|b| \leq |bh|$.

Suppose $bh = b'$ for some $b,b' \in B$ and some $h \in H$. Then $|b| = |b'|$ since $|b|$ and $|b'|$ are both minimal in bH by the first part. Therefore

$$2d(b,h^{-1}) = |h| = 2d(b',h) .$$

Moreover by part (i) of Lemma 1

$$2d(b,h^{-1}) \leq r .$$

By hypothesis $2d(b,a), 2d(b',a) > r$ so by A4

$$2d(b,b') > r .$$

Applying A4 again

$$2d(b',h^{-1}) = 2d(b,h^{-1})$$
$$= 2d(b',h) = |h| \text{ from above,}$$

which using A4 again gives

$$2d(h,h^{-1}) \geq |h|$$

that is $h = e$.

Let G be finitely generated, then applying the results so far to a finite set of generators, we have that G has finitely many elements of any given length ℓ, and there is an algorithm to find all such elements.

Algorithm 3 [7]. *If g is finitely generated, and if X is a finite subset of G, then there is an algorithm*

(i) *to decide if H is of finite index in G*

(ii) *to decide if H is normal in G, and*

(iii) _if_ H _is normal and of finite index, to construct the multiplication table for_ G/H.

Proof. Let Z be a progressive set of generators for G and let $t = \max\{|z|, z \in Z\}$. As before, assume that X is progressive and let $r = \max\{|x|, x \in X\}$. If every coset of H in G contains an element with $2|a| \leq r+t$, then H is of finite index. Suppose conversely that there is a coset with no such element, let $a = (z_1 z_2 \ldots z_\ell)^{-1}$, $z_1 \ldots z_\ell$ reduced with $z_i \in Z^{\pm 1}$, satisfy

(α) $2|a| > r+t$,

(β) $|a|$ is minimal in aH, and

(γ) ℓ is minimal subject to (α) and (β).

Let $b = z_1 z_2 \ldots z_{\ell+1} \ldots z_m$ also be reduced with $z_i \in Z^{\pm 1}$. Since

$$|z_1 \ldots z_\ell| \leq |z_1 \ldots z_\ell z_{\ell+1}| \leq \ldots \leq |z_1 \ldots z_m|,$$

part (i) of Lemma 1 gives

$$2d(a^{-1}, (ab)^{-1}) \leq t, \text{ that is}$$

$$|a| + |ab| - |b| \leq t, \text{ and so}$$

$$2d(b^{-1}, a) = |b| + |a| - |ab|$$

$$\geq 2|a| - t > r.$$

Since there are infinitely many such b, H has infinite index by Lemma 5.

Suppose that $2|z_1 z_2 \ldots z_{\ell-1}| > r+t$ then

$$2d((z_1 z_2 \ldots z_{\ell-1})^{-1}, a) > r+t - |z_\ell| \geq r$$

and by Lemma 5 $|z_{\ell-1}^{-1} \ldots z_1^{-1}|$ is minimal in $z_{\ell-1}^{-1} \ldots z_1^{-1} H$ contradicting the choice of ℓ. By (1) $|z_1 \ldots z_\ell| \leq |z_1 \ldots z_{\ell-1}| + |z_\ell| \leq (r+3t)/2$. Thus H has infinite index if and only if there is an element $a \in G$ with $|a|$ minimal in aH and $r+t < 2|a| \leq r+3t$. This is decidable by Algorithm 2.

To decide if H is normal, use Algorithm 2 to decide if $gxg^{-1} \in H$ for each $x \in X$ and each generator g of G. To construct the multiplication table for G/H, use Algorithm 2 again to list the elements of G/H and to find the coset $aHbH = abH$.

§3. We now apply these results to double cosets. Suppose X and Y are two finite progressive subsets of G and let $\langle X\rangle = H$ and $\langle Y\rangle = K$. Given any a in G we can use Algorithm 2 alternately on left cosets of H and right cosets of K to find $a' \in KaH$ such that $|a'|$ is minimal in Ka' and in $a'H$.

Lemma 6. *If a is minimal in Ka and in aH and if $2|a| > |x| + |y|$ for all $x \in X$ and $y \in Y$, then*

(i) $|a|$ is minimal in KaH,

(ii) $aHa^{-1} \cap K = e$ *and the cosets* kaH *for* $k \in K$ *and* Kah *for* $h \in H$ *are all distinct.*

Algorithm 4. *If a satisfies the conditions of Lemma 6 then there is an algorithm to find all elements of KaH of given length.*

Proofs. Let $r = \max\{|x| : x \in X\}$ and $s = \max\{|y| : y \in Y\}$. By part (i) of Lemma 1, $2d(a,h^{-1}) \leq r$ and $2d(a^{-1},k) \leq s$ for all $h \in H$ and $k \in K$. Thus

$$|ka| + |ah| \geq |k| + |a| - |r| + |a| + |h| - s$$
$$> |k| + |h|, \text{ since } 2|a| > r + s .$$

Therefore by Proposition 1

$$|kah| = |ka| + |ah| - |a| \geq |a|$$

which proves (i) of Lemma 6.

If $k \in aHa^{-1} \cap K$ then $kah = a$ for some $h \in H$ and so from above $|ka| = |ah| = |a|$. Thus

$$2d(k,a^{-1}) = |k| + |a| - |ka| = |k| \text{ and}$$
$$2d(k^{-1},a^{-1}) = |k| + |a| - |k^{-1}a| = |k|, \text{ since } k^{-1}a = ah .$$

Therefore by A4, $2d\{k,k^{-1}\} \geq |k|$, and so $k = e$. Thus $aHa^{-1} \cap K = e$, which implies that all the cosets are distinct. Since $|kah| = |ka| + |ah| - |a|$, to find all elements kah, with $|kah| \leq \ell$ we need only consider those k and h such that $|ka| + |ah| \leq \ell + |a|$, which we can find by Algorithm 2.

Algorithm 5 (Moldavanskii [6]). *Suppose G has only finitely many elements of any given length. If X and Y are finite subsets of G, then there is*

<u>an algorithm to decide whether</u>

(i) $aHa^{-1} \leq K$

(ii) $aHa^{-1} = k$

<u>for some</u> a <u>in</u> G, <u>and to find one such</u> a.

<u>Proof</u>. Transform X and Y to progressive sets by Algorithm 1. If there exists some a such that $aHa^{-1} \cap K$ is non-trivial, then by Lemma 6 there is some a with $2|a| \leq r + s$. We can use Algorithm 2 to test for all such $a \in G$, and for all $x \in X$ whether $axa^{-1} \in K$. This gives an algorithm for (i). Interchanging X and Y will give (ii).

We now consider what happens in cosets KaH when $2|a| \leq r + s$. In this case it is possible for $|kah|$ to be small for arbitrarily large $|h|$ and $|k|$, making it difficult to enumerate all elements of the coset. We show however that every kah can be approached by a sequence with a bounded part followed by an increasing part. We first need to consider some properties of kah.

<u>Lemma 7</u>. <u>Let</u> X <u>and</u> Y <u>be finite progressive subsets of</u> G <u>and let</u> $w = y_m^{-1} \ldots y_2^{-1} y_1^{-1} a x_1 \ldots x_n$ <u>where</u> $x_1 \ldots x_n$ <u>and</u> $y_1 \ldots y_m$ <u>are reduced words in</u> X <u>and</u> Y <u>respectively</u>, with $m,n \geq 0$

(i) <u>If</u> $|y_m^{-1} \ldots y_2^{-1} y_1^{-1} a x_1 \ldots x_{i-1}| < |y_m^{-1} \ldots y_1^{-1} a x_1 \ldots x_i|$ <u>then</u> $|y_m^{-1} \ldots y_1^{-1} a x_1 \ldots x_s|$ <u>is a non-decreasing function of</u> s <u>for</u> $s \geq i$.

(ii) <u>If</u> $|y_j^{-1} \ldots y_2^{-1} y_1^{-1} a x_1 \ldots x_n| > |y_{j-1}^{-1} \ldots y_2^{-1} y_1^{-1} a x_1 \ldots x_n|$ <u>then</u> $|y_t^{-1} \ldots y_2^{-1} y_1^{-1} a x_1 \ldots x_n|$ <u>is a non-decreasing function of</u> t <u>for</u> $t \geq j$.

(iii) <u>If</u> $v = y_j^{-1} \ldots y_2^{-1} y_1^{-1} a x_1 \ldots x_i$ <u>where</u> $i,j \geq 1$ <u>satisfies</u>

$$2|v| > |x_i| + |y_j|$$

$$2d(v,x_i) > |x_i|$$

$$2d(v^{-1},y_j) > |y_j|$$

<u>then</u> $|y_t^{-1} \ldots y_2^{-1} y_1^{-1} a x_1 \ldots x_s|$ <u>is a non-decreasing function of</u> s <u>and</u> t <u>for</u> $s \geq i$ <u>and</u> $t \geq j$.

<u>Moreover in each case any increase of</u> $2^{|X|}$ <u>in</u> s <u>and any increase of</u> $2^{|Y|}$ <u>in</u> t <u>gives a strict increase in the length</u>.

<u>Proof</u>. Putting $y_m^{-1}\ldots y_2^{-1}y_1^{-1}ax_1\ldots x_{i-1} = b$ in the Corollary to Lemma 4 gives the first part. The second part is similar with $b = x_n^{-1}\ldots x_1^{-1}a^{-1}y_1\ldots y_j$.

Suppose that v satisfies the conditions of part (iii). Since X and Y are progressive $|x_i x_{i+1}\ldots x_s| \geq |x_{i+1}\ldots x_s|$ and $|y_j y_{j+1}\ldots y_t| \geq |y_{j+1}\ldots y_t|$, so

$$2d(x_i, h_s^{-1}) \leq |x_i| \text{ and}$$
$$2d(y_j, k_t^{-1}) \leq |y_j|$$

where $h_s = x_{i+1}\ldots x_s$ and $k_t = y_{j+1}\ldots y_t$.

By hypothesis $2d(v,x_i) > |x_i|$ and $2d(v^{-1},y_j) > |y_j|$. Therefore by A4

$$\left.\begin{aligned} 2d(v,h_s^{-1}) &= 2d(x_i,h_s^{-1}) \leq |x_i| \text{ and} \\ 2d(v^{-1},k_t^{-1}) &= 2d(y_j,k_t^{-1}) \leq |y_j| . \end{aligned}\right\} \qquad (2)$$

Adding gives

$$|v| + |h_s| - |vh_s| + |v| + |k_t| - |k_t^{-1}v| \leq |x_i| + |y_j| .$$

By hypothesis $2|v| > |x_i| + |y_j|$, so $|k_t^{-1}v| + |vh_s| > |k_t| + |h_s|$.

Thus by Proposition 1

$$|k_t^{-1}vh_s| = |k_t^{-1}v| + |vh_s| - |v| .$$

Moreover from (2) we also have

$$|v| + |h_s| - |vh_s| = |x_i| + |h_s| - |x_i h_s| \text{ and}$$
$$|v| + |k_t| - |k_t^{-1}v| = |y_j| + |k_t| - |y_j k_t| .$$

Thus

$$|k_t^{-1}vh_s| = |y_j k_t| - |y_j| + |x_i h_s| - |x_i| + |v|$$

and (iii) follows because $|x_i h_s|$ and $|y_j k_t|$ are non-decreasing functions of s and t.

The last part follows directly from Lemma 4.

We now assume that $|a|$ is minimal in Ka and in aH and we construct for each word $w = y_m^{-1}\ldots y_1^{-1}ax_1\ldots x_n$ a subword which is either w itself or satisfies one of the conditions of Lemma 7, where $m,n \geq 0$.

Put $a_0^{-1} = a$ and put $a_1 = a_0^{-1}x_1\ldots x_{i_1}$ such that

$$|a_0^{-1}| \geq |a_0^{-1}x_1| \geq |a_0^{-1}x_1x_2| \geq \ldots \geq |a_0^{-1}x_1\ldots x_{i_1-1}| < |a_1|$$

if such an integer $i_1 \geq 1$ exists, and put $a_1 = a_0^{-1}x_1\ldots x_n$ otherwise. [In fact equalities hold since $|a|$ is minimal in aH.]

Put $a_2 = a_1^{-1}y_1\ldots y_{j_1}$ such that

$$|a_1^{-1}| \geq |a_1^{-1}y_1| \geq \ldots \geq |a_1^{-1}y_1\ldots y_{j_1-1}| < |a_2|$$

if such a $j_1 \geq 1$ exists, and put $a_2 = a_1^{-1}y_1\ldots y_m$ otherwise.

We continue this process, defining $a_3 = a_2^{-1}x_{i_1+1}\ldots x_{i_2}$, $a_4 = a_3^{-1}y_{j_1+1}\ldots y_{j_2}$, etc. and stop when we reach some $a_i^{\pm 1}$ satisfying one of the conditions of Lemma 7 or we reach $a_i = w$.

<u>Lemma 8</u>. If $|a| \leq r$, where $r = \max\{|x|,|y|: x \in X,\ y \in Y\}$ <u>then</u>

$$|a_0|,|a_1|,\ldots,|a_{i-1}| \leq 2r .$$

<u>Proof</u>. By assumption $|a_0| = |a| \leq r$. Since $a_1 = a_0^{-1}$ (if $n = 0 \neq m$) or $|a_0| \geq |a_0x_1| \geq \ldots \geq |a_0x_1\ldots x_{i_1-1}|$ and $|x_{i_1}| \leq r$, $|a_1| \leq 2r$. Let j be the least integer, if any, less than i such that $|a_j| > 2r$, then $|a_j| > |a_{j-1}|$. Suppose j is odd then $j = 2\ell+1$, $\ell \geq 1$, and by construction $a_j = a_{j-1}^{-1}x_{i_\ell+1}\ldots x_{i_{\ell+1}}$ where $i_\ell < i_{\ell+1} \leq n$ and

$$|a_{j-1}^{-1}| \geq |a_{j-1}^{-1}x_{i_\ell+1}| \geq \ldots \geq |a_jx_{i_{\ell+1}}^{-1}| < |a_j| .$$

Moreover a_j is not $y_m^{-1}\ldots y_1^{-1}ax_1\ldots x_{i_{\ell+1}}$, otherwise it satisfies condition (i) of Lemma 7. Therefore $a_{j-1} = a_{j-2}^{-1}\ldots y_{j_\ell}$ where

$$|a_{j-2}| \geq \ldots \geq |a_{j-1}y_{j_\ell}^{-1}| < |a_{j-1}| .$$

Thus we have

$$2d(a_{j-1},y_{j_\ell}) = |a_{j-1}| + |y_{j_\ell}| - |a_{j-1}y_{j_\ell}^{-1}| > |y_{j_\ell}| \quad \text{and}$$

$$2d(a_j,x_{i_{\ell+1}}) = |a_j| + |x_{i_{\ell+1}}| - |a_jx_{i_{\ell+1}}^{-1}| > |x_{i_{\ell+1}}| .$$

Now by (1), $|a_jx_{i_{\ell+1}}^{-1}| \geq |a_j| - |x_{i_{\ell+1}}|$,

$$> r \quad \text{since} \quad |a_j| > 2r \quad \text{and} \quad |x_{i_{\ell+1}}| < r .$$

Therefore

$$2d(x_{i_{\ell+1}}a_j^{-1},a_j^{-1}) = |x_{i_{\ell+1}}a_j^{-1}| + |a_j^{-1}| - |x_{i_{\ell+1}}| > 2r .$$

Moreover by the Corollary to Lemma 1, proved below,

$$2d(a_{j-1},x_{i_{\ell+1}}a_j^{-1}) \geq 2|a_jx_{i_{\ell+1}}^{-1}| - \max\{|x_{i_\ell+1}| \ldots\} > r .$$

Therefore by A4

$$2d(a_{j-1},a_j^{-1}) > r \geq |y_{j_\ell}| ,$$

and applying A4 again

$$2d(a_j^{-1},y_{j_\ell}) > |y_{j_\ell}| .$$

Thus $a_j = y_{j_\ell}^{-1}\ldots y_1^{-1}ax_1\ldots x_{i_{\ell+1}}$ satisfies condition (iii) of Lemma 7 contradicting the choice of j. The proof for j even is the same with X and Y interchanged. It remains to prove the following.

<u>Corollary to Lemma 1</u>. <u>If</u> $|x_0| \geq |x_0x_1| \geq \ldots \geq |x_0x_1\ldots x_n|$ <u>and</u> $|x_1| \leq |x_1x_2| \leq \ldots \leq |x_1\ldots x_n|$, $n \geq 0$, <u>then</u>

$$2d(x_0^{-1},(x_0x_1\ldots x_n)^{-1}) \geq 2|x_0x_1\ldots x_n| - \max\{|x_1|,\ldots,|x_n|\} .$$

<u>Proof</u>. Clear for $n = 0$. Let $v = x_0\ldots x_n$, then

$$|v| \leq |vx_n^{-1}| \leq \ldots \leq |vx_n^{-1}\ldots x_1^{-1}| \quad \text{and}$$

$$|x_n^{-1}\ldots x_1^{-1}| \geq \ldots \geq |x_1^{-1}| ,$$

so applying part (i) of Lemma 1

$$2d(v,x_1\ldots x_n) \leq \max\{|x_1|,\ldots,|x_n|\} .$$

Therefore

$$2d(x_0^{-1}, v^{-1}) = 2|v| - 2d(v, x_0^{-1}v)$$
$$\geq 2|v| - \max\{|x_1|, \ldots, |x_n|\} .$$

Algorithm 6. Suppose G has only finitely many elements of any given length then there is an algorithm

(i) to decide whether a and b are in the same (K,H) double coset

(ii) to find all elements of any given length ℓ in any (K,H) double coset.

Proof. It suffices to prove (ii) since (i) follows. Take any double coset KaH, where $|a|$ is minimalized in Ka and in aH by Algorithm 2. Apply Algorithm 4 if $|a| > v$, if not suppose that b is any element of KaH of length ℓ and suppose that b can be expressed as a product

$$b = y_m^{-1} \ldots y_1^{-1} a x_1 \ldots x_n$$

with m+n minimal. Consider the sequence of words

$$a_0^{-1}, a_0^{-1}x_1, \ldots, a_0^{-1}x_1 \ldots x_{i_1-1}$$

$$a_1, y_1^{-1}a_1, \ldots, y_{j_1-1}^{-1} \ldots y_1^{-1}a_1 ,$$

$$\vdots$$

$$a_i^{\varepsilon}$$

where $\varepsilon = (-1)^{i+1}$ and $a_0, a_1, \ldots, a_i$ are as defined for Lemma 8. Each term of the sequence is a subword of $y_m^{-1} \ldots y_1^{-1} a x_1 \ldots x_n$ and is also a subword of all succeeding terms. Therefore by the minimality of $m + n$ no two of these give the same element of G, and so there are at most M of them before a_i, where M is the number of elements of G of length $\leq 2r$. Now a_i satisfies the conditions for Lemma 7 or $a_i = b$ so

$$\ell \geq (m + n - M)/2^{\mu}, \text{ that is}$$

$$m + n \leq M + 2^{\mu}\ell$$

where $\mu = \max\{|X|, |Y|\}$.

We now extend Lemma 5 to double cosets.

Lemma 9. Suppose X and Y are progressive and $\max\{|x|, x \in X\} = r$ and $\max\{|y|, y \in Y\} = s$. Suppose also that $|a|$ is minimal in Ka and in aH and $2|a| > r+s$. Let B be any subset of G with $2d(b,a) > r$, $2d(b^{-1},a^{-1}) > s$, and $2|b| > r+s$, then the double cosets KbH are distinct for all $b \in B$.

Proof. Suppose $KbH = Kb'H$, for $b,b' \in B$, then $b'H = kbH$ for some $k \in K$. By hypothesis $|a^{-1}|$ is minimal in $a^{-1}K$ and, using part (i) of Lemma 1 and Lemma 4, $2d(k,a^{-1}) \leq s$ so by A4, $2d(k,b^{-1}) = 2d(k,a^{-1})$. Now

$$\begin{aligned} 2d(kb,b) &= |kb| + |b| - |k| \\ &= 2|b| - 2d(k,b^{-1}) \\ &> r . \end{aligned}$$

Therefore by A4, $2d(kb,a) > r$. Applying Lemma 5 to b' and kb gives $b' = kb$, i.e. $Kb' = Kb$, and by Lemma 5 again $b' = b$.

Algorithm 7. If G is finitely generated, and X and Y are finite subsets of G, then there is an algorithm to decide whether G has finitely many (K,H) double cosets.

Proof. Let Z be a progressive set of generators for G and let $t = \max\{|z|, z \in Z\}$. As before suppose that X and Y are progressive with $r = \max\{|x|, x \in X\}$ and $s = \max\{|y|, y \in Y\}$. If every double coset contains an element a with $|a|$ minimal in aH and in Ka and with $2|a| \leq r+s+4t$, then there are only finitely many double cosets. Suppose conversely that there is some element a in G with $|a|$ minimal in aH and in Ka and with $2|a| > r+s+4t$. Suppose $a = (z_1 \ldots z_\ell)^{-1}$ where $z_1 \ldots z_\ell$ is a reduced word in $Z^{\pm 1}$. Let i be the least integer such that $|z_1 \ldots z_i| > (r+t)/2$ and j the greatest integer such that $|z_j \ldots z_\ell| > (t+s)/2$. Then $j \geq i+1$, otherwise

$$\begin{aligned} |z_1 \ldots z_\ell| &\leq |z_1 \ldots z_i| + |z_{i+1} \ldots z_\ell| \\ &\leq |z_1 \ldots z_{i-1}| + |z_i| + |z_{j+1} \ldots z_1| \\ &\leq (r+t)/2 + t + (t+s)/2 \\ &= (r+s+4t)/2. \end{aligned}$$

Now there are infinitely many distinct reduced words $b = (z_1 \ldots z_i z'_1 \ldots z'_m z_j \ldots z_\ell)^{-1}$, $m \geq 0$, and at least one with $m = 1$. For each such b we have by part (i) of Lemma 1

$$2d(z_1 \ldots z_i, (z'_1 \ldots z_\ell)^{-1}) \leq t, \quad \text{i.e.}$$

$$|z_1 \ldots z_i| + |z'_1 \ldots z_\ell| - |b| \leq t .$$

Therefore $|b| > (r+t)/2 + (s+t)/2 - t = (r+s)/2$ and

$$2d(b, (z_1 \ldots z_i)^{-1}) \geq 2|z_1 \ldots z_i| - t > r .$$

Similarly

$$2d(b^{-1}, z_j \ldots z_\ell) > s .$$

In particular this is true for $b = a$ and so using A4

$$2d(b,a) > r \quad \text{and}$$

$$2d(b^{-1}, a^{-1}) > s .$$

Therefore by Lemma 9 there are infinitely many (K,H) double cosets.

If $m = 1$, then

$$\begin{aligned} |z_1 \ldots z_i z'_1 z_j \ldots z_l| &\leq |z_1 \ldots z_{i-1}| + |z_i| + |z'_1| + |z_j| + |z_{j-1} \ldots z_1| \\ &\leq (r+t)/2 + 3t + (t+s)/2 \\ &= (t+s+8t)/2 . \end{aligned}$$

Thus G has infinitely many (K,H) double cosets if and only if G has an element a with $r+s+4t < 2|a| \leq r+s+8t$ and with $|a|$ minimal in Ka and in aH. This is decidable by Algorithm 2.

REFERENCES

1. H. Federer and B. Jónsson, "Some properties of free groups," Trans. Amer. Math. Soc. 68 (1950), 1-27.

2. A. H. M. Hoare, "On length functions and Nielsen methods in free groups," J. London Math. Soc. (2), 14 (1976), 188-192.

3. A. H. M. Hoare, "Nielsen methods in groups with a length function," Math. Scand., 48 (1981), 153-164.

4. A. H. M. Hoare, "Length functions," in Journées Groupes et Langages. Amiens, 1980.

5. R. C. Lyndon, "Length functions in groups," Math. Scand., 12 (1963), 209-234.

6. D. I. Moldavanskii, "Conjugacy of subgroups of a free group," Algebra i Logika, 8 (1969), 691-694.

7. J. Petresco, "Algorithmes de décision et de construction dans les groups libres," Math. Zeit., 79 (1962), 32-43.

DEPARTMENT OF MATHEMATICS
UNIVERSITY OF BIRMINGHAM
P. O. BOX 363
BIRMINGHAM
B15 2TT

Received October 21, 1982

Contemporary Mathematics
Volume 33, 1984

SPLIT EXTENSIONS OF ABELIAN GROUPS WITH IDENTICAL SUBGROUP STRUCTURES

Charles Holmes

1. Introduction

This paper begins with the observation that the known examples of pairs of non-isomorphic groups with identical subgroup structures [3,4,7] are semidirect products of an abelian group A by two cyclic subgroups B, C of $\operatorname{Aut} A$ of the same prime power order. Theorem 1 shows that the semidirect products $A \circ B$ and $A \circ C$ under consideration are isomorphic if and only if B and C are conjugate in $\operatorname{Aut} A$. In the known examples A has exactly two maximal B, C-invariant subgroups. In this paper we investigate $A \circ B$ and $A \circ C$ where A is abelian, B and C are two cyclic subgroups of $\operatorname{Aut} A$ having the same order, and A has one (possibly trivial) maximal B, C-invariant subgroup. In Theorem 2 we show that if A is B, C-irreducible, then B and C are conjugate so that $A \circ B$ and $A \circ C$ are isomorphic and do not yield an example of non-isomorphic groups with identical subgroup structures.* Theorem 3 has the same conclusion when A has one non-trivial maximal B, C-invariant subgroup and the order of B and C is relatively prime to the order of A. The remaining cases based on the number of maximal B, C-invariant subgroups seem to yield examples of non-isomorphic groups with identical subgroup structures. The methods used in those cases differ considerably from the methods used in this paper, and will not be gone into here.

---

*After this paper was submitted Roland Schmidt proved alternative versions of Theorems 1 and 2 as well as other interesting results.

© 1984 American Mathematical Society
0271-4132/84 $1.00 + $.25 per page

We now establish the notation and definitions for this paper. By $H \leq G$ we mean that H is a subgroup of the group G. It is well known that $L(G) = \{H: H \leq G\}$, the subgroup lattice of G, is a lattice with respect to the usual operations. Two groups G and H have identical subgroup structures [7] if and only if there is a lattice isomorphism $p: L(G) \to L(H)$ which preserves indices, conjugacy, and isomorphism. Here p preserves indices means that the indices $[pB:pA]$ and $[B:A]$ are equal whenever A is a subgroup of B. By p preserves conjugacy we mean that pA is conjugate to pB in pC if and only if A is conjugate to B in C for appropriate A, B and C. And p preserves isomorphism means that pA is group isomorphic to A for any proper subgroup A of G. Of course any two isomorphic groups have identical subgroup structures.

As usual $\mathrm{Aut}\, G$ means the group of all automorphisms of G. If $b \in \mathrm{Aut}\, G$ and $g \in G$, then the image of g under b is g^b. When $B \leq \mathrm{Aut}\, G$ we denote the semidirect product of G with B by $G \circ B$, denote the elements of this group by gb, and define the product of $g'b'$ by gb to be $g'g^{b'}b'b$. The set of elements $\overline{G} = \{g1_B \in G \circ B\}$ is a normal subgroup of $G \circ B$ isomorphic to G and $\overline{B} = \{1_G b \in G \circ B\}$ is a non-normal subgroup of $G \circ B$ isomorphic to B. As usual $G \circ B = \overline{G}\overline{B}$ and $\{1_G 1_B\} = \overline{G} \cap \overline{B}$. We usually identify $\overline{G}$ with G and $\overline{B}$ with B. If H is a B-invariant subgroup of G, we write $H \circ B$ for $[H,B]$ the subgroup of $G \circ B$ generated by H and B.

If $G \circ B$ and H are groups with identical subgroup structures then $H = G \circ C$ and the following properties of B and C hold:

1. B is isomorphic to C,
2. the lattice of B-invariant subgroups of G is lattice isomorphic to the lattice of C-invariant subgroups of G,
3. the groups $M \circ B$ and $M \circ C$ are isomorphic for every proper maximal B, C-invariant subgroup M in G.

These properties guide the development of all results.

The first examples of non-isomorphic groups with identical subgroup structures were given by A. Rottlander [4] in 1927. In that paper the abelian group A is isomorphic to the vector space $Z_p \oplus Z_p$ and B and C are generated by automorphisms b and c of prime order q. The smallest order admitting non-isomorphic $A \circ B$ and $A \circ C$ is 605. Working independently of one another Honda [3] and Yff [7] created examples where A is a cylic group $Z_p \oplus Z_q$, p,q primes, and B and C are cyclic of prime power order. The smallest order admitting an example is 260. In 1976 Holmes [2] constructed an example where $A = Z_{p^2} \oplus Z_q$, p an odd prime dividing $q-1$ and B and C have order p. The smallest order admitting such an example is 189. Theorem 1 shows how B and C must be chosen for these examples.

The non-isomorphic groups with identical subgroup structures are of interest in several ways. They have been used in the study of the subgroup lattice of direct products $L(G \times G)$ by Schmidt [5]. On a broader scale the rise of lattice theory and universal algebra in the 1930's came almost as a response to Rottlander's examples limiting the importance of lattices in group theory. The thrust of this response was to affirm the importance of lattices in general algebraic processes. Still the examples remain and are made more troublesome by the fact that "usually" these groups will be the same in most other ways; i.e. same automorphism groups, same varieties, same character tables, etc. This paper is a small step toward understanding and bounding this bad behavior.

2. $G \circ B$ and $G \circ C$ are isomorphic

The crucial element in constructing non-isomorphic semi-direct products $G \circ B$ and $G \circ C$ having identical subgroup structures is to know what conditions on B and C force the groups to be non-isomorphic. The following simple theorem does this. It may be proved elsewhere, but not knowing where we give the proof.

<u>Theorem 1</u>. Let G be a group and B, C subgroups in Aut G. There is an isomorphism $f\colon G\circ B \to G\circ C$ such that $f(G) = G$ and $f(B) = C$ if and only if B and C are conjugate in Aut G.

<u>Proof</u>. Suppose first that f is an isomorphism. Let $r = f|_G \in \text{Aut } G$ and $s = f|_B$. Then $(gb)f = g^r(b)s$. Now we compare $(g'b'gb)f$ and $(g'b')f(gb)f$,

$$(g'b'gb)f = (g'g^{b'}b'b)f = (g'g^{b'})f(b'b)f = g'^r g^{b'r}(b'b)s$$

$$(g'b')f(gb)f = g'^r(b')sg^r(b)s = g'^r g^{r(b')s}(b'b)s \ .$$

Therefore

$$b'r = r(b')s, \quad (b')s = r^{-1}b'r$$

and B is conjugate to C in Aut G.

Conversely, it is easy to see that when $C = r^{-1}Br$ and s is conjugation by r the mapping $f\colon G\circ B \to G\circ C$ defined by $(gb)f = g^r(b)s$ is an isomorphism.

<u>Corollary</u>. Suppose $G\circ B$ and H are non-isomorphic groups with identical subgroup structure, and C is the subgroup of H corresponding to B under the lattice isomorphism. Then considered as subgroups of Aut G the groups B and C are not conjugate.

3. G is B-irreducible in $G\circ B$

Suppose now that G has a composition series and that G is B-irreducible in $G\circ B$. Then G is a minimal normal subgroup of $G\circ B$ and G is a direct product of isomorphic simple groups. (See for example M. Hall [1], Thm. 8.6.1, p. 131.) Thus if G is a finite abelian group, G must be elementary abelian of order p^n, say. Now if B were a p-group, B would have to centralize some subgroup for any p-group has non-trivial center. But then the B-irreducibility of G forces B to centralize all of G and G to be a cyclic group of prime order. Thus if A is elementary abelian of order p^n with $n > 1$ there is no p-group B in Aut A such that A is B-irreducible.

Suppose now that A is elementary abelian of order p^n, B is an abelian subgroup of Aut A of order r, and A is B-irreducible. As is well known, A can be thought of as a vector space over Z_p and Aut A as the general linear group $GL(n,p)$. From Schur's lemma and the commutativity of B, we have that $\mathrm{Hom}_B(A,A)$ is a finite field K containing B and Z_p. Thus K has order p^m for some m, B is cyclic, and r divides $p^m - 1$ so that r is prime to p. We let $B = [b]$ where the order of b is r. Since A is B-irreducible, b has an irreducible minimal polynomial $m_b(x)$ which is also the characteristic polynomial of b so that the degree of $m_b(x)$ is n. From $b^r = 1$ we have that $m_b(x)$ divides $x^r - 1$.

Now let d be a root of $m_b(x)$ in an extension field F of degree n over Z_p. Thus $F^* = F\backslash\{0\}$, the multiplicative group of F, has an element, d, of order r. Hence r divides $p^n - 1$. From the fact that F^* is a cyclic group satisfying the converse of Lagrange's Theorem, we have that n the degree of the irreducible polynomial $m_b(x)$ is the smallest integer k such that r divides $p^k - 1$. Thus p has multiplicative order n modulo r and n divides $\varphi(r)$ where φ is the Euler function. The $\varphi(r)$ elements $\{d^i: i \text{ prime to } r\}$ in F^* are primitive roots of $x^r - 1$, and $\varphi(r)$ is the degree of cyclotomic polynomial $\Phi_r(x) = \pi(x - d^i)$ where the product runs over all i prime to r.

Suppose $m_b(x)$ is monic. Then $m_b(x) = (x-d)(x-d^p)\dots(x-d^{p^{n-1}})$, since $m_b(x)$ is invariant under the Frobenius automorphism. Hence the irreducible polynomial of d^i, a primitive r-th root, is given by $m_i(x) = (x-d^i)(x-d^{ip})\dots(x-d^{ip^{n-1}})$. In fact $m_i(x)$ is the minimal polynomial for b^i. Moreover any irreducible polynomial of degree n dividing $x^r - 1$ has roots which are of multiplicative order r modulo p and must split in $Z_p(d)$. Thus any irreducible polynomial of degree n which divides $x^r - 1$ must be one of the $m_i(x)$. We summarize this background information in the following lemma.

<u>Lemma</u>. $\Phi_r(x)$ is the product of $s = \frac{\varphi(r)}{n}$ irreducible polynomials of degree n, i.e. $\Phi_r(x) = p_1(x)p_2(x)\dots p_s(x)$. If d is a root of one of these, then for each i there is a k such that d^k is a root of $p_i(x)$.

Before stating and proving the next theorem, the author wishes to thank his colleague, Stanley Payne, for several suggestions which were helpful in the development of the preceding lemma.

<u>Theorem 2</u>. Let A be a finite abelian group. Also let B and C be cyclic subgroups of Aut A of order r. If A is B-irreducible and C-irreducible, then $A \circ B$ is isomorphic to $A \circ C$.

<u>Proof</u>. As above A is an elementary abelian group of order p^n, and r is prime to p. Let $B = [b]$ and $C = [c]$ with b and c having irreducible polynomials $m_b(x)$ and $m_c(x)$, respectively. Then by the lemma some power c^i has minimal polynomial (and characteristic polynomial) $m_b(x)$ so that b is conjugate to c^i in Aut A. Therefore, $A \circ B$ is isomorphic to $A \circ C$ by Theorem 1.

4. A has at most one B-maximal subgroup

In this section we study a finite abelian group A with a subgroup B in Aut A for which A has just one maximal B-invariant subgroup. If there are two distinct primes which divide the order of A, then by the primary decomposition theorem A would have more than one maximal B-invariant subgroup. Thus A is a primary abelian group of order p^n, say, and in addition is B-indecomposable. It might be tempting to think that B-indecomposable is equivalent to the existence of just one B-maximal subgroup. However, $A = Z_{3^2} + Z_3$ and $b = \begin{bmatrix} 1 & 3 \\ 0 & 1 \end{bmatrix}$ in Aut A gives a counterexample. For $B = [b]$ has order 3, and there are two B-maximal subgroups generated by $\begin{bmatrix} 1 \\ 0 \end{bmatrix}$ and $\begin{bmatrix} 3 \\ 0 \end{bmatrix}, \begin{bmatrix} 0 \\ 1 \end{bmatrix}$ respectively, but A is still B-indecomposable.

We now consider the case where the order of B is relatively prime to p. The subgroups $A^{p^i} = \{c \in A : c = a^{p^i} \text{ some } a \in A\}$, for $i = 0,1,2,\dots$ are B-invariant and form a descending chain

$$A = A^{p^0} \geq A^p \geq A^{p^2} \geq \dots \geq A^{p^r} = 1 .$$

Furthermore A^p must be the maximal B-invariant subgroup. Otherwise there is a maximal B-group M properly containing A^p. Let B operate on A/A^p. By Maschke's theorem M/A^p has a complement in A/A^p which pulls back to a B-invariant group $N \geq A^p$ so that $MN = A$. Thus there is no unique maximal B-invariant subgroup which contradicts our basic assumption. Therefore A^p is the maximal B-invariant subgroup of A.

Furthermore A^{p^2} must be the maximal B-subgroup in A^p. Otherwise by an argument similar to the above we have $A^p = MN$ where $M \cap N = A^{p^2}$. Now let $\mathfrak{M} = \{a \in A: a^p \in M\}$ and $\mathfrak{N} = \{a \in A: a^p \in N\}$. These are B-subgroups of A, and by maximality of A^p and the fact that $\mathfrak{M} \not\leq A^p$ we have $\mathfrak{M}A^p = A$. Again there is a maximal B-subgroup of A containing $\mathfrak{M}$ which contradicts our basic assumption. Continuing the argument in this way we have that A^{p^i} is the maximal B-subgroup in $A^{p^{i-1}}$ for all $i = 1,\dots,r$.

However, $A_p = \{g \in A: \text{order of } g \leq p\}$ is also a B-subgroup of A containing $A^{p^{r-1}}$. If A_p properly contains $A^{p^{r-1}}$, then there is some k such that $A_p \leq A^{p^k}$ but $A_p \not\leq A^{p^{k+1}}$. Thus A^{p^k} would contain an additional maximal B-subgroup containing A_p. From this contradiction it follows that $A_p = A^{p^{r-1}}$. By the basis theorem we know that A has a direct sum decomposition $A = M_1 + M_2 + \dots + M_r$ where M_i is a direct sum of cyclic groups of order p^i. A routine computation shows that $A \neq M_r$ implies $A_p \neq A^{p^{r-1}}$. Hence $A = M_r$ or in other words A has a direct sum decomposition

$$A = Z_{p^r} \oplus Z_{p^r} \oplus \dots \oplus Z_{p^r} .$$

We have now proved the following lemma.

<u>Lemma</u>. Let A be a finite abelian group and $B \leq \text{Aut } A$ such that A has exactly one maximal B-subgroup. Then A is a primary abelian group of order p^n. If the order of B is relatively prime to p, then A is isomorphic to

$Z_{p^r} \oplus \ldots \oplus Z_{p^r}$ and every B-invariant subgroup of A is an element of the chain $A \geq A^p \geq A^{p^2} \geq \ldots \geq A^{p^r} = 1$, so that A is B-indecomposable.

We are now ready to state and prove our last theorem.

Theorem 3. Let A be an abelian group and let B and C be isomorphic cyclic groups in Aut A such that the order of B and the order of A are relatively prime. If A contains exactly one maximal B-subgroup and exactly one maximal C-subgroup, then $A \circ B$ is isomorphic to $A \circ C$.

Proof. The structure of A and its B, C-invariant subgroups are determined by the lemma. The notation is as in the lemma, but in addition let N designate the subgroup of Aut A consisting of all automorphisms which are congruent to the identity modulo p, i.e.

$$N = \{f \in \text{Aut } A : f(a) = a + b_a \text{ where } b_a \in A^p\} .$$

It is well known that N is a group of order p^{rn} where n is the number of summands in Z_{p^r} in A, and that (Aut A)/N is isomorphic to $GL(n,p)$ and to $\text{Aut}(A/A^p)$. See for example Speiser [6].

Let $f \in \text{Aut } A$ and $\bar{f}$ the element in $\text{Aut}(A/A^p)$ such $\bar{f}(aA^p) = f(a)A^p$. Define a homomorphism $h: \text{Aut } A \to \text{Aut}(A/A^p)$ by $h(f) = \bar{f}$. Then $\ker h = N$. By Theorem 2 $h(B)$ and $h(C)$ are conjugate in $\text{Aut}(A/A^p)$. But this means that BN and CN are conjugate in (Aut A)/N. That is, there is fN in (Aut A)/N such that $fN(B)f^{-1}N = fBf^{-1}N = CN$. But then fBf^{-1} and C are two subgroups of the solvable group of [C,N] in Aut A. By a result of Philip Hall extending the Sylow theorems fBf^{-1} and C are conjugate in [C,N]. Therefore B and C are conjugate in Aut A, and by Theorem 1, $A \circ B$ is isomorphic to $A \circ C$.

REFERENCES

[1] M. Hall, The Theory of Groups. New York, 1959.

[2] C. Holmes, Groups of order p^3q with Identical Subgroup Structures. Atti Accad. Sci. Ist. Bologna Cl. Sci. Fis. Rend. 3 (1976), 113-123.

[3] K. Honda, On Finite Groups whose Sylow Groups are all Cyclic. Commentarii Math. Univ. Sancti. Pauli, 1 (1952), 5-39.

[4] A. Rottlander, Nachweis der Existenz nicht-isomopher Gruppen von gleicher Situation der Untergruppen. Math. Zeitschrift, 28 (1928), 641-653.

[5] R. Schmidt, Der Untergruppenverband des directen Productes zweierisomorpher Gruppen. Journal of Algebra, 73 (1981), 264-272.

[5a] R. Schmidt, Untergruppenverbände endlicher Gruppen mit elementarabelschen Hallschen Normalteilern. Journal Reine Angew. Math. 334 (1982), 116-140.

[6] A. Speiser, Die Theorie der Gruppen von endlicher Ordnung. Berlin, 1937.

[7] P. Yff, Groups with Identical Subgroup Structures. Math. Zeitschrift, 99 (1967), 178-181.

DEPARTMENT OF MATHEMATICS AND STATISTICS
MIAMI UNIVERSITY
OXFORD, OH 45045

Received April 12, 1982

Contemporary Mathematics
Volume 33, 1984

NON-PLANARITY OF PLANAR GROUPS WITH REFLECTIONS

Johannes Huebschmann

To Roger Lyndon

A planar net (4.1 of [7]) is a 2-complex whose underlying space is (homeomorphic to) the standard plane. A group of automorphisms of a planar net will here be called a planar group. Planar groups arise in complex analysis as discontinuous groups of motions of the Euclidean or Non-Euclidean plane where they are often called Fuchsian groups (occasionally only those that do not contain reflections). If G is a planar group, i.e. a group of automorphisms of a planar net E, there is a classical construction due to Poincaré [6] of reading off a presentation (X;R) of G from E. If G is infinite, finitely generated and does not contain reflections, it is known (III.7 of [5], 4.13 of [7], or [3]) that things may be arranged so that the (combinatorial) Cayley complex C(X;R) (details will be given below) is the dual of E, and hence it is again a planar net. If G contains reflections, this does not seem to be possible unless the Cayley complex is suitably modified in a way already suggested by Cayley [1]; the result of this modification is usually called the modified Cayley complex (III.7 of [5], 4.13 of [7], or [3]). The purpose of this note is to show that the modification of the Cayley complex is essential. More precisely, it will be shown that a finitely generated infinite planar group with reflections cannot have a planar net as (ordinary) Cayley complex.

© 1984 American Mathematical Society
0271-4132/84 $1.00 + $.25 per page

We shall employ the concept of combinatorial asphericity. In order to explain this we first recall briefly the construction of the (ordinary) combinatorial Cayley complex.

Let $(X;R)$ be a presentation of a group G, and denote by F the free group of X and by $p: F \to G$ the obvious surjection. Then the corresponding (combinatorial) _Cayley complex_ $C(X;R)$ can be described as follows: The set (that underlies) G constitutes the set of vertices. Next, for each $g \in G$ and $x \in X$, $C(X;R)$ has an (oriented) edge (g,x) joining g and $gp(x)$; the _inverse_ of (g,x) is then written $(gp(x),x^{-1})$. Finally, for each $r \in R$, write $r = s^m$, $m \geq 1$, where s is not a proper power in F, and denote by $W_G(r)$ the subgroup of G generated by $p(s) \in G$; now $C(X;R)$ has, for each $r \in R$ and each left coset $gW_G(r)$ defined by some $g \in G$, a face $(gW_G(r),r)$ which is attached, starting at say g, by the path

$$(g_1,x_1^{e_1})(g_2,x_2^{e_2})\ldots(g_n,x_n^{e_n})$$

where $r = x_1^{e_1}\ldots x_n^{e_n}$ and $g_1 = g$, $g_{i+1} = g_i p(x_i^{e_i})$, $e_i = \pm 1$. We mention that one may first attach, for each $g \in G$ and $r \in R$, a face (g,r) and then identify all faces

$$(g,r),(gp(s),r),\ldots,(gp(s^{m-1}),r)$$

where $r = s^m$ as above. More details may be found in III.4 of [5] and in [2].

A presentation $(X;R)$ has been called _combinatorially aspherical_ in [2] if $C(X;R)$ is aspherical (in the topological sense). In particular, if $C(X;R)$ is a planar net its underlying space is contractible and so $(X;R)$ is then combinatorially aspherical. A group G is called _combinatorially aspherical_ if it has a combinatorially aspherical presentation.

PROPOSITION. _Let G be a group. Suppose that either G contains two involutions (i.e. elements of order 2) (say) c_1 and c_2 whose product has order > 1, or that G contains an involution (say) c and some non-trivial element (say) e that commutes with c. Then G is not combinatorially aspherical._

Proof. The hypothesis means that G either contains a finite dihedral group or the direct product of a cyclic group with a cyclic group of order 2. Now neither a finite dihedral group nor the direct product of two cyclic groups with one factor of finite order are combinatorially aspherical. This follows from the formulas for the cohomology of a combinatorially aspherical group given in [4]. For example, if the group $\mathbb{Z} \times \mathbb{Z}/k$ were combinatorially aspherical its third integral cohomology group would be zero which is not the case; likewise, the formulas for the cohomology of a combinatorially aspherical group imply for example that, if the group is finite, it must be cyclic. For clarity we note that the combinatorially aspherical groups have been called aspherical in [4]. Now G cannot be combinatorially aspherical either since this property is inherited by subgroups (Proposition 2.5 of [2]). q.e.d.

We can now state and prove the main result of this note.

THEOREM. *Let* G *be a finitely generated, infinite planar group with reflections. Then* G *cannot have a planar net as (ordinary) Cayley complex.*

Proof. A finitely generated, infinite planar group satisfies the hypothesis of the above Proposition. This may be seen from the canonical presentations as given in 4.5.6 and 4.11.5 of [7] for the compact- and non-compact fundamental domain cases, respectively. Hence such a group cannot be combinatorially aspherical and therefore cannot have a planar net as (ordinary) Cayley complex.

Remark. We restricted attention to finitely generated groups to avoid considerations of ends: If G is a planar group that is not finitely generated, the (perhaps modified) Cayley complex C(X;R) of a suitable presentation (X;R) may be "embedded" in the plane without the vertices only; the vertices can be considered as "ends" on the boundary (4.13 of [7] or [3]). Thus C(X;R) will not then be (homeomorphic to) the ordinary plane. Further, we did not consider finite planar groups. For in that case, the reasonable question to ask is whether the combinatorial Cayley complex is a net on a sphere rather than on the plane. However, there are only very few finite planar groups, and the question whether or not such a group admits a net on a sphere as Cayley complex can be settled by direct inspection.

Finally, we should like to mention that the papers [2] and [4], and the present note also, were inspired by the seminal work of Roger Lyndon (see [5] and the additional references to his work there).

REFERENCES

1. A. Cayley: The theory of groups: graphical representations. Amer. J. Math. 1 (1878), 174-176.
2. I. M. Chiswell, D. J. Collins and J. Huebschmann: Aspherical group presentations. Math. Z. 178 (1981), 1-36.
3. E. Gramberg and H. Zieschang: Order reduced Reidemeister-Schreier subgroup presentations and applications. Math. Z. 168 (1979), 53-70.
4. J. Huebschmann: Cohomology theory of aspherical groups and of small cancellation groups. J. of Pure and Applied Algebra 14 (1979), 137-143.
5. R. C. Lyndon and P. E. Schupp: Combinatorial group theory. Ergebnisse der Math. Vol. 89. Berlin-Heidelberg-New York: Springer 1977.
6. H. Poincaré: Théorie des groupes fuchsiens. Acta. Math. 1 (1882), 1-62.
7. H. Zieschang, E. Vogt, and H. D. Coldewey: Surfaces and planar discontinuous groups. Lecture Notes in Mathematics, 835. Berlin-Heidelberg-New York: Springer 1980 (Revised and expanded translation of: Flaechen und ebene diskontinuierliche Gruppen, Lecture Notes in Mathematics, 122. Berlin-Heidelberg-New York: Springer 1970).

MATHEMATISCHES INSTITUT
IM NEUENHEIMER FELD 288
D-6900 HEIDELBERG
W. GERMANY

Received November 20, 1982

Contemporary Mathematics
Volume 33, 1984

A SURVEY OF THE CONJUGACY PROBLEM

R. Daniel Hurwitz

I. Introduction

The conjugacy problem is one of the three group theoretic decision problems which were formulated and then solved in certain cases by Max Dehn [31]. Dehn called them " . . . fundamental problems, the solution of which is very important and, indeed, not possible without a penetrating study of the material." These three problems are: the word problem (Identitätsproblem), the conjugacy problem (Transformationsproblem), and the isomorphism problem (Isomorphieproblem). A solution of the word problem for a given group presentation is an algorithm which determines whether or not two arbitrary words on the generators represent the same element. A solution to the conjugacy problem for a given presentation is an algorithm which determines whether or not the elements represented by two arbitrary words on the generators are conjugate. A solution to the isomorphism problem is an algorithm which determines for two given presentations whether or not the groups presented are isomorphic.

Dehn stated in his introduction, "It is to be assumed that the solution of these problems forms the natural foundation of a methodical presentation of the theory of infinite groups." Indeed, the study of these and other decision problems has been important to logicians and to topologists as well. It has been evident for the last thirty years that a "solution" for group presentations is seldom a simple matter and in many cases impossible. This

*This work was supported by a research grant from Skidmore College.

© 1984 American Mathematical Society
0271-4132/84 $1.00 + $.25 per page

realization has generated extensive work on the solvability and unsolvability of Dehn's decision problems, their relation to each other, and to other questions.

This survey concentrates on the conjugacy problem. Compendia of what was known about the conjugacy problem have been included in several other surveys (for example, Miller [74], [75], Schupp [93], and Baumslag [12]). This survey attempts to state or to refer to as many known results on the conjugacy problem as possible, including the considerable number which have been discovered in recent years. No doubt there have been omissions, for which the author apologizes. Nothing has been omitted intentionally.

Work of Roger Lyndon [69], to whom this Festschrift is dedicated, concentrated on the papers by Dehn in which the decision problems were formulated. Lyndon's results and methods played an important role in the progress made in the study of these problems, especially in the area of small cancellation theory.

II. Some Fundamental Results and Relationships to Other Decision Problems

Dehn solved the conjugacy problem for the fundamental groups of closed orientable 2-manifolds. For years after the formulation of the problems, the only groups for which the solvability of the word problem was known also had solvable conjugacy problem. An important turning point was the discovery of groups in which the conjugacy problem was unsolvable.

A. Word and Conjugacy Problems. One construction which is extremely useful in the study of group theoretic decision problems is the Higman-Neumann-Neumann (HNN) extension. If $G = \langle X;R \rangle$ is a group having subgroups A and B which are isomorphic under $\varphi: A \to B$, then the HNN extension of G with associated subgroups A and B is the group G^* with presentation $G^* = \langle X,t;R,t^{-1}at = a\varphi,\ a \in A\rangle$. The element t, which does not occur in the set X of generators of G, is called a stable letter. The group G is canonically embedded in G^*. An important result giving a normal form to the elements of an HNN extension is Britton's Lemma, which states that any word equal to the identity in G^* and having occurrences of the stable letter t must have a subword of the form $t^{-1}at$ for some $a \in A$ or tbt^{-1} for some $b \in B$.

A group having unsolvable word problem clearly also has unsolvable conjugacy problem. Thus the famous works by Boone, Novikov, and Britton gave finite presentations of groups with unsolvable conjugacy problem as well. Earlier work by Novikov [78], [79] gave a somewhat simpler presentation of a group with unsolvable conjugacy problem. Fridman [38] then showed that the word problem was solvable in Novikov's example. Bokut' [15] also did this, using Britton's Lemma.

A very valuable example of this type was given by Miller [74] in his survey. Miller's group was an HNN extension with several stable letters of a free group. The word problem is thus solvable using Britton's Lemma. The group was so constructed, however, that the conjugacy problem was reducible to the word problem of another finitely presented group, and this word problem was unsolvable.

B. <u>Conjugacy Separability</u>. Miller's example was also a residually finite group. A group G is called residually finite if for every non-trivial element $g \in G$ there is a homomorphism φ of G onto a finite group H such that $g\varphi$ is not trivial. A strengthening of the property of residual finiteness is called conjugacy separability (sometimes called residual finiteness under conjugacy or finite approximability with respect to conjugacy). G is conjugacy separable if for every non-conjugage pair of elements x and y in G, there is a homomorphism φ onto a finite group H such that $x\varphi$ and $y\varphi$ are not conjugate. The relationship between conjugacy separability and the conjugacy problem was established by Mostowski [77], who showed that finitely presented conjugacy separable groups have solvable conjugacy problem. The converse does not hold (for example, see Dyer [36] or Section IV on linear groups).

C. <u>Subgroups</u>. The relationship between groups and subgroups with respect to the conjugacy problem differs significantly from that relationship with respect to the word problem. Let G be a finitely generated group with finitely generated subgroup H. If G has solvable word problem, then H also has solvable word problem. On the other hand, Collins [24] constructed a finitely presented group G with unsolvable conjugacy problem which could be recursively embedded in a finitely generated group having solvable conjugacy problem. Later Collins and Miller [29] constructed two examples demonstrating this difference even more dramatically. The first example is a finitely presented group with unsolvable conjugacy problem having a subgroup of index 2 with solvable conjugacy problem. The second is a finitely presented group with solvable conjugacy problem having a subgroup of index 2 with unsolvable conjugacy problem.

D. Embedding. Another difference between these two decision problems has been established with respect to the Higman embedding theorem. Higman [56] proved that every finitely generated group with a recursively enumerable set of defining relations can be embedded in a finitely presented group. Clapham [22] showed that the solvability of the word problem was preserved under this embedding. In [28], Collins considered the analogous situation for the conjugacy problem, and gave examples of finitely generated groups having solvable conjugacy problem which may be embedded in groups with unsolvable conjugacy problem. Specific criteria were determined under which the Higman embedding in G of a finitely generated group C with solvable conjugacy problem insures that G has unsolvable conjugacy problem.

Sacerdote [88] considered this relationship with respect to the Boone-Higman Theorem [18]. This theorem states that a finitely generated group has solvable word problem if and only if it can be embedded into a simple subgroup of a finitely presented group. Sacerdote showed that a finitely generated group has solvable conjugacy problem if and only if $G * \langle x \rangle$ can be embedded into a finitely generated simple subgroup of a finitely presented group K, where another somewhat complicated restriction is placed on the embedding.

E. Number of Relations. One other type of comparison between groups with unsolvable word problems and conjugacy problems concerns the "size" of the presentation (i.e., the number of generators, the number of relators, or the sum of the two). The situation concerning groups having one relator will be discussed in the next section. In a work considering minimizing the number of relators, Collins [26] gave an example with unsolvable conjugacy problem having 11 defining relations.

III. Groups Defined by Presentations

A basic and well-known fact in combinatorial group theory is that free groups have solvable conjugacy problem (see, for example, Magnus - Karrass-Solitar [71], Thm. 1.3, or Lyndon and Schupp [70], Prop. 2.14). Several fairly recent works (Baumslag [11], Stebe [95], Remeslennikov [85], and Wehrfritz [106]) established that free groups are conjugacy separable as well. Dyer [35] extended this result to free-by-finite groups: any group having a free subgroup of finite index is conjugacy separable. In this section we will concentrate on groups whose presentations fall into several different classes: one-relator groups, small cancellation groups, Fuchsian groups, Artin and Coxeter groups, and groups arising in the study of the Burnside problem.

A. One-relator Groups. Probably the most interesting type of group for which the question of the solvability of the conjugacy problem has not completely been resolved is the group having one defining relator. There has been considerable work in the area. Gurevich [51], [52] and Fridman [39] solved the conjugacy problem for groups whose relation has stable letters in a certain form. These results are closely related to work in HNN extensions (see Section V). In fact, using a classification given by Pietrowski [83], Larsen's work [63] on cyclic HNN constructions and free products with amalgamation solved the conjugacy problem for all one-relator groups with non-trivial center. Carrying this one step further, Armstrong [10] and Dyer [36] showed that one-relator groups with non-trivial center are conjugacy separable.

B. Small Cancellation Groups. An important idea used by Dehn in singling out groups in which he could solve the conjugacy problem was to examine the amount of cancellation in certain products of elements. This idea led to the study of groups in which the cancellation in the products of cyclic conjugates of defining relators was relatively small. Such small cancellation groups have been studied using algebraic and geometric methods. The results have great significance in groups interesting to topologists (see Section VI).

We will begin with a short summary of the different cancellation conditions. A complete survey has been done by Schupp [93]. A piece relative to a symmetrized set of relators is a common initial subword of two distinct elements of that set. The three conditions on cancellation commonly considered are the following: (1) the metric $C'(\lambda)$: the length of any piece of a relator is less than the product of λ and the length of the relator, (2) non-metric $C(p)$: no relator is a product of fewer than p pieces, and (3) the generalized "triangle condition" $T(q)$: sequences of h relators having no adjacent inverse pairs will have a non-cancelling adjacent product if $3 \leq h < q$.

The first solutions of the conjugacy problem for small cancellation groups relied heavily on combinatorial arguments. Greendlinger [43] extended Dehn's algorithm to solve the conjugacy problem in $C'(1/8)$. This was followed by solutions in groups having both $C'(1/4)$ and $T(4)$ ([44] and [45]) and $C'(1/6)$ ([46] and [47]). Using the geometric method of annular diagrams, Schupp [91] solved the conjugacy problem for groups satisfying $C(6)$, $C(4)$ and $T(4)$, and $C(3)$ and $T(6)$. This geometric approach was again used by Schupp [92] to solve the conjugacy problem for a large class of quotient groups satisfying small cancellation conditions. These results and some applications to knot groups are also to be found in Lyndon and Schupp [70].

Lipschutz [65] showed that a free product of eighth groups (i.e., groups satisfying $C'(1/8)$) with a specific type of cyclic amalgamation has solvable conjugacy problem. Comerford and Truffault [30] generalized this to the solution of the conjugacy problem in free products of sixth groups with cyclic amalgamation and HNN extensions of sixth groups with cyclic associated subgroups. (For the general situation in such constructions, see Section V.)

One further extension of the theory has been pursued by Dô Long Vân [33], [34] and Palasin'ski [82]. They considered groups having so-called "non-homogeneous" small cancellation conditions, which essentially depend on left and right cancellation lengths. Under specific limitations of these lengths, such groups also have solvable conjugacy problem. The methods used again apply the cancellation diagrams of Lyndon and Schupp.

According to Gol'berg [41], an enlargement of the non-metric conditions $C(6)$, $C(4)$ and $T(4)$, or $C(3)$ and $T(6)$ will allow an isomorphic copy of every group, so there exist examples having slightly larger cancellation and unsolvable decision problems.

C. Fuchsian Groups. It is the case that many Fuchsian groups satisfy small cancellation conditions. A Fuchsian group G has presentation

$$G = \langle a_1, b_1, \ldots, a_g, b_g, x_1, \ldots, x_n, f_1, \ldots, f_k;$$
$$x_1^{m_1} = \ldots = x_n^{m_n} = f_1 \ldots f_k x_1 \ldots x_n \prod_{i=1}^{g} [a_i, b_i] = 1 \rangle, \quad m_i \geq 2 .$$

Dehn had used hyperbolic geometry to solve the conjugacy problem for finitely generated Fuchsian groups. In [30], using small cancellation theory, Comerford and Truffault solved the conjugacy problem for such presentations. Also, Stebe [98] showed that if $t \neq 0$ or $g \neq 0$, then the group is conjugacy separable.

D. Artin and Coxeter Groups. Another class of groups in which small cancellation techniques have been productive are the Artin and Coxeter groups. These groups have grown out of the study of braid groups (see Section VI). Briefly stated, an Artin group G has generators a_i, $i \in I$, and relations of the form $a_i a_j a_i \ldots = a_j a_i a_j \ldots$ where the length of the relation is fixed by the entries of a symmetric matrix (called the Coxeter matrix). The braid group B_n on n strings has a presentation of this type. From the Artin group G one defines the Coxeter group $\overline{G}$ by adding the relations $a_i^2 = 1$ for $i \in I$. G is said to be of finite type if $\overline{G}$ is finite.

Using the methods applied by Garside [40] to solve the conjugacy problem for braid groups, Brieskorn and Saito [19] and Deligne [32] solved the conjugacy problem for all Artin groups of finite type. Recently Appel and Schupp [8], [110] have used the methods of small cancellation theory to extend these results to certain classes of Artin and Coxeter groups which are not of finite type. They define large and extra-large type, and solve the conjugacy problem for all such groups which are finitely generated or, with an additional geometric hypothesis, recursively presented.

E. Burnside Problem. There are some results on the conjugacy problem which have been included in extensive works on the Burnside problem. Novikov and Adian [80] solved the conjugacy problem for $B(m,n)$ having m generators, $m \geq 2$, and odd exponent $n \geq 4381$. In [1], Adian improved this to n odd and $n \geq 665$. Also, referring to the work of Adian and Novikov as well as small cancellation theory, Ol'šanskiĭ [81] solved the conjugacy problem for his construction of an infinite group such that every proper subgroup has prime order and such that subgroups of the same order are conjugate.

IV. Algebraic Varieties

One of the "measures" of the elementary nature of the finitely presented groups discussed by Miller in his survey [75] is the algebraic class to which they belong. Baumslag also considers some of these classes in his survey [12] of infinite group-theoretic topics of interest to finite group theorists. In this section we make a general summary of results known in this area.

A. Metabelian Groups. One of the earliest algebraic classes in which the conjugacy problem was studied was the class of metabelian groups. A group is metabelian if its derived group is abelian. Gol'dina [42] and Matthews [73] both solved the conjugacy problem for finitely generated free metabelian groups. Matthews' method involved the embedding of the free metabelian group in a wreath product of abelian groups and solving certain decision problems in the factors. Using Matthews' work, Timošenko [104] proved that finitely generated free metabelian groups are conjugacy separable. At the other extreme, Wehrfritz [107], [108] has given three examples of finitely generated metabelian groups which are not conjugacy separable. Finally, in his survey Baumslag mentions a proof by Noskov that all finitely generated metabelian groups have solvable conjugacy problem: this work is unknown to this author.

B. Solvable Groups. Much of the work mentioned above is related closely to the general question of the conjugacy problem in solvable groups. One basic result in this classification was given by Blackburn [14], who proved that finitely generated nilpotent groups are conjugacy separable. Similarly, Seksenbaev [94] showed that Noetherian, almost-nilpotent groups are conjugacy separable.

Kargapolov and Remeslennikov [60] solved the conjugacy problem for free solvable groups. Kargapolov [59] also proved that super-solvable groups (finite normal series with cyclic factors) are conjugacy separable. Both this result and that of Blackburn also follow from later works by Remeslennikov [84] and Formanek [37], in which finite extensions of polycyclic groups are proved to be conjugacy separable. The general question about the conjugacy problem for solvable groups was answered negatively by Remeslennikov [86] who gave an example of a finitely presented solvable group with derived series of length 5 having unsolvable word problem.

C. Matrix Groups. There has been extensive study of conjugacy in groups which are linear, that is, isomorphic to matrix groups. Stebe [99] showed that GL(2,Z) and SL(2,Z) are conjugacy separable (Applegate and Onishi [9] have recently given an algorithm to solve the conjugacy problem in SL(2,Z) using continued fractions). Also in Stebe's work (and, in part, in that of Remeslennikov [85]), it is shown that GL(n,Z) and SL(n,Z) are not conjugacy separable for $n \geq 3$. Stebe went on to establish the conjugacy separability of the linear groups over the p-adic integers, but mentioned that this in itself does not solve the conjugacy problem unless the presentations are finite.

The conjugacy problem itself in these groups is considered in the work of Grunewald [48] and Sarkisjan [90]. In both of these works the conjugacy problem for groups of integral matrices over algebraic number fields is solved. Grunewald and Segal [49], [50] extended these results to prove that every arithmetic group has solvable conjugacy problem, where arithmetic groups are a certain type of algebraic subgroup of the general linear group.

In several of the special cases in these algebraic varieties (for example, SL(2,Z) and GL(2,Z)), the special algebraic structure (in this case as a free product with amalgamation) also gives an approach to the study of decision problems. These structures will be considered in the next section.

V. Group Theoretical Constructions

Early examples of groups with unsolvable decision problems were HNN extensions (then called Britton extensions) with several stable letters. Since such structures arise in various contexts, considerable attention has been given to HNN extensions as well as to free products with amalgamation and to other constructions. In this section we consider results known about group structures of these types.

A. Free Products and Wreath Products. The conjugacy problem in a free product is clearly reducible to the conjugacy problems of the factors. Stebe [95] and Remeslennikov [85] also have shown that free products of conjugacy separable groups are conjugacy separable. Matthews [73] reduced the conjugacy problem in a wreath product to the factors while adding an additional condition, and Remeslennikov [85] gave necessary and sufficient conditions for the conjugacy separability of discrete wreath products, extending the work of Timošenko [104].

B. Free Products with Amalgamation and HNN Extensions. The free product with amalgamation and the HNN extension are two closely related basic constructions

in combinatorial group theory. The HNN extension was defined in Section II. A free product with amalgamation may be defined as follows: if the groups G and H have presentations $G = \langle X;R\rangle$, $H = \langle Y;S\rangle$, and $A \subseteq G$, $B \subseteq H$ are subgroups which are isomorphic under the mapping $\varphi: A \to B$, then the free product of G and H amalgamating A and B has presentation
$\langle X,Y;\ R,S,\ a = a\varphi,\ a \in A\rangle$.

Two basic results on the structure of conjugate elements in these groups are Solitar's Theorem (Magnus-Karass-Solitar [71], Thm. 4.6) for free products with amalgamation and Collins' Lemma (Collins [25]) for HNN extensions. Each of them has served as a starting point for study of the conjugacy problem. Miller and Schupp [76] have established these results using the cancellation diagrams so important to the study of small cancellation groups, thus giving an important geometric approach to the constructions. There is no categorical positive answer for these groups: in [74], Miller gives examples of an HNN extension of a free group and a free product of free groups with finitely generated amalgamation having unsolvable conjugacy problem. The positive results which are known are related primarily to cyclic subgroups and to other decision problems.

C. Cyclic Subgroups. The conjugacy problem for a free product of free groups with infinite cyclic amalgamation was shown to be solvable by Lipschutz [66]. The corresponding result for HNN extensions of free groups with cyclic associated subgroups can be found in work by Anshel and Stebe [5] (one also finds this elsewhere, as in one of the cases considered by Fridman [39]). If the base group or groups satisfy small cancellation conditions, the corresponding results have already been mentioned in Section III.

Conditions satisfied by the generators of the amalgamated or associated cyclic subgroups which are sufficient to guarantee the solvability of the conjugacy problem have also been studied. In the free product with amalgamation case, Lipschutz [66] showed the sufficiency of having subgroups generated by semi-critical elements in the factors. Elements are semi-critical if they are non-self-conjugate, conjugate-power-solvable, and satisfy a double coset property. This result was established in corresponding HNN groups by Hurwitz [58] provided that the generators are mutually semi-critical. Lipschutz [67] extended this type of approach even further by considering conditions called semi-non-self-conjugacy and non-power elements (in the paper itself, the hyphens have been deleted). Thus, by considering the nature of the generators and by relating their properties to a system of equations in the group depending on the exponents of the cyclic subgroup generators (Anshel-Stebe [5]), some special cases with solvable conjugacy problem have been classified. This problem has been connected to the requirement that the exponents be relatively prime in pairs by Anshel [2], and then in turn to vector addition systems [3] and the word problem in commutative semigroups [4].

D. The General Structure. Other criteria for solvable conjugacy problem in these structures have also been found. Clapham [23] gave a long list of conditions under which a general free product with amalgamation would have solvable conjugacy problem. Included were the solvability of the generalized word problem in the factors, the generalized conjugacy problem in the amalgamated part, and a finite coset intersection property. Larsen [62] solved the conjugacy problem for a free product of isomorphic free groups with finitely generated amalgamated subgroups. Iterated free products with cyclic amalgamation of finitely generated free groups and cyclic HNN extensions of such groups have also been shown by Larsen [63] to have solvable conjugacy problem.

After relating it to a number theoretic decision problem for a difference equation on the exponents in a relator, Britton [20] solved the conjugacy problem for an HNN extension with one stable letter of a finitely presented abelian group.

E. Related Problems. The conjugacy separability of these constructions has also been considered. Stebe [96] proved that a certain type of free product with cyclic amalgamation of free groups is conjugacy separable. Dyer [36] extended this to several types of base groups: any free groups with cyclic amalgamation, finitely generated nilpotent groups with cyclic amalgamation, finite groups and certain constructions on conjugacy separable groups themselves.

Concerning a related construction, Hurwitz [57] solved the conjugacy problem in a free product of free groups with finitely generated commuting subgroups, and in a free product with commuting subgroups cyclic on semi-critical generators [58].

VI. Groups Connected to Topology

The first groups for which Dehn considered and solved the conjugacy problem included the fundamental groups of closed orientable 2-manifolds. As Haken [54] has pointed out in his survey of topological and group theoretical decision problems, the homotopy problem of closed curves in a connected complex is equivalent to the conjugacy problem in its fundamental group. In this section we survey results for braid groups, knot groups, and mapping class groups.

A. Braid Groups. The equivalence problem for two braids on n-threads is identical to the conjugacy problem in B_n, the braid group (see Stillwell [100]). This problem is solvable, as proved by Garside [40] and Makanin [72]. Garside's arguments were combinatorial and extended to several other topological groups including hypercubes, a truncated icosododecahedron, and a truncated cubohedron. Birman [13] gave a modified solution of the conjugacy problem for B_n which may simplify its inclusion in the link problem for braids.

B. Knot Groups. When considering knot groups, small cancellation theory turns out to be a valuable tool. Using non-metric conditions C(4) and T(4), Weinbaum [109] solved the conjugacy problem in the Dehn presentation of a tame, prime, alternating knot. This was extended by Appel and Schupp [7], who eliminated the condition that the knot be prime. Using Wirtinger presentations and small cancellation theory on the dual of conjugacy diagrams, Appel [6] then solved the conjugacy problem for several specific types of knots including some cable knots which are not alternating.

In a different approach to knot groups, Stebe [97] showed that hose knot groups are conjugacy separable. These presentations involve free products with cyclic amalgamation.

C. Mapping Class Groups. The homeomorphism problem for knots as well as more general 2- and 3-manifolds is directly related to the conjugacy problem. In [55], Hemion solved the conjugacy problem for the mapping class groups of compact orientable surfaces. Mapping class groups are quotient groups of the self-homeomorphisms of a surface by the normal subgroup of isotopies. Using this result and a method introduced by Haken [53], Waldhausen [105] showed how to classify a large class of 3-manifolds.

VII. Degrees and Complexity

A natural question from the field of recursive function theory concerning decision problems is their degree of unsolvability. In this section we summarize some results on recursively enumerable (r.e.) degrees, Turing reducibility, and the conjugacy problem.

A. R. E. Degrees. For any r.e. degree of unsolvability, Collins [25] gave a construction of a finitely presented group with solvable word problem and unsolvable conjugacy problem of that degree. Bokut' [16] and Miller [74] also established this result. Miller extended this by constructing finitely generated recursively presented examples which, for given r.e. degrees $D_1 \leq D_2$, had word problem of degree D_1 and conjugacy problem of degree D_2. Collins [27] then extended this further by giving an example of a finitely presented group with word problem of degree a and conjugacy problem of degree b, where a and b are arbitrary r.e. Turing degrees of unsolvability with $a \leq b$.

B. Complexity. Analogues to these results can be given in terms of the complexity of solution. In [68], Litvinceva considered the above type of examples in terms of a Turing machine with oracle. Using this approach, Tetruašvili [101], [102], [103] studied upper bounds on the complexity of solutions to the conjugacy problem for certain one-relator groups.

VIII. Problems

The following is a short list of open problems involving the conjugacy problem.

1. One-relator groups: Does there exist a one-relator group with trivial center having unsolvable conjugacy problem or is the conjugacy problem solvable for all one-relator groups?

2. Solvable groups (Remeslennikov-Romanovskiĭ [87]): Is the conjugacy problem solvable for solvable groups of derived length 2, 3, and 4? As mentioned in Section IV, for length 5 there is a counterexample.

3. Small cancellation groups: Are small cancellation groups conjugacy separable?

4. Cyclic subgroups (Anshel [MR 81k #20049]): Is there a free product with cyclic amalgamation (or an HNN extension with cyclic associated subgroups) with unsolvable conjugacy problem?

5. Higman embedding theorem (Collins [28]): What conditions are necessary and sufficient for the preservation of the solvability of the conjugacy problem in the Higman embedding?

6. Commutator problem (Comerford-Edmunds): The commutator problem for a group is to determine, given an element of the group, if it is a commutator. How is this problem related to the conjugacy problem?

The author would like to thank the referee for many helpful suggestions and to express his gratitude to Leo Comerford for his encouragement.

REFERENCES

1. Adian, S. I.: The Burnside Problem and Identities in Groups. Ergebnisse der Mathematik, Bd. 95 (Springer, Berlin-Heidelberg-New York, 1979). MR 55 #5753

2. Anshel, M.: Conjugate powers in HNN groups. Proc. Amer. Math. Soc. 54 (1976), 19-23. MR 52 #14059

3. Anshel, M.: Decision problems for HNN groups and vector addition systems. Math. Comput. 30 (1976), 154-156. MR 53 #626

4. Anshel, M.: The conjugacy problem for HNN groups and the word problem for commutative semigroups. Proc. Amer. Math. Soc. 61 (1976), no. 2, 223-224. MR 54 #10446

5. Anshel, M.; Stebe, P.: The solvability of the conjugacy problem for certain HNN groups. Bull. Amer. Math. Soc. 80 (1974), 266-270. MR 54 #7633

6. Appel, K. I.: On the conjugacy problem for knot groups. Math. Z. 138 (1974), 273-294. MR 50 #10090

7. Appel, K. I.; Schupp, P. E.: The conjugacy problem for the group of any tame alternating knot is solvable. Proc. Amer. Math. Soc. 33 (1972), 329-336. MR 45 #3530

8. Appel, K. I.; Schupp, P. E.: Artin groups and infinite Coxeter groups. Invent. math. 72 (1983), 201-220.

9. Applegate, H.; Onishi, H.: Continued fractions and the conjugacy problem. Comm. in Alg. 9 (1981), no. 11, 1121-1130. MR 82i #20052

10. Armstrong, S. M.: One relator groups with non-trivial centre. M. Phil. Thesis, Queen Mary College (1977).

11. Baumslag, G.: Residual nilpotence and relations in free groups. J. Algebra 2 (1965), 271-282. MR 31 #3487

12. Baumslag, G.: Problem areas in infinite group theory for finite group theorists. The Santa Cruz Conference on Finite Groups (Univ. California, Santa Cruz, Calif. 1979), pp. 217-223, Proc. Sympos. Pure Math. 37, Amer. Math. Soc., Providence, R.I., 1980. MR 82c #20063

13. Birman, J. S.: Braids, Links, and Mapping Class Groups. Ann. Math. Studies 82 (Princeton Univ. Press, Princeton, N.J.; Univ. of Tokyo Press, Tokyo, 1974). MR 51 #11477.

14. Blackburn, N.: Conjugacy in nilpotent groups. Proc. Amer. Math. Soc. 16 (1965), 143-148. MR 30 #3140

15. Bokut', L. A.: On the Novikov groups. Algebra i Logika Sem. 6 (1967), no. 1, 25-38. MR 36 #250

16. Bokut', L. A.: Degrees of unsolvability of the conjugacy problem for finitely presented groups. Algebra i Logika 7 (1968), no. 5, 4-70; ibid. 7 (1968), no. 6, 4-52. MR 41 #3574

17. Boler, J.: Conjugacy in abelian-by-cyclic groups. Proc. Amer. Math. Soc. 55 (1976), no. 1, 17-21. MR 53 #3117

18. Boone, W. W.; Higman, G.: An algebraic characterization of groups with solvable word problem. J. Austral. Math. Soc. 18 (1974), 41-53. MR 50 #10093

19. Brieskorn, E.; Saito, K.: Artin-Gruppen und Coxeter-Gruppen. Invent. Math. 17 (1972), 245-271. MR 48 #2263

20. Britton, J. L.: On the conjugacy problem and difference equations. J. London Math. Soc. (2) 17 (1978), no. 2, 240-250. MR 80k #20039

21. Britton, J. L.: The conjugacy problem for an HNN extension of an abelian group. Math. Sci. 4 (1979), no. 2, 85-92. MR 81h #20036

22. Clapham, C. R. J.: An embedding theorem for finitely generated groups. Proc. Lond. Math. Soc. 3 (1967), 419-430. MR 36 #5199

23. Clapham, C. R. J.: The conjugacy problem for a free product with amalgamation. Arch. Math. 22 (1971), 358-362. MR 46 #5464

24. Collins, D. J.: On embedding groups and the conjugacy problem. J. London Math. Soc. (2) 1 (1969), 674-682. MR 40 #5709

25. Collins, D. J.: Recursively enumerable degrees and the conjugacy problem. Acta Math. 122 (1969), 115-160. MR 39 #4001

26. Collins, D. J.: Word and conjugacy problems in groups with only a few defining relations. Z. Math. Logik Grund. Math. 15 (1969), 305-324. MR 41 #8502

27. Collins, D. J.: Representation of Turing reducibility for word and conjugacy problems in finitely presented groups. Acta Math. 128 (1972), no. 1-2, 73-90. MR 52 #13356

28. Collins, D. J.: Conjugacy and the Higman embedding theorem. Word Problems, II (Conf. on Decision Problems in Algebra, Oxford, 1976), pp. 81-85, Studies in Logic and Foundations of Math., 95, North-Holland, Amsterdam, 1980. MR 81m #20051

29. Collins, D. J.; Miller, C. F. III: The conjugacy problem and subgroups of finite index. Proc. Lond. Math. Soc. 34 (1977), no. 3, 535-556. MR 55 #8187

30. Comerford, L. P., Jr.; Truffault, B.: The conjugacy problem for free products of sixth-groups with cyclic amalgamation. Math. Z. 149 (1976), no. 2, 169-181. MR 53 #13418

31. Dehn, M.; Uber unendliche diskontinuerliche Gruppen. Math. Ann. 71 (1912), 116-144.

32. Deligne, P.: Les immeubles des groupes de tresses généralisés. Invent. Math. 17 (1972), 273-302. MR 54 #10659

33. Dô Long Vân: On identity and conjugacy problems for certain classes of finitely presented groups. Dokl. Akad. Nauk SSSR 241 (1978), no. 5, 1005-1008. (English translation: Soviet Math. Dokl. 19 (1978), no. 4, 938-941 (1979).) MR 80g #20046

34. Dô Long Vân: Problemes des mots et de conjugaison pour une classe de groupes de presentation finie. C. R. Acad. Sc. Paris 292 (1981), no. 17, 773-776. MR 82f #20061

35. Dyer, J. L.: Separating conjugates in free-by-finite groups. J. London Math. Soc. (2) 20 (1979), no. 2, 215-221. MR 81a #20041

36. Dyer, J. L.: Separating conjugates in amalgamated free products and HNN extensions. J. Austral. Math. Soc. Ser. A 29 (1980), no. 1, 35-51. MR 81f #20033

37. Formanek, E.: Conjugate separability in polycyclic groups. J. Algebra 42 (1976), no. 1, 1-10. MR 54 #7626

38. Fridman, A. A.: On the relation between the word problem and the conjugacy problem in finitely presented groups. Trudy Moskov. Mat. Obsc. 9 (1960), 329-356. MR 31 #1195

39. Fridman, A. A.: A solution of the conjugacy problem in a certain class of groups. Trudy Mat. Inst. Steklov 133 (1973), 233-242. MR 49 #424

40. Garside, F. A.: The braid group and other groups. Quart. J. Math. Oxford Ser. (2) 20 (1969), 235-254. MR 40 #2051

41. Gol'berg, A. I.: The impossibility of strengthening certain results of Greendlinger and Lyndon. Uspehi Mat. Nauk 33 (1978), no. 6 (204), 201-202. NR 80d #20037

42. Gol'dina, N. P.: Solution of some algorithmic problems for free and free nilpotent groups. Uspehi Mat. Nauk 13 (1958), no. 3 (81), 183-189. MR 20 #6455

43. Greendlinger, M. D.: On Dehn's algorithms for the conjugacy and word problems, with applications. Comm. Pure Appl. Math. 13 (1960), 641-677. MR 23 #A2327

44. Greendlinger, M. D.: Solutions of the word problem for a class of groups by Dehn's algorithm and of the conjugacy problem by a generalization of Dehn's algorithm. Dokl. Akad. Nauk SSSR 154 (1964), 507-509. MR 28 #3080

45. Greendlinger, M. D.: Solution by means of Dehn's generalized algorithm of the conjugacy problem for a class of groups which coincide with their anti-centers. Dokl. Akad. Nauk SSSR 158 (1964), 1254-1256. MR 30 #4819

46. Greendlinger, M. D.: On the word and conjugacy problems. Izv. Akad. Nauk SSSR Ser. Mat. 29 (1965), 245-268. MR 30 #4820

47. Greendlinger, M. D.: The problem of conjugacy and coincidence with an anti-center in the theory of groups. Sibirsk. Mat. Zh. 7 (1966), 785-803. MR 33 #7406

48. Grunewald, F. J.: Solution of the conjugacy problem in certain arithmetic groups. Word Problems, II (Conf. on Decision Problems in Algebra, Oxford, 1976), pp. 101-139, Studies in Logic and Foundations of Math., 95, North-Holland, Amsterdam, 1980. MR 81h #20054

49. Grunewald, F. J.; Segal, D.: The solubility of certain decision problems in arithmetic and algebra. Bull. Amer. Math. Soc. 1 (1979), no. 6, 915-918. MR 81b #10014

50. Grunewald, F. J.; Segal, D.: Some general algorithms. I: Arithmetic groups. Annals of Math. 112 (1980), 531-583. MR 82d #20048a

51. Gurevich, G. A.: On the conjugacy problem for groups with one defining relator. Dokl. Akad. Nauk SSSR 207 (1972), 18-20. (English translation: Soviet Math. Dokl. 13 (1972), 1436-1439.) MR 47 #5123

52. Gurevich, G. A.: On the conjugacy problem for groups with a single defining relation. Trudy Mat. Inst. Steklov 133 (1973), 109-120. MR 49 #2956

53. Haken, W.: Theorie der Normalflächen. Acta Math. 105 (1961), 245-375. MR 25 #4519

54. Haken, W.: Connections between topological and group theoretical decision problems. Word Problems: Decision Problems and the Burnside Problem in Group Theory (Conf. on Decision Problems in Group Theory, Univ. California, Irvine, Calif., 1969), pp. 427-441. Studies in Logic and the Foundations of Math., 71, North-Holland, Amsterdam, 1973. MR 53 #1594

55. Hemion, G.: On the classification of homeomorphisms of 2-manifolds and the classification of 3-manifolds. Acta Math. 142 (1979), no. 1-2, 123-155. MR 80f #57003

56. Higman, G.: Subgroups of finitely presented groups. Proc. Roy. Soc. London Ser. A 262 (1961), 455-475. MR 24 #A152

57. Hurwitz, R. D.: On the conjugacy problem in a free product with commuting subgroups. Math. Ann. 221 (1976), no. 1, 1-8. MR 54 #414

58. Hurwitz, R. D.: On cyclic subgroups and the conjugacy problem. Proc. Amer. Math. Soc. 79 (1980), no. 1, 1-8. MR 81k #20048

59. Kargapolov, M. I.: Finite approximability of supersolvable groups with respect to conjugacy. Algebra i Logika Sem. 6 (1967), no. 1, 63-68. MR 35 #6741

60. Kargapolov, M. I.; Remeslennikov, V.N.: The conjugacy problem for free solvable groups. Algebra i Logika Sem. 5 (1966), no. 6, 15-25. MR 34 #5905

61. Kargapolov, M. I.; Remeslennikov, V. N.; Romanovskiĭ, N. S.; Roman'kov, V. A.; Čurkin, V. A.: Algorithmic questions for σ-powered groups. Algebra i Logika 8 (1969), 643-659. MR 44 #293

62. Larsen, L.: The solvability of the conjugacy problem for certain free products with amalgamation. J. Algebra 43 (1976), no. 1, 28-41. MR 54 #12912

63. Larsen, L.: The conjugacy problem and cyclic HNN constructions. J. Austral. Math. Soc. Ser. A 23 (1977), no. 4, 385-401. MR 58 #909

64. Lipschutz, S.: Generalization of Dehn's result on the conjugacy problem. Proc. Amer. Math. Soc. 17 (1966), 759-762. MR 33 #5706

65. Lipschutz, S.: On the conjugacy problem and Greendlinger's eighth-groups. Proc. Amer. Math. Soc. 23 (1969), 101-106. MR 40 #4343

66. Lipschutz, S.: The conjugacy problem and cyclic amalgamations. Bull. Amer. Math. Soc. 81 (1975), 114-116. MR 52 #580

67. Lipschutz, S.: Groups with solvable conjugacy problems. Illinois J. Math. 24 (1980), no. 2, 192-195. MR 81k #20049

68. Litvinceva, Z. K.: The conjugacy problem for finitely presented groups. Dal'nevostočn. Mat. Sb. 1 (1970), 54-71. MR 52 #10400

69. Lyndon, R. C.: On Dehn's algorithm. Math. Ann. 166 (1966), 208-228. MR 35 #5499

70. Lyndon, R. C.; Schupp, P. E.: Combinatorial Group Theory. Ergebnisse der Mathematik, Bd. 89 (Springer, Berlin-Heidelberg-New York, 1977). MR 58 #28182

71. Magnus, W.; Karass, A.; Solitar, D.: Combinatorial Group Theory. Pure and Applied Math. 13 (Wiley, New York, 1965). MR 34 #7617

72. Makanin, G. S.: The conjugacy problem in the braid group. Dokl. Akad. Nauk SSSR 182 (1968), 495-496. MR 38 #2195

73. Matthews, J.: The conjugacy problem in wreath products and free metabelian groups. Trans. Amer. Math. Soc. 121 (1966), 329-339. MR 33 #1351

74. Miller, C. F. III: On Group Theoretic Decision Problems and Their Classification. Ann. Math. Studies 68 (Princeton Univ. Press, Princeton, N.J.; Univ. of Tokyo Press, Tokyo 1971). MR 46 #9147

75. Miller, C. F. III: Decision problems in algebraic classes of groups (a survey). Word Problems: Decision Problems and the Burnside Problem in Group Theory (Conf. on Decision Problems in Group Theory, Univ. California, Irvine, Calif., 1969), pp. 507-523. Studies in Logic and the Foundations of Math., 71, North-Holland, Amsterdam, 1973. MR 53 #3121

76. Miller, C. F. III; Schupp, P. E.: The geometry of Higman-Neumann-Neumann extensions. Comm. Pure Appl. Math. 26 (1973), 787-802. MR 49 #9091

77. Mostowski, A. W.: On the decidability of some problems in special classes of groups. Fund. Math. 59 (1966), 123-135. MR 37 #292

78. Novikov, P. S.: Unsolvability of the conjugacy problem in the theory of groups. Izv. Akad. Nauk SSSR Ser. Mat. 18 (1954), 485-524. MR 17 #706

79. Novikov, P. S.: The unsolvability of the problem of equivalence of words in a group and several other problems in algebra. Czechoslovak Math. J. 6 (81) (1956), 450-454. MR 22A #12133

80. Novikov, P. S.; Adian, S. I.: Commutative subgroups and the conjugacy problem in free periodic groups of odd exponent. Izv. Akad. Nauk SSSR Ser. Mat. 32 (1968), 1176-1190. MR 38 #2197

81. Ol'šanskiĭ, A. J.: Infinite groups with cyclic subgroups. Dokl. Akad. Nauk SSSR 245 (1979), no. 4, 785-787. (English translation: Soviet Math. Dokl. 20 (1979), no. 2, 343-346.) MR 80i #20013

82. Palasiński, M.: The identity and conjugacy problem for finitely presented groups. Z. Math. Logik Grundlag. Math. 26 (1980), no. 4, 311-326. MR 81m #20053

83. Pietrowski, A.: The isomorphism problem for one-relator groups with non-trivial centre. Math. Z. 136 (1974), 95-106. MR 50 #2344

84. Remeslennikov, V. N.: Conjugacy in polycyclic groups. Algebra i Logika 8 (1969), 712-725. MR 43 #6313

85. Remeslennikov, V. N.: Finite approximability of groups with respect to conjugacy. Sibirsk. Mat. Zh. 12 (1971), 1085-1099. (English translation: Siberian Math. J. 23 (1971), 783-792.) MR 45 #3539

86. Remeslennikov, V. N.: An example of a group, finitely presented in the variety $\mathfrak{A}^5$, with the unsolvable word problem. Algebra i Logika 12 (1973), no. 5, 577-602. (English translation: Algebra and Logic 12 (1973), no. 5, 327-346 (1975).) MR 51 #8266

87. Remeslennikov, V. N.; Romanovskiĭ, N. S.: Algorithmic problems for solvable groups. Word Problems, II (Conf. on Decision Problems in Algebra, Oxford, 1976), pp. 337-346, Studies in Logic and Foundations of Math., 95, North-Holland, Amsterdam, 1980. MR 81h #20044

88. Sacerdote, G. S.: The Boone-Higman theorem and the conjugacy problem. J. Algebra 49 (1977), no. 1, 212-221. MR 56 #8350

89. Sarkisjan, R. A.: Conjugacy in free polynilpotent groups. Algebra i Logika 11 (1972), no. 6, 694-710. (English translation: Algebra and Logic 11 (1972), 387-396 (1974).) MR 48 #11331

90. Sarkisjan, R. A.: The conjugacy problem for collections of integral matrices. Mat. Zametki 25 (1979), no. 6, 811-824. (English translation: Math. Notes 25 (1979), no. 5-6, 419-426.) MR 80k #20046

91. Schupp, P. E.: On Dehn's algorithm and the conjugacy problem. Math. Ann. 178 (1968), 119-130. MR 38 #5901

92. Schupp, P. E.: On the conjugacy problem for certain quotients of free products. Math. Ann. 186 (1970), 123-129. MR 41 #5475

93. Schupp, P. E.: A survey of small cancellation theory. Word Problems: Decision Problems and the Burnside Problem in Group Theory (Conf. on Decision Problems in Group Theory, Univ. California, Irvine, Calif. 1969), pp. 569-589. Studies in Logic and the Foundations of Math., 71, North-Holland, Amsterdam, 1973. MR 54 #415

94. Seksenbaev, K.: The finite approximability of the finite extension of a nilpotent group with respect to conjugacy. Algebra i Logika 6 (1967), no. 6, 29-31. MR 37 #2861

95. Stebe, P. F.: A residual property on certain groups. Proc. Amer. Math. Soc. 26 (1970), 37-42. MR 41 #5494

96. Stebe, P. F.: Conjugacy separability of certain free products with amalgamation. Trans. Amer. Math. Soc. 156 (1971), 119-129. MR 43 #360

97. Stebe, P. F.: Conjugacy separability of the groups of hose knots. Trans. Amer. Math. Soc. 159 (1971), 79-90. MR 44 #2808

98. Stebe, P. F.: Conjugacy separability of certain Fuchsian groups. Trans. Amer. Math. Soc. 163 (1972), 173-188. MR 45 #2030

99. Stebe, P. F.: Conjugacy separability of groups of integer matrices. Proc. Amer. Math. Soc. 32 (1972), 1-7. MR 44 #6854

100. Stillwell, J.: Classical Topology and Combinatorial Group Theory. Graduate Texts in Math. 72 (Springer, Berlin-Heidelberg-New York, 1980).

101. Tetruašvili, M. R.: The problem of conjugacy for groups with one defining relation and the complexity of Turing computations. Studies in mathematical logic and the theory of algorithms, pp. 29-43, Tbilis. Univ., Tbilisi. 1978. MR 80d #20038

102. Tetruašvili, M. R.: The problem of conjugacy for a class of groups and the complexity of computations. Tbilis. Gos. Univ. Inst. Prikl. Mat. Trudy 5/6 (1978), 250-258. MR 80e #68125

103. Tetruašvili, M. R.: The conjugacy problem for groups with one defining relation and the complexity of Turing calculations. Begriffsschrift-Jena Frege Conference (Friedrich-Schiller-Univ., Jena, 1979), pp. 477-482, Friedrich-Schiller-Univ., Jena, 1979. MR 82b #20050

104. Timošenko, E. I.: Conjugacy in free metabelian groups. Algebra i Logika Sem. 6 (1967), no. 2, 89-94. MR 35 #6740

105. Waldhausen, F.: Recent results on sufficiently large 3-manifolds. Proc. Symp. in Pure Math. 32 (1978), vol. 2, 21-38. MR 80e #57004

106. Wehrfritz, W. A. F.: Conjugacy separating representations of free groups. Proc. Amer. Math. Soc. 40 (1973), 52-56. MR 51 #10469

107. Wehrfritz, W. A. F.: Two examples of soluble groups that are not conjugacy separable. J. London Math. Soc. (2) 7 (1973), 312-316. MR 49 #2942

108. Wehrfritz, W. A. F.: Another example of a soluble group that is not conjugacy separable. J. London Math. Soc. (2) 14 (1976), 381-382. MR 54 #10418

109. Weinbaum, C. M.: The word and conjugacy problems for the knot group of any tame, prime, alternating knot. Proc. Amer. Math. Soc. 30 (1971), 22-26. MR 43 #4895

110. Appel, K. I.: On Artin Groups and Coxeter groups of large type, pp. 50-78 of this volume.

DEPARTMENT OF MATHEMATICS AND COMPUTER SCIENCE
SKIDMORE COLLEGE
SARATOGA SPRINGS, N.Y. 12866

Received June 2, 1982

Contemporary Mathematics
Volume 33, 1984

MAXIMAL ALGEBRAS OF BINARY RELATIONS

Bjarni Jónsson

(Dedicated to Roger Lyndon on the occasion of his 65th birthday.)

1. Introduction. In this note, the Galois connection between sets of binary relations on a set X and sets of permutations of X will be used to investigate certain subalgebras of $\mathcal{R}(X)$, the full algebra of binary relations on X. In particular, a large class of maximal proper subalgebras of $\mathcal{R}(X)$ will be exhibited.

The algebra $\mathcal{R}(X)$ has as its universe the family of all binary relations on X, i.e., the power set of the relation $V = X \times X$. Its basic operations are the set-theoretic union, intersection and complementation, the relative multiplication, and the conversion. We denote the relative product of two binary relations R and S by $R \circ S$, and the converse of R by $R^{\smile}$. In addition, we regard the universal relation V, the null relation $\emptyset$, and the identity relation E as distinguished elements. These relations and the diversity relation E^{-} therefore belong to every subalgebra of $\mathcal{R}(X)$. In fact, $\{V, \emptyset, E, E^{-}\}$ is the universe of the smallest subalgebra of $\mathcal{R}(X)$.

These investigations were inspired by a result obtained by G. Birkhoff: If G is a group of prime order, then the relations

$$\{(x,y) \in G \times G : x^{-1}y \in A\}$$

with $A \subseteq G$ form a maximal proper subalgebra of $\mathcal{R}(G)$. This will be proved

These investigations were supported by NSF Grant 7901735.

© 1984 American Mathematical Society
0271-4132/84 $1.00 + $.25 per page

below as a special case of Corollary 4, but Birkhoff's original proof was of course quite different.

The automorphism group of $\mathcal{R}(X)$ may be identified with the group $\Pi(X)$ of all permutations of X. More precisely, if we associate with each permutation ϕ of X the automorphism ϕ^* of $\mathcal{R}(X)$ such that $\phi^*(R) = \phi^{\smile} \circ R \circ \phi$ for all $R \in \mathcal{R}(X)$, then the correspondence $\phi \to \phi^*$ is an isomorphism from $\Pi(X)$ to the automorphism group of $\mathcal{R}(X)$. This gives rise to a Galois connection

$$A \to A^{\sigma}, \quad \phi \to \phi^{\rho},$$

between subsets A of $\Pi(X)$ and subsets $\mathcal{S}$ of $\mathcal{R}(X)$, with

$$A^{\sigma} = \{R \in \mathcal{R}(X) : \phi^*(R) = R \text{ for all } \phi \in A\},$$

$$S^{\rho} = \{\phi \in \Pi(X) : \phi^*(R) = R \text{ for all } R \in \mathcal{S}\}.$$

(For the general notion of Galois connection, see e.g. Birkhoff [1].) The Galois closed subsets of $\Pi(X)$ are obviously subgroups, but not every subgroup is Galois closed. It is easy to see (and is well known), that $G^{\sigma\rho}$, the Galois closure of a subgroup G of $\Pi(X)$, consists of all permutations ϕ of X with the property that, for all $x,y \in X$, there exists $\psi \in G$ with $\phi(x) = \psi(x)$ and $\phi(y) = \psi(y)$. The alternating group on a finite set with more than three elements is therefore an example of a permutation group that is not Galois closed.

The Galois closed subsets of $\mathcal{R}(X)$ are obviously subalgebras. In fact, they are closed subalgebras; i.e., they are closed under arbitrary unions and intersections as well as under complementation, relative multiplication and conversion. On the other hand, a closed subalgebra of $\mathcal{R}(X)$ need not be Galois closed. A counterexample will be constructed later.

Every closed subalgebra of $\mathcal{R}(X)$ is a complete, atomic Boolean algebra. The atoms are pairwise disjoint relations P_k, $k \in K$, whose union is the universal relation V. In addition, the following three conditions hold:

For all $k \in K$, $P_k^{\smile} = P_m$ for some $m \in K$.

For all $k \in K$, either $P_k \subseteq E$ or $P_k \subseteq E^-$.

For all $k,m,n \in K$, if $P_k \cap (P_m \circ P_n) \neq \emptyset$, then $P_k \subseteq P_m \circ P_n$.

Conversely, given a partitioning of V into relations with these three properties, these relations constitute the set of all atoms of a unique closed subalgebra of $\mathcal{R}(X)$. For a subgroup G of $\Pi(X)$, the atom of G^σ are the relations

$$G(x,y) = \{(\phi(x),\phi(y)) : \phi \in G\}$$

with $x,y \in X$.

No useful necessary and sufficient conditions are known for a subalgebra of $\mathcal{R}(X)$ to be Galois closed, but in the next section we obtain some simple sufficient conditions.

2. <u>A class of Galois closed subalgebras</u>. A subgroup G of $\Pi(X)$ is said to be semi-regular if, for all $\phi,\psi \in G$ and $x \in X$, the condition $\phi(x) = \psi(x)$ implies that $\phi = \psi$. In other words, to say that G is semi-regular means that no member of G except the identity permutation has a fixed point. If G is both semi-regular and transitive, then it is said to be regular.

<u>Theorem</u> 1. If $\mathcal{A}$ is a closed subalgebra of $\mathcal{R}(X)$ and every atom of $\mathcal{A}$ is a partial function, then $\mathcal{A}$ is Galois closed, and $\mathcal{A}^\rho$ is semi-regular. Conversely, if G is a semi-regular subgroup of $\Pi(X)$, then G is Galois closed, and every atom of G^σ is a partial function.

<u>Proof</u>. The converse of an atom is always an atom. Hence, under the hypothesis of the first part of the theorem, every atom of $\mathcal{A}$ is a bijection from a subset of X to a subset of X. Furthermore, the domains of any two atoms are either identical or else disjoint. Consequently, X may be partitioned into subsets X_k such that each atom of $\mathcal{A}$ is a bijection from some X_k to some X_m. For $x,y \in X$, let $P_{x,y}$ be the unique atom of $\mathcal{A}$ with $(x,y) \in P_{x,y}$.

We claim that the group $\mathcal{A}^\rho$ is transitive on each of the sets X_k. Given $u,v \in X_k$, let $\pi(x) = P_{u,x}(v)$. This function π is defined for all

$x \in X$, for $P_{u,x}$ has X_k as its domain. To see that π is a permutation of X, observe that the conditions

$$P_{u,x}(v) = y\ , \qquad P_{u,x} = P_{v,y}\ , \qquad P_{v,y}(u) = x$$

are equivalent. Consequently, for any $y \in X$ there is a unique $x \in X$ with $\pi(x) = y$, namely $x = P_{v,y}(u)$. Since $\pi(u) = v$, the proof of the claim will be completed if we show that $\pi \in \mathcal{A}^\rho$, i.e., that $\pi^*(P) = P$ for every atom P of $\mathcal{A}$.

The relation $\pi^*(P)$ consists of all ordered pairs $(\pi(x),\pi(y)) = (P_{u,x}(v),P_{u,y}(v))$ with $(x,y) \in P$, i.e., with $P_{x,y} = P$. Thus if $(s,t) \in \pi^*(P)$, then for some $(x,y) \in P$,

$$(v,s) \in P_{u,x} \text{ and } (v,t) \in P_{u,y}\ ,$$

and therefore

$$(s,t) \in P^\smile_{u,x} \circ P_{u,y} = P_{x,u} \circ P_{u,y} = P_{x,y} = P\ .$$

This shows that $\pi^*(P) \subseteq P$, and to establish the opposte inclusion we need only observe that if, in the definition of π, the roles of u and v are interchanged, the resulting permutation is $\pi^\smile$.

If $\mathcal{A}_1$ is a closed subalgebra of $\mathcal{R}(X)$ that properly contains $\mathcal{A}$, then some atom P_1 of $\mathcal{A}_1$ is properly contained in an atom P of $\mathcal{A}$. It follows that the domain Y_1 of P_1 is properly contained in the domain Y of P, and choosing $u,v \in Y$ with $u \in Y_1$ and $v \notin Y_1$, we use the transitivity of $\mathcal{A}^\rho$ on Y to obtain $\pi \in \mathcal{A}^\rho$ with $\pi(u) = v$. Noting that $\pi^*(P_1) \neq P_1$, hence $\pi \notin \mathcal{A}_1^\rho$, we conclude that $\mathcal{A}_1^\rho \neq \mathcal{A}^\rho$. Thus $\mathcal{A}$ is Galois closed. Finally, if $\phi,\psi \in \mathcal{A}^\rho$ then for all $x,y \in X$, $P_{x,y}(\phi(x)) = \phi(y)$ and $P_{x,y}(\psi(x)) = \psi(y)$, hence $\phi(x) = \psi(x)$ implies $\phi(y) = \psi(y)$. This shows that $\mathcal{A}^\rho$ is semi-regular.

Now suppose G is a semi-regular subgroup of $\Pi(X)$, and consider any $\phi \in G^{\sigma\rho}$. Fixing $x \in X$, we can, for each $y \in X$, find a member ψ_y of G such that

$$\phi(x) = \psi_y(x) \text{ and } \phi(y) = \psi_{(y)}(y)\ .$$

Since all the permutations ψ_y take on the same value at x, they must be equal to the same permutation ψ, in view of the semi-regularity of G. But this means that $\psi(y) = \phi(y)$ for all $y \in X$, so that $\phi = \psi \in G$. Consequently, G is Galois closed. Furthermore, by the semi-regularity of G, every atom of G^σ,

$$P_{x,y} = G(x,y) = \{(\phi(x),\phi(y)) : \phi \in G\},$$

is a partial function, for if two ordered pairs in $P_{x,y}$, $(\phi(x),\phi(y))$ and $(\psi(x),\psi(y))$ have the same first coordinate, $\phi(x) = \psi(x)$, then they have the same second coordinate, $\phi(y) = \psi(y)$.

This completes the proof of the theorem.

Corollary 2. If $\mathscr{A}$ is a closed subalgebra of $\mathcal{R}(X)$, and if every atom of $\mathscr{A}$ is a permutation of X, then $\mathscr{A}$ is Galois closed and $\mathscr{A}^\rho$ is regular. Conversely, if G is a regular subgroup of $\Pi(X)$, then G is Galois closed and every atom of G^σ is a permutation of X.

Theorem 3. Suppose G is a group. For any subgroup H of G, the relations

$$H\cdot(x,y) = \{(ax,ay) : a \in H\}$$

with $x,y \in G$ constitute the atoms of a Galois closed subalgebra $\mathscr{G}(H)$ of $\mathcal{R}(G)$. The correspondence $H \to \mathscr{G}(H)$ is a dual lattice isomorphism from the lattice of all subgroups of G to the lattice of all closed (or, equivalently, Galois closed) subalgebras of $\mathcal{R}(G)$ that contains $\mathscr{G}(G)$ as a subalgebra.

Proof. The relations $G\cdot(x,y)$ are the right translations $t \to R_b(t) = tb$ with $b \in G$, i.e.,

$$R_b = \{(t,tb) : t \in G\}.$$

In fact, $G\cdot(x,y) = R_b$, where $b = x^{-1}y$. Since the right translations form a regular permutation group, they are the atoms of a Galois closed subalgebra $\mathscr{G}(G)$ of $\mathcal{R}(G)$. For a subgroup H of G, $H\cdot(x,y)$ is the partial function obtained by restricting the domain of the permutation $G\cdot(x,y)$ to the coset Hx. The relations $H\cdot(x,y)$ are therefore partial functions, and they constitute a partitioning of the universal relation $G \times G$. To see that they are the atoms of a closed (and therefore Galois closed) subalgebra of $\mathcal{R}(G)$ it suffices to note that $H\cdot(x,y) \subseteq E$ if $x = y$, but $H\cdot(x,y) \subseteq E^-$ if $x \neq y$, that the converse of $H\cdot(x,y)$ is $H\cdot(y,x)$, and that $H\cdot(x,y) \circ H\cdot(u,v)$ is equal to $H(x,yu^{-1}v)$ if $yu^{-1} \in H$, but is the null relation otherwise.

For future reference observe that $\mathcal{B}(G)^{\rho}$ is the group of all left translations $t \to L_a(t) = at$, and that, for any subgroup H of G,

$$\mathcal{B}(H)^{\rho} = \{L_a : a \in H\} .$$

It is obvious that the map $H \to \mathcal{B}(H)$ is one-to-one, for if H and K are distinct subgroups of G, then the atom $H\cdot(x,y)$ of $\mathcal{B}(H)$ is distinct from the atom $K\cdot(x,y)$. To complete the proof, it suffices to show that every closed subalgebra $\mathcal{A}$ of $\mathcal{R}(G)$ that has $\mathcal{B}(G)$ as a subalgebra is of the form $\mathcal{B}(H)$. First observe that $\mathcal{A}$ is Galois closed, for every atom of $\mathcal{A}$ is a subset of an atom of $\mathcal{B}(G)$, and is therefore a partial function. The group $\mathcal{A}^{\rho}$ is a subgroup of $\mathcal{B}(G)^{\rho}$, the group of all left translations of G. Hence

$$\mathcal{A}^{\rho} = \{L_a : a \in H\} = \mathcal{B}(H)^{\rho}$$

for some subgroup H of G, and since both $\mathcal{A}$ and $\mathcal{B}(H)$ are Galois closed, we conclude that $\mathcal{A} = \mathcal{B}(H)$.

Corollary 4. In the notation of the preceding theorem, $\mathcal{B}(H)$ is a maximal proper subalgebra of $\mathcal{R}(G)$ if H is of prime order.

3. A class of maximal proper closed subalgebras. If X is a set of prime order p, and if G is a subgroup of $\Pi(X)$ of order p, then G is regular, and therefore Galois closed. From correspondence between Galois closed subgroups of $\Pi(X)$ and Galois closed subalgebras of $\mathcal{R}(X)$ it follows that G^{σ} is a maximal proper Galois closed subalgebra of $\mathcal{R}(X)$. Actually, G^{σ} turns out to be a maximal proper subalgebra of $\mathcal{R}(X)$. This is a special case of Theorem 6 below.

Theorem 5. Every finite, cyclic subgroup of $\Pi(X)$ is Galois closed.

Proof. Let $G = \langle \phi \rangle$ be a finite, cyclic subgroup of $\Pi(X)$. Since the order of ϕ is finite, each element $x \in X$ has an orbit of finite length m_x under ϕ, and the number of distinct orbit lengths m_x is also finite.

Consider any $\psi \in G^{\sigma\rho}$. For each $x \in X$, some power of ϕ, say ϕ^{k_x}, must agree with ψ at x, and for any integer k we therefore have

$$\psi(x) = \phi^k(x) \quad \text{iff} \quad k \equiv k_x \mod m_x .$$

For any two elements $x,y \in X$, there must be some power ϕ^k of ϕ that agrees with ψ at both x and y; in other words, the two congruences

$$k \equiv k_x \mod m_x , \quad k \equiv k_y \mod m_y$$

must have a common solution. Since the number of distinct moduli m_x is finite, we conclude by the Generalized Chinese Remainder Theorem that all the congruences have a common solution k, whence $\psi = \phi^k \in G$. Thus G is Galois closed.

<u>Theorem</u> 6. If G is a subgroup of $\Pi(X)$ of prime order, then G^σ is a maximal proper subalgebra of $\mathcal{R}(X)$.

<u>Proof</u>. Let the order of G be p. Each orbit Gx either has p elements, or else consists of the element x alone. If both Gx and Gy have p elements, then the atom $G(x,y)$ of G^σ is a bijection from Gx to Gy, but if either Gx or Gy consists of just one element, then $G(x,y) = Gx \times Gy$.

Consider a closed subalgebra $\mathcal{A}$ of $\mathcal{R}(X)$ that properly contains G^σ. Then some atom $G(x,y)$ of G^σ is not an atom of $\mathcal{A}$. We may assume that Gx has p elements, and it readily follows that $G(x,x)$ is not an atom of $\mathcal{A}$. Since $G(x,x)$ is finite, it is the union of finitely many atoms of $\mathcal{A}$, say $P_1, P_2, \ldots, P_n$. The relation $G(x,x)$ is the identity map on a p-element subset Y of X, and each P_j is the identity map on a subset Y_j of Y. For $j \leq n$ pick $y_j \in Y_j$. If $i,j \leq n$, then $G(y_i, y_j)$ is a permutation of Y, and $P_i \circ G(y_i,y_j) \circ P_j$ is therefore a bijection from a subset Z_i of Y_i to a subset Z_j of Y_j. But Z_i cannot be a proper subset of Y_i, for then P_i would not be an atom of $\mathcal{A}$. Similarly, Z_j cannot be a proper subset of Y_j. Consequently, the relation $P_i \circ G(y_i,y_j) \circ P_j$ is a bijection from Y_i to Y_j. This shows that all the sets Y_i have the same number of elements, say k.

Hence $p = kn$, and since p is a prime and $n > 1$, we must have $k = 1$. Each of the relations P_i thus consists of a single ordered pair (y_i, y_i).

We have shown that there exists $x \in X$ such that Gx has p elements but the relation $Q = \{(x,x)\}$ belongs to $\mathcal{J}$. For each $y \in X$, the relation $G(y,x) \circ Q \circ G(x,y) = \{(y,y)\}$ therefore belongs to $\mathcal{J}$, and we conclude that $\mathcal{J} = \mathcal{R}(X)$. Thus the only closed subalgebra of $\mathcal{R}(X)$ that properly contains G^σ is $\mathcal{R}(X)$ itself, as was to be shown.

4. Examples. For $n = 1,2,3$, all the subalgebras of $\mathcal{R}(n)$ are Galois closed, and so are the subgroups of S_n, the full symmetric group on n elements. This can easily be verified directly. We have checked the cases $n = 4$ and $n = 5$, and shall present the conclusions while sparing the reader the details.

Example 1. Subalgebras of $\mathcal{R}(4)$. All the subalgebras of $\mathcal{R}(4)$ are Galois closed, and every subgroup of S_4 except the alternating group is Galois closed. The group S_4 has 30 subgroups, which belong to 11 conjugate classes. Consequently, $\mathcal{R}(4)$ has 29 subalgebras, and they fall into 10 conjugate classes.

Example 2. Subalgebras of $\mathcal{R}(5)$. All the subalgebras of $\mathcal{R}(5)$ are Galois closed. The alternating group, A_5, is of course not Galois closed, and neither are the five copies of A_4. There are two other conjugate classes whose members are not Galois closed. One class consists of the subgroups of order 20, while each member of the other class is generated by a pair of permutations (i,j,k) and $(i,j)(m,n)$. It is now easy to count the subalgebras of $\mathcal{R}(5)$; there are 124.

Example 3. Subalgebras that are not Galois closed. Given $R,S,T \in \mathcal{R}(X)$, let $Q(R,S,T)$ be the relation consisting of all ordered pairs (x,y) such that, for some s,t,

$$xRsSy, \quad xStRy, \quad sTt .$$

Clearly every permutation of X that is compatible with R,S and T is also compatible with $Q(R,S,T)$. A subalgebra of $\mathcal{R}(X)$ that is not closed under Q therefore cannot be Galois closed.

Let $X = F \times F$, where F is some field. Regarding X as an affine plane over F, let R_0, R_1 and R_2 be the equivalence relations on X whose blocks are, respectively, the horizontal lines, the vertical lines, and the lines with slope 1. It is easy to check that the relation $R_3 = Q(R_0,R_1,R_2)$ is the equivalence relation whose blocks are the lines with slope -1. It is also easy to check that the subalgebra $\mathcal{A}$ of $\mathcal{R}(X)$ generated by the relations R_0, R_1 and R_2 has exactly five atoms, the identity relation E, the three relations $E^- \cap R_i$ with $i = 0,1,2$, and the relation $S = (R_0 \cup R_1 \cup R_2)^-$. If the characteristic of F is two, then $R_3 = R_2$, and if F is the three-element field, then $R_3 = E \cup S$, but in all other cases, $E \subset R_3 \subset E \cup S$ (strict inclusions), and therefore $R_3 \notin \mathcal{A}$.

REFERENCE

1. G. Birkhoff, Lattice theory, Amer. Math. Soc. Colloq. Publ., vol. 25, third ed., 1967.

DEPARTMENT OF MATHEMATICS
VANDERBILT UNIVERSITY
NASHVILLE, TN 37235

Received February 5, 1982

Contemporary Mathematics
Volume 33, 1984

ÜBER UNTERGRUPPEN EBENER DISKONTINUIERLICHER GRUPPEN

R. N. Kalia und G. Rosenberger

Herrn Professor R. C. Lyndon zum 65. Geburtstag gewidmet

Einleitung:

Ausgangspunkt dieser Arbeit ist ein Vortrag von H. Zieschang im Mai 1981 in Oberwolfach auf der Gruppentheorietagung. H. Zieschang berichtete dort über seine Untersuchungen von Untergruppen freier Produkte zyklischer Gruppen, die von Elementen endlicher Ordnung erzeugt werden, mit Hilfe eines Algorithmus, bei dem in jedem Schritt die erzeugenden Elemente endlicher Ordnung nur durch Elemente endlicher Ordnung ersetzt werden (vgl. auch [8]). Dieser Algorithmus ist naheliegend (vgl. zum Beispiel [1], [5], [8]) und ergibt sich in natürlicher Weise wegen seines topologischen Hintergrundes (vgl. [9]).
Hier erweitern wir die Untersuchungen von H. Zieschang auf die ebenen diskontinuierlichen Gruppen G mit kompaktem Fundamentalbereich. Wir betrachten Untergruppen H von G, die von Elementen x_j, $j \in J$, endlicher Ordnung erzeugt werden, und fragen insbesondere, unter welchen Bedingungen H schon das freie Produkt der gegebenen endlichen zyklischen Gruppen $\langle x_j \rangle$ ist (Satz 1 bis Satz 3). Die Beweise basieren auf einer modifizierten Nielsenschen Kürzungsmethode in freien Produkten mit Amalgam, bei der in jedem Kürzungsschritt die Elemente endlicher Ordnung nur durch Elemente endlicher Ordnung ersetzt werden. Wir bemerken noch, daß sich die Aussagen (2.4) bis (2.10) allgemeiner aussprechen und anwenden lassen.

© 1984 American Mathematical Society
0271-4132/84 $1.00 + $.25 per page

§1. Vorbemerkungen.

Diese Arbeit verwendet die Terminologie und Bezeichnungsweise von [1], [4] und [6], wobei $\langle \ldots | \ldots \rangle$ die Gruppenbeschreibung durch Erzeugende und Relationen bedeutet. Die verwendeten Begriffe und Sätze aus der kombinatorischen Gruppentheorie finden sich in [2] und [9].

Sie $G = H_1 \underset{A}{*} H_2$ freies Produkte der Gruppen H_1 und H_2 mit Amalgam $A = H_1 \cap H_2$.

Wir wählen in jedem H_i ein System L_i von Vertretern für die Linksrestklassen von H_i nach A (A werde durch 1 repräsentiert). Jedes $x \in G$ besitzt eine eindeutig bestimmte Darstellung

$$x = h_1 \ldots h_n a \ ,$$

wobei $a \in A$ und die h_i Linksrestklassenvertreter sind, die abwechselnd in verschiedenen Faktoren von G liegen und ungleich 1 sind.
Durch $L(x) = n$ wird eine Länge von x definiert.
Wir nehmen nun die Inversen der Linksrestklassenvertreter als System L_i^{-1} von Vertretern für die Rechtsrestklassen. Damit erhält jedes $x \in G$ die eindeutig bestimmte (symmetrische) Normalform

$$x = \ell_1 \ldots \ell_m k r_m \ldots r_1 \text{ , wobei gilt:}$$

(a) $m \geq 0$, $k \in H_1 \cup H_2$, $1 \neq \ell_j \in L_1 \cup L_2$, $1 \neq r_j \in L_1^{-1} \cup L_2^{-1}$;

(b) die ℓ_j bzw. r_j liegen abwechselnd in verschiedenen Faktoren;

(c) für $L(x) = 0$ ist $m = 0$ und $k \in A$;

(d) Für $L(x) = 2m$ $(m \geq 1)$ ist $k \in A$, und ℓ_m und r_m liegen in verschiedenen Faktoren H_i von G und

(e) für $L(x) = 2m+1$ ist $k \notin A$, $k \in H_i$ für einen Faktor H_i und $\ell_m, r_m \notin H_i$ falls $m \geq 1$.

Wir bezeichnen $\ell_1 \ldots \ell_m$ als die vordere Hälfte, $r_m \ldots r_1$ als die hintere Hälfte und k als den Kern von x.

Wir führen nun eine Ordnung auf G ein. Dazu ordnen wir erst die Menge $\{1,2\}$ (und damit die Faktoren H_1 und H_2) und in jedem Faktor H_i die Restklassenvertreter aus L_i vollständig (wobei 1 das erste Element sei).

Dabei werden ℓ und ℓ^{-1} nicht unterschieden. Nun ordnen wir für jedes m die Produkte $\ell_1 \ldots \ell_m$ von Linksrestklassenvertretern (wobei $\ell_j \in L_i$ für ein i und die ℓ_j abwechselnd aus verschiedenen Faktoren) lexikographisch. Diese Ordnung in G werde mit $<$ bezeichnet; sie hat folgende Eigenschaften:

(a) Aus $\ell_1 \ldots \ell_m \leq \ell'_1 \ldots \ell'_m$ und $\ell'_1 \ldots \ell'_m \leq \ell_1 \ldots \ell_m$ folgt $\ell_1 \ldots \ell_m = \ell'_1 \ldots \ell'_m$ und

(b) aus $\ell_1 \ldots \ell_m < \ell'_1 \ldots \ell'_m$ folgt $\ell_1 \ldots \ell_m \ell_{m+1} < \ell'_1 \ldots \ell'_m \ell'_{m+1}$ für beliebige zulässige ℓ_{m+1}, ℓ'_{m+1}.

Für die Untersuchung endlicher Systeme $\{x_1, \ldots, x_n\} \subset G$ und der von ihnen erzeugten Untergruppen von G können wir noch annehmen:

Die Ordnung $<$ in G genügt der Bedingung, daβ vor einem Produkt $\ell_1 \ldots \ell_m$ von Linksrestklassenvertretern (die ℓ_j abwechselnd aus verschiedenen Faktoren) nur endlich viele Produkte $\ell_1 \ldots \ell_{m-1} \ell$ liegen, wobei $\ell \in L_i$ falls $\ell_m \in L_i$.

Nun erweitern wir diese Ordnung $<$ auf die Menge der Paare $\{x, x^{-1}\}$, wobei die Bezeichnung so sei, daβ die vordere Hälfte von x bezüglich $<$ vor der von x^{-1} steht. Dann gelte $\{x, x^{-1}\} \prec \{x', x'^{-1}\}$, wenn entweder $L(x) < L(x')$ oder bei $L(x) = L(x')$ die vordere Hälfte von x echt vor der von x' steht oder falls bei $L(x) = L(x')$ diese übereinstimmen, die vordere Hälfte von x^{-1} vor der von x'^{-1} steht. Ist $\{x, x^{-1}\} \prec \{x', x'^{-1}\}$, so sagen wir, daβ x vor x' steht. Gilt $\{x, x^{-1}\} \prec \{x' x'^{-1}\}$ und $\{x', x'^{-1}\} \prec \{x, x^{-1}\}$, so unterscheiden sich x und x' nur im Kern.

Ein System $\{x_j\}_{j \in J}$ heiβt kürzer als ein System $\{x'_j\}_{j \in J}$, wenn $\{x_j, x_j^{-1}\} \prec \{x'_j, x'^{-1}_j\}$ für alle $j \in J$, aber für mindestens ein j nicht $\{x'_j, x'^{-1}_j\} \prec \{x_j, x_j^{-1}\}$ gilt.

Wir schreiben im folgenden $u_1 \ldots u_q \equiv v_1 \ldots v_n$ für die Gleilchheit zusammen mit der Tatsache, daβ die rechte Seite reduziert ist in dem Sinn, daβ $L(v_1 \ldots v_n) = L(v_1) + \ldots + L(v_n)$ gilt (die $u_i, v_j \in G$). Ist p Produkt von Restklassenvertretern, so bedeute die Schreibweise $p \equiv rs$ zusätzlich, daβ auch r und s Produkte von Restklassenvertretern sind.

Wir verwenden für die (symmetrische) Normalform eines Elementes x in $G = H_1 \underset{A}{*} H_2$ die Schreibweise $x \equiv p_x k_x q_x$, wobei p_x die vordere Hälfte,

q_x die hintere Hälfte und k_x der Kern von x sind (also $p_x \in G$, $q_x \in G$, $k_x \in A$ oder $k_x \in H_i \setminus A$, i = 1 oder 2, p_x Produkt von Linksrestklassenvertretern und q_x Produkt von Rechtsrestklassenvertretern). Es ist insbesondere

$$L(x) = L(p_x) + L(q_x) = 2L(p_x) \text{ falls } k_x \in A \text{ und}$$

$$L(x) = L(p_x) + L(q_x) + L(k_x) = 2L(p_x) + 1 \text{ falls } k_x \notin A.$$

Es ist x genau dann zu einem Element aus H_1 oder H_2 konjugiert, wenn $q_x = p_x^{-1}$ ist.

Es bezeichne:

$[a,b] = aba^{-1}b^{-1}$ den Kommutator von $a,b \in G$;

(α,β) den größten gemeinsamen Teiler von $\alpha,\beta \in \mathbb{Z}$;

$\mathfrak{N}_1 = \{p \mid p = 1$ oder $1 \neq p \equiv p'\ell$ Produkt von Linksrestklassenvertretern mit $L(\ell) = 1$ und $\ell \in H_2\}$;

$\mathfrak{N}_2 = \{p \mid p = 1$ oder $1 \neq p \equiv p'\ell$ Produkt von Linksrestklassenvertretern mit $L(\ell) = 1$ und $\ell \in H_1\}$.

Für eine nicht-leere Teilmenge $X \subset G$ von G bedeute:

$\langle X \rangle$ die von X erzeugte Untergruppe von G;

card X die Kardinalzahl von X;

$X_o = \{x \mid x \in X \text{ und } x \in A\}$;

$X_{1,p} = \{x \mid x \equiv pk_xp^{-1},\ k_x \in H_1 \setminus A,\ p \in \mathfrak{N}_1\}$, d.h. insbesondere $p_x \equiv p$ für $x \in X_{1,p}$;

$X_{2,p} = \{x \mid x \in X,\ x \equiv pk_xp^{-1},\ k_x \in H_2 \setminus A,\ p \in \mathfrak{N}_2\}$;

$X_3 = \{x \mid x \in X$, x ist nicht konjugiert zu einem Element aus H_1 oder $H_2\}$;

$\mathfrak{N}_1(X) = \{p \mid p \in \mathfrak{N}_1 \text{ und } X_{1,p} \neq \emptyset\}$;

$\mathfrak{N}_2(X) = \{p \mid p \in \mathfrak{N}_2 \text{ und } X_{2,p} \neq \emptyset\}$.

<u>Bemerkung</u>:

Es ist $X = X_0 \cup (\bigcup_{p \in \mathfrak{N}_1(X)} X_{1,p}) \cup (\bigcup_{p \in \mathfrak{N}_2(X)} X_{2,p}) \cup X_3$.

<u>Definition (1.1)</u>:

Eine nicht-leere Teilmenge $X = \{x_j\}_{j \in J} \subset G$ heißt <u>E-Menge</u>, wenn sie wie folgt zerlegt ist:

(a) $X = X_1 \cup X_2$, $X_1 \cap X_2 = \emptyset$;

(b) jedes $x_j \in X_1$ hat endliche Ordnung;

(c) jedes $x_j \in X_2$ hat unendliche Ordnung.

Wir ersetzen eine E-Menge $X = \{x_j\}_{j \in J}$ durch eine E-Menge $X' = \{x_j'\}_{j \in J'}$ aus G mittels einer (endlichen) Hintereinanderausführung folgender elementarer Transformationen:

(1) Ersetze ein $x_j \in X_1$ durch $x_j' = x_k^{\epsilon} x_j x_k^{-\epsilon}$, $k \neq j$, $\epsilon = \pm 1$, und lasse die x_i, $i \neq j$, unverändert.

(2) Ersetze ein $x_j \in X_2$ durch $x_j' = x_k^{\epsilon} x_j$ oder $x_j x_k^{\epsilon}$, $k \neq j$, $\epsilon = \pm 1$, und lasse die x_i, $i \neq j$, unverändert.

(3) Ersetze ein $x_j \in X$ durch $x_j' = x_j^{-1}$, und lasse die x_i, $i \neq j$, unverändert.

(4) Permutiere in X_1, und lasse X_2 unverändert.

(5) Permutiere in X_2, und lasse X_1 unverändert.

(6) Hat $x_j \in X_1$ die Ordnung α, $2 \leq \alpha < \infty$, so ersetze x_j durch $x_j' = x_j^{\gamma}$ mit $1 \leq \gamma < \alpha$ und $(\gamma, \alpha) = 1$, und lasse die x_i, $i \neq j$, unverändert.

(7) Erzeugen $x_j, x_k \in X$, $j \neq k$, $1 \neq x_j$, $1 \neq x_k$, eine zyklische Gruppe $\langle x_j' \rangle$, so ersetze x_j durch x_j' und x_k durch $x_k' = 1$, und lasse die x_i, $j \neq i \neq k$, unverändert.

(8) Ist $x_j \in X$, $x_j = 1$, so streiche x_j, und lasse die x_i, $i \neq j$, unverändert.

Bemerkungen:

1) Der für uns wichtigste Spezialfall einer elementaren Transformation (7) ist der Fall, daß x_j und x_k beide endliche Ordnung haben.

2) Die E-Menge X' kann durchaus mehr Elemente endlicher Ordnung enthalten als X selbst, d.h. es können (z.B. durch eine elementare Transformation (2)) neue Elemente endlicher Ordnung hinzukommen.
Dies kann aber nicht geschehen, wenn $X = X_1$ ist, d.h. wenn jedes Element aus X endliche Ordnung hat.

Definition (1.2):

(a) Eine (endliche) Folge von solchen elementaren Transformationen (1) bis (8) nennen wir E-Transformation.

(b) Wir nennen eine E-Menge X' herleitbar aus der E-Menge X, wenn es eine E-Transformation von X auf X' gibt.

(c) Eine E-Menge $X = \{x_j\}_{j \in J}$ heißt E-reduziert, wenn $x_j \neq 1$ für alle $j \in J$, und wenn es keine aus X herleitbare E-Menge $X' = \{x'_j\}_{j \in J'}$ gibt, für die eine der beiden folgenden Eigenschaften erfüllt ist:

(i) $x'_j = 1$ für ein $j \in J'$;

(ii) $J = J'$, $x'_j \neq 1$ für alle $j \in J'$, und es ist X' kürzer als X. □

Als unmittelbare Konsequenz erhalten wir

Lemma (1.3):

(a) Die E-Menge X' sei aus der E-Menge X herleitbar.
Dann ist $\langle X \rangle = \langle X' \rangle$.

(b) Sei X eine endliche E-Menge. Dann gibt es eine E-reduzierte, endliche E-Menge X', die aus X in endlich vielen Schritten herleitbar ist. □

§2. Von Elementen endlicher Ordnung erzeugte Untergruppen ebener diskontinuierlicher Gruppen

In diesem Paragraphen sei stets

$$G = \langle s_1,\dots,s_m,a_1,b_1,\dots,a_g,b_g \mid s_1^{\alpha_1} = \dots = s_m^{\alpha_m} =$$

$$= s_1\dots s_m[a_1,b_1]\dots[a_g,b_g] = 1\rangle ,$$

$0 \leq m$, $2 \leq \alpha_i$ $(i = 1,\dots,m)$, $0 \leq g$, $2g - 2 + \sum_{i=1}^{m} (1 - \alpha_i^{-1}) > 0$, sowie $3 \leq m$ falls $g = 0$ Fuchssche Gruppe mit kompaktem Fundamentalbereich. (Wir bemerken, daß wir die Ergebnisse dieser Arbeit direkt analog übertragen können auf die ebene diskontinuierliche Gruppe

$\langle s_1,\dots,s_m,a_1,\dots,a_g \mid s_1^{\alpha_1} = \dots = s_m^{\alpha_m} = s_1\dots s_m a_1^2\dots a_g^2 = 1\rangle$,

$0 \leq m$, $2 \leq \alpha_i$ $(i = 1,\dots,m)$, $1 \leq g$ sowie $2 \leq g$ falls $m \leq 1$).

Ist $m = 0$, so ist G torsionsfrei. Ist $m = 3$ falls $g = 0$, so ist G nicht zerlegbar in ein nicht-triviales freies Produkt mit Amalgam ([7]). Ist $m = 1$ falls $g = 1$, so ist G nicht zerlegbar in ein nicht-triviales freies Produkt mit zyklischem Amalgam ([7]); ist aber zusätzlich α_1 keine Potenz von 2, so ist G zerlegbar in ein nicht-triviales freies Produkt mit Amalgam ([3]; der Fall $\alpha_1 = 2^\nu$, $\nu \geq 1$, ist noch offen).

Wir treffen also im folgenden für G die weitere

<u>Voraussetzung (2.1)</u>:

Es sei $m \geq 1$ sowie $4 \leq m$ falls $g = 0$ und $2 \leq m$ falls $g = 1$.

Dann läßt sich G (meist) auf verschiedene Weise als nicht-triviales freies Produkt mit zyklischem Amalgam darstellen. Wir beschäftigen uns in dieser Note im wesentlichen mit zwei solchen Darstellungen:

(2.2): $G = H_{11} \underset{A_1}{*} H_{21}$ mit

$$H_{11} = \langle s_1,\ldots,s_{m'} | s_1^{\alpha_1} = \ldots = s_{m'}^{\alpha_{m'}} = 1\rangle,$$

$$H_{21} = \langle s_{m'+1},\ldots,s_m,a_1,b_1,\ldots,a_g,b_g | s_{m'+1}^{\alpha_{m'+1}} = \ldots = s_m^{\alpha_m} = 1\rangle,$$

$$A_{21} = \langle s_1\ldots s_{m'}\rangle = \langle (s_{m'+1}\ldots s_m[a_1,b_1]\ldots[a_g,b_g])^{-1}\rangle$$

mit $2 \leq m' \leq m$, $0 \leq g \leq 1$ sowie

$2 \leq m' \leq m-2$ falls $g = 0$ und $2 \leq m' = m$ falls $g = 1$.

(2.3): $G = H_{12} \underset{A_2}{*} H_{22}$ mit

$$H_{12} = \langle s_1,\ldots,s_m,a_1,b_1 | s_1^{\alpha_1} = \ldots = s_m^{\alpha_m} = 1\rangle,$$

$$H_{22} = \langle a_2,b_2,\ldots,a_g,b_g\rangle,$$

$$A_2 = \langle s_1\ldots s_m[a_1,b_1]\rangle = \langle ([a_2,b_2]\ldots[a_g,b_g])^{-1}\rangle$$

falls $m \geq 1$ und $g \geq 2$.

Die Länge L und die Ordnung beziehen sich auf die jeweilige Faktorisierung. Wollen wir beide Faktorisierungen gleichzeitig behandeln, so schreiben wir kurz $G = H_1 \underset{A}{*} H_2$ und verstehen darunter $H_j = H_{ji}$, $A = A_i$ für $j \in \{1,2\}$ und jeweils festes $i \in \{1,2\}$.

Sei also zunächst die Faktorisierung $G = H_1 \underset{A}{*} H_2$ in diesem Sinne fixiert (wir beziehen die Länge L, die Ordnung und die Verkürzungen auf diese Faktorisierung); auch werden wir meist nur G anstelle von $G = H_1 \underset{A}{*} H_2$ schreiben, beziehen uns aber auf die Faktorisierung $G = H_1 \underset{A}{*} H_2$.

In G ist ein Element x von endlicher Ordnung konjugiert zu einer Potenz von einem s_i, $i \in \{1,\ldots,m\}$, d.h. in G besitzt ein Element x von endlicher

Ordnung eine Normalform $x \equiv p_x k_x p_x^{-1}$, $p_x \in G$, $k_x \in H_\nu \backslash A$, $\nu = 1$ oder 2, wobei p_x Produkt von Restklassenvertretern und $k_x = \ell_x s_i^{\gamma_i} \ell_x^{-1}$, $\ell_x \in H_\nu$, $1 \leq \gamma_i < \alpha_i$; und die Ordnung von x ist ein Teiler von α_i.

Lemma (2.4):

Seien $x,y \in G$ mit $L(y) \leq L(x)$ und $y \equiv p_y k_y q_y$, $x \equiv p_x k_x p_x^{-1}$, $k_x \in H_\nu \backslash A$, $\nu \in \{1,2\}$.

Ist $L(xy^\epsilon) < L(x)$ oder $L(y^{-\epsilon}x) < L(x)$, $\epsilon = \pm 1$, so tritt einer der folgenden Fälle ein:

(a) $L(y^{-\epsilon}xy^\epsilon) < L(x)$;

(b) $q_y \equiv p_y^{-1} \equiv p_x^{-1}$, und es liegt $k_x k_y^\epsilon$ oder $k_y^{-\epsilon}k_x$ in A (d.h. xy^ϵ oder $y^{-\epsilon}x$ ist zu einem Element aus A konjugiert);

(c) $q_y \neq p_y^{-1}$, $k_y \in H_\nu \backslash A$, $L(x) = L(y)$, $L(y^{-\epsilon}xy^\epsilon) = L(x)$, und es liegt $k_x k_y^\epsilon$ oder $k_y^{-\epsilon}k_x$ in A. Insbesondere ist $L(xy^\epsilon) < L(y)$ oder $L(y^{-\epsilon}x) < L(y)$.

Beweis:

Sei etwa $L(xy) < L(x)$. Wegen $k_x \in H_\nu \backslash A$ ist $L(x) = 2L(p_x) + 1$.

Aus $L(y) \leq L(x)$ und $L(xy) < L(x)$ folgt natürlich notwendig, daß p_y ganz von p_x^{-1} gekürzt wird, d.h. es ist $p_x^{-1} \equiv r_x^{-1} p_y^{-1}$

(wäre $L(p_x^{-1}p_y) \geq L(p_x) - L(p_y) + 1$, so wäre $L(xy) \geq 2L(p_x) + 2 > L(x)$).

Ist $p_x^{-1} \equiv r_x^{-1} p_y^{-1}$ mit $r_x \neq 1$, d.h. $L(p_x) > L(p_y)$, so findet eine Reaktion zwischen r_x^{-1} und $k_y q_y$ statt, und wegen $L(y) \leq L(x)$ tritt Fall (a) ein.

Sei nun $p_x^{-1} \equiv p_y^{-1}$, d.h. $L(p_x) = L(p_y)$.

Ist $k_y \in A$, so findet eine Reaktion zwischen k_x und $k_y q_y$ statt, und es ist $L(y^{-1}xy) = L(q_y^{-1}k_y^{-1}k_x k_y q_y) < L(x)$, d.h. es tritt wieder Fall (a) ein.

Sei nun $k_y \notin A$. Dann ist $xy = p_x k_x k_y q_y$ und $k_x k_y \in A$ wegen $L(xy) < L(x)$.

Insbesondere ist $L(x) = L(y)$.

Ist $q_y \equiv p_x^{-1}$, so tritt Fall (b) ein.

Ist $q_y \neq p_x^{-1}$, so tritt Fall (c) ein. q.e.d.

Bemerkungen:

1) Sie $X \subset G$ eine E-Menge mit $X = X_1$. Seien $x,y \in X$, und es mögen für x,y die Voraussetzungen von Lemma (2.4) gelten. Dann tritt nur Fall (a) oder (b) ein.

2) Lemma (2.4) ist bekannt (wir haben aber in der Literatur keinen explizit durchgeführten Beweis gefunden) und wurde bereits beim Beweis von Korollar (2.10) aus [5] verwendet. Da wir dort die vielen Wiederholungen zum Beweis von Satz (2.2) aus [5] vermeiden wollten, begnügten wir uns beim Beweis von Korollar (2.10) aus [5] auf einen skizzenhaften Hinweis. Dieser ist nun (sei es Gedankenlosigkeit, Tipp-oder Druckfehler) etwas zu skizzenhaft geraten, und wir haben bei dem Hinweis ein Wort hinzuzufügen.

Richtig muß der Hinweis lauten:

Ist einmal $L(x_j) \leq L(x_j^{\varepsilon} c_i s_{\nu_i}^{\eta\beta_i} c_i^{-1}) < L(c_i s_{\nu_i}^{\beta_i} c_i^{-1})$, $i \in \{1,\dots,m-1\}$, $i \neq j$; $\varepsilon, \eta = \pm 1$, so ist such $L(x_j^{\varepsilon} c_i s_{\nu_i}^{\eta\beta_i} c_i^{-1}) < L(c_i s_{\nu_i}^{\eta\beta_i} c_i^{-1})$, oder es liegen x_j und $c_i s_{\nu_i}^{\beta_i} c_i^{-1}$ in einer zu H_1 oder H_2 konjugierten Untergruppe, und es ist $x_j^{\varepsilon} c_i s_{\nu_i}^{\eta\beta_i} c_i^{-1}$ zu einem von 1 verschiedenen Element aus A konjugiert. □

Es fehlt also lediglich der obige Zusatz über $L(x_j)$, der bei der ursprünglichen Version dabei war.

Ist aber $j \notin \{1,\dots,m-1\}$ d.h. insbesondere hat x_j unendliche Ordnung, und ist $L(x_j) > L(x_j^{\varepsilon} c_i s_{\nu_i}^{\eta\beta_i} c_i^{-1})$, so ersetzt man beim Beweis von Korollar (2.10) aus [5] natürlich x_j verkürzend durch $x_j^{\varepsilon} c_i s_{\nu_i}^{\eta\beta_i} c_i^{-1}$.

Der Fall zweier Elemente endlicher Ordnung ist einfach und wird vollständig durch Lemma (2.4) und die obige Bemerkung 1 behandelt (vgl. auch Lemma 2 aus [1]). □

Lemma (2.5):

Seien $x,y \in G$ mit $L(y) \leq L(x)$ und $x \equiv p_x k_x p_x^{-1}$, $k_x \in H_\nu \backslash A$, $\nu \in \{1,2\}$.

Ist $L(xy^{\varepsilon}) = L(x)$ oder $L(y^{-\varepsilon}x) + L(x)$, $\varepsilon = \pm 1$, so ist auch $L(y^{-\varepsilon}xy^{\varepsilon}) \leq L(x)$.

Beweis:

Sei etwa $L(xy) = L(x)$ und $y \equiv p_y k_y q_y$.

Ist $p_x = 1$, so ist auch $p_y = 1$, und es liegen x,y in demselben Faktor H_ν wegen $L(y) \leq L(x) = L(xy)$; d.h. aber insbesondere $L(y^{-1}xy) \leq L(x)$.

Sei nun $p_x \neq 1$.

Angenommen p_y wird nicht ganz von p_x^{-1} gekürzt. Wegen $L(k_x) = 1$ und $L(y) \leq L(x)$ ist dann $p_x^{-1} \equiv r_x \ell_1^{-1} s_x^{-1}$, $p_y \equiv s_x \ell_2 s_y$ mit $L(\ell_1) = L(\ell_2) = 1$, $L(\ell_1^{-1}\ell_2) \geq 1$. Es folgt

$L(xy) = L(p_x k_x r_x^{-1} \ell_1^{-1} \ell_2 s_y k_y q_y) \geq L(p_x) + 1 + L(r_x) + 1 + L(s_x) + 1 > L(x)$

im Widerspruch zu $L(xy) = L(x)$.

Also wird p_y ganz von p_x^{-1} gekürzt, d.h. $p_x^{-1} \equiv r_x^{-1} p_y^{-1}$, und wir erhalten $xy = p_x k_x r_x^{-1} k_y q_y$.

Ist $r_x \neq 1$, so ist wegen $L(xy) = L(x)$ notwendig $L(r_x^{-1} k_y q_y) = L(p_x)$, und es folgt $L(y^{-1}xy) = L(x)$ wegen $L(y) \leq L(x)$. Sei nun $r_x = 1$, d.h. $xy = p_x k_x k_y q_y$. Wegen $L(xy) = L(x)$ ist $L(k_x k_y) = 1$ und weiter $L(y^{-1}xy) = L(q_y^{-1} k_y^{-1} k_x k_y q_y) \leq L(x)$ wegen $L(y) \leq L(x)$. q.e.d.

<u>Korollar (2.6)</u>:

Es mögen die Voraussetzungen von Lemma (2.5) gelten. Ferner sei k_x in H_ν nicht konjugiert zu einem Element aus A. Dann ist $L(y^{-\varepsilon}xy^{\varepsilon}) = L(x)$. □

<u>Korollar (2.7)</u>:

Seien $x,y \in G$ mit $L(xy) \geq L(x), L(y)$ und $x \equiv p_x k_x p_x^{-1}$, $k_x \in H_\nu \backslash A$, $\nu = 1$ oder 2, sowie $y \equiv p_y k_y q_y$. Ferner stehe xy vor x.

Dann tritt einer der folgenden Fälle ein:

(a) Es ist $p_x = p_y = 1$; und x,y liegen beide in H_ν.

(b) Es ist $L(y) \leq L(x) = L(xy)$, und es steht auch $y^{-1}xy$ vor x.

Beweis:

Es ist notwendig $L(y) \leq L(x) = L(xy)$, da xy vor x steht und $L(x) < L(xy)$ ist.

Ist $p_x = p_y = 1$, so liegen damit x und y beide in H_ν (es kann dabei $k_y \in A$ sein).

Sei nun $p_x \neq 1$ oder $p_y \neq 1$. Es ist dann stets $p_x \neq 1$, denn aus $p_x = 1$ folgt $p_y = 1$ wegen $L(y) \leq L(x)$.

Wie in Lemma (2.5) folgt nun $p_x^{-1} \equiv r_x^{-1} p_y^{-1}$ und $xy = p_x k_x r_x^{-1} k_y q_y$.

Ist $r_x \neq 1$, so schreiben wir $r_x^{-1} \equiv s_x^{-1} \ell_1^{-1}$ mit $L(\ell_1) = 1$.

Ist $p_y \neq 1$, so schreiben wir $p_y \equiv r_y \ell_2$ mit $L(\ell_2) = 1$ sowie $q_y \equiv \ell_3 s_y$ mit $L(\ell_3) = 1$.

Wir beachten im folgenden stets, daß xy vor x steht und unterscheiden sechs Fälle:

a) $k_y \notin A$, $p_y \neq 1$.
Dann ist $L(r_x^{-1} k_y) = L(r_x)$ falls $r_x \neq 1$ und $L(k_x k_y) = 1$ falls $r_x = 1$.
Wir erhalten $q_y^{-1} \leq p_y$, also steht auch $y^{-1}xy$ vor x.

b) $k_y \notin A$, $p_y = 1$.
Dann ist $r_x \neq 1$ und $L(\ell_1^{-1} k_y) = 1$.
Sei t_1^{-1} der Rechtsrestklassenvertreter von $\ell_1^{-1} k_y$; es ist $t_1 \leq \ell_1$, also steht auch $y^{-1}xy$ vor x.

c) $k_y \in A$, $L(p_y) \geq 2$.
Dann ist $L(s_y) = L(r_y) \geq 1$ sowie $L(r_x^{-1} k_y \ell_3) = L(r_x)$ falls $r_x \neq 1$ und $L(k_x k_y \ell_3) = 1$ falls $r_x = 1$.
Wir erhalten $s_y^{-1} \leq r_y$, also steht auch $y^{-1}xy$ vor x.

d) $k_y \in A$, $L(p_y) = 1$, $r_x \neq 1$.
Dann ist $L(\ell_1^{-1} k_y p_y) = 1$. Sei t_2^{-1} der Rechtsrestklassenvertreter von $\ell_1^{-1} k_y p_y$; es ist $t_2 \leq \ell_1$, also steht $y^{-1}xy$ vor x.

e) $k_y \in A$, $L(p_y) = 1$, $r_x = 1$.
Dann ist $q_y^{-1} \leq p_y = p_x$, und es steht auch $y^{-1}xy$ vor x.

f) $k_y \in A$, $p_y = 1$.
Dann ist $r_x \neq 1$. Sei t_3^{-1} der Rechtsrestklassenvertreter von $\ell_1^{-1} k_y$; es ist $t_3 \leq \ell_1$, also steht auch $y^{-1}xy$ vor x.

Damit sind alle Möglichkeiten erfaßt. q.e.d.

Entsprechend erhalten wir

Korollar (2.8):

Sei $x,y \in G$ mit $L(yx) \geq L(x), L(y)$ und $x \equiv p_x k_x p_x^{-1}$, $k_x \in H_\nu \backslash A$, $\nu = 1$ oder 2, sowie $y \equiv p_y k_y q_y$. Ferner stehe yx vor x.

Dann tritt einer der folgenden Fälle ein:

(a) $p_x = p_y = 1$; und x,y liegen beide in H_ν.

(b) Es ist $L(y) \leq L(x) = L(xy)$, und es steht auch yxy^{-1} vor x.

Bemerkung:

Sei $X \subset G$ eine E-reduzierte E-Menge. Wegen Korollar (2.7) und Korollar (2.8) können wir Hilfssatz (1.7) aus [4] (vgl. auch [6]) direkt für Elemente $x,y,z \in X^{\pm 1}$ anwenden, falls für x,y,z die Voraussetzungen dieses Hilfssatzes erfüllt sind.

Das Analogon zu Hilfssatz (1.10) aus [4] ist gegeben durch

Lemma (2.9):

Sei $X \subset G$ eine E-reduzierte E-Menge und $x,y,z \in X^{\pm 1}$ mit $x \equiv p_x k_x q_x$, $y \equiv p_y k_y q_y$, $z \equiv p_z k_z q_z$.

Gilt $L(xyz) < L(x) - L(y) + L(z)$, so tritt einer der folgenden Fälle ein:

(a) y ist konjugiert zu einem Element aus A;

(b) $p_x \equiv p_y \equiv q_x^{-1} \equiv q_y^{-1}$, und es ist xy konjugiert zu einem Element aus A;

(c) $p_y \equiv p_z \equiv q_y^{-1} \equiv q_z^{-1}$, und es ist yz konjugiert zu einem Element aus A;

(d) $p_x \equiv p_y \equiv p_z \equiv q_x^{-1} \equiv q_y^{-1} \equiv q_z^{-1}$, und es ist xyz konjugiert zu einem Element aus A.

Beweis:

Ist $y \in A$, so ist nichts zu zeigen. Sei nun $L(y) \geq 1$.

Wir beachten stets, daß X E-reduziert ist.

Ist etwa $L(x) \leq L(y)$ und $L(xy) < L(y)$, so ist $q_y \equiv p_y^{-1}$ und $k_y \notin A$; und nach Lemma (2.4) ist $p_x \equiv p_y$ sowie xy konjugiert zu einem Element aus A.

Sei nun also $L(xy) \geq L(x), L(y)$ und $L(yz) \geq L(y), L(z)$.

Aus $L(xyz) < L(x) - L(y) + L(z)$ erhalten wir mit Hilfssatz (1.7) aus [4] sowie Korollar (2.7) und Korollar (2.8):

1) $L(x), L(z) \geq L(y)$;

2) $q_y \equiv p_y^{-1}$ und $k_y \notin A$;

3) beide Hälften von y werden gekürzt.

Ist $L(x), L(z) > L(y)$, so ist zusätzlich y konjugiert zu einem Element aus A.

Sei nun $L(x) = L(y)$ oder $L(z) = L(y)$, und etwa $L(x) = L(y)$;

d.h. insbesondere $L(xyz) < L(z)$ wegen $L(xyz) < L(x) - L(y) + L(z)$.

Angenommen $L(z) > L(y) = L(x) \geq 1$.

Dann ist $q_z \equiv p_z^{-1}$, $k_z \notin A$, d.h. insbesondere $L(y) \leq L(z) - 2$, und damit $L(xy) < L(z)$, denn wegen 2), 3) und $L(x) = L(y)$ ist $q_x \equiv p_y^{-1}$ sowie $xy = p_x k_x k_y p_y^{-1}$. Lemma (2.4) impliziert andererseits aber $L(xy) = L(z)$, was einen Widerspruch ergibt.

Also ist $1 \leq L(x) = L(y) = L(z)$. Dann ist aber $x \equiv p_y k_x p_y^{-1}$, $z \equiv p_y k_z p_y^{-1}$ und $k_x k_y k_z \in A$ wegen $L(xyz) < L(z)$; und es ist xyz zu einem Element aus A konjugiert.

q.e.d.

Lemma (2.10):

Sei $X = \{x_j\}_{j \in J} \subset G$ eine E-reduzierte E-Menge.

Dann tritt einer der folgenden Fälle ein:

(1) $X_o \neq \emptyset$.

(2) Für ein $p \in \mathfrak{N}_i(X)$, $i = 1$ oder 2, liegen n $(n \geq 1)$ der x_j in $X_{i,p}$, und ein Produkt in ihnen ist zu einem von 1 verschiedenen Element aus A konjugiert.

(3) $$\langle X \rangle = \left(\mathop{*}_{p \in \mathfrak{N}_1(X)} \langle X_{1,p} \rangle \right) * \left(\mathop{*}_{p \in \mathfrak{N}_2(X)} \langle X_{2,p} \rangle \right) * F(X_3),$$

wobei $F(X_3)$ freie Gruppe mit freiem Erzeugendensystem X_3 ist.

Beweis:

Es trete weder Fall (1) noch Fall (2) ein.

Wir berücksichtigen dies und beachten wieder stets, daß X E-reduziert ist.

Wir betrachten (frei gekürzge) Produkte

(*) $$w = \prod_{i=1}^{n} x_{\nu_i}^{\epsilon_i}, \quad n \geq 1, \quad \nu_i \in J, \quad \epsilon_i = \pm 1 \text{ und}$$

$$\epsilon_i = \epsilon_{i+1}, \text{ wenn } \nu_i = \nu_{i+1}.$$

Gibt es kein Produkt (*) mit $L(w) < L(x_{\nu_i})$ für ein i $(1 \leq i \leq n)$, so tritt natürlich Fall (3) ein.

Sei nun w ein Produkt (*) mit $L(w) < L(x_{\nu_i})$ für ein i $(1 \leq i \leq n)$. Wir betrachten ein Teilprodukt

$$w_1 = x_{\nu_{j+1}}^{\epsilon_{j+1}} \cdots x_{\nu_{j+t}}^{\epsilon_{j+t}} \quad (0 \leq j < j+t \leq n)$$

von w mit $L(w_1) < L(x_{\nu_{j+\ell}})$ für ein ℓ $(1 \leq \ell \leq t)$ und t minimal für alle möglichen Teilprodukte.

Wir setzen $x_{\nu_{j+i}}^{\epsilon_{j+i}} =. u_i$ $(i = 1,\dots,t)$ und schreiben $w_1 = u_1 \dots u_t$.

Behauptung (2.11):

Es ist $w_1 = 1$, und die $u_1,\dots,u_t$ liegen in $X_{\nu,p}$ für ein $p \in \mathfrak{N}_\nu(X)$, $\nu = 1$ oder 2.

Beweis von (2.11):

Es ist jedenfalls $t \geq 2$.

Ist einmal $L(u_i u_{i+1}) < \max\{L(u_i), L(u_{i+1})\}$ für $1 \leq i \leq t-1$, so gilt die Behauptung (2.11) nach Lemma (2.4), denn es ist dann notwendig $t = 2$, $u_1 = u_2$ und $u_1^2 = 1$ wegen der Minimalität von t, denn Fall (2.4(c)) tritt nicht ein.

Ist $L(u_i u_{i+1}) \geq L(u_i), L(u_{i+1})$ für alle i $(1 \leq i < t-1)$ und einmal $L(u_i u_{i+1} u_{i+2}) < L(u_i) - L(u_{i+1}) + L(u_{i+2})$ für $1 \leq i \leq t-2$, so gilt die Behauptung (2.11) nach Lemma (2.9), denn es ist dann notwendig $t = 3$, $u_1 = u_2 = u_3$ und $u_1^3 = 1$ wegen der Minimalität von t (vgl. auch Hilfssatz (1.11) aus [4]).

Es gelte nun stets

$L(u_i u_{i+1}) \geq L(u_i), L(u_{i+1})$ für $1 \leq i \leq t-1$ sowie
$L(u_i u_{i+1} u_{i+2}) \geq L(u_i) - L(u_{i+1}) + L(u_{i+2})$ für $1 \leq i \leq t-2$.

Wegen der Minimalität von t gilt sogar

$L(u_i u_{i+1} u_{i+2}) = L(u_i) - L(u_{i+1}) + L(u_{i+2})$ für alle i $(1 \leq i \leq t-2)$;
und mittels $L(u_1 \dots u_t) < L(u_j)$ für ein j $(1 \leq j \leq t)$, Hilfssatz (1.7) aus [4] sowie Korollar (2.7) und Korollar (2.8) erhalten wir damit:

(a) $L(u_2) = \dots = L(u_{t-1}) =. L$;

(b) $L(u_1)$, $L(u_t) \geq L$;

(c) $u_i \equiv pk_ip^{-1}$, $k_i \in H_{\nu_i} \backslash A$, $\nu_i = 1$ oder 2, $1 < i < t$;

(d) $\nu_2 = \ldots = \nu_{t-1} = . \nu$, d.h. alle k_i $(1 < i < t)$ liegen in einem und demselben Faktor H_ν;

(e) $u_1 \equiv p_1k_1q_1$, $q_1 \equiv r_1p^{-1}$, $k_1 \in H_{\nu_1}$, $\nu_1 = 1$ oder 2;

(f) $u_t \equiv p_tk_tq_t$, $p_t \equiv ps_t$, $k_t \in H_{\nu_t}$, $\nu_t = 1$ oder 2

(zur Begründung vergleiche etwa [4; pp. 6-7]).

Wir setzen $u . = u_2\ldots u_{t-1}$; wegen der Minimalität von t ist $u \neq 1$ und $L(u) = L$.

Aus $L(u_1), L(u_t) \geq L$ und $L(u_1uu_t) < L(u_j)$ für ein j $(1 \leq j \leq t)$ folgt $L(u_1uu_t) < L(u_1) - L(u) + L(u_t)$; ferner ist $L(u_1u) \geq L(u_1)$ sowie $L(uu_t) \geq L(u_t)$ wegen der Minimalität von t. Mittels Hilfssatz (1.7) aus [4] sowie Korollar (2.7) und Korollar (2.8) folgt nun $L(u_1) = L$ oder $L(u_t) = L$, denn für $L(u_1), L(u_t) > L$ ist $u_1 \neq u_i$ und $u_t \neq u_i$ für alle i mit $2 \leq i \leq t-1$. Analog wie beim Beweis von Lemma (2.9) erhalten wir

$L(u_1) = L(u_t) = L$, $p_1 \equiv p_t \equiv p \equiv q_1^{-1} \equiv q_t^{-1}$, $\nu_1 = \nu_t = \nu$,

d.h. $k_1, k_r \in H_\nu \backslash A$, und weiter $k_1\ldots k_t = 1$ wegen $L(u_1uu_t) < L$; damit ist $w_1 = u_1uu_t = 1$, und die Behauptung (2.11) bewiesen. □

Nun folgt Lemma (2.10) induktiv. q.e.d.

<u>Bemerkung</u>:

Im folgenden betrachten wir E-reduzierte E-Mengen mit $X = X_1$, d.h. X enthält nur Elemente endlicher Ordnung. Dann ist $X_o = \emptyset$ und $X_3 = \emptyset$.

<u>Lemma (2.12)</u>:

Sei $X \subset G$ eine E-reduzierte E-Menge mit $X = X_1$.

Sei $Y = \{y_i\}_{i \in I}$ eine nicht-leere Teilmenge von X mit $y_i \in H_\nu$, $\nu = 1$ oder 2 fest, für alle $i \in I$. Dann ist $\langle Y \rangle \cong \mathop{*}_{i \in I} \langle y_i \rangle$.

Beweis:

Als nicht-leere Teilmenge von X ist Y selbst eine E-Menge mit $Y = Y_1$. Wir beachten stets, daß alle y_i endliche Ordnung haben.

Angenommen $\langle Y \rangle \ncong \mathop{*}_{i \in I} \langle y_i \rangle$.

Dann gibt es endlich viele y_i, $i \in I$, etwa $y_1,\ldots,y_n$, mit $\langle y_1,\ldots,y_n\rangle \not\cong \langle y_i\rangle * \ldots * \langle y_n\rangle$, $n \geq 1$.

Andererseits ist $\langle y_1,\ldots,y_n\rangle$ als Untergruppe des freien Produktes H_ν zyklischer Gruppen selbst ein freies Produkt zyklischer Gruppen, etwa $\langle y_1,\ldots,y_n\rangle \cong \langle u_1\rangle * \ldots * \langle u_k\rangle$, und jedes y_i, $1 \leq i \leq n$, ist in H_ν zu einem u_j, $j \in \{1,\ldots,k\}$ konjugiert. Wegen $\langle y_1,\ldots,y_n\rangle \not\cong \langle y_1\rangle * \ldots * \langle y_n\rangle$ ist $k < n$. Daher gibt es eine aus Y herleitbare E-Menge $Y' = \{y_i'\}_{i \in I'}$ mit $y_i' = 1$ für ein $i \in I'$ (vgl. etwa Lemma 1 aus [1] oder [8]). Das ergibt aber einen Widerspruch dazu, daß X eine E-reduzierte E-Menge ist.

Also ist $\langle Y\rangle \cong \mathop{*}\limits_{i \in I} \langle y_i\rangle$. q.e.d.

<u>Bemerkung</u>:

Da sich bei der Formulierung der Hauptergebnisse dieser Arbeit bei den beiden Faktorisierungen (2.2) und (2.3) Unterschiede ergeben, behandeln wir ab sofort die beiden Faktorisierungen (2.2) und (2.3) getrennt; wir schreiben daher nun konsequent $G = H_{1i} \mathop{*}\limits_{A_i} H_{2i}$, $i = 1$ oder 2, und beziehen uns auf die jeweilige Faktorisierung.

<u>Satz 1</u>:

Sei $G = H_{11} \mathop{*}\limits_{A_1} H_{21}$ und $X = \{x_j\}_{j \in J} \subset G$ eine E-reduzierte E-Menge mit $X = X_1$. Ferner sei $\langle X\rangle \cap gA_1g^{-1} = 1$ für alle $g \in G$. Dann gilt:

(1) $\langle X\rangle = \mathop{*}\limits_{j \in J} \langle x_j\rangle$;

(2) Ist Y eine weitere E-reduzierte E-Menge mit $Y = Y_1$ und $\langle Y\rangle = \langle X\rangle$, so ist card X = card Y und $\langle X_{i,p}\rangle = \langle Y_{i,p}\rangle$ für alle $p \in \mathcal{N}_i$, $1 \leq i \leq 2$.

Beweis:

Wegen $X = X_1$ ist $X_o = \emptyset$ und $X_3 = \emptyset$; wegen $\langle X\rangle \cap gA_1g^{-1} = 1$ für alle $g \in G$ kann der Fall (2) aus Lemma (2.10) nicht eintreten. Damit folgt (1) aus Lemma (2.10) und Lemma (2.12). Die Aussage (2) ist klar, denn ein $p \in \mathcal{N}_i$ liegt genau dann in $\mathcal{N}_i(X)$, wenn es auch in $\mathcal{N}_i(Y)$ liegt; und card X = card Y folgt damit aus [8]. q.e.d.

Bemerkungen:

1) Für die Untergruppen $\langle X_{i,p}\rangle$, $p \in \mathfrak{N}_i$, $i = 1$ oder 2, können wir nun die Ergebnisse aus [8] anwenden, denn $X_{i,p}$ enthält nur Elemente endlicher Ordnung ($X = X_1$), und es ist $\langle X_{i,p}\rangle$ Untergruppe eines freien Produktes zyklischer Gruppen.

2) Die Bedingung $\langle X\rangle \cap gA_1g^{-1} = 1$ für alle $g \in G$ kann man (in vielen Fällen) sicherlich durch geeignete andere Bedingungen ersetzen; allerdings kann man natürlich nicht ganz auf eine geeignete Bedingung verzichten, wenn man die Aussage von Satz 1 haben möchte (G ist kein freies Produkt zyklischer Gruppen). Als Beispiel geben wir

Satz 2:

Sei $G = H_{11} \underset{A_1}{*} H_{21}$ und $X = \{x_j\}_{j \in J} \subset G$ eine E-reduzierte E-Menge mit $X = X_1$. Sei $\mathfrak{N}_2(X) = \emptyset$ und A_1 malnormal in G; ferner sei $|L(p_1) - L(p_2)| \geq 3$ falls $p_1, p_2 \in \mathfrak{N}_1(X)$ und $p_1 \neq p_2$. Dann gilt:

(1) $\langle X\rangle = \underset{j \in J}{*} \langle x_j\rangle$;

(2) Ist Y eine weitere E-reduzierte E-Menge mit $Y = Y_1$ und $\langle Y\rangle = \langle X\rangle$, so ist $\mathfrak{N}_2(Y) = 0$, card X = card Y und $\langle X_{1,p}\rangle = \langle Y_{1,p}\rangle$ für alle $p \in \mathfrak{N}_1$.

Beweis:

Wir beachten wieder stets, daß X E-reduziert ist und betrachten Produkte

$$(**) \qquad w = \prod_{i=1}^{n} p_i v_i p_i^{-1} \quad \text{mit } n \geq 1,\ 1 \neq v_i \in H_{11},$$

$p_i \in \mathfrak{N}_1(X)$ und $p_i v_i p_i^{-1} \in \langle X_{1,p_i}\rangle$ für $i = 1,\ldots,n$ sowie $p_i \neq p_j$ für $i \neq j$ $(1 \leq i,j \leq n)$.

Es ist $p_i = 1$ oder $p_i \equiv p_i' \ell_i$ mit $L(\ell_i) = 1$, $\ell_i \in H_{21} (1 \leq i \leq n)$.

Ist wirklich einmal $v_i \in A_1$ und $p_i \neq 1$, so ist $\ell_i v_i \ell_i^{-1} \notin A_1$, da A_1 malnormal in G ist.

Wird einmal p_i (bzw. p_i^{-1}) ganz gekürzt, so ist $p_{i-1} \equiv p_i r_{i-1}$, $L(r_{i-1}) \geq 3$ (bzw. $p_{i+1} \equiv p_i r_{i+1}$, $L(r_{i+1}) \geq 3$).

Wird nun wirklich einmal p_i und p_i^{-1} ganz gekürzt, so ist

$$p_{i-1}v_{i-1}p_{i-1}^{-1} p_i v_i p_i^{-1} p_{i+1} v_{i+1} p_{i+1}^{-1} = p_{i-1} v_{i-1} r_{i-1}^{-1} v_i r_{i+1} v_{i+1} p_{i+1}^{-1},$$

und es ist $r_{i-1}^{-1} v_i r_{i+1} \notin H_{11} \cup H_{21}$, da A_1 malnormal in G ist (vgl. [6]); diese Argumente gelten natürlich insbesondere für den Fall $p_i = 1$. Nach Durchführung aller möglichen Kürzungen und Zusammenfassungen von benachbarten Elementen aus demselben Faktor ergibt sich damit in (**) sofort $w \neq 1$ und weiter $\langle X \rangle = \ast_{j \in J} \langle x_j \rangle$ mit Hilfe von Lemma (2.12). Die Aussage (2) ist klar (vgl. den Beweis von Satz 1). q.e.d.

Bemerkungen:

1) Wir erwähnen, daß A_1 genau dann malnormal in G ist, wenn weder H_{11} noch H_{21} isomorph zur unendlichen Diedergruppe $\langle t_1, t_2 | t_1^2 = t_2^2 = 1 \rangle$ ist (vgl. [6]).

2) Auf die obige Voraussetzung $|L(p_1) - L(p_2)| \geq 3$ falls $p_1, p_2 \in \mathfrak{N}_1(X)$ und $p_1 \neq p_2$ können wir in Satz 2 nicht ohne weiteres verzichten, wenn wir die Aussage von Satz 2 haben möchten. Dies zeigt das folgende Beispiel.

Sei $G = H_{11} \ast_{A_1} H_{21}$ mit $H_{11} = \langle s_1, s_2 | s_1^2 = s_2^3 = 1 \rangle$, $H_{21} = \langle s_3, s_4 | s_3^2 = s_4^3 = 1 \rangle \cong H_{11}$, $A_1 = \langle s_1 s_2 \rangle = \langle (s_3 s_4)^{-1} \rangle$. Es ist A_1 malnormal in G. Sei $X = \{s_1, s_2, s_4^2 s_1 s_4, s_4^2 s_2 s_4\}$; X ist E-Menge mit $X = X_1$ und $\mathfrak{N}_2(X) = \emptyset$ (die Ordnung sei nun geeignet gewählt). Es gilt aber $(s_1 s_2)^{-1} (s_4^2 s_1 s_2 s_4)^{-1} = s_4^2$, d.h. $s_4 \in \langle X \rangle$. Damit ist $\langle X \rangle = G$.

In diesem Beispiel ist $\mathfrak{N}_1(X) = \{1, s_4^2\}$, $L(s_4^2) - L(1) = 1$. Wir schließen den Fall $|L(p_1) - L(p_2)| = 2$ falls $p_1, p_2 \in \mathfrak{N}_1(X)$, $p_1 \neq p_2$, deshalb aus, weil dann in einem Produkt (**) durchaus die folgende Möglichkeit eintreten kann: $p_i = 1$, $v_i = r_i a_i r_i^{-1}$, $a_i \in A_1$, $r_i \in H_{11} \backslash A_1$ und $p_{i+1} \equiv r_i s_{i+1}$ mit $L(s_{i+1}) = 1$, $s_{i+1} \in H_{21}$.

Wir werden in einer späteren Arbeit die Untersuchungen für $H_{11} \ast_{A_1} H_{21}$ weiterführen; insbesondere werden wir auch den Fall behandeln, wo $\langle X \rangle$ endlichen Index in G hat und damit insbesondere kein freies Produkt zyklischer Gruppen ist ($X \subset$ ist wieder E-reduzierte E-Menge mit $X = X_1$).

Satz 3:

Sei $G = H_{12} \underset{A_2}{*} H_{22}$ und $X = \{x_j\}_{j \in J} \subset G$ eine E-reduzierte E-Menge mit $X = X_1$.

Dann gilt:

(1) $\mathcal{N}_2(X) = \emptyset$;

(2) $\langle X \rangle = \underset{j \in J}{*} \langle x_j \rangle$;

(3) Ist Y eine weitere E-reduzierte E-Menge mit $Y = Y_1$ und $\langle Y \rangle = \langle X \rangle$, so ist $\mathcal{N}_2(Y) = \emptyset$, card X = card Y und $\langle X_{1,p} \rangle = \langle Y_{1,p} \rangle$ für alle $p \in \mathcal{N}_1$.

Beweis:

Sei $\langle\langle X \rangle\rangle$ der normale Abschluß von $\langle X \rangle$ in G und $\varphi: G \twoheadrightarrow G/_{\langle\langle X \rangle\rangle}$ der kanonische Epimorphismus. Wegen $X = X_1$ ist $\varphi(a) \neq 1$ für alle $a \in A_2$ mit $a \neq 1$. Also mit $\langle X \rangle \cap gA_2g^{-1} = 1$ für alle $g \in G$ und $\mathcal{N}_2(X) = \emptyset$, da H_{21} freie Gruppe ist. Nun folgt Satz 3 analog wie Satz 1. q.e.d.

Wir danken dem Referenten für einige wertvolle Hinweise, durch die unsere Arbeit verständlicher wurde.

Literatur:

[1] R. C. Lyndon; Quadratic equations in free products with amalgamation. Houston J. Math. 4 (1978), 91-103.

[2] R. C. Lyndon, P. E. Schupp; Combinatorial group theory. Ergebnisse der Mathematik 189. Berlin-Heidelberg-New York: Springer, 1977.

[3] G. Rosenberger, Bemerkungen zu einer Arbeit von H. Zieschang. Archiv der Math. 29 (1977), 623-627.

[4] G. Rosenberger; Gleichungen in freien Produkten mit Amalgam. Math. Z. 173 (1980), 1-12; Berichtigung. Math. Z. 178 (1981), 579.

[5] G. Rosenberger; Automorphismen ebener diskontinuierlicher Gruppen. Proc. of the Stony Brook Conf. on Riemann Surfaces and related topics 1978; Annals of Math. Studies 97 (1981), 439-455, Princeton University Press.

[6] H. Zieschang; Über die Nielsensche Kürzungsmethode in freien Produkten mit Amalgam. Invent. Math. 10 (1970), 4-37.

[7] H. Zieschang; On decompositions of discontinuous groups of the plane. Math. Z. 151 (1976), 165-188.

[8] H. Zieschang; On subgroups of free products of cyclic groups (Russian). (to appear)

[9] H. Zieschang, E. Vogt, H.-D. Coldewey; Surfaces and planar discontinuous groups. Lecture Notes in Math. 835. Berlin-Heidelberg-New York: Springer, 1980.

DEPARTMENT OF MATHEMATICS AND C.S.
SOUTHWEST STATE UNIVERSITY
MARSHALL, MINNESOTA 56258
U.S.A.

Received June 28, 1982

ABTEILUNG MATHEMATIK
UNIVERSITÄT DORTMUND
POSTFACH 500500
D-4600 DORTMUND 50
FED. REP. OF GERMANY

Contemporary Mathematics
Volume 33, 1984

AUTOMORPHISMS OF A FREE PRODUCT WITH AN AMALGAMATED SUBGROUP

Abe Karrass, Alfred Pietrowski, and Donald Solitar*

Dedicated to Roger Lyndon on his 65th birthday

Introduction

In this paper we show that for many interesting cases the automorphism group of an amalgamated product $G = \langle A*B;U\rangle$ where $A \neq U \neq B$ is itself such an amalgamated product, and can be effectively computed modulo computations within the automorphism groups of the factors. Precisely, if the amalgamated subgroup is conjugate maximal, i.e., maximal along all its conjugates in G, and if in addition the two factors are mapped by the automorphisms of G onto conjugates of themselves or each other, then the automorphism group of G is the proper free product of two factors with an amalgamated subgroup. The condition on the amalgamated subgroup U automatically holds if it is, e.g., finite, normal, or even malnormal, or has finite index in the factors. The condition on the factors holds, if, e.g., they are finite (in which case the two factors for the automorphism group are also finite) or indecompsable as amalgamated products and have no infinite cyclic factor groups. However, even if the factors are decomposable, they and their conjugates may play such a special role in G that they must be mapped among themselves by each automorphism of G.

Artin's braid group on four strings B_4 is an amalgamated product of two factors which are decomposable but which do play such a special role. Moreover,

*This research was supported by NSERC grants A5602, A9108, and A5614.

© 1984 American Mathematical Society
0271-4132/84 $1.00 + $.25 per page

the amalgamated subgroup in this case is normal in B_4. Our results allow us to compute the automorphism group of B_4, as an amalgamated product of two factors (see [3]). (Moreover, this amalgamated product is a splitting extension of the inner automorphisms by the automorphism of order two given by mapping the standard braid group generators into their inverses. The authors' conjecture that the automorphism group of any braid group is such a splitting extension was proved by Dyer and Grossman [2].)

If the amalgamated subgroup is conjugate maximal in G but no condition is imposed on the factors, then the subgroup of all the automorphisms which do map conjugates of factors among themselves still has the form of a proper amalgamated product with two factors. Moreover, the amalgamated subgroup of this product is also conjugate maximal in it and therefore one can form a series of such automorphism subgroups (in fact, one can also form a series in which each of the factors is mapped into a conjugate of itself). For the amalgamated product of two finite groups this series will consist of amalgamated products of two finite groups and will be the same as its series of automorphism groups. If G has a trivial centre then so has Aut G and G can be embedded in Aut G as Inn G; hence, the automorphism series will be a tower. In the case of finite groups, Wielandt [8] showed that the automorphism tower is constant after finitely many terms. This need not be so for amalgamated products of two finite factors, although it can happen.

Results

Def. If U is a subgroup of H then we call U conjugate maximal (in H) if $hUh^{-1} < U$ implies $hUh^{-1} = U$ for any $h \in H$. □

Clearly, if U is finite, or more generally, U contains no proper isomorphic subgroup, then U is conjugate maximal in any containing group H. Moreover, if U is a normal, or malnormal subgroup of a group H, or has finite index in H then U is conjugate maximal in H. It immediately follows from standard cancellation agruments that U is conjugate maxiaml in $G = \langle A*B;U \rangle$ iff U is conjugate maximal in A and in B.

Lemma 1. Let $G = \langle A*B;U\rangle$, $A \neq U \neq B$, where U is conjugate maximal in G. If $gp(A,gBg^{-1}) = G$, and $A \cap gBg^{-1} = kUk^{-1}$, then for some $a \in A$,

$$gBg^{-1} = aBa^{-1} \text{ and } kUk^{-1} = aUa^{-1} .$$

Proof. Obviously if $gBg^{-1} = aBa^{-1}$ then

$$kUk^{-1} = A \cap aBa^{-1} = a(A \cap B)a^{-1} = aUa^{-1} .$$

We prove our result by induction on the syllable length of g. If this is 0 or 1 then the result is immediate. If g has syllable length > 1 and ends in B or begins in A then g can be shortened (A and U may have to be replaced by their conjugates by an element of A) and the result follows by inductive assumption. We now show that assuming that g has a reduced form

$$g = b_1 a_1 \ldots b_n a_n ,$$

where $n > 1$ and $b_i \in B \backslash U$, and $a_i \in A \backslash U$, leads to a contradiction. Indeed,

$$gbg^{-1} = b_1 a_1 \ldots b_n a_n b a_n^{-1} b_n^{-1} \ldots a_1^{-1} b_1^{-1} ,$$

where $b \in B$, has syllable length at most 1 implies that $b \in U$. Thus

$$B \cap gBg^{-1} < gUg^{-1}, \; kUk^{-1} = A \cap gBg^{-1} < gUg^{-1} ,$$

and so since U is conjugate maximal, we have

$$A \cap gBg^{-1} = gUg^{-1} .$$

On the other hand, since g begins in B, if gbg^{-1} has syllable length at most 1, it must be in B. Thus

$$gUg^{-1} = A \cap gBg^{-1} < A \cap B = U$$

and so by the conjugate maximality of U

$$B \cap gBg^{-1} < gUg^{-1} = A \cap gBg^{-1} = U .$$

We can now show that no element $x \in B\backslash U$ can be in $gp(A,gBg^{-1})$. For clearly x is not in A or in gBg^{-1}. Moreover, if x is a product of at least two elements alternating out of $A\backslash U$ and $g(B\backslash U)g^{-1}$, then x has syllable length > 1 in $\langle A*B;U\rangle$. This completes our proof. □

As is customary, for any group H we denote the centre of H by $Z(H)$, the automorphism group of H by Aut H, and the inner automorphism group of H by Inn H.

Defs. Let $G = \langle A*B;U\rangle$, where $A \neq U \neq B$. We define certain subgroups of Aut G relative to this decomposition.

A conjugating automorphism of G is one which takes A onto a conjugate of A, B onto a conjugate of B, and U onto a conjugate of U. All conjugating automorphisms of G form a subgroup denoted by Con(G).

A reverse conjugating automorphism of G is one which takes A onto a conjugate of B, B onto a conjugate of A, and U onto a conjugate of U. If G has a reverse conjugating automorphism we call G reversible. All reverse conjugating and conjugating automorphisms of G form a subgroup denoted by RCon(G).

A pairing automorphism of G is one which takes A onto A, and B onto B. All pairing automorphisms of G form a subgroup denoted by Pair(G).

A reverse pairing automorphism of G is one which takes A onto B, and B onto A. All reverse pairing and pairing automorphisms of G form a subgroup denoted by RPair(G).

If $H < G$ then XtnH(G) denotes the subgroup of conjugating automorphisms of G which take H onto H (extensions from H); moreover, Inn H(G) denotes the subgroup of Inn G determined by conjugation of G by elements of H. □

Lemma 2. Let $G = \langle A*B;U\rangle$, $A \neq U \neq B$, where U is conjugate maximal in G. Then any automorphism of G which takes A onto a conjugate of A and B onto a conjugate of B is a conjugating automorphism. Moreover, any automorphism of G which takes A onto a conjugate of B and B onto a conjugate of A is a reverse conjugating automorphism.

Proof. By applying a suitable inner automorphism of G we may assume that A and B are taken onto A and gBg^{-1} in some order, where g does not end in B. Let V be the image of U under the automorphism of G; we show that V is contained in a conjugate of U. Indeed, since V is just A intersected with gBg^{-1}, it follows by a standard reduction argument that $V < gUg^{-1}$. Now under the inverse automorphism of G, the factors A and gBg^{-1} are taken into their conjugates A and B in some order. Hence by the first part,

$$U < hVh^{-1} < hgUg^{-1}h^{-1} ,$$

for some $h \in G$, and so by the conjugate maximality of U in G, V is a conjugate of U. □

<u>Lemma</u> 3. Let $G = \langle A*B;U\rangle$, $A \neq U \neq B$, where U is conjugate-maximal. Then the following hold:

$$\text{XtnA}(G) = \text{Inn } A(G)\cdot\text{Pair}(G) ;$$

$$\text{Inn } A(G) \lhd \text{XtnA}(G) ;$$

$$\text{Inn } A(G) \cap \text{Pair}(G) = \text{Inn } U(G) ;$$

$$\text{Inn } G \cap \text{XtnA}(G) = \text{Inn } A(G) ;$$

$$\text{Inn } G \cap \text{Pair}(G) = \text{Inn } U(G) ;$$

$$\text{Inn } A(G) \cong A/(U \cap Z(A) \cap Z(B)) = A/Z(G) ;$$

$$\text{Inn } U(G) \cong U/Z(G) .$$

<u>Proof</u>. We first show that $\text{XtnA}(G) = \text{Inn } A(G)\cdot\text{Pair}(G)$. For, a given automorphism in $\text{XtnA}(G)$ will take A onto A, B onto a conjugate of B, and U onto a conjugate of U. Application of Lemma 1 shows that this conjugating automorphism differs from a pairing by an inner automorphism determined by an element of A. Thus $\text{XtnA}(G) = \text{Inn } A(G)\cdot\text{Pair}(G)$.

Since an automorphism of G conjugates an inner automorphism of G by being applied to the element determining the inner automorphism, and $\text{Pair}(G)$ takes A onto A, it is clear that $\text{Inn } A(G)$ is normal in $\text{XtnA}(G)$. Moreover, since $\text{Pair}(G)$ takes B onto B and $aBa^{-1} = B$ implies that $a \in U$, it is clear that $\text{Inn } A(G) \cap \text{Pair}(G) = \text{Inn } U(G)$. Similarly $\text{Inn } G \cap \text{XtnA}(G) = \text{Inn } A(G)$ and therefore, $\text{Inn } G \cap \text{Pair}(G) = \text{Inn } U(G)$. The last two results follow easily since conjugation by an element of G defines the identity on A and on B if the element is in $Z(G) = U \cap Z(A) \cap Z(B)$. □

Theorem 1. Let $G = \langle A*B;U\rangle$, $A \neq U \neq B$, where U is conjugate-maximal. Then the following hold:

(1) $\mathrm{Con}(G) = \langle \mathrm{XtnA}(G)*\mathrm{XtnB}(G);\mathrm{Pair}(G)\rangle$;

$\mathrm{Inn}\, G = \langle \mathrm{Inn}\, A(G)*\mathrm{Inn}\, B(G);\mathrm{Inn}\, U(G)\rangle$.

(2) $[\mathrm{XtnA}(G):\mathrm{Pair}(G)] = [A:U]$, $[\mathrm{XtnB}(G):\mathrm{Pair}(G)] = [B:U]$;

$[\mathrm{Con}(G):\mathrm{Inn}\, G] = [\mathrm{XtnA}(G):\mathrm{Inn}\, A(G)] = [\mathrm{XtnB}(G):\mathrm{Inn}\, B(G)]$

$= [\mathrm{Pair}(G):\mathrm{Inn}\, U(G)]$.

(3) Moreover, if G is reversible then

$\mathrm{RCon}(G) = \langle \mathrm{XtnA}(G)*\mathrm{RPair}(G);\mathrm{Pair}(G)\rangle$, and

$\mathrm{RPair}(G) = \mathrm{gp}(\rho,\mathrm{Pair}(G))$, for any reverse pairing automorphism ρ of G, and

$[\mathrm{RPair}(G):\mathrm{Pair}(G)] = 2 = [\mathrm{RCon}(G):\mathrm{Con}(G)]$.

Proof. Clearly $\mathrm{Inn}\, G \lhd \mathrm{Con}(G)$. We show that $\mathrm{Con}(G) = \mathrm{gp}(\mathrm{Inn}\, G,\mathrm{Pair}(G))$. For, a given conjugating automorphism when multiplied by a suitable inner automorphism of G will take A onto A, B onto a conjugate of B, and U onto a conjugate of U and therefore be in $\mathrm{XtnA}(G)$. Since $\mathrm{Inn}\, A(G) < \mathrm{Inn}\, G$, our result follows immediately from Lemma 3.

We can now prove (1). First since $\mathrm{Inn}\, G \cong G/Z(G)$ and $Z(G)$ is normal in A, B, and U, clearly $\mathrm{Inn}\, G = \langle \mathrm{Inn}\, A(G)*\mathrm{Inn}\, B(G);\mathrm{Inn}\, U(G)\rangle$.

We have already established that $\mathrm{Con}(G) = \mathrm{gp}(\mathrm{Inn}\, G,\mathrm{Pair}(G))$. But $\mathrm{Inn}\, G = \mathrm{gp}(\mathrm{Inn}\, A(G),\mathrm{Inn}\, B(G))$. Thus by Lemma 3 applied to A and B, we have $\mathrm{Con}(G) = \mathrm{gp}(\mathrm{XtnA}(G),\mathrm{XtnB}(G))$. Moreover, an automorphism in $\mathrm{XtnA}(G)$ and $\mathrm{XtnB}(G)$ must take A onto A and B onto B and so be in $\mathrm{Pair}(G)$. Suppose now that

$$x_1 y_1 \ldots x_n y_n$$

is a product of automorphisms alternating out of $\mathrm{XtnA}(G)\setminus\mathrm{Pair}(G)$ and $\mathrm{XtnB}(G)\setminus\mathrm{Pair}(G)$ which defines the identity. From Lemma 3 applied to A and B we can replace such a product by

$$z_1w_1 \ldots z_nw_np$$

where $p \in \mathrm{Pair}(G)$ and the factors z_i, w_i, alternate out of $\mathrm{Inn}\ A(G)\backslash\mathrm{Inn}\ U(G)$ and $\mathrm{Inn}\ B(G)\backslash\mathrm{Inn}\ U(G)$. But this last product cannot take A onto A. This contradiction completes the proof of (1).

To prove (2) use Lemma 3 and that for any two subgroups P and Q of a group H, $[P\cdot Q:Q] = [P:P \cap Q]$.

Finally, we prove (3). If G is reversible then there exists a reverse conjugating automorphism of G. By multiplying it by a suitable inner automorphism of G one obtains a reverse conjugating automorphism taking A onto B. But then applying Lemma 1 with A and B interchanged shows that if this new reverse conjugating automorphism is multiplied by a suitable inner automorphism determined by an element of B, a reverse pairing automorphism is obtained. Let ρ be any reverse pairing automorphism of G.

Clearly $\mathrm{RCon}(G) = \mathrm{gp}(\rho,\mathrm{Con}(G))$, $[\mathrm{RCon}(G):\mathrm{Con}(G)] = 2$, $\mathrm{RPair}(G) = \mathrm{gp}(\rho,\mathrm{Pair}(G))$, and $[\mathrm{RPair}(G):\mathrm{Pair}(G)] = 2$. Moreover, $\rho\mathrm{Inn}\ A(G)\rho^{-1} = \mathrm{Inn}\ B(G)$, and also $\rho\mathrm{Inn}\ U(G)\rho^{-1} = \mathrm{Inn}\ U(G)$. Thus $\mathrm{RCon}(G) = \mathrm{gp}(\rho,\mathrm{Inn}\ A(G),\mathrm{Pair}(G)) = \mathrm{gp}(\rho(\mathrm{XtnA}(G)) = \mathrm{gp}(\mathrm{XtnA}(G),\mathrm{RPair}(G))$. Clearly $\mathrm{XtnA}(G) \cap \mathrm{RPair}(G) = \mathrm{Pair}(G)$. Suppose now

$$x_1y_1 \ldots x_ny_n$$

is a product of automorphisms alternating out of $\mathrm{XtnA}(G)\backslash\mathrm{Pair}(G)$ and $\mathrm{RPair}(G)\backslash\mathrm{Pair}(G)$ which defines the identity. Since $[\mathrm{RCon}(G):\mathrm{Con}(G)] = 2$, n must be even. Such a product can be replaced by a product

$$z_1\rho z_2\rho^{-1} \ldots \rho z_n\rho^{-1}p$$

where $p \in \mathrm{Pair}(G)$ and the factors $z_i \in \mathrm{Inn}\ A(G)\backslash\mathrm{Inn}\ U(G)$. Now the automorphism $\rho z_i\rho^{-1} \in \mathrm{Inn}\ B(G)\backslash\mathrm{Inn}\ U(G)$. But this last product cannot take A onto A. This contradiction completes the proof of (3), and of Theorem 1. □

<u>Corollary</u>. If $G = \langle A*B;U\rangle$, $A \neq U \neq B$, where U has finite index in A and in B then $\mathrm{Con}(G)$ and $\mathrm{RCon}(G)$ as given in Theorem 1 have the form

$$\langle C*D;V\rangle\ ,\ C \neq V \neq D\ ,$$

where $[C:V] = [A:U]$ and $[D:V] = [B:U]$ or $[D:V] = 2$, respectively. Moreover, if every automorphism of G maps U into a conjugate of U, then $\text{Aut } G = \text{Con}(G)$ or $\text{RCon}(G)$.

<u>Proof</u>. Since U is of finite index in A and in B, U is conjugate maximal in A and in B and hence in G. Theorem 1 applies to yield the statement on indices.

Suppose now that U is mapped into a conjugate of itself by every automorphism of G. We first show that A (and therefore by symmetry B) is mapped onto a conjugate of A or B. Clearly we may assume that U is mapped onto itself. Now $\alpha(U) = U$ has finite index in $\alpha(A)$ and $A \cap \alpha(A)$ is a vertex group in the treed HNN group presentation for $\alpha(A)$ as a subgroup of $G = \langle A*B;U\rangle$. It follows that this presentation must reduce to a single vertex with no stable letters (see Karrass and Solitar [5]). Hence, $\alpha(A)$ is a conjugate of A or B. Thus $\text{Aut } G = \text{Con}(G)$ or $\text{Aut } G = \text{RCon}(G)$. □

We note that if A and B are finite extensions of a central group U and $G = \langle A*B;U\rangle$ then U is the centre of G and the preceding Corollary applies; therefore one can immediately compute the automorphism group of $G(r,s) = \langle a,b;a^r = b^s\rangle$ where r and s are positive integers > 1 (see Schreier [7]).

If $r \neq s$ then there is no reverse pairing since U has different indices in A and in B. On the other hand, if $r = s$ then interchanging A and B is a reverse conjugating automorphism of G. Then $\text{Aut } G = \text{Con}(G)$ if $r \neq s$ and $\text{Aut } G = \text{RCon}(G)$ if $r = s$.

Now

$$\text{Pair}(G) = \langle \epsilon\colon a \to a^{-1},\ b \to b^{-1}$$

$$\text{Inn } A(G) = \langle I_a : I_a{}^r\rangle$$

$$\text{Inn } B(G) = \langle I_b ; I_b{}^s\rangle$$

$$\text{XtnA}(G) = \langle \epsilon, I_a ; \epsilon^2, I_a{}^s,\ \epsilon I_a \epsilon^{-1} = I_a^{-1}\rangle$$

$$\text{XtnB}(G) = \langle \epsilon, I_b ; \epsilon^2, I_b{}^s,\ \epsilon I_b \epsilon^{-1} = I_b^{-1}\rangle ;$$

moreover, if $r = s$ then

$$\mathrm{RPair}(G) = \langle \varepsilon, \rho; \varepsilon^2, \rho^2, \varepsilon\rho = \rho\varepsilon \rangle \quad \text{where} \quad \rho: a \to b,\ b \to a .$$

Thus $\mathrm{Aut}\, G = \langle C*D; V \rangle$ where C is dihedral of order $2r$, V is cyclic of order 2, and if $r \neq s$ then D is dihedral of order $2s$ while if $r = s$ then D is the Klein four-group.

<u>Theorem</u> 2. If $G = \langle A*B; U \rangle$, $A \neq U \neq B$, where U is conjugate maximal in G, then $\mathrm{Pair}(G)$ is conjugate maximal in $\mathrm{Con}(G)$ and $\mathrm{RCon}(G)$.

<u>Proof</u>. To show that $\mathrm{Pair}(G)$ is conjugate maximal in $\mathrm{Con}(G)$ it suffices to show that $\mathrm{Pair}(G)$ is conjugate maximal in $\mathrm{XtnA}(G)$, since by symmetry it will then be conjugate maximal in $\mathrm{XtnB}(G)$. Moreover, since $\mathrm{Pair}(G)$ is normal in $\mathrm{RPair}(G)$ it will also follow that $\mathrm{Pair}(G)$ is conjugate maximal in $\mathrm{RCon}(G)$.

Since $\mathrm{XtnA}(G) = \mathrm{gp}(\mathrm{Inn}\, A(G), \mathrm{Pair}(G))$ and $\mathrm{Inn}\, A(G)$ is normal in $\mathrm{XtnA}(G)$, we must show that

$$I_a \mathrm{Pair}(G) I_a^{-1} < \mathrm{Pair}(G)$$

for some $a \in A$ implies that

$$I_a \mathrm{Pair}(G) I_a^{-1} = \mathrm{Pair}(G) .$$

Suppose the containment holds. We first show that a normalizes U.

Since $\mathrm{Inn}\, U(G) < \mathrm{Pair}(G)$ we have

$$I_a \mathrm{Inn}\, U(G) I_a^{-1} = \mathrm{Inn}\, aUa^{-1}(G) < \mathrm{Pair}(G) .$$

But $\mathrm{Inn}\, A(G) \cap \mathrm{Pair}(G) = \mathrm{Inn}\, U(G)$ and $Z(G) < U$. Thus $aUa^{-1} < U$ which implies that $aUa^{-1} = U$.

Let $\alpha \in \mathrm{Pair}(G)$. Then

$$I_a \alpha I_a^{-1} = (I_a \alpha I_a^{-1} \alpha^{-1})\alpha = I_{a\alpha(a)^{-1}} \alpha \in \mathrm{Pair}(G) \quad \text{iff} \quad a\alpha(a)^{-1} \in U .$$

Hence $\alpha(a) \in Ua = aU$ so that $a^{-1}\alpha(a) \in U$; therefore

$$I_{a^{-1}\alpha(a)}\, \alpha = I_a^{-1} \alpha I_a \in \mathrm{Pair}(G)$$

and so

$$I_a^{-1}\mathrm{Pair}(G)I_a < \mathrm{Pair}(G)$$

which implies

$$I_a\mathrm{Pair}(G)I_a^{-1} = \mathrm{Pair}(G) .$$

This completes the proof of Theorem 2. □

Theorem 3. If $G = \langle A*B;U\rangle$ where A and B are finite then $\mathrm{Aut}\ G = \langle C*D;V\rangle$ where C and D are finite.

Proof. Clearly we may assume that $A \neq U \neq B$. Since every finite subgroup of G is contained in a conjugate of A or of B, and neither factor is contained in a conjugate of the other, it follows that the conjugates of A and B are precisely the maximal finite subgroups of G. Hence they are mapped among themselves by an automorphism of G. Moreover, since U is finite it is conjugate maximal in G and therefore Theorem 1 applies to yield the desired result. □

Corollary. If $G = \langle A*B;U\rangle$ where A and B are indecomposable as proper amalgamated products and have no infinite cyclic factor group, and U is conjugate maximal in A and in B then $\mathrm{Aut}\ G = \mathrm{Con}(G)$ or $\mathrm{RCon}(G)$.

Proof. Since A and B are indecomposable as proper amalgamated products and have no infinite cyclic factor groups they are absolutely indecomposable in the sense of Karrass and Solitar [6]. Moreover, any automorphism of G maps A and B onto absolutely indecomposable groups which are the factors in an amalgamated product description for G. Hence, by Karrass and Solitar [6], every automorphism of G maps A and B onto conjugates of themselves and Theorem 1 applies to yield the desired result. □

We now consider two analogues of the automorphism series for $G = \langle A*B;U\rangle$, $A \neq U \neq B$ (relative to the given decomposition of G), where U is conjugate-maximal in G.

Defs. The Con series for G is obtained by starting with G and following each term H by $Con(H)$, while in the RCon series for G each term H is followed by $RCon(H)$ (which is the same as $Con(H)$ if H is not reversible).

We call $[A:U]$ the left index of G and $[B:U]$ the right index of G. □

The terms of each series are amalgamated free products with decompositions given by Theorem 1; Theorem 2 implies that the amalgamated subgroup of each such decomposition is conjugate maximal in that product.

Theorem 4. Let $G = \langle A*B;U\rangle$, $A \neq U \neq B$, where U is conjugate-maximal in G. If U does not have index two in both A and B then the RCon series for G has at most one reversible term. Moreover, in the Con series for G the left index of its terms is constant; similarly, the right index of its terms is constant. In the RCon series the left index of its terms is also constant; the right index of its terms is either constant or is constant for finitely many terms till a term is reversible and then the right index becomes and remains two.

Proof. By (1) and (2) of Theorem 1, the results on the constant indices in the Con series are immediate. The results for the constant left indices of the terms in the RCon series are also immediate from (2) and (3) of Theorem 1; before a reversible factor is reached in the RCon series the right index is constant, and once a reversible factor is reached (if ever) the right index becomes and remains two. Suppose a reversible term is reached; then the left and right index are the same. Since the left index is constant while the right index remains two from then on, to have another reversible term the left index of each term must be two. Therefore the right index which was previously constant or 2 must also be 2, and so the amalgamated subgroup of G has index 2 in both factors of G. □

Corollary 1. If $Z(G) = 1$ then the Con series and the RCon series for G are towers (where as usual, we must identify each term H with Inn H). Moreover, in the Con series, the left factors, the right factors, and the amalgamated subgroups form towers in which the indices of any term in its successor is the same as the index of the corresponding term in the Con series in its successor. In the RCon series the same holds true for the left factors and the amalgamated subgroups.

Proof. If $Z(H) = 1$, then H is naturally embedded in Inn H, which is contained in Con(H) which is itself contained in RCon(H), and this in turn is contained in Aut G. Hence, the Aut series, the RCon series, and the Con series are all towers.

By (2) of Theorem 1 both the left factors, the right factors, and the amalgamated subgroups of the terms of the Con series for G form a tower in which the indices of any term in its successor is the same as the index of Inn H in Con H. Moreover, in the RCon series for G, the same holds true for the left factors and the amalgamated subgroups. □

Corollary 2. Suppose the factors of $G = \langle A*B;U\rangle$, $A \neq U \neq B$, are finite and also $Z(G) = 1$. If U is not of index two in both A and B, or more generally, if the Aut series for G has only finitely many reversible terms then two terms in the Aut series of G are isomorphic iff the Aut series for G is identically constant from some term on.

Proof. If the factors of $G = \langle A*B;U\rangle$, $A \neq U \neq B$, are finite then by Theorem 3 the same will be true for Aut G = Con(G) or RCon(G). Hence, the results of Theorem 4 and Corollary 1 hold for the automorphism series of G. Suppose in addition to G having finite vertices, $Z(G) = 1$. Since the vertices of an amalgamated product of two finite factors are unique, and the left and right factors form a tower, it follows that two terms of the Con series can be isomorphic iff they are identical and identical to all terms between them. But if the Aut series of G has only finitely many reversible terms it becomes a Con series after finitely many terms. □

The group $G = Z_2 * Z_2$ has an Aut series in which each term is reversible and isomorphic to G, but no two terms are identical.

There are Aut series known in which no two terms are isomorphic. In fact, $G = Z*Z_3$ has such an Aut series (see J. Dyer [1]). Moreover, $\mathrm{Aut}(Z*Z_3)$ is the free product of two finite factors with an amalgamated subgroup (see [4]).

The automorphism series of groups without finite factors may also be computable using the previous results. For example, B_3 is isomorphic to $G(2,3) = \langle a,b; a^2 = b^3\rangle$, and this has as its automorphism group the amalgamated product of two finite dihedral groups. For the automorphism series of both B_3 and B_4 and a presentation of B_4 as an amalgamated product with two factors, see [3].

BIBLIOGRAPHY

1. J. Dyer, Free products and long automorphism sequences. Comm. in Alg., 6 (6), 1978, 583-609.
2. J. Dyer and E. Grossman, The automorphism groups of the braid groups. Amer. J. of Math., 103 (6), 1981, 1151-1169.
3. A. Karrass, A. Pietrowski, and D. Solitar, Some remarks on braid groups, to be published.
4. A. Karrass, A. Pietrowski, and D. Solitar, Automorphisms of an HNN group, to be published.
5. A. Karrass and D. Solitar, The subgroups of a free product of two groups with an amalgamated subgroup. Trans. Amer. Math. Soc., 150 (1), 1970, 227-255.
6. A. Karrass and D. Solitar, Uniqueness of the vertices in a treed HNN decomposition, to be published.
7. O. Schreier, Ueber die Gruppen $A^a B^b = 1$, Abh. Math. Sem. Univ. Hamburg, 3, 1924, 167-169.
8. H. Wielandt, Eine Verallgemeinerung der invarianten Untergruppen. Math. Z., 45, 1939, 209-244.

MATHEMATICS DEPARTMENT
YORK UNIVERSITY
DOWNSVIEW, ONTARIO

Received July 20, 1982

Contemporary Mathematics
Volume 33, 1984

SOME REMARKS ON BRAID GROUPS

Abe Karrass, Alfred Pietrowski, and Donald Solitar*

Introduction

In this paper we show that the Artin braid group B_n

$$B_n = \langle \sigma_1, \sigma_2, \ldots, \sigma_{n-1}; \sigma_i\sigma_{i+1}\sigma_i = \sigma_{i+1}\sigma_i\sigma_{i+1}, \sigma_i\sigma_j = \sigma_j\sigma_i \ (|i-j| \geq 2) \rangle$$

is decomposable as a proper free product with amalgamated subgroup iff $n = 3$ or 4. The group B_3 has the well-known presentation $B_3 = \langle a, b; a^3 = b^2 \rangle$. We show that the group B_4 has the presentation

$$B_4 = \langle A*B; N \rangle, \tag{1}$$

where

$$A = \langle a, u; a^4 = u^2 \rangle,\ B = \langle b, v; b^3 = v^2 \rangle,\ N = \langle x, y, z; x^2 = y^2 = z^2 \rangle,$$
$$x = a^2 = b^{-1}vb,\ y = u = v, \text{ and } z = aua^{-1} = bvb^{-1},$$

and in terms of the standard generators σ_i, $i = 1,2,3$,

$$a = \sigma_1\sigma_2\sigma_3\ ,\ b = \sigma_1\sigma_2\sigma_3\sigma_1 .$$

Moreover, N is characteristic in B_4, and if $B_4 = \langle C*D; V \rangle$, $C \neq V \neq D$, then $V = N$ and the pair C, D is, in some order, the pair gAg^{-1}, gBg^{-1} for some $g \in B_4$.

Furthermore, we show that $\mathrm{Aut}\, B_4 \cong \mathrm{Aut}\, F_2$, where F_2 is the free group on two generators, and hence, $\mathrm{Aut}\, B_4$ is complete. Using results in [4], we compute the automorphism series for B_3 and B_4.

---

*This research was supported by NSERC grants A5602, A9108, and A5614.

© 1984 American Mathematical Society
0271-4132/84 $1.00 + $.25 per page

Results

We shall make use of the presentation for B_n given in [1]

(2) $$B_n = \langle a,b; a^n = b^{n-1}, [b^{-1}a, a^{-j}(b^{-1}a)a^j], 2 \leq j \leq n/2 \rangle ,$$

where

$$a = \sigma_1\sigma_2\cdots\sigma_{n-1} , \ b = a\sigma_1, \text{ and } [x,y] = x^{-1}y^{-1}xy .$$

Lemma 1. If $B_n = \langle C*D;V \rangle$ where $C \neq V \neq D$, and $n \geq 4$, then also $B_n = \langle X*Y;T \rangle$ where $X \neq T \neq Y$ such that a is in $X \setminus T$, b is conjugate to an element of $Y\setminus T$ and begins and ends in $Y\setminus T$ in its reduced form in $\langle X*Y;T \rangle$, and the three elements a^2, ba^2b^{-1}, and $b^{-1}a^2b$ are all in T. Moreover, X, Y, T, are obtained from C, D, V, by conjugation by some element of B_n preceeded if needed by an interchange of C and D.

Proof. Since $a^n = b^{n-1}$ is in the centre of B_n which is in V, it follows by standard cancellation arguments in amalgamated products that a and b are in conjugates of C or D.

Consider now all presentations $B_n = \langle X*Y;T \rangle$ which arise from conjugating the presentations $B_n = \langle C*D;V \rangle$ or $\langle D*C;V \rangle$ by each element of B_n. Choose from among these one such that

(3) $$a \in X$$

(4) $$b = p_1\cdots p_r f p_r^{-1}\cdots p_1^{-1}$$

where $p_1, p_2, \ldots, p_r$, f alternate out of $X\setminus T$ and $Y\setminus T$ and r is minimal over all such presentations satisfying (3).

Since all the considered presentations are proper amalgamated products, a and b cannot be in the same factor X or Y. If $r = 0$ then $f \in Y\setminus T$; if $r > 0$ then $p_1 \in Y\setminus T$, since r is minimal. In any case b begins and ends in $Y\setminus T$; moreover, $a \in X\setminus T$, again since r is minimal.

Now, for $n \geq 4$, B_n has the relator $[b^{-1}a, a^{-2}(b^{-1}a)a^2]$ which is conjugate to W where

(5) $$W = a^{-3}ba^2b^{-1}a^{-1}b^{-1}a^2b .$$

We first show that $a^2 \in T$. For if otherwise, then

$$ba^2b^{-1}a^{-1}b^{-1}a^2b$$

has a reduced syllable length in $\langle X*Y;T\rangle$ of at least seven; thus W in (5) cannot be a relator. Moreover, since $a \in X\backslash T$ and $a^2 \in T$, it follows that $a^m \in X\backslash T$ for odd m.

We show now that ba^2b^{-1} and $b^{-1}a^2b$ are also in T. Clearly by standard cancellation arguments each of ba^2b^{-1} and $b^{-1}a^2b$ is in T or in reduced form begins and ends in $Y\backslash T$. If neither is in T then in (5) W has reduced syllable length at least four. If exactly one is in T then W has reduced syllable length at least one. In any event W could not be a relator unless both are in T. □

<u>Theorem</u> 1. The Artin braid group B_n is indecomposable as a proper tree product if $n \geq 5$.

<u>Proof</u>. If B_n were decomposable as a proper tree product, since it is finitely generated it would have a presentation of the form $\langle X*T;T\rangle$ as described in Lemma 1. Now $a^n = b^{n-1}$ is in the centre of B_n and so must be in T which also contains a^2. But if n is odd this implies that $a \in T$, a contradiction. Hence, for odd $n \geq 5$, B_n is indecomposable.

If n is even then $n \geq 6$, and so we can make use of an additional relator in (2) with $j = 3$, i.e., $[b^{-1}a, a^{-3}(b^{-1}a)a^3]$, which is conjugate to S where

$$(6) \qquad S = a^3(b^{-1}a^{-2}b)b^{-2}a^3(ba^{-4}b^{-1})b^2 .$$

Since n is even and $b^{n-1} \in T$ while $b \notin T$, it follows that $b^2 \notin T$; hence, in reduced form b^2 begins and ends in $Y\backslash T$. Since $b^{-1}a^{-2}b$ and $ba^{-4}b^{-1}$ are in T it follows that S in (6) has reduced syllable length ≥ 4 and cannot be a relator. □

<u>Corollary</u> 1. For $n \geq 5$, Inn B_n is absolutely indecomposable, i.e., is not a proper tree product and has no infinite cyclic factor group.

Proof. If $Z(B_n)$ is the centre of B_n then $\text{Inn } B_n \cong B_n/Z(B_n)$. If the factor group $\text{Inn } B_n$ of B_n is a proper tree product so is B_n itself, contrary to Theorem 1.

On the other hand, $Z(B_n) = \text{gp}(a^n)$. Hence, $B_n/Z(B_n)$ can be presented by adding the relator a^n expressed in terms of the standard generators for B_n to the standard presentation for B_n. Now, all of the standard generators are conjugate and therefore under a mapping onto an infinite cyclic factor group they would all have to go into the generator or all go into its inverse. But then the relator a^n would imply that the generator had finite order, and so $B_n/Z(B_n)$ cannot have an infinite cyclic factor group. □

Corollary 2. The mapping class group $M(n,0)$ is absolutely indecomposable if $n \neq 4$.

Proof. According to Theorem N9 of Magnus, Karrass, and Solitar [6, 3.7], $M(n,0) \cong B_n/T_n$ where $T_n > Z(B_n)$, so that $M(n,0)$ is a factor group of $\text{Inn } B_n$. If $n < 4$ then $M(n,0)$ is finite. If $n > 4$ then any presentation for $M(n,0)$ as a proper amalgamated free product would result (upon taking preimages) in a proper amalgamated free product presentation for $\text{Inn } B_n$, contrary to Corollary 1. Similarly, an infinite cyclic factor group of $M(n,0)$ would also be a factor group of $\text{Inn } B_n$. □

Theorem 2. The Artin braid group B_4 has the presentation $B_4 = \langle A*B;N\rangle$ where

$$A = \langle a,u;a^4 = u^2\rangle,\ B = \langle b,v;b^3 = v^2\rangle,\ N = \langle x,y,z;x^2 = y^2 = z^2\rangle,$$
$$x = a^2 = b^{-1}vb,\ y = u = v, \text{ and } z = aua^{-1} = bvb^{-1},$$

and in terms of the standard generators σ_i, $i = 1,2,3$,

$$a = \sigma_1\sigma_2\sigma_3,\ b = \sigma_1\sigma_2\sigma_3\sigma_1 .$$

Moreover, N is characteristic in B_4, and if $B_4 = \langle C*D;V\rangle$ where $C \neq V \neq D$ then $V = N$, and the pair C, D is, in some order, the pair gAg^{-1}, gBg^{-1} for some $g \in B_4$.

Proof. If $\langle X*Y;T\rangle$ is a presentation for B_4 given as in Lemma 1, then T contains

$$N = gp(a^2, ba^2, b^{-1}, b^{-1}a^2b) .$$

We first show that N is normal in B_4. To do this note that $a^4 = b^3$ and that the relator W in (5) implies that

$$a(ba^2b^{-1})a^{-1} = a^4(b^{-1}a^{-2}b) = b^{-1}a^2b ,$$

and therefore, also that

$$a(b^{-1}a^2b)a^{-1} = a^2(a^{-1}(b^{-1}a^2b)a)a^{-2} = a^2(ba^2b^{-1})a^{-2} .$$

Hence, $aNa^{-1} = N$. Clearly the conjugate of a^2 and of $b^{-1}a^2b$ by b are in N. Moreover,

$$b(ba^2b^{-1})b^{-1} = b^3(b^{-1}a^2b)b^{-3} = b^{-1}a^2b ,$$

so that $bNb^{-1} = N$. Thus N is normal in B_4, and contained in any amalgamated subgroup V for which $B_4 = \langle C*D;V\rangle$ where $C \neq V \neq D$.

We now show that $B_4 = \langle A*B;N\rangle$. For, the second defining relator for B_4 given in (2) is equivalent to the cyclic permutation of W in (5)

(7) $\quad aba^2b^{-1}a^{-1}b^{-1}a^2ba^{-4}$,

and since $a^4 = b^3$ is in the centre of B_4, we can present B_4 by

(8) $\quad B_4 = \langle a,b;a^4 = b^3, aba^2b^{-1}a^{-1} = b^{-1}a^2b\rangle .$

We now express the relators in the presentation (8) in terms of a, b, and $u = ba^2b^{-1}$ to obtain

(9) $\quad B_4 = \langle a,b,u;a^4 = u^2 = b^3, u = ba^2b^{-1}, a^2 = b^{-1}ub,$

$$b^{-1}a^2b = bub^{-1} = aua^{-1}\rangle .$$

Now $u = ba^2b^{-1}$, and $b^{-1}a^2b = bub^{-1}$ are derivable from the remaining defining relations in (9). Therefore,

(10) $\quad B_4 = \langle a,b;a^4 = u^2 = b^3, a^2 = b^{-1}ub, aua^{-1} = bub^{-1}\rangle .$

But (10) suggests that B_4 is the amalgamated product of the two factors

$$A = \langle a,u;a^4 = u^2\rangle,\ B = \langle b,v;b^3 = v^2\rangle$$

with $gp(a^2,u,aua^{-1})$ and $gp(b^{-1}vb,v,bvb^{-1})$ amalgamated. As can be shown using the standard presentations for subgroups of A and B each of these last two subgroups has the presentation $\langle x,y,z;x^2 = y^2 = z^2\rangle$. It follows that $B_4 = \langle A*B;N\rangle$.

Finally we show the uniqueness of this decomposition for B_4. Suppose $B_4 = \langle C*D;V\rangle$ where $C \neq V \neq D$. We already know that $N < V$. Therefore, $B_4/N = \langle C/N*D/N;V/N\rangle = A/N*B/N = \langle q,s;q^2,s^3\rangle$ where $q = aN$ and $s = bN$. We may assume without loss of generality that $q \in C/N$ and $s \in gEg^{-1}$ for some $g \in B_4$ where $E = C/N$ or $E = D/N$. We now choose a presentation $\langle X*Y;T\rangle$ as in Lemma 1, but replace B_4, a, b, C, D, V, by B_4/N, aN, bN, C/N, D/N, and V/N, respectively. It is clear from standard cancellation arguments that if aN does not generate X, or $r > 0$, or bN does not generate Y, then aN, bN do not generate B_4/N. Hence, the pair X, Y is, in some order, the pair A/N, B/N, and therefore, the pair C, D is, in some order, the pair gAg^{-1}, gBg^{-1}. Clearly then, $V = N$, and of course, N is characteristic. □

<u>Corollary</u>. The mapping class group $M(4,0)$ is isomorphic to $\langle D_4*A_4;D_2\rangle$ where $D_4 = \langle a,y;a^4,y^2,(ay)^2\rangle$, $A_4 = \langle b,y;b^3,y^2,(by)^3\rangle$, $D_3 = \langle x,y;x^2,y^2,(xy)^2\rangle$, and $x = a^2 = b^{-1}yb$. Moreover, any presentations for $M(4,0)$ as a proper amalgamated free product are conjugate.

<u>Proof</u>. According to Theorem N9 of Magnus, Karrass, and Solitar [6], $M(4,0) \cong B_4/T_4$ where T_4 is the normal subgroup of B_4 generated by the elements $xzy = (a^2b^{-1})^3$, $(xy)^2 = (a^2ba^2b^{-1})^2$, and $a^4 = b^3$ where we have converted the defining relators given in terms of the standard generators for B_4 into those given in terms of a, b, x, y, and z in Theorem 2. Since $T_4 < N$, we have that $M(4,0) \cong \langle A/T_4*B/T_4;N/T_4\rangle$ which is easily shown to have the presentation given in the statement of this corollary. □

It follows easily from the preceding corollary that $M(4,0)$ is an extension of a free group of rank 2 by the symmetric group of degree 4.

Theorem 3. Aut B_4 is a splitting extension of degree two over Inn B_4 with outer automorphism α given by $a \to a^{-1}$ and $b \to b^{-1}$, where a and b are the generators for B_4 in the presentation (2).

Proof. Our proof makes extensive use of the terminology and results in [4]. It is clear from Theorem 2 that N is characteristic in B_4 and has index two and three in A and in B, respectively. Hence, by the Corollary to Theorem 1 in [4] Aut B_4 = Con(B_4) or RCon(B_4). Since N has different indices in A and in B, B_4 is not reversible. Therefore, Aut B_4 = Con(B_4). Now Pair(B_4) contains Inn N(B_4) as a normal subgroup. Moreover, the mapping $a \to a^{-1}$, $b \to b^{-1}$ maps $x \to x^{-1}$, $y = u = v = ba^2b^{-1} \to b^{-1}a^{-2}b = b^{-2}v^{-1}b^2 = bv^{-1}b^{-1} = z^{-1}$, and $z = bvb^{-1} \to b^{-1}z^{-1}b = y^{-1}$, and hence defines a pairing automorphism α of B_4. Now the elements x, y, z, x^{-1}, y^{-1}, z^{-1} define distinct conjugate classes of N (this is easily seen by considering the homomorphisms of N onto $\langle x,y,z;x^2,y^2,z^2\rangle$ and onto $\langle x,y,z;x = y = z\rangle$). Thus $\alpha \notin$ Inn N(B_4). Moreover, $\alpha^2 = 1$. We now show that Pair(B_4) = gp(α,Inn N(B_4)).

It is clear that Aut A = gp(α_1,Inn A) = gp(α_1,C_a,Inn N(A)), where α_1 is the automorphism of A induced by α and given by $a \to a^{-1}$, $u \to au^{-1}a^{-1}$ and C_a is conjugation by the element a acting on A. Since [Inn A:Inn N(A)] = 2, each element of Aut A is a product $r_1q_1p_1$ where $p_1 = 1$ or $p_1 = C_a$, $q_1 = 1$ or $q_1 = \alpha_1$, and $r_1 = C_n$ for some $n \in N$. Similarly each element of Aut B is a product $r_2q_2p_2$ where $p_2 = 1$ or $p_2 = D_b$ or $p_2 = D_b^{-1}$, $q_2 = 1$ or $q_2 = \alpha_2$, and $r_2 = D_m$ for some $m \in N$, and where α_2 is the automorphism of B induced by α and given by $b \to b^{-1}$, $v \to bv^{-1}b^{-1}$ and D_b is conjugation by the element b acting on B. Clearly the pair of automorphisms α_1 of A and α_2 of B determine the pairing automorphism α of B_4. Moreover, C_a maps $x \to x$, $y \to z$, and $z \to xyx^{-1}$, while D_b maps $x \to y$, $y \to z$, and $z \to x$.

In order to examine the possible pairing automorphisms of B_4, consider now any pairing automorphism ω and the automorphisms ω_1 of A and ω_2 of

B that it induces. Let $\omega_i = r_i q_i p_i$ as above, where $i = 1$ or $i = 2$. Since $\mathrm{Inn}\, N(B_4) < \mathrm{Pair}(B_4)$, we may assume that $r_2 = 1$. Moreover, in order for ω_1 and ω_2 to agree on N we must have the images of x, y, and z under $\omega_3 = q_1 p_1$ conjugate in N to the images of x, y, and z under ω_2 respectively, in which case we say ω_3 <u>matches</u> ω_2. We consider six cases depending on whether p_2 is 1 or D_b or D_b^{-1}, and whether q_2 is 1 or α_2.

Case 1: $p_2 = 1$ and $q_2 = 1$; $x \to x$, $y \to y$, $z \to z$

(a) $p_1 = 1$ and $q_1 = 1$; agree on x, y, z; match

(b) $p_1 = C_a$ and $q_1 = 1$; $y \to z$; no match

(c) $p_1 = 1$ and $q_1 = \alpha_1$; $x \to x^{-1}$; no match

(d) $p_1 = C_a$ and $q_1 = \alpha_1$; $x \to x^{-1}$; no match

Case 2: $p_2 = D_b$ and $q_2 = 1$; $x \to y$

ω_3 maps $x \to x$ or x^{-1}; no match

Case 3: $p_2 = D_b^{-1}$ and $q_2 = 1$; $x \to z$

ω_3 maps $x \to x$ or x^{-1}; no match

Case 4: $p_2 = 1$ and $q_2 = \alpha_2$; $x \to x^{-1}$, $y \to z^{-1}$, $z \to y^{-1}$

(a) $p_1 = 1$ and $q_1 = 1$; $x \to x$; no match

(b) $p_1 = C_a$ and $q_1 = 1$; $x \to x$; no match

(c) $p_1 = 1$ and $q_1 = \alpha_1$; agree on x, y, z; match

(d) $p_1 = C_a$ and $q_1 = \alpha_1$; $y \to y^{-1}$; no match

Case 5: $p_2 = D_b$ and $q_2 = \alpha_2$; $x \to z^{-1}$

ω_3 maps $x \to x$ or x^{-1}; no match

Case 6: $p_2 = D_b^{-1}$ and $q_2 = \alpha_2$; $x \to y^{-1}$

ω_3 maps $x \to x$ or x^{-1}; no match

In the cases of the matches, namely Case 1(a), and Case 4(c), $\omega_3 = \omega_2$ and therefore r_1 defines the identity on N and n is in $Z(N) = Z(A)$, the centre of N and of A. Hence, modulo $\mathrm{Inn}\, N(B_4)$, $\omega = 1$ or $\omega = \alpha$. Thus, $\mathrm{Pair}(B_4) = \mathrm{gp}(\alpha, \mathrm{Inn}\, N(B_4))$ and $\mathrm{Aut}\, B_4 = \mathrm{gp}(\mathrm{Pair}(B_4), \mathrm{Inn}\, B_4) = \mathrm{gp}(\alpha, \mathrm{Inn}\, B_4)$. □

Corollary. For $n \geq 5$, Aut B_n is absolutely indecomposable, i.e., is not a proper tree product and has no infinite cyclic factor group.

Proof. The results of Theorem 3 hold for any B_n (Dyer and Grossman [3]). Thus Inn B_n has index 2 in Aut B_n, and is absolutely indecomposable by Corollary 1 to Theorem 1. Clearly then, Aut B_n has no infinite cyclic factor group. Moreover, by Karrass and Solitar [5], Inn B_n is contained in a conjugate of some proper factor of Aut B_n if it is a proper amalgamated product, contrary to $[\text{Aut } B_n : \text{Inn } B_n] = 2$. □

Theorem 4. Aut $B_4 = \langle C*D;V\rangle$ where $C = \langle A_1*B_1;V_1\rangle$, $D = \langle A_2*B_2;V_2\rangle$, and

$$A_1 = \langle a,p;a^4,p^2,(ap)^2\rangle \simeq D_4 ,$$

$$B_1 = \langle p,u;p^2,u^2,(pu)^2\rangle \simeq D_2 ,$$

$$V_1 = \langle p;p^2\rangle \simeq Z_2 ,$$

$$A_2 = \langle b,q;b^3,q^2,(bq)^2\rangle \simeq D_3 ,$$

$$B_2 = \langle q,u;q^2,u^2,(qu)^2\rangle \simeq D_2 ,$$

$$V_2 = \langle q;q^2\rangle \simeq Z_2 ,$$

$$V = gP(ap,a^2,u) = gp(bq,b^{-1}ub,u) \simeq Z_2*Z_2*Z_2 ,$$

with a, b, $u = ba^2b^{-1}$ being inner automorphisms of B_4 determined by the corresponding elements of B_4, and $p = \alpha a$, $q = \alpha b$.

Proof. Clearly Inn $B_4 = \langle \text{Inn } A(B_4)*\text{Inn } B_4);\text{Inn } N(B_4)\rangle$ and conjugation by α maps each factor and the amalgamated subgroup in this presentation onto itself. If we let a, b, and u denote the inner automorphisms of B_4 determined by the corresponding elements of B_4, then

$$gp(\alpha,\text{Inn } A(B_4)) = \langle \alpha,a,u;\alpha^2,a^4,u^2,(\alpha a)^2,\alpha u\alpha^{-1} = aua^{-1}\rangle , \tag{11}$$

$$gp(\alpha,\text{Inn } B(B_4)) = \langle \alpha,b,u;\alpha^2,b^3,u^2,(\alpha b)^2,\alpha u\alpha^{-1} = bub^{-1}\rangle , \tag{12}$$

$$gp(\alpha,\text{Inn } N(B_4)) = \langle \alpha,x,y,z;\alpha^2,x^2,y^2,z^2,\alpha x\alpha^{-1} = x,\alpha y\alpha^{-1} = z,\alpha z\alpha^{-1} = y\rangle , \tag{13}$$

where $x = a^2$, $y = u$, and $z = aua^{-1} = bub^{-1}$.

If we define new generators $p = \alpha a = a^{-1}\alpha$ and $q = \alpha b = b^{-1}\alpha$ and replace α by $ap = bq$ then the relators in (11) become

$$(14) \quad (ap)^2, a^4, u^2, p^2, pup^{-1} = u$$

while those in (12) become

$$(15) \quad (bq)^2, b^3, u^2, q^2, quq^{-1} = u .$$

The relation $aua^{-1} = bub^{-1}$ is equivalent to $\alpha aua^{-1}\alpha^{-1} = \alpha bub^{-1}\alpha^{-1}$ and therefore becomes $pup^{-1} = quq^{-1}$ which is derivable from the last relations in (14) and (15).

The relations $\alpha a = a^{-1}\alpha$ and $\alpha b = b^{-1}\alpha$ become $apa = p$ and $bqb = q$ which are derivable from the first and fourth relators in (14) and (15).

The relation $u = ba^2b^{-1}$ is equivalent to $a^2 = b^{-1}ub$.

The relations in (13) are all derivable from those in (14) if x, y, z are replaced by their definitions in terms of $a, b,$ and u.

The relation $ap = bq$ must be added to the rest.

Thus $\mathrm{Aut}\ B_4$ has the presentation

$$(16) \quad \mathrm{Aut}\ B_4 = \langle a,p,u,b,q; a^4, p^2, u^2, b^3, q^2, (ap)^2, (pu)^2, (bq)^2, (qu)^2, ap = bq, a^2 = b^{-1}ub \rangle .$$

Using the definitions for $A_1, B_1, V_1, A_2, B_2, V_2,$ and V given in the statement of this theorem, the results of this theorem are immediate. □

<u>Lemma 2</u>. Let $\omega: P \to Q$ be a homomorphism of P into Q such that a normal subgroup L of P is mapped isomorphically onto a normal subgroup M of Q, and the induced homomorphism of P/L into Q/M is also an onto isomorphism. Then ω is an isomorphism of P onto Q.

<u>Proof</u>. If $\omega(p) = 1$ for some $p \in P$ then $\omega(pL) = M$. Thus $p \in L$ and so $p = 1$. Clearly ω is onto. □

<u>Theorem</u> 5. $\mathrm{Aut}\ B_4$ is isomorphic to $\mathrm{Aut}\ F_2$ where F_2 is the free group on two generators. More specifically, the elements $s = b^{-1}a^2ba^{-2}$ and

$t = ba^2b^{-1}a^{-2}$ freely generate a free normal subgroup F_2 of Aut B_4, and Aut B_4 acts naturally as the automorphism group of F_2 by means of conjugation.

Proof. The elements $x = a^2$, $y = ba^2b^{-1}$, and $z = b^{-1}a^2b$ generate Inn $N(B_4)$ with the presentation $\langle ,xy,z;x^2,y^2,z^2\rangle$, and therefore the elements $s = zx$ and $t = yz$ freely generate a free group F_2, namely, the subgroup of words of even length in Inn $N(B_4)$. Since a, b, and α conjugate x, y, and z into words of odd length in x, y, and z, it follows that F_2 is normal in Aut B_4. Specifically, acting by conjugation in Aut B_4,

$$a: x \to x,\ y \to z,\ z \to xyx^{-1}$$

$$b: x \to y,\ y \to z,\ z \to x$$

$$\alpha: x \to x,\ y \to z,\ z \to y$$

and so $a: s \to t^{-1}, t \to s$; $b: s \to t^{-1}, t \to st^{-1}$; $\alpha: s \to t, t \to s$.

In order to show that this action of Aut B_4 on $F_2 = \langle s,t\rangle$ is an isomorphism onto Aut F_2 we observe that $b^{-1}a^2ba^{-2}$ acts on F_2 by conjugating it by s, while $ba^2b^{-1}a^{-2}$ acts on F_2 by conjugating it by t. It is well known that the natural homomorphism from Aut$\langle s,t\rangle$ onto Aut$\langle s,t;st = ts\rangle$ has kernel generated by I_s and I_t (Nielsen [7]) which freely generate a free group. Moreover, Inn B_4 is generated by a and b and has index two in Aut B_4, and the determinant of the transformations induced on $Ab_2 = \langle s,t;st = ts\rangle$ is 1 for a and b, and -1 for α. It follows that the group Aut F_2^+, the automorphisms of F_2 inducing those of Ab_2 with determinant 1, are generated by a and b. Thus the given action of Aut B_4 on F_2 induces a homomorphism of Inn B_4/F_2 onto Aut $F_2^+/\text{Inn } F_2 = Ab_2^+$. Again, it is well known that Ab_2^+ has the presentation $\langle c,d;c^4,d^6,c^2 = d^3\rangle$. But Inn B_4/F_2 has the presentation $\langle a,b;a^4,b^3,ba^2 = a^2b\rangle$, and if one replaces b using a new generator $e = ba^2$, then the presentation becomes $\langle a,e;a^4,e^6,a^2 = e^3\rangle$. Thus under the action of Aut B_4 on F_2, F_2 is mapped isomorphically onto Inn F_2, and Inn B_4/F_2 is mapped isomorphically onto

$\mathrm{Aut}\ F_2^+/\mathrm{Inn}\ F_2$. Hence from Lemma 2 it follows that $\mathrm{Inn}\ B_4$ is mapped isomorphically onto $\mathrm{Aut}\ F_2^+$. Moreover, since $\mathrm{Aut}\ F_2/\mathrm{Aut}\ F_2^+$ is Z_2 which is isomorphic to $\mathrm{Aut}\ B_4/\mathrm{Inn}\ B_4$, it follows again by Lemma 2 that $\mathrm{Aut}\ B_4$ is isomorphic to $\mathrm{Aut}\ F_2$. □

We shall now compute the automorphism series for the braid groups B_n, $n \leq 4$. This is easily done for B_2 which is infinite cyclic, and therefore has a three term series consisting of itself, Z_2, and 1. For B_3, using [4], we can show that its automorphism series also has three terms. These are $B_3 = \langle a,b;a^3 = b^2\rangle$, $\langle D_3 * D_2; Z_2\rangle$, and $\langle (Z_2 \times D_3) * D_4; D_2\rangle$, where the last amalgamation has trivial centre and therefore, the nontrivial central element of $Z_2 \times D_3$ is amalgamated with a non-central element of D_4, and vice versa. For B_4, since $\mathrm{Aut}\ B_4 \cong \mathrm{Aut}\ F_2$ and $\mathrm{Aut}\ F_2$ is complete (Dyer and Formanek [2]), it follows that $\mathrm{Aut}\ B_4$ is complete. J. Dyer and E. Grossman [3] show that $\mathrm{Aut}\ B_n$ is complete for $n \geq 4$.

BIBLIOGRAPHY

1. H. S. M. Coxeter and W. O. J. Moser, Generators and relations for discrete groups. Ergebnisse der Mathematik, Bd. 14, 3rd ed., Berlin-Heidelberg-New York, Springer, 1972.

2. J. Dyer and E. Formanek, The automorphism group of a free group is complete. J. London Math. Soc., (2) 11, 1975, 181-190.

3. J. Dyer and E. Grossman, The automorphism groups of the braid groups. Amer. J. of Math., 103 (6), 1981, 1151-1169.

4. A. Karrass, A. Pietrowski, and D. Solitar, Automorphisms of a free product with an amalgamated subgroup, to appear.

5. A. Karrass and D. Solitar, The subgroups of a free product of two groups with an amalgamated subgroup. Trans. Amer. Math. Soc., 150 (1), 1970, 227-255.

6. W. Magnus, A. Karrass, and D. Solitar, Combinatorial Group Theory. 2nd rev. ed., New York, Dover, 1976.

7. J. Nielsen, Die Isomorphismengruppe der allegemeinen unendlichen Gruppe mit zwei Erzeugenden, Math. Ann., 78, 1918, 385-397.

MATHEMATICS DEPARTMENT
YORK UNIVERSITY
DOWNSVIEW, ONTARIO

Received July 20, 1982

Contemporary Mathematics
Volume 33, 1984

GEOMETRIC REALIZATION OF GROUP EXTENSIONS BY THE SEIFERT CONSTRUCTION

Kyung Bai Lee and Frank Raymond*

Dedicated to Roger C. Lyndon, teacher, colleague and friend

§0. Introduction

Mathematical constructions which blend several disciplines of mathematics can be expected to exhibit interesting connections between them. The Seifert construction translates geometric, topological and group theoretic data into a group action on a topological space and certain associated mappings. This space and mappings faithfully display both the geometric and group theoretic data and have intrinsic topological interest.

The classical 3-dimensional Seifert manifolds [Seifert] are singular fiberings over 2-dimensional surfaces with typical fiber the circle. These fiberings predated the theory of fiber bundles but Seifert ingeniuously anticipates many aspects of bundle theory. In [CR1] the study of transformation groups on aspherical manifolds was developed from initial contributions of Conner and Montgomery, [CM] and [C]. In the latter half of [CR1] the Seifert construction pertaining to compact group actions, as we now envision it, was initiated. (For the classical Seifert fiberings this reduces the topological study of these 3-dimensional manifolds to the cohomology of Fuchsian groups.)

This theory was extensively developed in a series of papers. Kodaira [K] and Holmann [H] had introduced holomorphic singular fiberings and [CR6]

* Supported in part by the National Science Foundation.

© 1984 American Mathematical Society
0271-4132/84 $1.00 + $.25 per page

is an extension of the earlier Seifert construction to embrace these more extensive phenomena. In [CR5] the deformation theory of Seifert manifolds, which was implicit in some of the earlier papers, was developed. K. B. Lee [L1] refined certain aspects of [CR5] still further enabling him to show that certain infranilmanifolds possess an affine structure. One innovation was the construction of a "universal group" for the Seifert construction and its deformation theory. This unifying point of view almost has a functorial character and will prove to be very useful here.

Roughly speaking, the Seifert construction provides a homomorphism of a group E into the group of homeomorphisms of a contractible manifold $\mathbb{R}^m \times W$ so that a certain normal series of E acts as "translations" on the corresponding factors of $\mathbb{R}^m$. If E is torsion-free, then it acts as covering transformations on $\mathbb{R}^m \times W$ so that $E\backslash(\mathbb{R}^m \times W)$ is a $K(E,1)$-manifold. This gives a nice way of constructing compact aspherical manifolds with given fundamental groups. The case where E is not necessarily torsion-free is also important. For example, when E contains a torsion-free normal subgroup π, then the quotient group E/π will act on the $K(\pi,1)$-manifold $\pi\backslash\mathbb{R}^m \times W$. We have a uniqueness and rigidity theorem for the construction so that we can compare two manifolds if they are of the form $\pi\backslash(\mathbb{R}^m \times W)$ as above. This yields rigidity theorems for certain manifolds. The rigidity is also used to show that for certain manifolds the maximal toral action characterizes all the compact Lie group actions on the manifold.

We now describe the main results of this paper. Section 1 starts by recalling the Seifert construction where the typical fiber is a k-torus. Given a properly discontinuous action of a group Q on a manifold W and a group extension $1 \to \mathbb{Z}^k \to P \to Q \to 1$, a Seifert construction is a homomorphism ψ of P into $H^F(\mathbb{R}^k \times W)$, a group of (fiber-preserving) self-homeomorphisms of $\mathbb{R}^k \times W$. This construction is unique and satisfies a rigidity criterion. If a group E has a filtration which yields a series of group extensions with free abelian groups as kernels, we find a homomorphism

ψ of E into $\mathcal{U}(\mathbb{R}^m \times W;K)$, a certain group of self-homeomorphisms of $\mathbb{R}^m \times W$. This is achieved by iterating the construction with abelian kernels. The iterated Seifert construction also satisfies the uniqueness and rigidity properties (1.8) which are used throughout the paper.

In Section 2, we construct for a given extension $1 \to \pi \to E \to Q \to 1$ where π is a finitely generated, torsion-free virtually poly-$\mathbb{Z}$ group and Q acts on a contractible W with compact orbit space, a closed $K(E,1)$-manifold which admits a series of Seifert fiberings with typical fiber tori (2.5).

Section 3 is devoted to an analysis of the construction when π is nilpotent. In this case, the fibers are infranilmanifolds (3.8). We shall also prove a generalization of the classical Bieberbach theorem: An isomorphism between almost crystallographic groups for a nilpotent Lie group L is conjugation by a diffeomorphism of L (3.7). This corrects and sharpens [A1, Theorem 2].

More constructions are discussed in Section 4. By combining the iterated construction with that of [RW] and results of Auslander [A2], we obtain in (4.4) a construction whose typical fiber is a closed aspherical manifold homotopy equivalent to a double coset space of a Lie group, $\Gamma\backslash G/K$, where K is a maximal compact subgroup of G and Γ is a uniform lattice in G. Related to this construction is the construction of Johnson [J] for poly-$\mathbb{L}_+$ groups. We are able to get rid of the extra assumptions that were made in [J] proving that for a torsion-free poly-$\mathbb{L}_+$ group G, there exists a closed smooth $K(G,1)$-manifold, see (4.6).

Section 5 is an application of our construction and rigidity to the theory of transformation groups. If M is a closed aspherical manifold, then the only connected compact Lie groups that can act effectively on M are tori T^s with $s \leq$ rank of the center of $\pi_1(M)$. A toral action (T^k,M) is called maximal if k = rank of the center of $\pi_1(M)$. In (5.3), we show that many of the manifolds we have created by the Seifert construction admit maximal toral actions. For example, if M is a closed $K(\pi,1)$-manifold with

π a virtually nilpotent or a special Mostow-Wang group, M admits a "standard" maximal toral action. Furthermore, every smooth toral action can be topologically conjugated into the standard toral action, see (5.6) and (5.9). The theorems in Section 5 are stated in considerable more generality than stated here. In fact, a special feature is the treatment of general compact (not necessarily connected) Lie group actions.

The last section is concerned with the geometric realization problem of a finite group of homotopy classes of self-homotopy equivalences of a closed aspherical manifold M as a group of homeomorphisms (or diffeomorphisms). This is an important question in geometry and topology. The authors have solved this problem for flat Riemannian manifolds in [LR1]. This section treats and solves the problem for nilmanifolds, infranilmanifolds (and hence flat Riemannian manifolds) and virtually poly-$\mathbb{Z}$ manifolds. Moreover, the constructed geometric realizations all have a geometric character in that they are achieved as groups of fiber-preserving diffeomorphisms. These realizations then provide homotopy theoretic models for all possible topological actions.

We would like to thank Yoshinobu Kamishima for a preprint of his independent and related work on Seifert fiberings with nilmanifolds as typical fiber. In a subsequent publication with Y. Kamishima, the authors will develop the theory of iterated Seifert constructions where the terms of the filtration have nilpotent (rather than abelian) kernels.*

§1. The Seifert Constructions

1.1. An extensive generalization of the Seifert fiber construction with typical fiber a k-torus [CR1] was alluded to in a paper by Raymond and Wigner, [RW, p. 421]. This generalization is obtained by iterating the earlier torus fiber construction to yield a Seifert fibering with typical fiber a poly-$\mathbb{Z}$ manifold. In this section we shall describe this iterative procedure in detail and establish the uniqueness and rigidity of the resulting construction.

*Added in proof: See [LKR] for this development as well as a sharpening of some of the results of this paper.

1.2. We first recall the Seifert fiber construction (with typical fiber a torus) and describe some of its known facts. Suppose that we have (Q,W) satisfying

(*) The group Q acts on W, a connected and simply connected paracompact ANR space; properly discontinuously via a homomorphism $\rho: Q \to \mathcal{H}(W)$. $\mathcal{H}(W)$ denotes the group of all self-homeomorphisms of W. (The constructions work for very general (Q,W). But most of our interest is centered on W being a manifold (or manifold factor) and $Q\backslash W$ being compact so that the construction yields a closed manifold.)

Given an extension of $\mathbb{Z}^k$ by Q, $1 \to \mathbb{Z}^k \to P \to Q \to 1$, there exists an action of Q on $T^k \times W$ which acts on the W factor via ρ, and is compatible with the translational T^k action on the first factor. That is,

$$\alpha(t\cdot(s,w)) = \phi(\alpha)(t)\cdot(\alpha\cdot(s,w))$$

for all $t \in T^k$, $(s,w) \in T^k \times W$ and $\alpha \in Q$, where $\phi: Q \to \mathrm{Aut}(T^k)$ is the homomorphism induced by the exact sequence. We note, in passing, that we are really defining an action of the semi-direct product of T^k by Q, $T^k \circ Q$, on $T^k \times W$ by $(t,\alpha)(s,w) = t\cdot(\alpha\cdot(s,w))$. The group law is $(t,\alpha)(u,\beta) = (t + \phi(\alpha)u, \alpha\beta)$. Now, lifting the Q action to the covering space $\mathbb{R}^k \times W$ of $T^k \times W$, we obtain an action of P on $\mathbb{R}^k \times W$ which is compatible with the translational $\mathbb{R}^k$ action. That is, we lift the $T^k \circ Q$ action to an $\mathbb{Z}^k\backslash(\mathbb{R}^k \circ P)$ action.

1.3. Let us describe this construction and lifting in more detail. Let $\mathcal{M}(W,\mathbb{R}^k)$ be the real vector space of all continuous maps of W into $\mathbb{R}^k$. The group $\mathrm{Aut}(\mathbb{R}^k) \times \mathcal{H}(W)$ acts on $\mathcal{M}(W,\mathbb{R}^k)$ by

$$(g,h)\cdot\lambda = g \circ \lambda \circ h^{-1}$$

making $\mathcal{M}(W,\mathbb{R}^k)$ an $\mathrm{Aut}(\mathbb{R}^k) \times \mathcal{H}(W)$-space. The subgroup $\mathbb{R}^k$ of all constant maps of W into $\mathbb{R}^k$ is an $\mathrm{Aut}(\mathbb{R}^k) \times \mathcal{H}(W)$-subspace (trivial as an $\mathcal{H}(W)$-space). Form the semi-direct product $\mathcal{M}(W,\mathbb{R}^k) \circ (\mathrm{Aut}(\mathbb{R}^k) \times \mathcal{H}(W))$ which we denote by $\mathcal{H}^F(\mathbb{R}^k \times W)$ for simplicity. Of course, the group law is

$(\lambda,g,h)(\lambda_1,g_1,h_1) = (\lambda + g\lambda_1 h^{-1}, gg_1, hh_1)$. Then $\mathcal{H}^F(\mathbb{R}^k \times W)$ is naturally a subgroup of $\mathcal{H}(\mathbb{R}^k \times W)$ via

$$(\lambda,g,h)(x,w) = (gx + \lambda hw, hw) .$$

The extension sequence $1 \to \mathbb{Z}^k \to P \to Q \to 1$ induces a representation $\phi\colon Q \to \mathrm{Aut}(\mathbb{Z}^k)$. Suppose we have an embedding $\epsilon\colon \mathbb{Z}^k \to \mathbb{R}^k$. It gives rise to an embedding $\epsilon\colon \mathrm{Aut}(\mathbb{Z}^k) \to \mathrm{Aut}(\mathbb{R}^k)$, and hence a homomorphism $\phi_\epsilon\colon Q \to \mathrm{Aut}(\mathbb{Z}^k) \to \mathrm{Aut}(\mathbb{R}^k)$. Also there is induced an embedding $\epsilon\colon \mathbb{Z}^k \to \mathbb{R}^k \to \mathcal{M}(W,\mathbb{R}^k)$. A <u>Seifert construction</u>, for the extension sequence $1 \to \mathbb{Z}^k \to P \to Q \to 1$ together with $\rho\colon Q \to \mathcal{H}(W)$ satisfying (*) and an embedding $\epsilon\colon \mathbb{Z}^k \to \mathbb{R}^k$, is a homomorphism $\psi\colon P \to \mathcal{H}^F(\mathbb{R}^k \times W)$ which makes the following diagram with exact rows commutative:

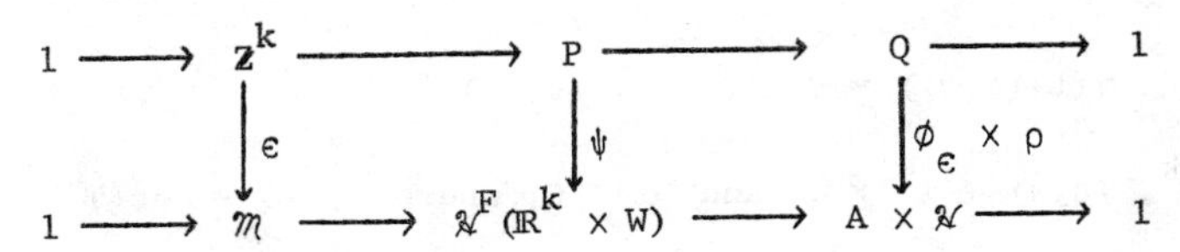

where $\mathcal{M} = \mathcal{M}(W,\mathbb{R}^k)$, $A = \mathrm{Aut}(\mathbb{R}^k)$ and $\mathcal{H} = \mathcal{H}(W)$. Such a homomorphism ψ always exists (see [CR6] or [L1]. Consequently, P acts on $\mathbb{R}^k \times W$ via $\psi\colon P \to \mathcal{H}^F(\mathbb{R}^k \times W) \subset \mathcal{H}(\mathbb{R}^k \times W)$. This description of the P action on $\mathbb{R}^k \times W$ is exactly equivalent to imposing the compatibility of the Q-action on $T^k \times W$ with the translational T^k action in (1.2). If W is a smooth manifold and if the Q action on W is smooth (i.e., $\rho\colon Q \to \mathrm{Diff}(W)$), then the construction can be done smoothly so that $\psi\colon P \to C(W,\mathbb{R}^k) \circ (\mathrm{Aut}(\mathbb{R}^k) \times \mathrm{Diff}(W)) = \mathrm{Diff}^F(\mathbb{R}^k \times W) \subset \mathrm{Diff}(\mathbb{R}^k \times W)$, where $C(W,\mathbb{R}^k)$ is the space of all smooth maps of W into $\mathbb{R}^k$. (Note that $C(W,\mathbb{R}^k)$ is $\mathrm{Aut}(\mathbb{R}^k) \times \mathrm{Diff}(W)$-invariant.)

<u>Notation</u>. Let A be a group. Then μ denotes the <u>conjugation</u> map of A into $\mathrm{Aut}(A)$ so that $\mu(a)x = axa^{-1}$ for all $a \in A$ and $x \in A$. $z(A)$ will always denote the <u>center</u> of A. If B is a subgroup of A, then the <u>centralizer</u> of B in A will be denoted by $C_A(B)$.

Let Q_0 = kernel of ρ, and Q_w = stabilizer of Q at $w \in W$. Let P_0, P_w be the complete inverse images of Q_0, Q_w, respectively, under the homomorphism $P \to Q$. Then the P action on $\mathbb{R}^k \times W$ is effective if and only if $C_{P_0}(\mathbb{Z}^k)$ is torsion-free. The P action on $\mathbb{R}^k \times W$ is free if and only if P_w is torsion-free for all $w \in W$.

1.4. Given an extension sequence $1 \to \mathbb{Z}^k \to P \to Q \to 1$ together with $\rho: Q \to \mathcal{H}(W)$ and $\epsilon: \mathbb{Z}^k \to \mathbb{R}^k$, suppose we have two constructions $\psi_1, \psi_2: P \to \mathcal{H}^F(\mathbb{R}^k \times W)$. Then there exists $\lambda \in \mathcal{M}(W,\mathbb{R}^k)$ so that $\psi_2 = \mu(\lambda) \circ \psi_1$, where $\lambda = (\lambda,1,1) \in \mathcal{H}^F(\mathbb{R}^k \times W)$.

Therefore, the Seifert construction for $1 \to \mathbb{Z}^k \to P \to Q \to 1$ and fixed ϵ and ρ is unique up to conjugation by elements of $\mathcal{M}(W,\mathbb{R}^k)$.

We may also ask what happens if, for a fixed extension $1 \to \mathbb{Z}^k \to P \to Q \to 1$ and ρ, we change ϵ. Let $\psi, \psi': P \to \mathcal{H}^F(\mathbb{R}^k \times W)$ be two constructions for $1 \to \mathbb{Z}^k \to P \to Q \to 1$, $\rho: Q \to \mathcal{H}(W)$ and $\epsilon, \epsilon': \mathbb{Z}^k \to \mathbb{R}^k$, respectively. Then there exists $g \in \mathrm{Aut}(\mathbb{R}^k)$ for which $\epsilon' = g \circ \epsilon$. Then $\mu(g)$, the conjugation by $g = (0,g,1) \in \mathcal{H}^F(\mathbb{R}^k \times W)$, is an automorphism of $\mathcal{H}^F(\mathbb{R}^k \times W)$. Let $\psi'' = \mu(g) \circ \psi$. Then ψ'' is another construction for $1 \to \mathbb{Z}^k \to P \to Q \to 1$, and ρ and ϵ'. We can compare ψ' and ψ''. By the previous case, there exists $\lambda \in \mathcal{M}(W,\mathbb{R}^k)$ such that $\psi' = \mu(\lambda) \circ \psi''$. Consequently, $\psi' = \mu(\lambda,g,1) \circ \psi$. Observe that $(\lambda,g,1)$ is a global linear transformation on $\mathbb{R}^k$ followed by an "equivariant" isotopy moving just in each "fiber" $\mathbb{R}^k \times W$. (In fact, any element of $\mathcal{M}(W,\mathbb{R}^k) \subset \mathcal{H}^F(\mathbb{R}^k \times W)$ moves only along the fibers and each element of $\mathcal{H}^F(\mathbb{R}^k \times W)$ is fiber-preserving. The superscript F in $\mathcal{H}^F(\mathbb{R}^k \times W)$ refers to "fiber-preserving" homeomorphisms.) Thus, we have seen that the choice of $\epsilon: \mathbb{Z}^k \to \mathbb{R}^k$ does not make too much difference. Sometimes we delete ϵ from our data, understanding that we are using some ϵ and that the result is still unique up to conjugation by elements of $\mathcal{M}(W,\mathbb{R}^k) \circ \mathrm{Aut}(\mathbb{R}^k)$, because most of the time we will work up to conjugation. For more details the reader is referred to [CR1], [CR5] and [L1]. For a quick sketch of a connection between the approaches of

(1.2) and (1.3), consult the Appendix. (The Appendix will not be used until Section 5.) We also remark that we have adopted left actions of Q on W instead of the right actions of [CR1], and so the notation is closer to that of [L1].

The following is crucial for the rigidity of the iterated Seifert construction and hence for the rigidity of infranilmanifolds. Let $1 \to \mathbb{Z}^k \to P \to Q \to 1$, $\rho: Q \to \mathcal{H}(W)$; $1 \to \mathbb{Z}^k \to P' \to Q' \to 1$, $\rho' = Q' \to \mathcal{H}(W)$ be two sets of data. Suppose there exists an isomorphism $\theta: P \to P'$ inducing an isomorphism $\hat{\theta}: \mathbb{Z}^k \to \mathbb{Z}^k$ (and hence, an isomorphism $\overline{\theta}: Q \to Q'$). Further suppose that there exists $h \in \mathcal{H}(W)$ so that $\rho' \circ \overline{\theta} = \mu(h) \circ \rho$. Namely, the following two diagrams are commutative:

$$\begin{array}{ccccccccc} 1 & \longrightarrow & \mathbb{Z}^k & \longrightarrow & P & \longrightarrow & Q & \longrightarrow & 1 \\ & & \downarrow \hat{\theta} & & \downarrow \theta & & \downarrow \overline{\theta} & & \\ 1 & \longrightarrow & \mathbb{Z}^k & \longrightarrow & P' & \longrightarrow & Q' & \longrightarrow & 1 \end{array} \qquad \begin{array}{ccc} Q & \xrightarrow{\rho} & \mathcal{H}(W) \\ \downarrow \overline{\theta} & & \downarrow \mu(h) \\ Q' & \xrightarrow{\rho'} & \mathcal{H}(W) \end{array}$$

Let ψ, ψ' be Seifert construction for $1 \to \mathbb{Z}^k \to P \to Q \to 1$, ϵ and ρ; $1 \to \mathbb{Z}^k \to P' \to Q' \to 1$, ϵ' and ρ', respectively.

Rigidity. There exists $\hat{h} \in \mathcal{H}^F(\mathbb{R}^k \times W)$ such that $\psi' \circ \theta = \mu(\hat{h}) \circ \psi$.

Proof. This is essentially proved in [CR1, Theorem (8.5)]. From the definition of ψ', the following is commutative.

$$\begin{array}{ccccccccc} 1 & \longrightarrow & \mathbb{Z}^k & \longrightarrow & P' & \longrightarrow & Q' & \longrightarrow & 1 \\ & & \downarrow \epsilon' & & \downarrow \psi' & & \downarrow \phi'_{\epsilon'} \times \rho' & & \\ 1 & \longrightarrow & \mathcal{M} & \longrightarrow & \mathcal{M} \circ (A \times \mathcal{H}) & \longrightarrow & A \times \mathcal{H} & \longrightarrow & 1 \end{array}$$

where $\phi'_{\epsilon'}: Q' \xrightarrow{\phi'} \mathrm{Aut}(\mathbb{Z}^k) \xrightarrow{\epsilon'} \mathrm{Aut}(\mathbb{R}^k)$. Let $\psi'' = \mu(h^{-1}) \circ \psi' \circ \theta$, where $h = (0,1,h) \in \mathcal{H}^F(\mathbb{R}^k \times W)$. We claim that ψ'' is another construction for $1 \to \mathbb{Z}^k \to P \to Q \to 1$, $\epsilon'' = \epsilon' \cdot \hat{\theta}$ and ρ. For, $\mu(h^{-1}) \circ \phi'_{\epsilon'} \circ \overline{\theta} = \phi'_{\epsilon'} \circ \overline{\theta} = \phi_{\epsilon' \circ \hat{\theta}} = \phi_{\epsilon''}$ and $\mu(h^{-1}) \circ \rho' \circ \overline{\theta} = \rho$ since $\rho' \circ \overline{\theta} = \mu(h) \circ \rho$ by the hypothesis. Therefore, $(\phi'_{\epsilon'} \times \rho') \circ \overline{\theta} = \phi_{\epsilon''} \times \rho$. Consequently, we have two constructions ψ and ψ'' for $1 \to \mathbb{Z}^k \to P \to Q \to 1$

and $\rho: Q \to \mathcal{U}(W)$. They use different ϵ, ϵ''. However, as we have seen already, they are same up to conjugation by elements of $\mathcal{M}(W,\mathbb{R}^k) \circ \mathrm{Aut}(\mathbb{R}^k)$. More precisely, there exists $(\lambda,g) \in \mathcal{M}(W,\mathbb{R}^k) \circ \mathrm{Aut}(\mathbb{R}^k)$ such that $\psi'' = \mu(\lambda,g,1) \circ \psi$. As $\psi'' = \mu(h)^{-1} \circ \psi' \circ \theta$, $\psi' \circ \theta = \mu(h) \circ \psi'' = \mu(\lambda h^{-1},g,h) \circ \psi$. Put $(\lambda h^{-1},g,h) = \hat{h} \in \mathcal{U}^F(\mathbb{R}^k \times W)$. Note that g is determined by $\epsilon' \circ \theta = g \circ \epsilon$. □

1.5. Let m be a positive integer. $K = (k_0,k_1,\ldots,k_r)$ is called a <u>partition</u> of m if $\Sigma k_i = m$ and $k_i > 0$ for all $i = 0,1,\ldots,r$. Let W be a space. We will define a subgroup of $\mathcal{U}(\mathbb{R}^m \times W)$ which depends only on K and W. Let $m_i = k_i + \ldots + k_r$ (partial sum). Then $K_i = (k_i,k_{i+1},\ldots,k_r)$ is a partition of m_i (with $K_0 = K$ and $m_0 = m$). Let $\mathcal{U}(W;K_{r+1}) = \mathcal{U}(w)$. (Note that K_{r+1} is an empty set.) Inductively, we let

$$\mathcal{U}(\mathbb{R}^{m_i} \times W;K_i) = \mathcal{M}(\mathbb{R}^{m_{i+1}} \times W,\mathbb{R}^{k_i}) \circ (\mathrm{Aut}(\mathbb{R}^{k_i}) \times \mathcal{U}(\mathbb{R}^{m_{i+1}} \times W;K_{i+1})) .$$

We define an action of $\mathcal{U}(\mathbb{R}^m \times W;K)$ on $\mathbb{R}^m \times W$ by induction again. Suppose we have $\mathcal{U}(\mathbb{R}^{m_{i+1}} \times W;K_{i+1}) \subset \mathcal{U}(\mathbb{R}^{m_{i+1}} \times W)$. Then

$$(\lambda,g,h)(x,w) = (gx + \lambda hw,hw)$$

for $(\lambda,g,h) \in \mathcal{U}(\mathbb{R}^{m_i} \times W;K_i)$ and $(x,w) \in \mathbb{R}^{k_i} \times (\mathbb{R}^{m_{i+1}} \times W) = \mathbb{R}^{m_i} \times W$. This also explains the group structure of $\mathcal{U}(\mathbb{R}^{m_i} \times W;K_i)$ as an iterated semi-direct product. Therefore, $\mathcal{U}(\mathbb{R}^{m_i} \times W;K_i) \subset \mathcal{U}(\mathbb{R}^{m_i} \times W)$. By induction, we have defined $\mathcal{U}(\mathbb{R}^m \times W;K)$ by $\mathcal{U}(\mathbb{R}^{m_0} \times W;K_0)$. Observe that $\mathcal{U}(\mathbb{R}^m \times W;K)$ <u>depends only on the partition</u> K <u>and the space</u> W.

When we talk about a fixed $K = (k_0,k_1,\ldots,k_r)$, it is convenient to abbreviate as

$$\mathcal{U}(\mathbb{R}^{m_i} \times W;K_i) = \mathcal{U}_i$$

$$\mathcal{M}(\mathbb{R}^{m_{i+1}} \times W,\mathbb{R}^{k_i}) = \mathcal{M}_i$$

$$\mathrm{Aut}(\mathbb{R}^{k_i}) = A_i$$

$$\mathcal{M}_i \circ \mathcal{M}_{i+1} \circ \ldots \circ \mathcal{M}_r = \hat{\mathcal{M}}_i$$

$$A_i \times A_{i+1} \times \ldots \times A_r = \hat{GL}_i \ .$$

Then $\hat{\mathcal{M}}_i$ is easily seen to be a normal subgroup of $\mathcal{N}_i$. The quotient $\mathcal{N}_i/\hat{\mathcal{M}}_i$ is $\hat{GL}_i \times \mathcal{N}(W)$. In fact, $\mathcal{N}_i = \hat{\mathcal{M}}_i \circ (\hat{GL}_i \times \mathcal{N}(W))$. We delete the subscript 0 so $\hat{\mathcal{M}} = \hat{\mathcal{M}}_0 = \hat{\mathcal{M}}(\mathbb{R}^m \times W;K)$, $\hat{GL} = \hat{GL}_0 = \hat{GL}(\mathcal{M};K)$ and $\mathcal{N} = \mathcal{N}_0$, etc. Then

$$\mathcal{N}(\mathbb{R}^m \times W;K) = \hat{\mathcal{M}} \circ (\hat{GL} \times \mathcal{N}(W)) = \hat{\mathcal{M}}(\mathbb{R}^m \times W;K) \circ (\hat{GL}(\mathcal{M};K) \times \mathcal{N}(W)) \ .$$

When W is a point, $\mathcal{N}(\mathbb{R}^m \times \text{point};K)$ is isomorphic to $\mathcal{N}(\mathbb{R}^m;K)$, which is a subgroup of $\mathcal{N}(\mathbb{R}^m)$. Note that $\mathcal{N}(\mathbb{R}^m;K)$ <u>depends only on the partition</u> $K = (k_0,k_1,\ldots,k_r)$. Elements of $\mathcal{N}(\mathbb{R}^m \times W;K)$ preserves the fibers in the sense that $\mathbb{R}^m \times w$ is mapped onto $\mathbb{R}^m \times w'$ by an element of $\mathcal{N}(\mathbb{R}^m \times W;K)$. An element of $\hat{\mathcal{M}} \circ \hat{GL}$ moves only along the fibers, i.e., $\mathbb{R}^m \times w$ is mapped onto itself.

1.6. Suppose we have a sequence of extensions

$$\mathfrak{F}(E): \quad \begin{array}{l} 1 \to \mathbb{Z}^{k_0} \to E \to E_1 \to 1 \\ 1 \to \mathbb{Z}^{k_1} \to E_1 \to E_2 \to 1 \\ \quad\vdots \\ 1 \to \mathbb{Z}^{k_r} \to E_r \to E_{r+1} \to 1 \\ 1 \to F \to E_{r+1} \to Q \to 1 \end{array}$$

where F is a finite group. It determines a partition $K = (k_0,k_1,\ldots,k_r)$ of $m = \Sigma k_i$. Suppose also that Q acts on a space W, via $\rho: Q \to \mathcal{N}(W)$, satisfying (*) of (1.2). As soon as we are given a set of embeddings $\epsilon = (\epsilon_0,\epsilon_1,\ldots,\epsilon_r)$, $\epsilon_i: \mathbb{Z}^{k_i} \to \mathbb{R}^{k_i}$, we can construct an action of E on $\mathbb{R}^m \times W$ as follows:

The group E_{r+1} acts on W via $\rho_{r+1}: E_{r+1} \to Q \overset{\rho}{\to} \mathcal{N}(W) = \mathcal{N}(\mathbb{R}^0 \times W;K_{r+1})$. Suppose we have constructed $\psi_{i+1}: E_{i+1} \to \mathcal{N}(\mathbb{R}^{m_{i+1}} \times W;K_{i+1})$ satisfying (*). Using the extension sequence $1 \to \mathbb{Z}^{k_i} \to E_i \to E_{i+1} \to 1$ together with $\epsilon_i: \mathbb{Z}^{k_i} \to \mathbb{R}^{k_i}$ and $\psi_{i+1}: E_{i+1} \to \mathcal{N}(\mathbb{R}^{m_{i+1}} \times W;K_{i+1})$ (given by the induction hypothesis), we can do the construction of (1.3). The result is a homomorphism $\psi_i: E_i \to \mathcal{U}(\mathbb{R}^{m_i} \times W; K_i)$ which satisfies (*) and makes the following diagram

$$\begin{array}{ccccccccc} 1 & \longrightarrow & \mathbb{Z}^{k_i} & \longrightarrow & E_i & \longrightarrow & E_{i+1} & \longrightarrow & 1 \\ & & \downarrow \epsilon_i & & \downarrow \psi_i & & \downarrow \phi_i \times \psi_{i+1} & & \\ 1 & \longrightarrow & \mathcal{M}_i & \longrightarrow & \mathcal{U}_i & \longrightarrow & A_i \times \mathcal{U}_{i+1} & \longrightarrow & 1 \end{array}$$

commutative. (Recall $A_i = \mathrm{Aut}(\mathbb{R}^{k_i})$ and ϕ_i is induced by $\epsilon_i: \mathbb{Z}^{k_i} \to \mathbb{R}^{k_i}$ so that $\phi_i = \phi_{\epsilon_i}$.) This finishes the induction step. Consequently, we have constructed a homomorphism $\psi: E \to \mathcal{U}(\mathbb{R}^m \times W; K)$ as we desired. Such a construction is called an (iterated) <u>Steifert fiber construction for</u> $\mathcal{F}(E)$, ρ <u>and</u> ϵ. <u>Thus we have shown</u>

<u>Existence</u>. For a set of extension sequences

$$1 \to \mathbb{Z}^{k_0} \to E \to E_1 \to 1$$
$$1 \to \mathbb{Z}^{k_1} \to E_1 \to E_2 \to 1$$
$$\vdots$$
$$1 \to \mathbb{Z}^{k_r} \to E_r \to E_{r+1} \to 1$$
$$1 \to F \to E_{r+1} \to Q \to 1$$

(F = finite), a set of embeddings $\epsilon = (\epsilon_0, \epsilon_1, \ldots, \epsilon_r)$: $(\mathbb{Z}^{k_0}, \mathbb{Z}^{k_1}, \ldots, \mathbb{Z}^{k_r}) \to (\mathbb{R}^{k_0}, \mathbb{R}^{k_1}, \ldots, \mathbb{R}^{k_r})$ and an action of Q on W, $\rho: Q \to \mathcal{U}(W)$, satisfying (*) of (1.2), there exists a Seifert fiber construction $\psi: E \to \mathcal{U}(\mathbb{R}^m \times W; K)$, where $K = (k_0, k_1, \ldots, k_r)$.

1.7. Let π be the kernel of the natural quotient homomorphism $E \to Q$ so that

$$1 \to \pi \to E \to Q \to 1$$

is exact. π contains the poly-$\mathbb{Z}$ group Γ = kernel $(E \to E_{r+1})$ of rank m and of finite index. Let Q_0 = kernel (ρ), Q_w = the stabilizer of Q at $w \in W$

as before. Also let E_0, E_w be the complete inverse image of Q_0, Q_w, respectively, in E. Note that Q_0 and Q_w are finite subgroups of Q because of (*). Clearly $Q_0 \subset Q_w$ for all $w \in W$.

<u>Theorem</u>. The action of E on $\mathbb{R}^m \times W$, via $\psi: E \to \mathcal{U}(\mathbb{R}^m \times W;K)$, obtained by the Seifert fiber construction for $\mathcal{J}(E)$, ρ and ϵ, is effective if and only if $C_{E_0}(\Gamma)$ is torsion-free. The action of E on $\mathbb{R}^m \times W$ is free if and only if E_w is torsion-free for all $w \in W$. Moreover, $E\backslash(\mathbb{R}^m \times W)$ is compact if and only if $Q\backslash W$ is compact.

<u>Proof</u>. E_0 = kernel $(E \to \rho(Q))$ maps homomorphically onto E^0_{r+1} = kernel $(E_{r+1} \to Q \to \rho(Q))$ with kernel Γ. Since E^0_{r+1} is finite, $C_{E_0}(\Gamma)$ is free abelian if and only if it is torsion-free (see [LR1]). The group E_0 acts properly discontinuously on $\mathbb{R}^m \times W$ since E does, and, in fact, E_0 acts properly discontinuously on $\mathbb{R}^m \times w$ for each $w \in W$, since E^0_{r+1} acts trivially on W.

For each $w \in W$, the group Γ acts freely on $\mathbb{R}^m \times w$ with $C_{E_0}(\Gamma)$ commuting with the Γ action. Consequently if $C_{E_0}(\Gamma)$ has non-trivial p-torsion say $\mathbb{Z}/p$, then $\mathbb{Z}/p$ fixes a $\mathbb{Z}/p$-acyclic subset of $\mathbb{R}^m \times w$. Since Γ is torsion-free, $C_{E_0}(\Gamma)$ contains $\Gamma \times \mathbb{Z}/p$. The restriction to the $\mathbb{Z}/p$ action on $\mathbb{R}^m \times w$ is trivial for each $w \in W$, by [CR2, A-11]. Therefore $C_{E_0}(\Gamma)$ does not act effectively and hence E would not act effectively.

Conversely now suppose that E does not act effectively, then E contains a finite normal subgroup A which is the stabilizer of all $\mathbb{R}^m \times W$. The group A projects isomorphically into E^0_{r+1} since Γ is torsion-free. Let E_A denote the complete inverse image of A in E. Then $E_A \cong \Gamma \circ A$, a semi-direct product with the splitting A normal in E_A. Consequently A commutes with Γ so that $A \subset C_{E_0}(\Gamma)$. Therefore we have shown that the E action is effective if and only if $C_{E_0}(\Gamma)$ has trivial torsion.

The claim in the Theorem that E acts freely if and only if each extension E_w is torsion-free is similar to the Γ = abelian case, [CR1]. In fact, for each $w \in W$, Q_w stabilizes $w \in W$ and $(E_{r+1})_w$ is the stabilizer of the E_{r+1} action on W. From the commutative diagram of extensions

$$\begin{array}{ccccccccc} & & 1 & & 1 & & & & \\ & & \downarrow & & \downarrow & & & & \\ & & \Gamma & \overset{=}{\to} & \Gamma & & & & \\ & & \downarrow & & \downarrow & & & & \\ 1 & \to & \pi & \to & E_w & \to & Q_w & \to & 1 \\ & & \downarrow & & \downarrow & & \downarrow = & & \\ 1 & \to & F & \to & (E_{r+1})_w & \to & Q_w & \to & 1 \\ & & \downarrow & & \downarrow & & & & \\ & & 1 & & 1 & & & & \end{array}$$

we see that E_w acts properly discontinuously on $\mathbb{R}^m \times w$ by our construction. Moreover, the image of the E-orbit of any (x,w) corresponds to the image of the E_w-orbit through (x,w). E acts freely on the E-orbit of (x,w) if and only if E_w acts freely on $E(x,w)$ for each w. Since E_w is acting properly discontinuously on $\mathbb{R}^m \times w$, it acts freely, if and only if, it is torsion-free by the Smith Theorems. Finally, $E\backslash(\mathbb{R}^m \times W)$ is compact if and only if $Q\backslash W$ is compact follows easily from the construction since each $\pi\backslash(\mathbb{R}^m \times w)$ is compact. This completes the proof of the theorem. □

Remarks. If W is a smooth manifold and if the Q action on W is smooth (i.e., $\rho: Q \to \mathrm{Diff}(W)$), then the construction can be done smoothly. Given $K = (k_0, k_1, \ldots, k_r)$, a partition of m, we define $\mathrm{Diff}(W; K_{r+1}) = \mathrm{Diff}(W)$ and

$$\mathrm{Diff}(\mathbb{R}^{m_i} \times W; K_i) = C(\mathbb{R}^{m_{i+1}}, \mathbb{R}^{k_i}) \circ (\mathrm{Aut}(\mathbb{R}^{k_i}) \times \mathrm{Diff}(\mathbb{R}^{m_{i+1}} \times W; K_{i+1}))$$

inductively. Then clearly, $\mathrm{Diff}(\mathbb{R}^{m_i} \times W; K_i) \subset \mathscr{H}(\mathbb{R}^{m_i} \times W; K_i) \cap \mathrm{Diff}(\mathbb{R}^{m_i} \times W)$. In particular $\mathrm{Diff}(\mathbb{R}^m \times W; K)$ is a subgroup of $\mathrm{Diff}(\mathbb{R}^m \times W)$ which depends only on K and W. If ρ has the image in $\mathrm{Diff}(W)$, then $\mathfrak{F}(E)$ and $\varepsilon = (\varepsilon_0, \varepsilon_1, \ldots, \varepsilon_r)$ give rise to a $\psi: E \to \mathrm{Diff}(\mathbb{R}^m \times W; K)$. This smooth construction also satisfies the theorem in (1.7).

Our construction automatically restricts to a free action of Γ on $\mathbb{R}^m \times w$ for each $w \in W$, with quotient a closed $K(\Gamma,1)$-manifold $\Gamma\backslash(\mathbb{R}^m \times w)$.

When Γ is nilpotent, it is isomorphic to a uniform lattice in a connected, simply connected nilpotent Lie group L, diffeomorphic to $\mathbb{R}^m$. We shall observe the uniqueness of the construction and properties of nilmanifolds, to be discussed in Section 2, enable us to show that the constructed fiber $\Gamma\backslash(\mathbb{R}^m \times w)$ is diffeomorphic to the nilmanifold $\Gamma\backslash L$.

1.8. We turn now to studying the uniqueness and the rigidity of our construction. Let $\mathfrak{F}(E)$ be the series of extension sequences in (1.6), $\rho: Q \to \mathcal{H}(W)$ and $\epsilon = (\epsilon_0, \epsilon_1, \ldots, \epsilon_r)$ as before.

Uniqueness (1). Given $\mathfrak{F}(E)$, ρ and ϵ, the construction $\psi: E \to \mathcal{H}(\mathbb{R}^m \times W;K)$ is unique up to conjugation by elements of $\hat{\mathcal{M}}(\mathbb{R}^m \times W;K)$. In fact, one construction can be deformed to another via conjugations by elements of $\hat{\mathcal{M}}(\mathbb{R}^m \times W;K)$.

If we vary ϵ, we still get uniqueness. This time we have to use a bigger group $\hat{\mathcal{M}} \circ \hat{GL}$.

Uniqueness (2). Given $\mathfrak{F}(E)$ and ρ, let $\psi, \psi': E \to \mathcal{H}(\mathbb{R}^m \times W;K)$ be constructions for $(\mathfrak{F}(E), \rho, \epsilon)$, $(\mathfrak{F}(E), \rho, \epsilon')$, respectively, where ϵ and ϵ' are embeddings of $(\mathbb{Z}^{k_0}, \mathbb{Z}^{k_1}, \ldots, \mathbb{Z}^{k_r})$ into $(\mathbb{R}^{k_0}, \mathbb{R}^{k_1}, \ldots, \mathbb{R}^{k_r})$. Then there exists $(\lambda, G) \in \hat{\mathcal{M}} \circ \hat{GL}$ so that $\psi' = (\lambda, G, 1) \circ \psi$. $G = (g_0, g_1, \ldots, g_r)$ is determined by $\epsilon_i' = g_i \circ \epsilon_1$ $(i = 0, 1, \ldots, r)$.

Now we want to change even the group E, and the action of Q on W. Let E, E' be groups with filtrations $\mathfrak{F}(E)$, $\mathfrak{F}(E')$ as in (1.6). Suppose an isomorphism $\theta: E \to E'$ induces isomorphisms $\hat{\theta}_i: \mathbb{Z}^{k_i} \to \mathbb{Z}^{k_i}$, $\theta_i: E_i \to E_i'$ for $i = 0, 1, \ldots, r+1$ and $\bar{\theta}: Q \to Q'$. Let $\rho, \rho': Q \to \mathcal{H}(W)$ be two actions satisfying (*) and $\rho' = \mu(h) \circ \rho$ for some $h \in \mathcal{H}(W)$.

Rigidity. Let $\psi: E \to \mathcal{H}(\mathbb{R}^m \times W;K)$, $\psi': E \to \mathcal{H}(\mathbb{R}^m \times W;K)$ be constructions for $(\mathfrak{F}(E), \rho, \epsilon)$, $(\mathfrak{F}(E'), \rho', \epsilon')$, respectively. Then there exists $\hat{h} = (\lambda, G, h) \in \mathcal{H}(\mathbb{R}^m \times W;K)$ so that $\psi' \circ \theta = \mu(\hat{h}) \circ \psi$.

Proof (of rigidity). As E_{r+1}, E_{r+1}' act on W via $\psi_{r+1}: E_{r+1} \to Q \xrightarrow{\rho} \mathcal{H}(W)$, $\psi_{r+1}': E_{r+1}' \to Q' \xrightarrow{\rho'} \mathcal{H}(W)$, respectively, and both squares in

$$\begin{array}{ccccc} E_{r+1} & \longrightarrow & Q & \xrightarrow{\rho} & \mathcal{U}(W) \\ \downarrow \theta_{r+1} & & \downarrow \bar{\theta} & & \downarrow \mu(h) \\ E'_{r+1} & \longrightarrow & Q' & \xrightarrow{\rho'} & \mathcal{U}(W) \end{array}$$

commute, we have $\psi'_{r+1} \circ \theta_{r+1} = \mu(h) \circ \psi_{r+1}$. Suppose that we have $\psi_{i+1} \circ \theta_{i+1} = \mu(h_{i+1}) \circ \psi_{i+1}$ for some $h_{i+1} \in \mathcal{U}(\mathbb{R}^{m_{i+1}} \times W;K_{i+1})$. Then we have commutative diagrams

$$\begin{array}{ccccccccc} 1 & \longrightarrow & \mathbb{Z}^{k_i} & \longrightarrow & E_i & \longrightarrow & E_{i+1} & \longrightarrow & 1 \\ & & \downarrow \hat{\theta}_i & & \downarrow \theta_i & & \downarrow \theta_{i+1} & & \\ 1 & \longrightarrow & \mathbb{Z}^{k_i} & \longrightarrow & E'_i & \longrightarrow & E'_{i+1} & \longrightarrow & 1 \end{array} \qquad \begin{array}{ccc} E_{i+1} & \xrightarrow{\psi_{i+1}} & \mathcal{U}(\mathbb{R}^{m_{i+1}} \times W;K_{i+1}) \\ \downarrow \theta_{i+1} & & \downarrow \mu(h_{i+1}) \\ E'_{i+1} & \xrightarrow{\psi'_{i+1}} & \mathcal{U}(\mathbb{R}^{m_{i+1}} \times W;K_{i+1}) \end{array}$$

This is exactly what we need in order to apply the rigidity of (1.4). Therefore we can find $h_i = (\lambda_i, g_i, h_{i+1}) \in \mathcal{M}(\mathbb{R}^{m_{i+1}} \times W;\mathbb{R}^{k_i}) \circ (\mathrm{Aut}(\mathbb{R}^{k_i}) \times \mathcal{U}(\mathbb{R}^{m_{i+1}} \times W;K_{i+1}))$ for which $\psi'_i \circ \theta_i = \mu(h_i) \circ \psi_i$. Note that g_i is determined by $\epsilon'_i \circ \hat{\theta}_i = g_i \circ \epsilon_i$. This finishes the induction step. At the last step, we find $\hat{h} = (\lambda, G, h) \in \hat{\mathcal{M}} \circ (\hat{GL} \times \mathcal{U}(W)$ so that $\psi' \circ \theta = \mu(\hat{h}) \circ \psi$, where $G = (g_0, g_1, \ldots, g_r)$. □

The proof given above proves Uniqueness (2) and the first statement of Uniqueness (1). For the second statement, note that the conjugation $\psi' = \mu(\lambda) \circ \psi$ is by an element λ of $\hat{\mathcal{M}}$. Let λ_r be the last slot of λ as an element of $\hat{\mathcal{M}} = \mathcal{M}_0 \circ \mathcal{M}_1 \circ \cdots \circ \mathcal{M}_r$. Observe that $\mathcal{M}_r = \mathcal{M}(W, \mathbb{R}^{k_r})$ is naturally a subgroup of $\mathcal{U}(\mathbb{R}^m \times W;K)$. Using this extended λ_r, we change ψ to $\psi^{(r)} = \mu(\lambda_r) \circ \psi\colon E \to \mathcal{U}(\mathbb{R}^m \times W;K)$. $\psi^{(r)}$ is easily seen to be a new Seifert construction for $(\mathfrak{F}(E), \rho, \epsilon)$ so that we may compare it with ψ'.

By definition $\psi^{(r)}$ and ψ' induce the same homomorphisms $\psi_r^{(r)} = \psi'_r$ on E_r. By applying the rigidity, we know $\psi' = \mu(**, \lambda_{r-1}, 0) \circ \psi^{(r)}$. Put $\psi^{(r-1)} = \mu(\lambda_{r-1}) \circ \psi^{(r)}$ (again using the extended $\lambda_{r-1} \in \mathcal{U}(\mathbb{R}^m \times W;K)$). Then $\psi'_i = \psi_i^{(r-1)}$ for $i = r-1, r$. We continue this process to get $\psi^{(0)}$. Then

$\psi_i^{(0)} = \psi_i'$ for $i = 0,1,\ldots,r$. This means $\psi^{(0)} = \psi'$. Therefore we have $\psi' = \mu(\lambda_0) \circ \mu(\lambda_1) \circ \ldots \circ \mu(\lambda_r) \circ \psi$ for $\lambda_i \in \mathcal{M}(\mathbb{R}^{m_{i+1}} \times W; \mathbb{R}^{k_i})$.

At each stage the conjugation by λ_i can be parametrized by $t\lambda_i$, $0 \leq t \leq 1$, a 1-parameter family of conjugations $\mu(t\lambda_i)$, so that $\mu(t\lambda_i) \circ \psi^{(i+1)}$ is $\psi^{(i+1)}$ if $t = 0$, $\psi^{(i)}$ if $t = 1$. Let us define

$$\lambda(t) = (r+1)(t-(r-1)/(r+1))\lambda_i \circ \ldots \circ \lambda_r \in \hat{\mathcal{M}}_i$$
$$\text{if } (r-i)/(r+1) \leq t \leq (r-i+1)/(r+1) .$$

Then $\lambda(t)$ $(0 \leq t \leq 1)$ is a 1-parameter family in $\hat{\mathcal{M}}$. Put $\psi(t) = \mu(\lambda(t)) \circ \psi$. Then for each $0 \leq t \leq 1$, $\psi(t): E \to \mathcal{A}(\mathbb{R}^m \times W;K)$ is a Seifert construction for $(\mathfrak{F}(E),\rho,\epsilon)$, and clearly $\psi(0) = \psi$, $\psi(1) = \psi'$. □

Any Seifert construction for $(\mathfrak{F}(E),\rho)$ (using some ϵ) will be called a <u>Seifert construction associated with</u> $(\mathfrak{F}(E),\rho)$. By Uniqueness (1) and (2), we do not have to specify which ϵ we use, as long as we view two constructions which are conjugate to each other by an element of $\hat{\mathcal{M}} \circ \hat{GL}$ as the "same".

As we mentioned in (1.7), the smooth versions of the existence (1.6) and Theorem (1.7) are true. With exactly the same modification (replacing $\mathcal{M}$ by C, etc.), the uniqueness and rigidity are true for smooth constructions also.

§2. The Existence of the Seifert Construction for $1 \to \pi \to E \to Q \to 1$ (π = torsion-free virtually poly-$\mathbb{Z}$).

2.1. First we shall analyze the map $X = E\backslash(\mathbb{R}^m \times W) \xrightarrow{\rho} Q\backslash W$ induced from our Seifert construction for the given data $(\mathfrak{F}(E),\rho,\epsilon)$ as in (1.6). This mapping is called a <u>Seifert fibering</u> with typical fiber $\pi\backslash\mathbb{R}^m$. Since π is virtually poly-$\mathbb{Z}$, $\pi\backslash\mathbb{R}^m$, the typical fiber, will be a closed $K(\pi,1)$-manifold if π is torsion-free. We shall observe that $X \to Q\backslash W$ actually factors through a series of Seifert fiberings having tori as typical fibers. This sequence of course reflects the family $\mathfrak{F}(E)$. Of special interest is the case when (Q,W) is a smooth action on a smooth manifold. This will insure that the constructed action $(E,\mathbb{R}^m \times W)$ is smooth. If the E action is free, X

will be a smooth manifold, and moreover if W is contractible $M(E) = X = E\backslash(\mathbb{R}^m \times W)$ is a $K(E,1)$-manifold.

On the other hand, if we are just given simply an exact sequence $1 \to \pi \to E \to Q \to 1$, with $\rho: Q \to \mathcal{H}(W)$ defining a properly discontinuous action on W and π is a virtually poly-$\mathbb{Z}$ group, we produce various families of filtrations $\mathfrak{F}$ which yield actions of E on $\mathbb{R}^m \times W$. The filtrations arise from nilpotent normal (in E) subgroups N of π for which π/N is virtually free abelian. It is then shown how one finds such N and that the construction only depends on the extension, ρ and the choice of N.

The goal of this section then is achieved in the Theorem of (2.5) where to each such extension with E torsion-free and W contractible, a $K(E,1)$-manifold is constructed.

2.2. For given data $\mathfrak{F}(E)$, ρ and ε as in (1.6), we have constructed an action of E on $\mathbb{R}^m \times W$, $\psi: E \to \mathcal{H}(\mathbb{R}^m \times W;K)$, called a Seifert construction associated with $(\mathfrak{F}(E),\rho,\varepsilon)$. Let $\pi = \text{kernel } (E \to Q)$ as in (1.7). There is a commutative diagram of equivariant maps

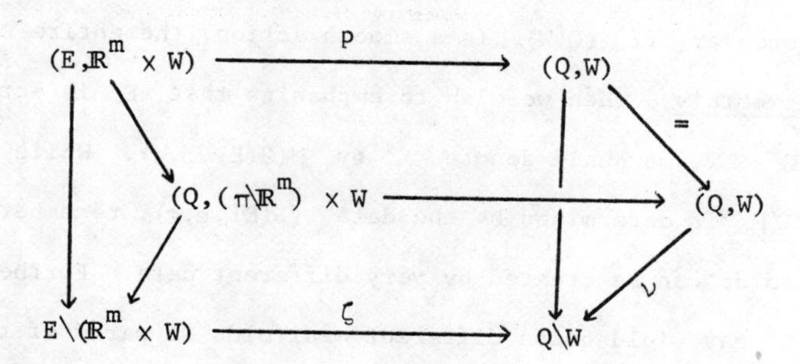

Let us denote $E\backslash(\mathbb{R}^m \times W)$ by X. For each $x \in X$, we may choose $w \in W$ so that $\nu(w) = w^*$. Then $\zeta^{-1}(w^*)$ is called the <u>fiber</u> over w^*. It can be described as $E_w\backslash(\mathbb{R}^m \times w) = Q_w\backslash(\pi\backslash(\mathbb{R}^m \times w))$. That is, $\zeta^{-1}(w^*)$ is the quotient of $\mathbb{R}^m \times w$ by the action of the group E_w, or equivalently, the quotient of the closed $K(\pi,1)$ manifold $\pi\backslash(\mathbb{R}^m \times w)$ (provided π is torsion-free) by the finite group Q_w. If the action of E_w is free, then $\zeta^{-1}(w^*)$ is a closed manifold which is finitely covered by $\pi\backslash(\mathbb{R}^m \times w)$. The mapping ζ is called a <u>Seifert fibering</u> with <u>typical fiber</u> the $K(\pi,1)$ manifold $\pi\backslash\mathbb{R}^m$.

The space X together with $\zeta: X \to Q\backslash W$ is called a Seifert fiber space, and $Q\backslash W$ is called its base. If $Q_w \neq 1$, the fiber $\zeta^{-1}(w^*)$ is called a singular or exceptional fiber and if $Q_w = 1$, the fiber $\zeta^{-1}(w^*)$ is called an ordinary or typical fiber. It is quite possible that there are no typical fibers (that is, all fibers are exceptional).

Associated with X is a sequence of Seifert fiberings:

$$X = X_0 \xrightarrow{\zeta_0} X_1 \xrightarrow{\zeta_1} X_2 \xrightarrow{\zeta_2} \cdots \longrightarrow X_r \xrightarrow{\zeta_r} E_{r+1}\backslash W$$

where $\zeta_0, \zeta_1, \ldots, \zeta_r$ all have tori as typical fibers. These fiberings are a result of the iterative procedure and the composite $\zeta = \zeta_0 \circ \cdots \circ \zeta_r$ has the compact $K(\Gamma,1)$-manifold $\Gamma\backslash\mathbb{R}^m$ as a typical fiber. However, as $E_{r+1}\backslash W = Q\backslash W$, we can say $X \xrightarrow{\zeta} Q\backslash W$ has the compact $K(\pi,1)$-manifold $\pi\backslash\mathbb{R}^m$ as a typical fiber when π is torsion-free.

Certain situations of the Seifert construction are of special interest to us. If W is a manifold or manifold factor (i.e., $\mathbb{R}^m \times W$ is a topological manifold) and the action of E is free (unbranched) then $E\backslash(\mathbb{R}^m \times W)$ is a manifold. Moreover, if (Q,W) is a smooth action, the entire construction can be done smoothly. When we wish to emphasize that E is acting freely on a manifold $\mathbb{R}^m \times W$, we shall denote X by $M(\mathcal{F}(E),\rho,\epsilon)$. While the construction of M is determined by the data $(\mathcal{F}(E),\rho,\epsilon)$, remember that diffeomorphic manifolds can be created by very different data. Furthermore, the same group E may yield quite different manifolds if parts of the data such as $\rho: Q \to \mathcal{H}(W)$ or the set of extension sequences $\mathcal{F}(E)$ is altered (for a fixed group E).

2.3. We examine the series of extension sequences $\mathcal{F}(E)$ in (1.6):

$$\mathcal{F}(E) \qquad \begin{array}{l} 1 \to \mathbb{Z}^{k_0} \to E \to E_1 \to 1 \\ 1 \to \mathbb{Z}^{k_1} \to E_1 \to E_2 \to 1 \\ \qquad \vdots \\ 1 \to \mathbb{Z}^{k_r} \to E_r \to E_{r+1} \to 1 \\ 1 \to F \to E_{r+1} \to Q \to 1 \end{array}$$

Let π = kernel $(E \to Q)$. Then π is a virtually poly-$\mathbb{Z}$ group. (A group is <u>virtually</u> $\mathcal{P}$ if it contains a normal subgroup of finite index which is $\mathcal{P}$), and $1 \to \pi \to E \to Q \to 1$ is exact. Let Γ^i = kernel $(E \to E_1 \to \dots \to E_i)$ for $i = 1,2,\dots,r+1$, and let $\Gamma^0 = 1$. Then

$$\mathcal{F}(E) \qquad 1 = \Gamma^0 \subset \Gamma^1 \subset \Gamma^2 \subset \dots \subset \Gamma^{r+1} \subset \pi \subset E$$

is a filtration of E such that Γ^i and π are normal in E and Γ^{i+1}/Γ^i is free abelian $(\cong \mathbb{Z}^{k_i})$, π/Γ^{r+1} is finite $(\cong F)$, $E/\pi = Q$.

Conversely if we have such a filtration, $1 = \Gamma^0 \subset \Gamma^1 \subset \Gamma^2 \subset \dots \subset \Gamma^{r+1} \subset \pi \subset E$, we obviously get a series of extension sequences. Therefore we denote the filtration of E by $\mathcal{F}(E)$ again.

2.4. For a Seifert construction for $1 \to \pi \to E \to Q \to 1$ (where π is a torsion-free, virtually poly-$\mathbb{Z}$ group) and $\rho\colon Q \to \mathcal{H}(W)$, what we need is a filtration for E as above. However, we shall see that a nilpotent normal (in E) subgroup N of π for which π/N is virtually free abelian is enough to get a filtration.

<u>Lemma</u>. If N is as above, there exists a normal (in E) subgroup Γ of π such that π/Γ is finite, $N \subset \Gamma$ and Γ/N is free abelian.

<u>Proof</u>. Let $\mathbb{Z}^s \subset \pi/N$ be a free abelian subgroup with $(\pi/N)/\mathbb{Z}^s = F$ being finite, say of order q. Let $\pi(q)$ be the characteristic group of π generated by $\{x^q \mid x \in \pi\}$. Then $\pi(q)/\pi(q) \cap N \cong \mathbb{Z}^s$. Now let $\Gamma = \pi(q)\cdot N$, the subgroup of π generated by $\pi(q)$ and N. Clearly Γ is normal in E, π/Γ is finite, $N \subset \Gamma$ and $\Gamma/N \cong \pi(q)/\pi(q) \cap N \cong \mathbb{Z}^s$. $\square$

Therefore, if we have a nilpotent normal (in E) subgroup N of π for which π/N is virtually free abelian, we have a Γ as above. Consequently, we have

$$1 \to N \to E \to E/N \to 1$$
$$1 \to \Gamma/N \to E/N \to E/\Gamma \to 1$$
$$1 \to \pi/\Gamma \to E/\Gamma \to Q \to 1$$

where $\Gamma/N \cong Z^s$, $\pi/\Gamma \cong F$ (finite). For N, we can always use its ascending central series. More precisely, let $\mathbb{Z}^0(N) = 1$, $\mathbb{Z}^{i+1}(N)$ be the subgroup of N corresponding to the center of $N/\mathbb{Z}^i(N)$. As N is nilpotent, $N/\mathbb{Z}^{r-1}(N)$ is free abelian for some r. The higher centers $\mathbb{Z}^i(N)$ of N are characteristics in N and hence normal in E. Thus we have

$$1 = \mathbb{Z}^0(N) \subset \mathbb{Z}^1(N) \subset \ldots \subset \mathbb{Z}^{r-1}(N) \subset N \subset \Gamma \subset \pi \subset E$$

satisfying all the conditions for $\mathfrak{F}(E)$.

There are possibly many Γ's which satisfy the above lemma. We would like to examine the Seifert constructions for the filtration $\mathfrak{F}(E) : N \subset \Gamma \subset \pi \subset E$. It should be understood that we are using the ascending central series for N, below N, in order to have a complete filtration $\mathfrak{F}(E)$.

<u>Proposition</u>. Let $1 \to \pi \to E \to Q \to 1$ be exact, where π is torsion-free, virtually poly-$\mathbb{Z}$ and $\rho: Q \to \mathcal{N}(W)$ be given. The Seifert construction for $N \subset \Gamma \subset \pi \subset E$ is independent of Γ.

<u>Proof</u>. It is enough to show that the Seifert construction for E/N is independent of the choice of Γ, because once we know the action of E/N on $\mathbb{R}^s \times W$ (s = rank of E/N), then the construction for $1 \to N \to E \to E/N \to 1$ is determined by the ascending central series of N. We first show that the subgroup $C_{\pi/N}(\Gamma/N)$ of π/N is characteristic and independent of Γ. Let Γ' be another normal subgroup of π for which Γ'/N is free abelian. Then $A = (\Gamma \cap \Gamma')/N$ is a free abelian normal subgroup of π/N of rank s. We wish to see that $C_{\pi/N}(A) = C_{\pi/N}(\Gamma/N)$. (Then it will follow that $C_{\pi/N}(\Gamma/N) = C_{\pi/N}(\Gamma'/N)$.) Since $A \subset \Gamma/N$, $C_{\pi/N}(A) \supset C_{\pi/N}(\Gamma/N)$. Suppose $e \in C_{\pi/N}(A)$. Let p be the index of A in Γ/N. For any $z \in \Gamma/N$, $z^p \in A$. Now we have

$$z^p = ez^pe^{-1} = (eze^{-1})^p .$$

Since both z and eze^{-1} are elements of Γ/N, $z^p = (eze^{-1})^p$ implies that $z = eze^{-1}$. (Recall that Γ/N is abelian), so $e \in C_{\pi/N}(\Gamma/N)$. Therefore $C_{\pi/N}(A) = C_{\pi/N}(\Gamma/N)$.

Now we show that $C_{\pi/N}(A)$ is characteristic in π/N. Let f be any automorphism of π/N. We want $f(e) \in C_{\pi/N}(A)$ for any $e \in C_{\pi/N}(A)$. $[e,A] = 1$ implies $[f(e),f(A)] = 1$ so that $f(e) \in C_{\pi/N}(f(A))$. However, $f(A)$ is another free abelian normal subgroup of π/N of rank s, and hence $C_{\pi/N}(f(A)) = C_{\pi/N}(A)$. Consequently, we have $f(e) \in C_{\pi/N}(A)$.

By [LR1, Fact 2], the set of all torsion elements K of $C_{\pi/N}(A)$ is a finite characteristic subgroup of $C_{\pi/N}(A)$ (and hence characteristic in π/N) so that $C_{\pi/N}(A)/K$ is free abelian of rank s. Moreover, $(\pi/N)/K$ is an abstract crystallographic group. In fact, $C_{\pi/N}(A)/K$ is the unique maximal abelian subgroup of $(\pi/N)/K$. Note that K also does not depend on Γ.

By [L1, (1.4)] the subgroup K acts on $\mathbb{R}^s \times W$ trivially, whatever Γ we may use. Furthermore, for any Γ, Γ/N injects into $C_{\pi/N}(A)$ and the bigger group $C_{\pi/N}(A)$ maps into $\mathbb{R}^s \subset \mathcal{M}(W,\mathbb{R}^s)$. By the uniqueness in (1.3), a construction using Γ is the same as the construction via
$1 \to K \to E/N \to (E/N)/K \to 1$, $1 \to C_{\pi/N}(A)/K \to (E/N)/K \to (E/N)/C_{\pi/N}(A) \to 1$
together with $(E/N)/C_{\pi/N}(A) \to Q \to \mathcal{H}(W)$. This completes the proof of the proposition. □

In conclusion, to have a Seifert construction for $1 \to \pi \to E \to Q \to 1$ (π is a torsion-free virtually poly-$\mathbb{Z}$ group) and $\rho: Q \to \mathcal{H}(W)$, it is enough to specify the nilpotent subgroup N. We shall denote this as <u>the data</u> (N,π,E,Q,W), where N is a normal (in E) subgroup of π with E/N virtually free abelian.

2.5. <u>Model aspherical manifolds</u>. Let $1 \to \pi \to E \to Q \to 1$ (where π is a torsion-free virtually poly-$\mathbb{Z}$ group) be exact, and $\rho: Q \to \mathcal{H}(W)$ satisfy the condition (*). <u>If we have a nilpotent normal</u> (in E) <u>subgroup</u> N of π for which π/N is virtually free abelian, then the Seifert construction for the data (N,π,E,Q,W) enables us to construct an action of E on $\mathbb{R}^m \times W$ (m = rank of π). If W is a contractible manifold factor, then the action of E is free if and only if E is torsion-free. Consequently

$M(N,\pi,E,Q,W) = E\backslash(\mathbb{R}^m \times W)$ is a $K(E,1)$-manifold. It will be closed aspherical if $Q\backslash W$ is compact and W has empty boundary. The closed $K(E,1)$-manifold $M(N,\pi,E,Q,E)$ will be called a <u>model aspherical manifold associated with the data</u> (N,π,E,Q,W). The important point here is the existence of model aspherical manifolds for each possible torsion-free extension of a virtually poly-$\mathbb{Z}$ group π by Q. Moreover, the model manifolds exhibit the strong geometric property of being a Seifert fibering over Q/W. We emphasize that the model manifold $M(N,\pi,E,Q,W)$ is unique up to homeomorphism, for a given data (N,π,E,Q,W), by the uniqueness theorem (2) in (1.8). If we choose different N', then we cannot apply the uniqueness anymore, and therefore we do not know whether $M(N,\pi,E,Q,W)$ is homeomorphic to $M(N',\pi,E,Q,W)$ or not, except for some special cases (e,g., E itself is virtually nilpotent or a poly-$\mathbb{Z}$ group).

<u>Theorem</u>. Let $1 \to \pi \to E \to Q \to 1$ be exact, where π is a torsion-free, virtually poly-$\mathbb{Z}$ group. Suppose Q acts on a contractible manifold W properly discontinuously so that $Q\backslash W$ is compact. Then there exists a Seifert fibering $E\backslash(\mathbb{R}^m \times W) \to Q\backslash W$ with typical fiber a $K(\pi,1)$-manifold $\pi\backslash\mathbb{R}^m$ (m = rank of π). If E is torsion-free, $E\backslash(\mathbb{R}^m \times W)$ is a closed $K(E,1)$-manifold. If W is a smooth manifold and Q acts smoothly, then $E\backslash(\mathbb{R}^m \times W)$ can be constructed to be a smooth manifold.

<u>Proof</u>. The only thing that we need to do for our construction is to find a nilpotent normal (in E) subgroup N of π such that π/N is virtually free abelian. It is well-known that π contains a characteristic subgroup Γ of finite index which is an extension of a nilpotent group by a free abelian group and ${}^n\Gamma = {}^n\pi$ (nilradicals). We use $N = {}^n\pi$ as our N. As N contains any nilpotent normal subgroup of Γ, Γ/N is abelian. The free abelian part of Γ/N may not be normal in π/N. However, by an argument similar to the proof of Lemma (2.4), one can show that π/N contains a free abelian normal subgroup of finite index. Thus we have shown that the nilradical of π satisfies the conditions for N. □

Corollary. For a torsion-free, virtually poly-$\mathbb{Z}$ group π, there exists a smooth closed $K(\pi,1)$-manifold. □

Remarks. (1) Auslander and Johnson [AJ] have obtained the same result earlier as the Corollary using a "predivisible" subgroup of π. At this moment, the present authors do not know if the manifold constructed here is diffeomorphic to the one by [AJ]. However, we will treat the geometric realization problem for the manifolds constructed in [AJ] in (6.5).

(2) With the notation of (2.4), the manifold constructed in the Theorem can be denoted by $M({}^n\pi,\pi,E,Q,W)$. If there is no confusion likely, we denote it simply by $M(E)$, the model aspherical manifold. However, one should remember that N does not have to be the nilradical of π. In fact, when π is a uniform lattice of a connected, simply connected solvable Lie group R, and if we want a Seifert construction so that its typical fiber is the homogeneous space $\pi\backslash R$, we are forced to use N different from ${}^n\pi$. See (4.5).

(3) Finally we make the obvious but nonetheless important observation that finding $N = {}^n\Gamma = {}^n\pi$ did not depend upon the contractibility of W nor on the π being torsion-free. Consequently, the existence of $1 \to \pi \to E \to Q \to 1$ with π virtually poly-Z, and $\rho: Q \to \mathcal{H}(W)$ enables one to find families $\mathcal{F}(E)$ depending only upon a choice of N, which yields a Seifert construction and a Seifert fibering $X \to Q\backslash W$. Thus $\rho: Q \to \mathcal{H}(W)$ just satisfying (*) of Section 1, yields a Seifert construction for the data (N,π,E,Q,W).

§3. A Generalization of Bieberbach's Theorem to Infranilmanifolds

3.1. One of Bieberbach's theorems states that homotopy equivalent flat Riemannian manifolds are affinely diffeomorphic. In this section we use the rigidity of the Seifert construction to generalize the above result to: Homotopy equivalent infranilmanifolds are diffeomorphic. A crucual ingredient is that the affine group $A(L) = L \circ \mathrm{Aut}(L)$ of a connected, simply connected nilpotent Lie group is really a subgroup of $\mathrm{Diff}(\mathbb{R}^n;K)$. The result also applies to show that the typical fiber of the Seifert construction for virtually nilpotent normal subgroups are homeomorphic to infranilmanifolds.

3.2. A subgroup E of $E(n) = \mathbb{R}^n \circ O(n)$, the group of rigid motions of $\mathbb{R}^n$, which operates without accumulation points and with compact fundamental domain is called an n-dimensional crystallographic group. A torsion-free crystallographic group is a Bieberbach group. Therefore, a Bieberbach group acts on $\mathbb{R}^n$ freely and properly discontinuously. Thus, Bieberbach groups are exactly the fundamental groups of flat Riemannian manifolds. These groups are completely understood algebraically by the celebrated work of Bieberbach.

Theorem A. Let E be an n-dimensional crystallographic group. Then $E \cap \mathbb{R}^n$ is a uniform lattice of $\mathbb{R}^n$ and $E/E \cap \mathbb{R}^n$ is finite.

Theorem B. An abstract group is isomorphic to a crystallographic group if and only if it contains a free abelian normal subgroup of finite index which is maximal abelian.

Theorem C. Let $\theta: E \to E'$ be an isomorphism between two crystallographic groups (say, of dimension n). Then θ is the conjugation by an element of $A(n) = \mathbb{R}^n \circ GL(n,\mathbb{R})$, the group of affine motions.

3.3. Let us review our Seifert fiber construction in the special case $(N,\pi,E,Q,W) = (\mathbb{Z}^n,\mathbb{Z}^n,E,Q,\text{point})$. Then Q is necessarily a finite group. (Recall the condition (*) of (1.2).) The Seifert fiber construction maps E into $\mathcal{A}^F(\mathbb{R}^n \times \text{point}) = \mathbb{R}^n \circ \text{Aut}(\mathbb{R}^n) = A(n)$, the group of affine motions. By a proper choice of basis for $\mathbb{R}^n$, we can arrange that E maps into $E(n)$, and so the image of E acts on $\mathbb{R}^n$ as a group of isometries of $\mathbb{R}^n$. The action is effective if and only if $C_E(\mathbb{Z}^n)$ is torsion-free. When $C_E(\mathbb{Z}^n)$ is torsion-free, it is a free abelian (see [LR1]) normal subgroup which is maximal abelian. This recovers Theorem B above. The action of E is free if and only if E is torsion-free. In this case E becomes a Bieberbach group, and $E\backslash\mathbb{R}^n$ is a closed flat Riemannian manifold, and all flat Riemannian manifolds arise in exactly this manner.

A consequence of this fact is that in a Seifert construction with typical fiber a torus, each of the fibers is a quotient of the torus by the action of some Q_w. This finite group acts on $T^n \times w$ as a group of affine motions and after choosing a basis for $\mathbb{R}^n$ as a group of euclidean motions. The orbit space $Q_w \backslash (T^n \times w)$ is the (perhaps exceptional) fiber over the image of w in $Q \backslash W$.

Let E be a crystallographic group (of dimension n). Then by Theorem A, $E \cap \mathbb{R}^n$ is a free abelian normal subgroup of E, of finite index. Certainly $E \hookrightarrow E(n)$ is a Seifert construction for the exact sequence $1 \to \mathbb{Z}^n \to E \to E/\mathbb{Z}^n \to 1$ (where $\mathbb{Z}^n = E \cap \mathbb{R}^n$) and $(E/\mathbb{Z}^n, \text{point})$ as our (Q,W). Let $\theta: E \to E'$ be an isomorphism of E onto another crystallographic group E'. Since W is a point, $\rho: E/E \cap \mathbb{R}^n \to \mathcal{H}(\text{point})$ and $\rho': E'/E' \cap \mathbb{R}^n \to \mathcal{H}(\text{point})$ automatically satisfy the condition $\rho' = \mu(1) \circ \rho$. Then, by the rigidity (1.8), there exists $h \in \mathcal{H}^F(\mathbb{R}^n \times \text{point}) = A(n)$ such that $\psi' = \mu(h) \circ \psi$ ($\psi: E \to E(n)$ and $\psi': E' \hookrightarrow E(n)$ are original embeddings of E and E'.) As $\psi' = \theta \circ \psi$ from the beginning, we have $\theta = \mu(h)$. Consequently, we have shown that any isomorphism between two crystallographic groups is conjugation by an affine motion, which is the Theorem C above.

3.4. In 1960, L. Auslander generalized these theorems to nilpotent groups. For convenience, we define some terminology. Let L be a connected and simply connected nilpotent Lie group. $A(L) = L \circ \mathrm{Aut}(L)$ is called the _affine group_ of L, and it acts on L by $(z,\alpha)\cdot x = z\cdot\alpha(x)$. Note that the L-factor acts on L as left translations. Consider the linear connection D on L defined by the left invariant vector fields. It is known [KT] that $A(L)$ is the group of connection preserving self-diffeomorphisms of L. A discrete uniform subgroup E of $L \circ C$, where C is any compact subgroup of $\mathrm{Aut}(L)$, is called an A.C. group (= _almost crystallographic group_). If we denote the projection $A(L) \to \mathrm{Aut}(L)$ by c, then $c(E)$ is, in fact, finite (see A' below). A torsion-free A.C. group is an A.B. group (= _almost_

Bieberbach group). Thus the A.B. groups are exactly the fundamental groups of infranilmanifolds. The term "almost" is motivated from the fact that the class of infranilmanifolds is identical with the class of almost flat manifolds up to diffeomorphisms. See [G], [FH] and [Ruh]. The following have been claimed by Auslander [A1].

A'. Let E be an A.C. group. Then $E \cap L$ is a uniform lattice in L and $E/E \cap L$ is finite.

B'. $E \cap L$ is the maximal normal nilpotent subgroup of E.

C'. Let $\theta: E \to E'$ be an isomorphism between two A.C. groups. Then θ can be uniquely extended to a continuous automorphism θ^* of $L \circ C$ onto itself, where C is any compact subgroup of $\mathrm{Aut}(L)$ containing $c(E)$ and $c(E')$.

Unfortunately, C' is not true as it stands, even for Bieberbach groups. We shall state and prove a correct generalization of Theorem C in (3.7).

3.5. We want to characterize A.C. groups algebraically. An essential part of the following has been proved by Auslander [A1].

<u>Proposition</u>. For a finitely generated group E, the following are equivalent.

(a) E is isomorphic to an A.C. group,

(b) E contains a (unique) torsion-free maximal normal nilpotent subgroup of finite index which is maximal nilpotent, and

(c) E contains a torsion-free nilpotent normal subgroup Δ of finite index such that $C_E(\Delta)$ is torsion-free.

<u>Proof</u>. (a $\Rightarrow$ b) is essentially proved by Auslander [A_1, proposition 2]. (b $\Rightarrow$ c) is easy. Observe that such a subgroup in (b) is necessarily characteristic in E. For (c $\Rightarrow$ a), recall the following well-known construction. Let L be a connected, simply connected nilpotent Lie group containing Δ as a uniform lattice. (Such an L exists by Malcev.) Using the unique automorphism extension property of the pair (Δ, L) (again by Malcev), we form a "pushout" diagram

$$\begin{array}{ccccccccc} 1 & \to & \Delta & \to & E & \to & F & \to & 1 \\ & & \cap & & \downarrow & & \downarrow & & \\ 1 & \to & L & \to & L\circ F & \to & F & \to & 1 \end{array}$$

Note that the bottom sequence splits because $F = E/\Delta$ is finite and L is divisible. This splitting yields a homomorphism $F \to \mathrm{Aut}(L)$, whence a homomorphism $f: E \to L \circ F \to L \circ \mathrm{Aut}(L) = A(L)$.

It remains to show that f is injective. This is a consequence of the following useful easily checked fact.

<u>Lemma</u>. Let $1 \to A \to B \to C \to 1$ be exact, where A is torsion-free and F is finite. Let f be a homomorphism of B into a group H such that $f|_A$ is injective. If $C_B(A)$ is torsion-free, then f itself is injective. □

3.6. Let L be a connected, simply connected nilpotent Lie group of dimension n. We would like to show that the action of $A(L)$ on L is via a Seifert construction. More precisely, we want to show that $A(L) \subset \mathrm{Diff}(\mathbb{R}^n;K)$ (for some partition K of n which depends only on L). Let

$$1 = Z^0(L) \subset \ldots \subset Z^r(L) \subset Z^{r+1}(L) = L$$

be the ascending central series for L. Then $Z^{i+1}(L)/Z^i(L)$ is an abelian Lie group, say of dimension k_i. Therefore the above filtration gives rise to a partition $K = (k_0, k_1, \ldots, k_r)$ of n.

Let $L_i = L/Z^i(L)$. Then L_i is again a connected, simply connected nilpotent Lie group diffeomoprhic to R^{n_i} $(n_i = k_i + k_{i+1} + \ldots + k_r)$. Therefore,

$$\hat{C}(L) = C(L_1, z(L)) \circ C(L_2, z(L_1)) \circ \ldots \circ C(L_{r+1}, z(L_r))$$

and consequently,

$$\mathrm{Diff}(L;K) = \mathrm{Diff}(\mathbb{R}^n;K) = \hat{C}(L) \circ \hat{GL}(K)$$

where $\hat{GL}(K) = \mathrm{Aut}(\mathbb{R}^{k_0}) \times \mathrm{Aut}(\mathbb{R}^{k_1}) \times \ldots \times \mathrm{Aut}(\mathbb{R}^{k_r})$.

<u>Lemma</u>. $L_i \subset \hat{C}(L_i)$ for each i. In particular, $L \subset \hat{C}(L)$.

<u>Proof</u>. We use induction. Certainly $L_r = C(L_{r+1}, z(L_r))$, since $L_{r+1} = 1$ and $z(L_r) = L_r$. Suppose that $L_{i+1} \subset \hat{C}(L_{i+1})$. The exact sequence $1 \to z(L_i) \to L_i \to L_{i+1} \to 1$ represents an element of $H^2(L_{i+1}, z(L_i))$. We consider L_i as $z(L_i) \times L_{i+1}$ with group multiplication

$$(x,w)(x',w') = (x + x' + f(w,w'), ww')$$

where $f\colon L_{i+1} \times L_{i+1} \to z(L_i)$ is a 2-cocycle representing the element of $H^2(L_{i+1}, z(L_i))$ mentioned above. We use $\ell\colon L_j \hookrightarrow \mathrm{Diff}(L_j)$ for the action of L_j on itself by left translations, for all j. By the induction hypothesis, $\ell(w) \in \hat{C}(L_{i+1})$. The multiplication of $L_i = z(L_i) \times L_{i+1}$ confirms that

$$\ell(x,w) = (\ell(x) + f(w,-) \circ \ell(w^{-1}), 1, \ell(w))$$

as an element of $C(L_{i+1}, z(L_i)) \circ (\mathrm{Aut}(z(L_i)) \times \hat{C}(L_{i+1}))$. (Recall that (λ, g, h) acts on (x',w') by $(\lambda,g,h)(x',w') = (g(x') + \lambda h(w'), h(w'))$.) This shows that $L_i \subset \hat{C}(L_i)$, completing the induction step. □

<u>Proposition</u>. $A(L) \subset \mathrm{Diff}(L;K)$. Moreover,

$$\begin{array}{ccccccccc}
1 & \longrightarrow & L & \longrightarrow & A(L) & \longrightarrow & \mathrm{Aut}(L) & \longrightarrow & 1 \\
 & & \cap & & \cap & & \downarrow & & \\
1 & \longrightarrow & \hat{C}(L) & \longrightarrow & \mathrm{Diff}(L;K) & \longrightarrow & \hat{GL}(K) & \longrightarrow & 1
\end{array}$$

is commutative.

<u>Proof</u>. As $A(L)$ and $\mathrm{Diff}(L;K)$ are subgroups of $\mathrm{Diff}(L)$, it suffices to identity them. We use induction again. Certainly, $\mathrm{Aut}(L_r) \subset \mathrm{Diff}(L_r) = \mathrm{Diff}(L_r; K_r)$. Assume that $\mathrm{Aut}(L_{i+1}) \subset \mathrm{Diff}(L_{i+1}; K_{i+1})$. Let $\alpha \in \mathrm{Aut}(L_i)$. Then α induces unique automorphisms of $z(L_i)$ and of L_{i+1}, say $\hat{\alpha}$ and $\bar{\alpha}$, respectively. We still condier L_i as $z(L_i) \times L_{i+1}$, with group multiplication as before (using the 2-cocycle f). We define a map $\lambda\colon z(L_i) \times L_{i+1} \to z(L_i)$ by

$$\alpha(x,w) = (\hat{\alpha}(x) + \lambda(x,w), \bar{\alpha}(w))$$

for all $(x,w) \in L_i$. As α restricts to $\hat{\alpha}$ on $z(L_i)$, $\lambda(x,1) = 0$ for all $x \in z(L_i)$. This readily implies that λ is independent of x (assuming that f is normalized so that $f(1,w) = 0 = f(w,1)$ for all $w \in L_{i+1}$) so that $\lambda: L_{i+1} \to z(L_i)$. The triple $\lambda, \hat{\alpha}, \bar{\alpha}$ satisfy the relation

$$f(\bar{\alpha}(w_1), \bar{\alpha}(w_2)) = \lambda(w_1 w_2) - (\lambda(w_1) + \lambda(w_2)) + \hat{\alpha}(f(w_1, w_2)) .$$

Conversely, any triple satisfying this relation determines an automorphism $\alpha \in \mathrm{Aut}(L_i)$ uniquely.

By the induction hypothesis, $\bar{\alpha} \in \mathrm{Aut}(L_{i+1}) \subset \mathrm{Diff}(L_{i+1};K_{i+1})$. Therefore, $\alpha = (\lambda \circ \bar{\alpha}^{-1}, \hat{\alpha}, \bar{\alpha}) \in C(L_{i+1}, z(L_i)) \circ (\mathrm{Aut}(z(L_i)) \times \mathrm{Diff}(L_{i+1};K_{i+1})) = \mathrm{Diff}(L_i;K_i)$. Thus we have proved that $\mathrm{Aut}(L_i) \subset \mathrm{Diff}(L_i;K_i)$.

As $L_i \subset \hat{C}(L_i) \subset \mathrm{Diff}(L_i;K_i)$ by the lemma, we have $A(L_i) = L_i \circ \mathrm{Aut}(L_i) \subset \mathrm{Diff}(L_i;K_i)$. We can describe this more precisely. Let $(z,\alpha) \in L_i \circ \mathrm{Aut}(L_i) = A(L_i)$. Then $z = (x,w) \in z(L_i) \times L_{i+1}$ uniquely. Let $\lambda, \hat{\alpha}, \bar{\alpha}$ be as above. As the action of (z,α) on L_i is $(z,\alpha)z' = z \circ \alpha(z')$, it can be verified that

$$(z,\alpha) = (x + f(w,-) \circ \ell(w^{-1}) + \lambda\bar{\alpha}^{-1}\ell(w^{-1}), \hat{\alpha}, \ell(w) \circ \bar{\alpha})$$

is an identification of $(z,\alpha) \in A(L_i)$ as an element of $\mathrm{Diff}(L_i;K_i)$. This proves the induction step, and our claim has been proved. □

3.7. In (3.2) we have seen how we can prove Bieberbach's rigidity theorem using Seifert constructions. A similar argument enables us to generalize the rigidity theorem to the nilpotent case. The Proposition (3.6) is crucial for our argument.

<u>Theorem</u>. Let $\theta: E \to E'$ be an isomorphism between two A.C. groups. Then θ is the conjugation by an element of $\mathrm{Diff}(L;K)$, and hence it extends to a continuous automorphism of $\mathrm{Diff}(L;K)$ (and hence, of $\mathrm{Diff}(L)$).

<u>Proof</u>. As E and E' are subgroups of $A(L)$, they are in $\mathrm{Diff}(L;K)$, thanks to (3.6). We will claim something more. Let $N = E \cap L$, $N' = E \cap L$. Then (3.6) tells us that $\psi: E \hookrightarrow \mathrm{Diff}(L;K)$, $\psi': E' \hookrightarrow \mathrm{Diff}(L;K)$ are Seifert

constructions for the data $(N,N,E,E/N,\text{point})$, $(N',N',E',E'/N',\text{point})$. This is a consequence of the fact that $Z^i(N) = Z^i(L) \cap N$ and $Z^i(N') = Z^i(L) \cap N'$.

As θ induces an isomorphism of N onto N', it induces isomorphisms of the ascending central series of N onto that of N', $Z^i(N) \overset{\cong}{\to} Z^i(N')$ for each $i = 0,1,2,\ldots,r+1$. Furthermore, $Q = E/N$ and $Q' = E'/N'$ act on $W = \text{point}$ trivially. These facts are enough to be able to apply the rigidity of Seifert construction (1.8). Consequently, there exists $\hat{h} \in \mathrm{Diff}(L;K)$ for which $\theta = \mu(\hat{h})$. This completes the proof of the theorem. □

Corollary. A homotopy equivalence between two infranilmanifolds is homotopic to a diffeomorphism.

Proof. Let $M = \pi\backslash L$, $M' = \pi'\backslash L'$ be infranilmanifolds and let $f: M \to M'$ be a homotopy equivalence. As L is diffeomorphic to L' we may assume $L = L'$. Let us denote the isomorphism of π onto π' induced by f, by f_*. The theorem above says that $f_* = \mu(\hat{h})$ for some $\hat{h} \in \mathrm{Diff}(L;K)$. (Recall that E,E' are A.C. groups.) This $\hat{h}$ then induces a fiber preserving diffeomorphism of M onto M' which is certainly homotopic to f. □

3.8. Proposition (3.6) also enables us to identify the fibers of Seifert fibering for $1 \to N \to E \to Q \to 1$, where N is nilpotent. By Malcev [M], the finitely generated torsion-free nilpotent group N is isomorphic to a uniform lattice in a connected, simply connected nilpotent Lie group $L(N)$, which is diffeomorphic to $\mathbb{R}^n$. Let $\phi: E \to \mathcal{H}(\mathbb{R}^n \times W;K)$ be a Seifert construction for (N,N,E,Q,W). The typical fiber is $N\backslash\mathbb{R}^n$ where N acts on $\mathbb{R}^n$ as a subgroup of $\mathcal{H}(\mathbb{R}^n \times W;K)$. In fact, N is embedded in $\hat{\mathcal{M}}(\mathbb{R}^n;K)$. On the other hand, N acts on $L(N)$ as left translations and (3.6) shows that $N \subset \hat{C}(L(N)) \subset \hat{\mathcal{M}}(L(N))$. Clearly $\hat{\mathcal{M}}(\mathbb{R}^n,K) = \hat{\mathcal{M}}(L(N);K)$ after we identify $L(N)_i$ with $\mathbb{R}^{n_i}$. Now by the rigidity of construction (1.8), $N\backslash\mathbb{R}^n$ is homeomorphic to $N\backslash L(N)$. If the Q action on W is smooth, one can do the smooth construction so that $N\backslash\mathbb{R}^n$ is diffeomorphic to $N\backslash L(N)$. Thus we have proved the first half of the

Theorem. Let E be an extension of a finitely generated torsion-free nilpotent group N by Q. Let $\rho: Q \to \mathcal{H}(W)$ be a properly discontinuous action of Q on a space W satisfying (*). Then the typical fiber of the Seifert fibering $E\backslash(\mathbb{R}^n \times W) \to Q\backslash W$ is homeomorphic to a (closed) nilmanifold. The singular fibers are homeomorphic to the quotient of a simply connected Lie group by almost crystallographic groups.

In particular if the E action is a covering action (for example, if E is torsion-free), then the singular fibers are homeomorphic to infranilmanifolds. If the Q action on W is smooth, then one can do the construction smoothly. In this case the fibers are diffeomorphic to nill or infranilmanifolds when the restriction to a fiber is free.

Proof. Singular fibers come from Q_w (= stabilizer of Q at $w \in W$) which is non-trivial. As the Q action is properly discontinuous, Q_w has to be finite. $1 \to N \to E_w \to Q_w \to 1$ is exact and E_w acts on $L(N) \times w$. Let K be the kernel of this action. K is a normal torsion subgroup of E_w which centralizes N. The group E_w acts effectively as E_w/K on $L(N) \times w$. It contains N and $C_{E_w/K}(N)$ is torsion-free since E_w/K acts effectively. By (3.5) Proposition, E_w/K can be embedded into $A(L(N))$ as an A.C. group. However, we know $E_w \to E_w/K \subset A(L(N)) \subset \mathrm{Diff}(L(N);K)$ is also a Seifert construction by (3.6). Now using rigidity (1.8), we have a homeomorphism of $E_w\backslash(\mathbb{R}^n \times w)$ onto $E_w\backslash L(N)$. The rest of the theorem is a consequence of the construction. □

Corollary. If E in the theorem is torsion-free and W is a contractible manifold, then E acts freely and $E\backslash(\mathbb{R}^n \times W)$ is a $K(E,1)$-manifold with typical fiber homeomorphic to the nilmanifold $N\backslash L(N)$ and singular fibers homeomorphic to the infranilmanifolds $E_w\backslash L(N)$.

Remarks. (1) When E is torsion-free virtually nilpotent (so that Q is finite and W is a point), the Seifert construction yields a closed manifold diffeomorphic to an infranilmanifold, because the Q action on W is smooth already.

(2) When E is torsion-free virtually poly-$\mathbb{Z}$, we take $N = {}^nE$ as before. Then $M = M(N,E,E,1,\text{point}) = M(N,N,E/N,\mathbb{R}^s)$ (where $s = \text{rank}(E/N)$) is a smooth closed aspherical manifold, with a predetermined Seifert structure with typical fiber diffeomorphic to the closed nilmanifold $M({}^nE)$ and with base a compact orbit space of a euclidean space by a crystallographic group.

§4. A Seifert Construction for Uniform Lattices in a Lie Group.

4.1. We have seen that there are Seifert constructions with typical fibers tori, nilmanifolds and $K(\pi,1)$-manifolds when π is virtually poly-$\mathbb{Z}$. In [RW] a construction was made when the typical fiber was a locally symmetric space. In this section we take a torsion-free uniform lattice Γ in a simply connected, connected Lie group L without compact normal factors. We combine both techniques to describe a Seifert construction over $Q\backslash W$. The typical fiber is a closed manifold homotopically equivalent to the homogeneous manifold $\Gamma\backslash L/K$, where K denotes a maximal compact subgroup of L.

4.2. Let L be a connected and simply connected Lie group with radical R (= the maximal connected normal solvable subgroup of L), and let S denote the simply connected semi-simple quotient L/R. It is well-known that L splits (Levi-decomposition):

$$1 \to R \to L \underset{\rho}{\overset{\theta}{\leftrightarrows}} S \to 1 .$$

We will write L as $R \circ_\theta S$. This splitting induces a homomorphism of S into the group of continuous automorphisms of R via conjugation. Let C denote the center of S. Then $S/C = S^*$ is the adjoint form of S, and it is centerless. C is discrete in S. As usual, we work with L for which S has <u>no normal compact direct factor</u>.

Let Γ be a uniform lattice in L. <u>Assume</u> also that if S^* contains any 3-dimensional factors, then the image of Γ in S^* is dense in each of these factors. Given an extension of Γ, $1 \to \Gamma \to E \to Q \to 1$ with (Q,W) satisfying (*) as before, we shall construct a properly discontinuous action of E on $(L/K) \times W$ for which the mapping $E\backslash((L/K) \times W) \to Q\backslash W$ is a Seifert fibering with typical fiber homotopy equivalent to the double coset space $\Gamma\backslash L/K$, where K is a maximal compact subgroup of S.

4.3. For an arbitrary group H, we will let $r(H)$, $n\text{-}r(H)$ denote its <u>discrete</u> <u>radical</u>, <u>discrete</u> <u>nilradical</u>, resp. [A2] if they exist. These are the maximal normal solvable subgroup, the maximal normal nilpotent subgroup of H, resp. If H is a subgroup of any Lie group, $r(H)$ always exists.

Let L, R and Γ be as above. Then $\Gamma \cap R$, $\rho(\Gamma)$ are uniform lattices in R, S, respectively. In [A2], it is shown that

(i) $R = (r(L))_0$, the connected component of the identity,

(ii) $\rho(\Gamma) \cap \rho(r(L))$ is of finite index in $\rho(r(L))$ and hence,

(iii) the cosets $r(L)/r(L) \cap \Gamma$ have a finite number of connected components each of which is compact.

We wish to observe that $r(L) = R \circ C$. First we claim that $r(S) = C$. For, $C \subset r(S)$ and Auslander shows that $r(S)/C$ is finite. Therefore $r(S)$ is a discrete normal subgroup, and consequently $S \to S/r(S)$ is a covering map. But $S \to S/C$ is the maximal normal covering map, so $r(S) = C$. Now observe that $R \circ C \subset r(L)$. Therefore $\rho(r(L))$ is a normal solvable subgroup of S containing C. However, $r(S) = C$. Therefore $\rho(r(L)) = C$ and hence, $r(L) = R \circ C$.

Next we would like to show that $r(\Gamma) = r(L) \cap \Gamma$. Let ρ^* denote the composite $L \xrightarrow{\rho} S \to S^* = L/r(L)$. Surely $r(L) \cap \Gamma$, which is solvable and normal in Γ, is contained in $r(\Gamma)$. Therefore, $\rho^*(r(\Gamma))$ is a normal solvable subgroup of $\Gamma^* = \rho^*(\Gamma)$. But if this subgroup is non-trivial, it would have a discrete nilradical and consequently a non-trivial center. This center will be then a normal abelian subgroup of Γ^*. However, Γ^* is a uniform lattice in S^* since $\rho(\Gamma) \cap C$ has finite index in C and $\rho(\Gamma)$ is a uniform lattice in S. This contradicts the fact that there are no normal abelian subgroups for uniform lattices in S^*, see [CR5, (6.4)] for a proof. Therefore, $r(\Gamma) = r(L) \cap \Gamma$. Thus we have the

<u>Proposition</u>. Let Γ be a uniform lattice in a connected, simply connected Lie group L whose Levi-decomposition is $R \circ S$, where S contains no normal non-trivial compact direct factors. Then $r(L) = R \circ C$, where C is the center of S. Moreover, $r(\Gamma) = \Gamma \cap (R \circ C)$, and $\Gamma/r(\Gamma)$ is a uniform lattice in $S/C \cap r(\Gamma)$. $\Gamma/r(\Gamma)$ projects isomorphically onto $\rho^*(\Gamma)$; a centerless uniform lattice in $S/C = S^*$. $\square$

4.4. Let W be a simply connected space and (Q,W) a properly discontinuous action. Let Γ be a uniform lattice of a Lie group L as in the proposition above. Also <u>assume</u> that if $L/r(L)$ contains any 3-dimensional factors, then the image $\Gamma^* = \Gamma/r(\Gamma)$ is dense in each of these factors. See [RW] for a discussion about the last condition. Assume that Γ is torsion-free. Let K be the maximal compact subgroup of L (hence of S). Then the quotient space $\Gamma\backslash L/K$ is a closed $K(\Gamma,1)$-manifold.

<u>Theorem</u>. Let $1 \to \Gamma \to E \to Q \to 1$ be any extension. Then on $L/K \times W$ there exists a properly discontinuous action of E with quotient $X = E\backslash(L/K \times W)$ and a Seifert fiber map $q\colon X \to Q\backslash W$ with typical fiber $M(\Gamma)$, a closed manifold homotopically equivalent to $\Gamma\backslash L/K$. If E is torsion-free and W is a contractible manifold factor with $Q\backslash W$ compact, then X is a closed (smooth if the Q action on W is smooth) $K(E,1)$-manifold.

<u>Proof</u>. L/K is diffeomorphic to euclidean space $\mathbb{R}^n$, where $n = \dim L - \dim K$. We use the exact sequence $1 \to r(\Gamma) \to \Gamma \to \Gamma^* \to 1$, to induce $1 \to r(\Gamma) \to E \to E/r(\Gamma) \to 1$ ($r(\Gamma)$ is characteristic in Γ) and $1 \to \Gamma^* \to E/r(\Gamma) \to Q \to 1$. Γ^* is a lattice in $S^* = S/C$, and so acts properly discontinuously on S^*/K^*, where K^* is a maximal compact subgroup of S^*. S^*/K^* is diffeomorphic to $\mathbb{R}^{n-t-m}$, where $t = \operatorname{rank} C$ and $m = \dim R$. We apply [RW] to construct an action of $E/r(\Gamma)$ on $(S^*/K^*) \times W \approx \mathbb{R}^{n-t-m} \times W$. This yields a Seifert fibering on $(E/r(\Gamma))\backslash((S^*/K^*) \times W)$ over $Q\backslash W$ with typical fiber the compact locally symmetric space $\Gamma^*\backslash(S^*/K^*)$. $r(\Gamma)$ is torsion-free and solvable and has rank equal to rank R + rank $C = m + r$.

We now apply the Theorem (2.5), Remark 3, to the extension sequence $1 \to r(\Gamma) \to E \to E/r(\Gamma) \to 1$ and the action $(E/r(\Gamma), (S^*/K^*) \times W)$ to obtain a properly discontinuous action of E on $\mathbb{R}^{m+t} \times (\mathbb{R}^{n-t-m} \times W) = \mathbb{R}^n \times W$. The typical fiber is the closed aspherical manifold $\Gamma\backslash\mathbb{R}^n$ which has the same homotopy type as $\Gamma\backslash L/K$.

We have fixed the action of Q on W as a part of our data. The choices of K and K^* are unique up to conjugation in S and S^* and the action of Γ^* on S^*/K^* is rigid and so the action of $E/r(\Gamma)$ on $(S^*/K^*) \times W$ is determined. The remaining part of our construction which utilizes (2.5) is canonical and so the construction of our E action on $\mathbb{R}^n \times W$ is canonical. When W is a contractible manifold factor and $Q\backslash W$ is compact, E torsion-free, then $M(E) = E\backslash(\mathbb{R}^n \times W)$ is a closed $K(E,1)$-manifold. □

As it stands, it is not clear as to whether the restriction of the E action to Γ produces on each $\mathbb{R}^n \times w$ a manifold diffeomorphic to $\Gamma\backslash L/K$. The following seems to be the best concrete result for the typical fiber to be a homogeneous space.

4.5. Let $1 \to \Gamma \to \pi \to Q \to 1$ be a torsion-free extension, where Γ is a discrete uniform lattice of a connected, simply connected solvable Lie group R, and let (Q,W) be a properly discontinuous action of Q on a contractible manifold factor W such that $Q\backslash W$ is compact. Let $\mathfrak{N}$ be the nilradical of R.

<u>Theorem</u>. If $\mathfrak{N} \cap \Gamma$ is normal in π, then there exists a Seifert construction for $(\mathfrak{N} \cap \Gamma, \Gamma, \pi, Q, W)$. The closed $K(\pi,1)$-manifold $M = M(\mathfrak{N} \cap \Gamma, \Gamma, \pi, Q, W)$ has a fibering onto $Q\backslash W$ with typical fiber homeomorphic (resp. diffeomorphic) to $\Gamma\backslash R$ (resp. if Q action is smooth).

<u>Proof</u>. As we noted at the beginning of (2.5), it suffices for the first statement that $\mathfrak{N} \cap \Gamma$ is normal in π and $\Gamma/\mathfrak{N} \cap \Gamma$ is virtually free

abelian. This is easy to check. It remains to identify the typical fiber for $M \to Q\backslash W$.

We would like to show first that the group R of left translation of R is a Seifert construction. More precisely, let $\mathfrak{n}$ be the nilradical of R. Let $K = (k_0, k_1, \ldots, k_r)$ be the partition of $n = \dim \mathfrak{n}$, induced by the ascending central series of $\mathfrak{n}$, see (3.7). Let $R/\mathfrak{n} = \mathbb{R}^s$, and $K' = (k_0, k_1, \ldots, k_r, k_s)$. Then by an argument similar to Lemma (3.7), one can show $R \subset \mathrm{Diff}(\mathbb{R}^{n+s}; K')$. Note that R is not a subgroup of $\hat{C}(\mathbb{R}^{n+s}; K')$ in general. This is due to the fact that the $\mathbb{R}^s$ part does not act trivially on $z(\mathfrak{n})$ etc., in general. Consequently, $\Gamma \subset R \subset \mathrm{Diff}(R; K')$.

On the other hand, the typical fiber of $M \to Q\backslash W$ is $\Gamma\backslash\mathbb{R}^{n+s}$. We may compare $(\Gamma, \mathbb{R}^{n+s})$ with (Γ, R), because both of them are Seifert constructions for the same filtration which is the restriction of $1 = Z^0(\mathfrak{n}) \subset Z^1(\mathfrak{n}) \subset \ldots \subset Z^r(\mathfrak{n}) \subset Z^{r+1}(\mathfrak{n}) = \mathfrak{n} \subset R$, to Γ. By the uniqueness of (1.8), the typical fiber $\Gamma\backslash\mathbb{R}^{n+s}$ is homeomorphic to $\Gamma\backslash R$. If W is smooth and the Q action on W is smooth, then construction can be done smoothly so that $\Gamma\backslash\mathbb{R}^{n+s}$ is diffeomorphic to $\Gamma\backslash R$. □

Remarks. (1) When $\mathfrak{n} \cap \Gamma$ is not normal in π, it is not always possible to put E into $\mathcal{A}(R \times W; K')$. For example, let $R = \mathbb{R}^2 \circ \mathbb{R}^1$ be the universal covering group of $\mathbb{R}^2 \circ SO(2) \subset E(2)$. $\Gamma = \mathbb{Z}^3$ be the uniform lattice of R, $\mathbb{Z}^2$ from the standard orthonormal basis for the translational part $\mathbb{R}^2$ of R, and a remaining $\mathbb{Z}$ is from $\mathbb{R}^1$ acting trivially on $\mathbb{R}^2$. Let $Q = \mathbb{Z}$ be generated by α, acting on $W = \mathbb{R}$ as translations. Let π be a torsion-free extension of Γ by Q, where α acts on $\Gamma = \langle t_1, t_2, t_3 \rangle$ by $\alpha(t_1) = t_2$, $\alpha(t_2) = t_3$, $\alpha(t_3) = t_1$. A 4-dimensional Bieberbach group with holonomy $\cong \mathbb{Z}_3$ and center $\cong \mathbb{Z}$ is such a group. In this case,

$$\mathrm{Diff}(R \times W; K') = (C(R \times W, \mathbb{R}^2) \circ C(W, \mathbb{R})) \circ ((GL(2,\mathbb{R}) \times GL(1,\mathbb{R})) \times \mathrm{Diff}(W)),$$

and the automorphism of $\mathbb{Z}^3$ defined by α cannot be factored into $GL(2,\mathbb{R}) \times GL(1,\mathbb{R})$. In other words, the Q action mixes the natural fiber structure of R. This shows that the condition for $\Gamma \cap \mathfrak{n}$ being normal in

π is a very natural one.

(2) Observe, once again, that it is neither necessary that W be a contractible manifold factor nor that π be torsion-free for us to determine that π acts properly discontinuously on $R \times W$ so that the typical fiber in the fibering $X = \pi\backslash(R \times W) \to Q\backslash W$ is homeomorphic to the homogeneous space $\Gamma\backslash R$.

(3) If we return to the Theorem (4.4) it is possible to describe still a different Seifert construction. For each uniform lattice Γ we factored it into the extension $1 \to r(\Gamma) \to \Gamma \to \Gamma^* \to 1$. However, there are many extensions of $r(\Gamma)$ by Γ^* some of which will surely be torsion-free. In any case $r(\Gamma)$ will be the discrete radical of the extension, for otherwise Γ^* would have a non-trivial solvable subgroup. Thus $r(\Gamma)$ is normal in E' (where E' is an extension of Γ' by Q). If we follow the argument of (4.4) then we can construct a properly discontinuous action of E' on $\mathbb{R}^n \times W$ and a Seifert fibering $E'\backslash(\mathbb{R}^n \times W)$ over $Q\backslash W$ with typical fiber $\Gamma'\backslash\mathbb{R}^n$.

4.6. In [J], Johnson introduces the notion of a poly $\mathbb{L}_+$-group and shows (Theorem 3) that if G is a torsion-free poly-$\mathbb{L}_+$-group, then there exists a closed smooth $K(G,1)$-manifold provided that two additional hypotheses on G hold:

(i) $r(G)$ is a poly-$\mathbb{Z}$ group and

(ii) $G/r(G)$ is torsion-free.

Recall $r(G)$ is the (discrete) radical of G.

Johnson intimated that he believed these extra conditions were actually unnecessary. However, he was unable to remove them. Essentially a poly-$\mathbb{L}_+$-group G is a group having a filtration

$$1 = G_0 \subset \ldots \subset G_k = G$$

so that G_i is normal in G_{i+1} and $G_{i+1}/G_i \in \mathbb{L}_+$. A group is in $\mathbb{L}$ if it is a uniform discrete lattice in a connected finite covering group of the adjoint form S^* of S as in (4.3). G_i then is defined to be in $\mathbb{L}_+$ if either G_i is in $\mathbb{L}$ or is virtually poly-$\mathbb{Z}$.

Johnson shows in his Proposition 2 that if G is a poly-$\mathbb{L}_+$-group then (a) $r(G)$ is virtually poly-$\mathbb{Z}$ and (b) $G/r(G)$ is a virtual $\mathbb{L}$-group.

We shall show that his main realization result (Theorem 3) is true <u>without</u> the extra hypotheses of (i) and (ii).

We have $1 \to r(G) \to G \to G/r(G) \to 1$ exact. It suffices to show that $G/r(G)$ acts smoothly and properly discontinuously on a contractible manifold W with a compact quotient $(G/r(G))\backslash W$. Let us denote $G/r(G)$ by Γ. Then Γ contains Γ_1, a normal (actually characteristic according to (b) above) $\mathbb{L}$-subgroup of finite index. Put $F = \Gamma/\Gamma_1$. We obtain the commutative diagram with exact rows:

$$
\begin{array}{ccccccccc}
 & & 0 & & 1 & & 1 & & \\
 & & \downarrow & & \downarrow & & \downarrow & & \\
0 & \longrightarrow & z(\Gamma_1) & \longrightarrow & C_\Gamma(\Gamma_1) & \longrightarrow & F_1 & \longrightarrow & 1 \\
 & & \downarrow & & \downarrow & & \downarrow & & \\
1 & \longrightarrow & \Gamma_1 & \longrightarrow & \Gamma & \longrightarrow & F & \longrightarrow & 1 \\
 & & \downarrow & & \downarrow & & \downarrow & & \\
1 & \longrightarrow & \Gamma_1/z(\Gamma_1) & \longrightarrow & \Gamma/C_\Gamma(\Gamma_1) & \longrightarrow & F/F_1 & \longrightarrow & 1 \\
 & & \cap & & \cap & & \cap & & \\
1 & \longrightarrow & \mathrm{Inn}\,\Gamma_1 & \longrightarrow & \mathrm{Aut}\,\Gamma_1 & \longrightarrow & \mathrm{Out}\,\Gamma_1 & \longrightarrow & 1
\end{array}
$$

Now, $\Gamma^* = \Gamma_1/z(\Gamma_1)$ is a lattice in S^* (Γ_1 is a lattice in a finite central covering of S^*) and $\Gamma/C_\Gamma(\Gamma_1)$ is then a lattice in $\mathrm{Aut}(S^*)$, see [RW, Corollary 4]. Therefore, $\Gamma/C_\Gamma(\Gamma_1)$ acts smoothly on $\mathrm{Aut}(S^*)$ as a uniform lattice. Let K^{**} be a maximal compact subgroup of $\mathrm{Aut}(S^*)$. Now, $\mathrm{Aut}(S^*)/K^{**} = S^*/K^* \approx W$, a simply connected non-compact symmetric space, which is diffeomorphic to some $\mathbb{R}^m$ (K^* is a maximal compact subgroup of S^*). So we may define a smooth action of Γ on W via $\rho\colon \Gamma \to \Gamma/C_\Gamma(\Gamma_1) \to \mathrm{Diff}(W)$. Because $C_\Gamma(\Gamma_1)$ is finite, the Γ action on W (which is possibly ineffective) is smooth with compact quotient. We are now ready to apply the Theorem (2.5) since $r(G)$ is a torsion-free virtually poly-$\mathbb{Z}$ group. Thus we obtain a smooth model aspherical manifold $M({}^n r(G), r(G), G, G/r(G), W)$.

§5. The Maximal Toral Actions.

5.1. For a closed aspherical manifold M, it is known [CR1, Theorem 5.6] that the only connected compact Lie groups that act effectively on M are toral groups T^s with $s \leq$ rank of $z(\pi_1 M)$. If T^s acts effectively on M, let $ev^x: (T^s,1) \to (M,x)$ be the evaluation map, $ev^x(t) = tx$. Then $ev^x_*: \pi_1(T^s,1) \to \pi_1(M,x)$ is injective and has image in $z(\pi_1(M,x))$. We will denote the image of ev^x_* by $\pi_1(T^s)$ if there is no confusion likely. In [CR5] it was conjectured that if M is a closed aspherical manifold, then

(1) the center $z(\pi_1 M)$ is finitely generated, say of rank k, and

(2) there exists a toral group T^k acting effectively on M so that $\pi_1(T^k) = z(\pi_1 M)$.

These may be verified in many interesting cases. For example, (a) if M is a smooth manifold admitting a Riemannian metric with non-positive sectional curvature, then (1) and (2) can be derived from the work of Lawson and Yau [LY], or (b) if M is a nilmanifold, then (1) and (2) hold [CR5, 5.7(d)]. For these and other classes of examples see [CR5, (9.3)]. As further evidence in support of these conjectures it was shown

Proposition [CR4]. Let (T^s,M) be an effective toral action on a closed aspherical manifold M. Then the image of evaluation map, $\pi_1(T^s)$, is a direct summand of $z(\pi_1 M)$.

5.2. We say that a closed aspherical manifold admits a maximal toral action if there exists an effective T^k action on M, where $k = \text{rank } z(\pi_1 M)$. (Note that such a T^k is the biggest connected compact Lie group that can act on M effectively.) Then necessarily $\pi_1(T^k) = z(\pi_1 M)$ by Proposition (5.1).

Now suppose that M admits a maximal toral action (T^k,M). If (T^r,M) is any toral action, Proposition (5.1) says that $\pi_1(T^r)$ is a direct summand of $\pi_1(T^k)$. In other words, the T^r-action is the restriction of the maximal toral action $T^k = T^r \times T^{k-r}$ to the first factor on the homotopy level.

Unfortunately, examples readily show that the homotopy interpretation cannot be improved to the more exact level of transformation group equivalence.

See [CR1, Example 7.6]. On the other hand, if we place restrictions on M or on the actions, then in some situations the T^r action is the restriction of the T^k action up to equivalence. We state this problem as a conjecture.

(3) Let M be a closed aspherical manifold admitting a maximal toral action (T^k, M). For any homotopically trivial free action of a compact Lie group G on M, there exists a self homeomorphism h of M so that $\mu(h)$ (= conjugation by h) maps G into T^k.

In this section, we will exploit the Seifert construction to verify Conjectures (1), (2) and (3) for certain classes of manifolds.

5.3. Let (π, E, Q, W) be given data as in Section 2 with π a virtually poly-$\mathbb{Z}$ group. We may assume that Q acts effectively on W. (Note that this is not a real restriction because when we extend π by Q_0 = kernel of $\rho: Q \to \mathcal{N}(W)$, it will still be a virtually poly-Z group.) Assume that E is torsion-free and W is a contractible manifold. Then by Theorem (2.5), E acts on $\mathbb{R}^m \times W$, via $E \to \mathcal{N}(\mathbb{R}^m \times W; K)$, and $M({}^n\pi, \pi, E, Q, W) = E\backslash(\mathbb{R}^m \times W)$ is a closed $K(E,1)$-manifold.

Theorem. If Q is centerless, then $M({}^n\pi, \pi, E, Q, W)$ admits a maximal toral action. This maximal toral action is a factor of the principal Seifert fibering corresponding to the center of ${}^n\pi = N$.

Proof. As Q is centerless, $z(E)$ is contained in π. This easily implies that $z(E) \subset z(N)$. We claim that $z(E)$ is a direct summand of $z(N)$. Suppose $c \in z(N)$ with $c^p \in z(E)$ for some $p > 0$. For any $x \in E$, $c^p = xc^px^{-1} = (xcx^{-1})^p$. Since both c and xcx^{-1} belong to $z(N)$ and $z(N)$ is torsion-free, $c = xcx^{-1}$ for all $x \in E$. Thus $c \in z(E)$. Consequently we have shown that $z(E)$ is a direct summand of $z(N)$.

$M = M({}^n\pi, \pi, E, Q, W)$ fibers over M_1 (see (2.2)) with typical fiber T^{k_0} (k_0 = rank of $z(N)$). We shall now use our second version of the Seifert construction. This will be in terms of injective Seifert fiberings initially sketched in (1.2) and expanded somewhat in the Appendix. On the covering space corresponding to the image of $\pi_1(T^{k_0})$ in $\pi_1(M)$, we have a splitting $T^{k_0} \times \tilde{M}_1$. For any $\alpha \in E/z(N)$ and $t \in T^{k_0}$ we have

$$\alpha(t(x,\tilde{m})) = \phi(\alpha)(t)(\alpha(s,\tilde{m})) .$$

The summand T^k of T^{k_0} corresponding to $z(E)$ acts on $T^{k_0} \times \tilde{M}_1$ as a subgroup of T^{k_0}. If $t \in T^k$, $\phi(\alpha)(t) = t$. This implies that T^k commutes with the action of $E/z(N)$ and so induces an effective action of T^k on the quotient $(E/z(N))\backslash(T^{k_0} \times \tilde{M}_1) = M$. We also observe that Q need not be centerless to obtain the result above. What is only used is that $z(E) \subset \pi$. □

5.4. C. T. C. Wall has studied a class of aspherical manifolds whose fundamental groups are poly-$\mathbb{Z}$ groups. He showed [W] that if $f: M_1 \to M_2$ is a map between closed aspherical manifolds with poly-$\mathbb{Z}$ fundamental groups and if f induces an isomorphism on their fundamental groups, then f is homotopic to a homeomorphism provided that dimension of $M_i \neq 3,4$.

Recently F. T. Farrell and W. C. Hsiang generalized the special case of nilpotent fundamental groups to virtually nilpotent groups. Namely, they proved [FH] that any homotopy equivalence between closed aspherical manifolds of dimension $\neq 3,4$ with virtually nilpotent fundamental groups is homotopic to a homeomorphism.

Therefore, Theorem (5.3) together with Wall's and Farrell-Hsiang's results yields the following.

Corollary. A closed aspherical manifold of dimension $\neq 3,4$ whose fundamental group is either poly-$\mathbb{Z}$ or virtually nilpotent admits a maximal toral action. □

5.5. Let π be a torsion-free, virtually poly-$\mathbb{Z}$ group. Then the model aspherical manifold $M(\pi)$ constructed in (2.5) admits a maximal toral action described in Theorem (5.4). This maximal toral action is unique in the following sense. For a fixed model $M(\pi)$, the toral action is unique. If $M'(\pi)$ is another model, then there exists a fiber-preserving diffeomorphism of $M(\pi)$ onto $M'(\pi)$ and under the fiber-preserving diffeomorphism, the maximal toral action of $M(\pi)$ maps onto that of $M'(\pi)$ isomorphically. This is a consequence of the uniqueness of the construction, (1.8). We call this particular maximal toral action <u>the standard maximal toral action on</u> $M(\pi)$. However, we should remember that this maximal toral action is always subject to a reparametrization of the torus which corresponds to an automorphism of $z(\pi)$.

For nilmanifolds or infranilmanifolds we have a different description for the standard maximal toral action. Let $M = \pi\backslash L$ be an infranilmanifold so that L is a connected, simply connected nilpotent Lie group, and $\pi \subset A(L)$. Let C be the subgroup of L centralizing π (in $A(L)$). Then C is easily seen to be a factor of $z(L)$, of dimension equal to the rank of $z(\pi)$. The action of C on L (by left translations) gives rise to an action of the torus $C/z(\pi)$ on M. This toral action is the standard one on M.

5.6. We turn our attention to the third problem. We will restrict ourselves to the closed aspherical manifolds whose fundamental groups are either virtually nilpotent or poly-$\mathbb{Z}$ group whose quotient by its nilradical is free abelian.

<u>Theorem</u>. Let M be an infranilmanifold, and let G be a compact Lie group acting freely, smoothly and homotopically trivially on M. Assume that $\dim M - \dim G \neq 3,4$. Then the action of G is topologically equivalent to a restriction of the standard maximal toral action.

<u>Proof</u>. The connected component G_0 of the identity of G is a torus (see (5.1)), say of dimension s. Let $\pi = \pi_1 M$. By Proposition (5.1), $\mathbb{Z}^s = \pi_1(G_0)$ is a direct summand of $z(\pi)$. Therefore, $\pi/\mathbb{Z}^s$ is again virtually nilpotent and torsion-free since the action of G_0 is free. See [CR1, Theorem (7.4)]. We prove the theorem for G_0 first.

As $\mathbb{Z}^s$ is a direct summand of $z(\pi)$, we have a toral action T^s which is a factor of the standard maximal toral action, corresponding to $\mathbb{Z}^s$. Since we may alter the maximal toral action by simply reparametrizing the torus, let us choose G_0 to be our fixed s-torus and reparametrize T^k so that the identity map $G_0 \to T^s$ is compatible with the evaluation map. That is,

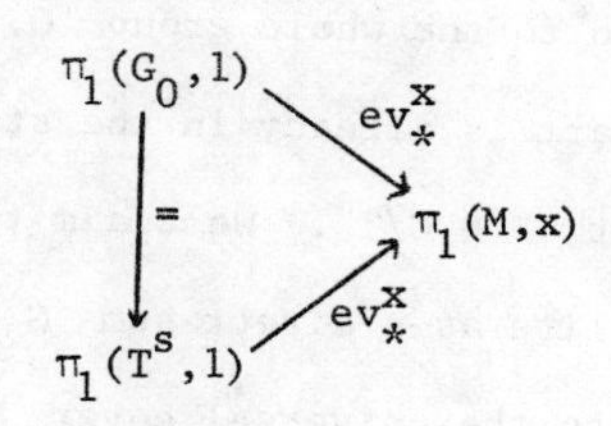

commutes. All this does is to change our original standard action to a weakly equivalent one. We have two principal fiber structures on M,

$$G_0 \to M \to M/G_0 \quad \text{and} \quad T^s \to M \to M/T^s .$$

Let BT^s be the classifying space for T^s. Then all the principal T^s-bundles over a space Y are classified by the homotopy classes $[Y, BT^s]$. Since BT^s is $K(\mathbb{Z}^s, 2)$, $[Y, BT^s] \cong H^2(Y, \mathbb{Z}^s)$. If Y is a $K(Q,1)$-manifold, then $H^2(Y, \mathbb{Z}^s) \cong H^2(Q, \mathbb{Z}^s)$. All these isomorphisms are natural. An element of $H^2(Q, \mathbb{Z}^s)$ is the equivalence class of group extensions of $\mathbb{Z}^s$ by Q. A homeomorphism $h: Y \to Y'$ induces an isomorphism $h^*: H^2(Q', \mathbb{Z}^s) \cong [Y', BT^s] \to [Y, BT^s] \cong H^2(Q, \mathbb{Z}^s)$. Therefore, if $h^*(a') = a$, then the resulting bundles will be bundle isomorphic.

We apply the above fact to $Y = M/G_0$ and $Y' = M/T^s$. Let $f: Y \to BT^s$ and $f': Y' \to BT^s$ be the classifying maps for the principle T^s-bundles. We seek a homeomorphism $h: Y \to Y'$ so that $f' \circ h$ is homotopic to f. Now f and f' determine group extensions $a, a' \in H^2(Y \text{ or } Y'; \mathbb{Z}^s)$ which were already known to be isomorphic inducing the identity on $\pi_1(T^s)$. (Note that $\pi_1(Y) \cong \pi_1(Y')$ since they are isomorphic to quotients of $\pi_1(M)$ by the same group.) The actions arise from bundles equivalent to the original actions. Since G_0 (and T^s) act smoothly, Y and Y' are (smooth) manifolds with isomorphic fundamental groups. By [FH] we may choose a homeomorphism $h: Y \to Y'$ so that $h^*(a') = a$ which means $f' \circ h \simeq f$. By the argument above, two bundles are isomorphic. In other words, there exists an equivariant homeomorphism $\hat{h}: M \to M$ which covers h. This completes the proof of the theorem for G_0.

Now we turn our attention to the whole group G. By the above argument, we may assume that the G_0 part is already in the standard maximal toral action (we use $G_0 = T^s$). Let $F = G/T^s$. We claim that F is a finite abelian group and that G splits as a direct sum $G = T^s \oplus F$. Let E be the group of all liftings of G to the universal cover $\tilde{M}$ of M, so that $1 \to \pi \to E \to G \to 1$ is exact. Since G acts homotopically trivially on M, the induced map $G \to \pi_0\mathcal{E}(M) \cong \text{Out}\,\pi$ is trivial, (where $\mathcal{E}(M)$ is the H-space of all self-homotopy equivalences of M). Therefore, the exact sequence above contains $1 \to z(\pi) \to C_E(\pi) \to G \to 1$. The inverse image of $T^s \subset G$ is $\mathbb{R}^s$ and $z(\pi) = \mathbb{Z}^s \oplus \mathbb{Z}^{k-s}$ (k = rank of $z(\pi)$). Thus we have $1 \to \mathbb{Z}^{k-s} \to C_E(\pi)/\mathbb{R}^s \to F \to 1$ exact. However, it is known that a torsion-free central extension of a free abelian group is again free abelian [LR1, p. 261] so that $C_E(\pi)/\mathbb{R}^s$ is free abelian, say $'\mathbb{Z}^{k-s}$. We look at the exact sequence $1 \to \mathbb{R}^s \to C_E(\pi) \to {}'\mathbb{Z}^{k-s} \to 1$. Since $C_E(\pi)$ is 2-step nilpotent, it is readily shown that $'\mathbb{Z}^{k-s}$ splits back into $C_E(\pi)$ so that $C_E(\pi) = \mathbb{R}^s \oplus {}'\mathbb{Z}^{k-s}$. Consequently, $G = T^s \oplus F$.

Since F is a direct summand of G, the F action on M induces an action of F on $Y = M/T^s$. We have another action of F on Y denoted by (F', Y) which is induced by $'\mathbb{Z}^{k-s}/\mathbb{Z}^{k-s}$. This is embedded in T^{k-s}, a part of the standard maximal toral action of T^k on M. Now both actions F and F' induce isomorphic group extensions with the isomorphism being the identity on π and $\mathbb{Z}^s$. Since both F and F' are free on Y, this means that $\pi_1(Y/F)$ and $\pi_1(Y/F')$ are isomorphic with the isomorphism being the identity on $\pi_1(Y)$. We can find a homeomorphism k by virtue of [FH] which induces this isomorphism between $\pi_1(Y/F)$ and $\pi_1(Y/F')$. This homeomorphism k lifts to a homeomorphism $\hat{k}: Y \to Y$ inducing the identity $\hat{k}_*: \pi_1(Y) \to \pi_1(Y)$. This tells us that the F action on Y is equivalent to the (pull back of) F' action on Y. But this F' action is embedded in part of the standard action on Y (coming from the embedding of F' in the T^{k-s} action on M). This implies that the lifting $\hat{k}$ of k of M conjugates G into the standard maximal toral action, i.e., for any $g \in G$, $\hat{k} \circ g \circ \hat{k}^{-1}$ is an element of the standard maximal toral action T^k on M. □

5.7. For a nilmanifold, the standard maximal toral action is necessarily free as shown in the following lemma. Therefore, we can drop the condition that G acts freely in Theorem (5.6) in this case.

<u>Lemma</u>. Let M be a closed aspherical manifold with a free maximal toral action. If a compact Lie group acts effectively and homotopically trivially on M, then it acts freely.

<u>Proof</u>. Let $\pi = \pi_1 M$ and let $1 \to \pi \to E \to G \to 1$ be the lifting sequence of the compact Lie group action G on M. Then $1 \to c_E(\pi) \to E \to \pi/z(\pi) \to 1$ is exact since G maps trivially into $\mathrm{Out}\,\pi$ as G acts homotopically trivially. Since G acts effectively, $c_E(\pi)$ is torsion-free [LR1, Lemma 1]. Since M admits a free maximal toral action, by (5.1) and [CR1, (7.4)], $\pi/z(\pi)$ is torsion-free. Therefore we have shown that E is torsion-free.

The isotropy group G_x at any point x of M must be finite since the connected component of the identity G_0 is a torus that acts injectively [CR3]. The lifting sequence of the action (G_x, M), $1 \to \pi \to E_x \to G_x \to 1$, then splits so that E_x contains G_x, a finite subgroup. Since E is torsion-free, G_x must be trivial, hence the G action is free. □

A useful <u>alternative</u> version of the lemma above <u>replaces</u> the hypothesized <u>free maximal toral action</u> with the requirement <u>that</u> $\pi_1 M/z(\pi_1 M)$ <u>is torsion-free</u>. The reason that this works just as well is that the free maximal toral action implied the torsion-freeness of $\pi_1 M/z(\pi_1 M)$ which was then used to obtain the conclusion.

5.8. <u>Theorem</u>. Let M be an infranilmanifold and G as in (5.6). If $\dim M - \dim G = 3$, assume that M/G_0 is irreducible. If $\dim M - \dim G = 4$, assume that M and M/G_0 are homeomorphic to nilmanifolds and that G is connected. Then the conclusion of the Theorem (5.6) is still true.

Proof. Let $\dim M - \dim G = 3$. Let us assume first that M is a nilmanifold since the argument for the infranil case depends upon some announced but unpublished work. Since $Y = M/G_0$ is homotopy equivalent to a nilmanifold, it has a non-trivial first betti number. By hypothesis Y is irreducible and as it finitely covers Y/F, Y/F is also irreducible. Moreover, Y/F is homotopy equivalent to a 3-dimensional nilmanifold and hence a Haken manifold. By Waldhausen's result, Y is homeomorphic to the nilmanifold $Y' = M/T^s$ and Y/F is homeomorphic to the nilmanifold Y'/F' and so the conclusion holds.

A recently announced result of Peter Scott enables one to extend the argument above to the infranilmanifold case. Once again $Y = M/G_0$ and Y/F are homotopy equivalent to infranilmanifolds (use (3.5)) and are irreducible. However, they may not have trivial first betti number and not be Haken. It is not hard to see that Y and Y/F are homotopy equivalent to closed Seifert 3-manifolds. Scott's result [S] then implies that the irreducible Y and Y/F are actually Seifert 3-manifolds and hence infranilmanifolds. The argument above then applies.

For dimension 4, M/G_0 is homeomorphic to the nilmanifold M/T^s. All we need for our conclusion then is that each homotopy equivalence can be deformed to a homeomorphism. This is a consequence of the Seifert fiber structure since M/G_0 fibers over a lower dimensional nilmanifold via the orbit mapping of a maximal toral action. We can then apply [CR5, Theorem 10]. □

5.9. By a <u>special Mostow-Wang group</u> Γ, we shall mean a poly-$\mathbb{Z}$ group Γ for which $\Gamma/{}^n\Gamma$ is free abelian, where ${}^n\Gamma$ is the nilradical of Γ. For such a Γ, we have a unique model manifold $M(\Gamma)$ with a standard maximal toral action, see (5.5). Note that these model manifolds exhaust all closed aspherical manifolds with such fundamental groups, up to homeomorphisms, in dimension $\neq 3,4$, by Wall. Note also that this class of manifolds contains all the nilmanifolds. By the same argument as in (5.6), we have

Theorem. Let Γ be a special Mostow-Wang group, and let G be a compact Lie group acting smoothly and homotopically trivially on $M(\Gamma)$. Assume that $\dim M(\Gamma) - \dim G \neq 3,4$. Then the action of G is topologically equivalent to a restriction of the standard maximal toral action.

Proof. First note that we have deleted the condition that G acts freely from the hypothesis. Nevertheless, it is a consequence that G does act freely by (5.7) after we confirm that $\Gamma/z(\Gamma)$ is torsion-free. First, $\Gamma/{}^n\Gamma$ is free abelian and since ${}^n\Gamma/z({}^n\Gamma)$ is torsion-free, $\Gamma/z({}^n\Gamma)$ is torsion-free. Because $z(\Gamma)$ is a direct summand of $z({}^n\Gamma)$ it follows that $\Gamma/z(\Gamma)$ is torsion-free and so the maximal toral action is free. This torsion-freeness enables us to utilize the argument of (5.6). We replace Farrell-Hsiang's theorem by Wall's to obtain our conclusion. □

Corollary 1. Let Γ be as above and let M be a closed manifold homotopy equivalent to $M(\Gamma)$. Assume that $\operatorname{rank}(\Gamma/z(\Gamma)) > 4$. Then any homotopically trivial smooth action of a compact Lie group on M is conjugate to a restriction of the standard maximal toral action. In particular, any two maximal toral actions on M are conjugate to each other.

Proof. We can work on the model aspherical manifold $M(\Gamma)$ and pull back the result by Wall's theorem. Again we remind the reader that we may have to alter our given maximal toral action by a reparametrization of our torus before we are able to conjugate the actions. □

Corollary 2. Let M be a closed $K(\Gamma,1)$-manifold of dimension $\neq 3,4$, where Γ is as in the theorem. Let K be the kernel of the natural homomorphism $\pi_0(\mathcal{H}(M)) \to \operatorname{Out}\Gamma$. Then no finite subgroup of K can be lifted back to $\mathcal{H}(M)$. □

This corollary states that no exotic finite group of homotopy classes of homeomorphisms homotopic to the identity can be realized as a group of homeomorphisms. It settles a conjecture of Lee and Raymond [LR2, §4] about closed aspherical manifolds in the affirmative for the class of closed nil-manifolds, and more generally, for the class of closed aspherical manifolds with special Mostow-Wang fundamental group. We mention in dimension 3 that if M is assumed to be irreducible (to avoid the Poincaré conjecture), then M is a Haken manifold and $\pi_0(\mathcal{H}(M))$ maps isomorphically onto Out Γ.

5.10. The hypothesis, while convenient, is stronger than is necessary to obtain the results (5.6)-(5.9). What is needed to carry through the arguments given is:

1. The closed aspherical manifold M admits a maximal toral action.
2. M/G_0 is a topological manifold (which holds if the action is smooth).
3. $\pi_1 M/\pi_1(G_0)$ and a torsion-free central extension of $\pi_1 M$ by a finite group F determines the topological type of M/G_0 and M/F. In our case, these were guaranteed by Wall's and Farrell-Hsiang's theorems.

5.11. The reason that we cannot claim (5.9) for the more general class of poly-$\mathbb{Z}$ groups Γ is that the standard maximal toral action on the model aspherical manifold $M(\Gamma)$ is not free in general. For example, the 3-dimensional flat Riemannian manifold $M(\mathcal{G}_2)$ with holonomy group $\mathbb{Z}_2$ has non-free maximal toral action. Observe that this example yields similar higher dimensional examples easily.

Even though we restrict ourselves to the case where the maximal toral actions are free, we cannot apply Wall's theorem anymore because $\Gamma/z(\Gamma)$ is not poly-$\mathbb{Z}$ in general.

<u>Examples</u>. There exists a closed aspherical manifold M with poly-$\mathbb{Z}$ fundamental group admitting a free maximal toral action T^k but $\pi_1(M/T^k)$ is not poly-$\mathbb{Z}$.

Let $\mathcal{G}_6$ be the 3-dimensional flat Riemannian manifold with holonomy $\mathbb{Z}/2 \oplus \mathbb{Z}/2$. $\pi_1(\mathcal{G}_6)$ has trivial center and 0 first betti number. Hence $\pi_1(\mathcal{G}_6)$ is not poly-$\mathbb{Z}$. $\mathcal{G}_6$ is the quotient of the 3-dimensional torus by the holonomy group $\mathbb{Z}/2 \oplus \mathbb{Z}/2$. We may factor this action through $\mathcal{G}_2$ by first dividing out a subgroup $\mathbb{Z}/2$ of $\mathbb{Z}/2 \oplus \mathbb{Z}/2$ to get $\mathcal{G}_2$ and then following by the quotient $\mathbb{Z}/2$ on $\mathcal{G}_2$. If we take the free diagonal action on $S^1 \times \mathcal{G}_2$ and form $M^4 = S^1 \times_{\mathbb{Z}/2} \mathcal{G}_2$ we obtain a flat Riemannian 4-manifold with an S^1-action by translating along the first factor. This action is free since $\mathbb{Z}/2$ on $\mathcal{G}_2$ is free and $M^4/S^1 = \mathcal{G}_6$. Then $z(\pi_1 M^4)$ is $\text{im}(ev^x_*: \pi_1(S^1,1) \to \pi_1(M^4,x))$ because $z(\pi_1\mathcal{G}_6) = 1$ and Proposition (5.1). Therefore $\pi_1(\mathcal{G}_6) = \pi_1 M^4/z(\pi_1 M^4)$, which is not poly-$\mathbb{Z}$.

It is now easy to obtain higher dimensional examples by just adding trivial toral factors to $\mathcal{G}_2$ and hence to $\mathcal{G}_6$.

5.12. Now we examine the faithfulness of the natural representation, $G \to \text{Out}\ \pi_1 M$. Let M be a closed aspherical manifold, G a compact Lie group acting effectively on M. The homomorphism $\phi_0: G \to \text{Out}\ \pi_1 M$ is not injective in general. Let $G_1 = \text{kernel}(\phi_0)$. Suppose G_1 acts freely on M and that M/G_1 is a topological manifold. Then the quotient group G/G_1 naturally acts on M/G_1. Let $G_2 = \text{kernel}(G \to G/G_1 \hookrightarrow \mathcal{H}(M/G_1) \to \text{Out}\ \pi_1(M/G_1))$. Suppose G_2 acts freely on M. Then M/G_2 is a manifold and G/G_2 acts on M/G_2. We could continue this process until $G/G_i \hookrightarrow \mathcal{H}(M/G_i) \to \text{Out}\ \pi_1(M/G_i)$ is faithful. In fact, however, if M admits a maximal toral action, this process surprisingly turns out to terminate at the second step always.

Lemma. Let F be a finite group acting effectively on a closed aspherical manifold M. Let $F_1 = \text{kernel}(F \hookrightarrow \mathcal{H}(M) \to \text{Out}\ \pi_1 M)$ and suppose that F_1 acts freely on M. Then $F/F_1 \to \mathcal{H}(M/F_1) \to \text{Out}\ \pi_1(M/F_1)$ is injective.

Proof. Let $1 \to \Gamma \to E \to F \to 1$ be the lifting sequence of (F,M), where $\Gamma = \pi_1 M$. The restriction of this exact sequence to F_1 is $1 \to \Gamma \to \Gamma_1 \to F_1 \to 1$, where $\Gamma_1 = C_E(\Gamma)\cdot\Gamma$, the subgroup of E generated by $C_E(\Gamma)$ and Γ. Note that $\Gamma_1 = \pi_1(M/F_1)$. Let $F_2 = \text{kernel}(F \to F/F_1 \hookrightarrow \mathcal{H}(M/F_1) \to \text{Out}\ \Gamma_1)$. Since the lifting sequence of

$(F/F_1, M/F_1)$ is $1 \to \Gamma_1 \to E \to F/F_1 \to 1$, the restriction of this sequence of F_2/F_1 is $1 \to \Gamma_1 \to \Gamma_2 \to F_2/F_1 \to 1$, where $\Gamma_2 = C_E(\Gamma_1)\cdot\Gamma_1$ as before. Certainly $\Gamma_1 = \Gamma_2$ so that $F_1 = F_2$. □

Theorem. Let M be a closed aspherical manifold with a special Mostow-Wang fundamental group Γ. Let G be a compact Lie group acting smoothly and effectively on M. Assume that $\dim M - \dim G \neq 3,4$. Let

$G_1 = \text{kernel}(G \hookrightarrow \mathcal{H}(M) \to \text{Out}\ \pi_1 M)$ and

$G_2 = \text{kernel}(G \to G/G_1 \hookrightarrow \mathcal{H}(M/G_1) \to \text{Out}\ \pi_1(M/G_1))$. Then G_2 acts freely on M and the G_1-action is conjugate to a restriction of a maximal toral action on M. On the topological manifold M/G_1, G/G_1 is equivalent to a restriction of a maximal toral action. On M/G_2, the natural homomorphism of the finite group G/G_2 into $\text{Out}\ \pi_1(M/G_2)$ is injective.

Proof. Let $F = G/G_1$ and apply the Lemma to the manifold $M_1 = M/G_1$. Since the finite group $F_1 = G_2/G_1$ acts homotopically trivially on M/G_1, the action is free by Lemma (5.7). This implies that G_2 acts freely on M. By the above Lemma, $G/G_2 \cong F/F_1 \hookrightarrow \mathcal{H}(M_1/F_1) = \mathcal{H}(M/G_2) \to \text{Out}\ \pi_1(M/G_2)$ is injective. □

Observe that $G_1 = G_0 \oplus (G_1/G_0)$ where G_0 is the connected component of the identity of G by (5.6).

Example. Let M_2 be the closed 3-dimensional nilmanifold with euler class 2. Then $0 \to \mathbb{Z} \to \pi_1(M_2) \to \mathbb{Z} \oplus \mathbb{Z} \to 1$ is the extension describing M_2 as a principal circle bundle over the 2-torus T^2. This central extension is represented by $2 \in \mathbb{Z} \cong H^2(Z \oplus Z, Z)$. If we take the subgroup $n\mathbb{Z} \oplus n\mathbb{Z}$ of $\mathbb{Z} \oplus \mathbb{Z}$ and take the natural $\mathbb{Z}_n \times \mathbb{Z}_n$-fold covering of this torus corresponding to $n\mathbb{Z} \oplus n\mathbb{Z}$, this circle bundle lifts to the circle bundle $M_{n^2\cdot 2}$ with euler class $n^2\cdot 2$. $M_{n^2\cdot 2}$ covers M_2 equivariantly with the group of regular covering transformations $\mathbb{Z}_n \times \mathbb{Z}_n$ commuting with the S^1-action. This group $\mathbb{Z}_n \times \mathbb{Z}_n$ embeds in $\text{Out}\ \pi_1(M_{n^2\cdot 2})$ and does not lift to a subgroup of $\text{Aut}\ \pi_1(M_{n^2\cdot 2})$, see [CR2, §8.2]. If we now consider $G = S^1 \times (\mathbb{Z}_n \times \mathbb{Z}_n)$ we see that $G_0 = S^1$, $G_1 = S^1$ and $G_2 = G$.

In a similar situation we can take $\mathbb{Z}_k \subset S^1$ and form $G = \mathbb{Z}_k \times (\mathbb{Z}_n \times \mathbb{Z}_n)$. Then $G_0 = 1$, $G_1 = \mathbb{Z}_k$ and $G_2 = G_1$. Note M/G_1 has euler class $2k\cdot n^2$, and M/G has euler class $2k$. Also if $k = n^2$, M/G non-trivially covers itself.

§6. The Geometric Realization Problem.

6.1. Recall that an <u>abstract kernel</u> is a homomorphism $\Psi\colon G \to \text{Out}\ \pi = \text{Aut}\ \pi/\text{Inn}\ \pi$. Each group extension E of π by the group G, $1 \to \pi \to E \to G \to 1$, defines an abstract kernel by conjugating the elements of π by the elements of E. To each abstract kernel we may attempt to find an extension which <u>algebraically realizes</u> this <u>abstract kernel</u>. There is an obstruction to constructing this extension in $H^3(G, z(\pi))$ which vanishes if and only if some extension, realzing the abstract kernel, exists. If the obstruction vanishes then all extensions realizing this kernel are classified, up to conjugacy, by the elements in $H^2(G, z(\pi))$.

6.2. If π is the fundamental group of an aspherical space M, then $\pi_0\mathcal{E}(M)$, the homotopy classes of the H-space of self-homotopy equivalences of M, is naturally isomorphic to $\text{Out}\ \pi$. Thus $\Psi\colon G \to \text{Out}\ \pi$ has a very significant geometric interpretation when π is the fundamental group of an aspherical space. There are natural homomorphisms $\theta\colon \mathcal{H}(M) \hookrightarrow \mathcal{E}(M) \to \pi_0\mathcal{E}(M) \xrightarrow{\cong} \text{Out}\ \pi$. Thus, the <u>geometric realization</u> (effective geometric realization, resp.) of an <u>abstract kernel</u> $\Psi\colon G \to \text{Out}\ \pi_1(M,x)$ is a lifting homomorphism (injective homomorphism, resp.) $\hat{\Psi}$ of Ψ to $\mathcal{H}(M)$ so that $\Psi = \theta \circ \hat{\Psi}$.

6.3. Let us recall some facts about admissible extensions from [LR1]. An extension of a torsion-free group π by a <u>finite</u> group G, $1 \to \pi \to E \to G \to 1$, is called <u>admissible</u> if $C_E(\pi)$ is torsion-free. We have seen this condition in (1.7) and (3.5) already. This concept is useful because of the following:

If G is an <u>effective</u> action on a closed aspherical manifold M whose fundamental group is π, let us define E to be the group of all lifts of the homeomorphisms in G to the universal covering $\tilde{M}$ of M. <u>Then</u> E <u>is an admissible extension of</u> π <u>by</u> G. Thus, the existence of an admissible extension of π by G is a necessary algebraic requirement for an abstract kernel $\Psi: G \to \text{Out}\,\pi$ to admit an effective geometric realization. So far, no examples are known where this requirement is not sufficient. The first examples exhibiting the failure of geometric realization are given in [RS]. These examples were constructed on nilmanifolds.

In [LR1] the geometric realization problem for flat Riemannian manifolds was posed and completely solved. We will treat the analogous situation for nilmanifolds, infranilmanifolds, and manifolds with poly-$\mathbb{Z}$ fundamental groups.

6.4. <u>Theorem</u>. Let π be a torsion-free, virtually poly-$\mathbb{Z}$ group and let $M(\pi)$ be a model smooth aspherical manifold. Then a finite abstract kernel can be geometrically realized (resp., effectively geometrically realized) as a group of fiber-preserving diffeomorphisms of $M(\pi)$ if and only if the abstract kernel admits an extension (resp., admissible extension).

<u>Proof</u>. Let $G \to \text{Out}\,\pi$ be the given abstract kernel and let $1 \to \pi \to E \to G \to 1$ be an extension realizing the abstract kernel. Recall that $M(\pi)$ is constructed by the Seifert construction using a characteristic subgroup Γ of finite index such that Γ contains the nilradical N of π and Γ/N is free abelian. More precisely, $\pi \subset \text{Diff}(\mathbb{R}^m;K)$ (where $m = \text{rank}\,\Gamma$ and K is the partition of m by the natural filtration of Γ) and $M(\pi) = \pi\backslash\mathbb{R}^m$. Using the extension sequence $1 \to \Gamma \to E \to E/\Gamma \to 1$ (recall that Γ is characteristic in π so that it is normal in E) and the natural filtration of Γ, one can map E into $\text{Diff}(\mathbb{R}^m;K)$, say $\psi: E \to \text{Diff}(\mathbb{R}^m;K)$, by our Seifert construction. Note that $\psi|_\pi$ is injective as π is torsion-free. $\psi(\pi)$ may be different from the original $\pi \subset \text{Diff}(\mathbb{R}^m;K)$. However, the uniqueness of the Seifert construction (1.8) tells us that we may assume that these two embeddings of π are identical (by conjugating by an element of

$\mathrm{Diff}(\mathbb{R}^m;K)$ if necessary). Therefore, $G = E/\pi$ acts on the model manifold $M(\pi)$ smoothly and fiber preservingly (as E maps into $\mathrm{Diff}(\mathbb{R}^m;K)$).

If E is admissible, we can apply Lemma (3.5) as ψ is injective on π. In this case, ψ itself is injective, whence G acts effectively on $M(\pi)$. □

Corollary 1. Let M be a closed aspherical manifold with virtually poly-$\mathbb{Z}$ fundamental group. Then a finite abstract kernel can be (effectively) geometrically realized on a closed smooth manifold M', homotopy equivalent to M, as a group of self-diffeomorphisms of M' if and only if the abstract kernel admits an (admissible) extension.

Proof. Let $\pi_1 M = \pi$. The model aspherical manifold $M(\pi)$ is homotopy equivalent to M. The extension given by the hypothesis gives rise to a geometric realization on $M(\pi)$ by the Theorem, as a group of fiber-preserving self-diffeomorphisms of $M(\pi)$. □

For the class of manifolds for which homotopy equivalences are homotopic to homeomorphisms, the Theorem gives a complete solution to the realization problem. By Wall's and Farrell-Hsiang's theorem, we have

Corollary 2. Let M be a closed aspherical manifold of dimension $\neq 3,4$ with fundamental group either virtually nilpotent or poly-$\mathbb{Z}$. Then a finite abstract kernel can be (effectively) geometrically realized as a group of self-homeomorphisms of M if and only if the abstract kernel admits an (admissible) extension. □

This corollary covers nilmanifolds, flat Riemannian manifolds, infranilmanifolds (or, almost flat Riemannian manifolds) and special solv-manifolds, etc. For infranilmanifolds we have a stronger result without appealing to the Farrell-Hsiang theorem. Furthermore, we do not need the dimension restriction. Simply recall that infranilmanifolds are our model aspherical manifolds because they are constructed by the Seifert construction, see (3.6). As a direct consequence of the theorem, we have

Corollary 3. Let M be an infranilmanifold. A finite abstract kernel can be (effectively) geometrically realized as a group of fiber-preserving self-diffeomorphisms of M if and only if it admits an (admissible) extension. □

6.5. For a torsion-free, virtually poly-$\mathbb{Z}$ group π, Auslander and Johnson [AJ] constructed a closed smooth $K(\pi,1)$-manifold M_π which is finitely covered by a special solv-manifold, as we mentioned in (2.5) Remark (1). If we follow their method closely, we see that their result does extend to also yield a smooth solution to the geometric realization problem for the $K(\pi,1)$-manifold M_π. We suspect that $M(\pi)$ and M_π are diffeomorphic and the actions are equivalent.

Proposition. A finite abstract kernel of π can be geometrically (effectively) realized as a group of self-diffeomorphisms of M_π if and only if it admits an (admissible) extension.

Proof. Choose the predivisible subgroup Γ of π which is used for constructing M_π. Use the same solvable Lie group $D(\Gamma)$ as in [AJ]. It is known that $D(\Gamma) = R(\Gamma) \circ C$, where $R(\Gamma)$ is a connected, simply connected solvable Lie group and C is a compact abelian group. Furthermore, Γ is a uniform lattice of $R(\Gamma)$, and $(\Gamma, D(\Gamma))$ has the unique automorphism extension property.

Let $1 \to \pi \to E \to G \to 1$ be an extension realizing the given abstract kernel. As Γ is characteristic in π, it is normal in E. Let $E/\Gamma = F$ (finite). Then $1 \to \Gamma \to E \to F \to 1$ is exact and one can form the "pushout" so that

$$\begin{array}{ccccccccc} 1 & \to & \Gamma & \to & E & \to & F & \to & 1 \\ & & \downarrow & & \downarrow & & \downarrow = & & \\ 1 & \to & D(\Gamma) & \to & ED(\Gamma) & \to & F & \to & 1 \end{array}$$

is commutative. Let H^* be a maximal compact subgroup of $ED(\Gamma)$ so that $D(\Gamma) \cap H^* = C$. Then E acts on the coset space $ED(\Gamma)/H^*$. Since π is normal in E, $G = E/\pi$ acts on $M_\pi = \pi\backslash ED(\Gamma)/H^*$, which is diffeomorphic to the model manifold M_π.

It remains to show that the constructed G action is effective if $C_E(\pi)$ is torsion-free. However, since $E \to \mathrm{Diff}(ED(\Gamma)/H^*) = \mathrm{Diff}(\tilde{M}_\pi)$ is injective on π already, it is injective on E if the extension is admissible, by Lemma (3.5). □

APPENDIX

A.1. We give, for the convenience of the reader, some more details on the relation between injective Seifert fiberings in say [CR5] and the Seifert constructions we have used here. They are really the same thing but the approach is a little different. For the section on maximal toral actions the point of view of injective Seifert fiberings is perhaps more useful. Some of the facts to be stated here are not obvious and the reader should consult [CR5] or [L1] for details.

For the (injective) Seifert construction with typical fiber a k-torus, one begins with a properly discontinuous action (Q,W) satisfying (*). The object is to construct all possible $T^k \circ Q$ action on $T^k \times W$. As mentioned in (1.2) this means by definition $(t,\alpha)\cdot(s,w) = t\cdot(\alpha\cdot(s,w))$ where $\alpha\cdot(t\cdot(s,w)) = \phi(\alpha)(t)\cdot(\alpha\cdot(s,w))$ for $t,s \in T^k$, $\alpha \in Q$, $w \in W$. Each such $T^k \circ Q$ action on $T^k \times W$ corresponds to a 1-cocycle $m \in Z^1(Q;\mathcal{M}(W,T^k))$. For each $\alpha \in Q$, $m(\alpha) \in \mathcal{M}(W,T^k)$, the abelian group of all continuous maps of W into $T^k = \mathbb{R}^k/\varepsilon(\mathbb{Z}^k)$. Cohomologous cocycles m' correspond to an equivalent $T^k \circ Q$ action obtained by conjugating by an element $\lambda = (\lambda,1,1) \in \mathcal{U}^F(T^k \times W)$ and conversely. Thus each 1-dimensional cohomology class corresponds exactly to a $T^k \circ Q$ action determined up to conjugation by an element of $\mathcal{M}(W,T^k)$. The coboundary $\delta\colon H^1(Q;\mathcal{M}(W,T^k)) \to H^2(Q;\mathbb{Z}^k)$ is a (surjective) isomorphism and $\delta[m]$ gives rise precisely to the conjugacy class of the extension $1 \to \mathbb{Z}^k \to P \to Q \to 1$.

A.2. The above presumes $\rho: Q \to \mathcal{H}(W)$, $\epsilon: \mathbb{Z}^k \to \mathbb{R}^k$, and a section $W \to T^k \times W$ all have been chosen. Then for each $T^k \circ Q$ action the cocycle is determined. (Altering the section corresponds to choosing cohomologous cocycles.)

The notation $G(T^k, X, \phi)$ was used in [CR5, §7] to denote the group of "fiber preserving homeomorphisms" of $Q\backslash(T^k \times W) = X$. This was actually defined as the quotient of $G(T^k, X', \phi)$, the group of homeomorphisms of $T^k \times W$ weakly equivariant with respect to the $T^k \circ Q$ action on $T^k \times W$ by the normal subgroup Q. The group $G(T^k, X', \phi)$ can be lifted to $\mathcal{H}(\mathbb{R}^k \times W)$ and, as such, it is a subgroup of $\mathcal{M}(W, \mathbb{R}^k) \circ (\mathrm{Aut}(\mathbb{R}^k) \times \mathcal{H}(W)) = \mathcal{H}^F(\mathbb{R}^k \times W)$. In fact, it is precisely the normalizer of $\psi(P)$ in $\mathcal{H}^F(\mathbb{R}^k \times W)$.

Theorem 10, one of the main results of [CR5, §8] gave conditions under which every outer automorphism of $\pi_1(X)$ can be geometrically realized by a fiber-preserving homeomorphism of X (i.e., $G(T^k, X, \phi) \to \mathrm{Out}\, \pi_1(X)$ is surjective). This, as the reader may easily discern, is a special case of what we have called rigidity in (1.4). It corresponds to (8.6) in [CR1] in the central case.

The rigidity statement (1.4) of present article, $\psi' \circ \theta \circ \psi^{-1}: \psi(P) \to \psi'(P')$ is given by $\mu(\hat{h})$ for some $\hat{h} \in \mathcal{H}^F(\mathbb{R}^k \times W)$, says that the subgroup $\psi(P)$ is conjugate to $\psi'(P')$ by $\hat{h} \in \mathcal{H}^F(\mathbb{R}^k \times W)$. Now $\hat{h}$ defines a fiber-preserving homeomorphism of $X = \psi(P)\backslash(\mathbb{R}^k \times W)$ to $X' = \psi'(P')\backslash(\mathbb{R}^k \times W)$.

If P acts as covering transformations on $\mathbb{R}^k \times W$ then covering space theory yields the converse to the statement above. Namely, if $X \to X'$ is a fiber-preserving homeomorphism then it is induced from $\mathcal{H}^F(\mathbb{R}^k \times W)$ as above.

A.3. The entire approach to the Seifert constructions arose from the study of injective toral actions [CR1]. There exists an induced T^k action on $X = Q\backslash(T^k \times W)$ only when Q acts trivially on $\mathbb{Z}^k$. This corresponds to the extension $\delta[m]$ being central. The action (T^k, X) is injective (see §5). In this particular situation the toral action on X lifts to the covering space Y associated to the image $ev_*^x(\pi_1(T^k,1)) \subset \pi_1(X,x)$, and splits. That is $(T^k,Y) = (T^k, T^k \times (T^k\backslash Y)) = (T^k, T^k \times W)$ with the lifted T^k action being just left translation. Moreover the covering transformations $Q \cong \pi_1(X,x)/\pi_1(Y,y) = \pi_1(X,x)/ev_*^x(\pi_1(T^k,1))$ commute with the T^k action on Y and so yields our $T^k \times Q$ action on Y. This projects to a properly discontinuous Q action on the simply connected space W and we have the <u>converse of our construction when the extension is central</u>.

To obtain an explicit description now of a Seifert construction as in (1.3) we can choose a cross section $\chi: W = T^k\backslash Y \to Y = T^k \times W$. We can identify $\chi(w)$ with $(1,w) \in T^k \times W$ and so $t\cdot\chi(w)$ would be written as (t,w). Then $\alpha\cdot(s,w) = s\cdot(\alpha\cdot(1,w)) = s\cdot(m(\alpha)(\alpha\cdot w),\alpha\cdot w) = (s+m(\alpha)(\alpha\cdot w),\alpha\cdot w) = (s+m(\alpha)(\alpha\cdot w))\cdot(1,\alpha\cdot w)$. Of course, lifts $\tilde{m}(\alpha): W \to \mathbb{R}^k$ yields our elements of $\mathcal{M}(W,\mathbb{R}^k)$. (The non-central case is very similar.)

A.4. It should be fairly clear now that the iterated Seifert construction could just as well take place on the T^k level.

One particular instance that is of interest in Section 5 concerns the construction associated with the ascending central series for a torsion-free, finitely generated nilpotent group N.

This has been called in §5 our <u>second version for the Seifert construction when the kernel is nilpotent</u>. We shall use "the" connected, simply connected nilpotent Lie group $L(N)$ that contains N as a uniform lattice. The center $z(L(N))$ is a vector space of positive dimension and $L(N)/z(L(N)) = L_1(N) = L(N_1) = L(N/Z^1(N))$, a connected, simply connected nilpotent Lie group of dimension rank N-rank $Z^1(N)$. The lattice $Z^1(N)$ in $z(L)$ is precisely $N \cap z(L)$.

If we look at $N\backslash L(N)$, then $L(N)$ acts transitively on the right and $z(L)$ commutes with the N action. This induces a left $z(L(N))/z(N) = T^{k_0}$ injective action on $N\backslash L(N)$. It lifts to $(T^{k_0}, \mathbb{Z}^{k_0}\backslash L(N))$ and splits to left translation $(z(N)\backslash z(L), (z(N)\backslash z(L)) \times z(L)\backslash L(N))$. Here, of course, we have a $T^{k_0} \times N_1 = (z(N)\backslash z(L(N)) \times (Z^1(N))\backslash N)$ action. If we divide out by the T^{k_0}

translations we get $(N_1, L(N_1))$, cf. with [CR5, (5.7)(d)]. For our situation we iterate this procedure as in (1.6) obtaining our family $\mathfrak{F}(E)$ associated to the ascending central series. At each stage we have a free injective $(T^{k_i}, N_i\backslash L(N_i))$ action.

For any extension, $1 \to \pi \to E \to Q \to 1$, where π is virtually poly-$\mathbb{Z}$, ((2.4) and Remark 3 in (2.5)) we choose the ascending central series for ${}^n\pi$. We then obtain the unique Seifert construction for $({}^n\pi, \pi, E, Q, W)$ by filtering $L(N)$ and taking the induced total actions $(T^{k_i}, N_i\backslash L(N_i))$.

REFERENCES

[A1] L. Auslander, Bieberbach's theorems on space groups and discrete subgroups of Lie groups, Ann. Math. 71 (1960), 579-590.

[A2] __________, On radicals of discrete subgroups of Lie groups, Amer. J. Math., 85 (1963), 145-150.

[AJ] L. Auslander and F. E. A. Johnson, On a conjecture of C. T. C. Wall, J. London Math. Soc. (2) 14 (1976), 331-332.

[C] P. E. Conner, Transformation groups on a $K(\pi,1)$, II, Michigan Math. J. 6 (1959), 413-417.

[CM] P. E. Conner and D. Montgomery, Transformation groups on a $K(\pi,1)$, I, Michigan Math. J. 6 (1959), 405-412.

[CR1] P. E. Conner and Frank Raymond, Actions of compact Lie groups on aspherical manifolds, Topology of Manifolds (Proc. Inst., Univ. of Georgia, Athens, 1969), Markham (1970), 227-264.

[CR2] __________, Manifolds with few periodic homeomorphisms, Proc. Second Conference on Compact Transformation Groups, Part II, Springer Lecture Notes in Math. #299 (1972), 1-75.

[CR3] __________, Injective actions of toral groups, Topology 10 (1971), 283-296.

[CR4] __________, Realizing finite groups of homeomorphisms from homotopy classes of self-homotopy equivalences, Manifold-Tokyo 1973, edited by A. Hattori, Tokyo 1975, 231-238.

[CR5] __________, Deforming homotopy equivalences to homeomorphisms in aspherical manifolds, Bull. AMS 83 (1977), 36-85.

[CR6] __________, Holomorphic Seifert fiberings, Proc. Second Conference on Compact Transformation Groups, Part II, Springer Lecture Notes in Math. #299 (1972), 124-204.

[FH] F. T. Farrell and W. C. Hsiang, Topological characterization of flat and almost flat Riemannian manifolds $M^n (n \neq 3,4)$, American J. Math., 105 (1983), 641-672.

[G] M. Gromov, Almost flat manifolds, J. Differential Geometry, 13 (1978), 231-241.

[H] H. Holmann, Seifertsche Faserräume, Math. Annalen 157 (1964), 138-166.

[J] F. E. A. Johnson, On Poincaré duality groups of poly-linear type and their realizations, Math. Z. 163 (1978), 145-148.

[K] K. Kodaira, Compact complex analytic surfaces, I, Amer. J. Math. 86 (1964), 751-798.

[KT] F. W. Kamber and Ph. Tondeur, Flat manifolds with parallel torsion, J. Differential Geometry, 2 (1968), 385-389.

[L1] K. B. Lee, Aspherical manifolds with virtually 3-step nilpotent fundamental group, American J. Math., 105 (1983), 1435-1453.

[L2] __________, Geometric realization of a finite subgroup of $\pi_0\mathcal{E}(M)$, II, Proc. AMS, 87 (1983), 175-178.

[LR1] K. B. Lee and Frank Raymond, Topological, affine and isometric actions on flat Riemannian manifolds, J. Differential Geometry 16 (1981), 255-269.

[LR2] __________, Ibid II, Topology and its Appl. 13 (1982), 295-310.

[LY] H. B. Lawson and S. T. Yau, Compact manifolds of non-positive curvature, J. Differential Geometry, 7 (1972), 211-228.

[M] A. I. Malcev, On a class of homogeneous spaces, AMS Transl. 39 (1951), 1-33.

[Rag] M. S. Raghunathan, Discrete subgroups of Lie groups, Ergebnisse der Math., vol. 68, Springer, 1972.

[R] Frank Raymond, The Nielson theorem for Seifert fibered spaces over locally symmetric spaces, J. Korean Math. Soc. 16 (1979), 87-93.

[Ruh] E. Ruh, Almost flat manifolds, J. Diff. Geometry, 17 (1982), 1-14.

[RS] Frank Raymond and L. Scott, The failure of Nielsen's theorem in higher dimensions, Arch. Math. 29 (1977), 643-654.

[RW] Frank Raymond and David Wigner, Construction of aspherical manifolds, Geometric Appl. of Homotopy Theory Proc. 1977, Springer Lecture Notes in Math. #657, 408-422.

[S] P. Scott, There are no fake Seifert spaces with infinite π_1, Ann. Math., 117 (1983), 35-70.

[Seifert] H. Seifert, Topologie drei-dimensionaler gefaserter Räume, Acta. Math. 60 (1933), 147-238.

[W] C. T. C. Wall, The topological space-form problems, Topology of Manifolds (Proc. Inst., Univ. of Georgia, Athens, 1969), Markham (1970), 319-331.

[LKR] Y. Kamishima, K. B. Lee and F. Raymond, The Seifert construction and its applications to infranilmanifolds, Quart. J. Math., Oxford (2), 34 (1983), 433-452.

UNIVERSITY OF OKLAHOMA
NORMAN, OK 73019

UNIVERSITY OF MICHIGAN
ANN ARBOR, MI 48109

Received April 1, 1982

Contemporary Mathematics
Volume 33, 1984

THE AUGMENTATION QUOTIENTS OF THE GROUPS OF ORDER 2^4

Gerald Losey and Nora Losey

Dedicated to Roger Lyndon on the occasion of his 65th birthday

1. Introduction

The goal of this paper is to determine the structure of the augmentation quotients of the groups of order 2^4.

Let G be a group, ZG the integral group ring of G and Δ the augmentation ideal of ZG. The abelian groups

$$Q_n(G) = \Delta^n/\Delta^{n+1}, \quad n \geq 1$$

are called the augmentation quotients of G. It is well known that $Q_1(G) \cong G/\gamma_2(G)$ for all G (where by $\gamma_i(G)$ we mean the ith term of the lower central series of G). The structure of $Q_2(G)$ is known for all finitely generated G ([1], [4]). In [12] Tahara has shown how to calculate $Q_3(G)$ for finite G.

In order to determine the structure of $Q_n(G)$ for finite G it suffices to consider the case that G is a finite p-group, p a prime. Passi ([8]) has shown that $Q_n(G) \cong G$ for G cyclic and has found the structure of $Q_n(G)$ in the case that G is an elementary abelian p-group. If G is nilpotent of class c then there exist natural numbers n_o and π, π dividing l.c.m.$\{1,2,\ldots,c\}$, such that $Q_n(G) \cong Q_{n+\pi}(G)$ for all $n \geq n_o$ ([2]). If G is abelian then $\pi = 1$ and in this case Singer ([9]) has called $Q_{n_0}(G)$ the augmentation terminal of G; he has determined the augmentation

© 1984 American Mathematical Society
0271-4132/84 $1.00 + $.25 per page

terminal for all finite abelian groups of exponent dividing 8 and found the order of $Q_{n_o}(G)$ for all finite abelian p-groups ([10], [11]). In [6] we found the structure of $Q_n(G)$ in the case that the lower central series of G satisfies $\gamma_i(G)^p \leq \gamma_{ip}(G)$ for all $i \geq 1$. In [7] the structure of $Q_n(G)$ for G a any group of order p^3 is determined.

There are 5 abelian groups of order 2^4 and the results of Passi ([8]) and Singer ([10]) yield the structure of $Q_n(G)$ in these case. There remain the 9 nonabelian groups of order 2^4 (cf. Burnside [3]). For each of these groups we determine the structure of $Q_n(G)$. The results are summarized in the table in Section 6.

2. <u>The Machinery</u>

In this section we review and refine the techniques used in [6] and [7].

Let G be a finite nilpotent group. For $x \in G$, $x \neq 1$, the <u>weight</u> $w(x)$ of x is the integer k such that $x \in \gamma_k(G) \setminus \gamma_{k+1}(G)$. A sequence $X = (x_1, x_2, \ldots, x_m)$ of elements of G is a <u>uniqueness basis</u> for G with respect to a sequence $R = (r_1, r_2, \ldots, r_m)$ of integers $r_i \geq 2$ if

i) $w(x_i) \leq w(x_j)$ if $i < j$

ii) every element $g \in G$ can be expressed uniquely in the form

$g = x_1^{e_1} x_2^{e_2} \ldots x_m^{e_m}$ where $0 \leq e_i < r_i$ for all i.

Let X be a fixed uniqueness basis for G with respect to the integer sequence R. A <u>proper sequence</u> $\alpha = (e_1, e_2, \ldots, e_m)$ is any m-sequence of non-negative integers; α is a <u>basic sequence</u> if $e_i < r_i$ for all i. The number of basic sequences is $r_1 r_2 \ldots r_m = |G|$. We order the set $\mathcal{P}$ of proper sequences lexicographically; $\mathcal{P}$ is then a well ordered set. The <u>weight</u> of the proper sequence α is defined to be

$$W(\alpha) = \Sigma_{i=1}^{m} e_i w(x_i) .$$

We set $i_\alpha = \max\{w(x_j) : e_j \neq 0\}$.

For a proper sequence $\alpha = (e_1, e_2, \ldots, e_m)$ we define the <u>proper product</u>

$$P(\alpha) = (x_1 - 1)^{e_1} (x_2 - 1)^{e_2} \ldots (x_m - 1)^{e_m} .$$

If α is basic then $P(\alpha)$ is called a <u>basic product</u>. If $W(\alpha) = n$ then we say that $P(\alpha)$ is a proper product of weight n. The unique proper product of weight 0 is $P(\underline{0}) = 1$ where $\underline{0} = (0,0,\ldots,0)$.

The following proposition generalizes slightly the corresponding result in [5] and is proved in essentially the same manner.

<u>Proposition 1</u>. Δ^n is spanned over Z by all proper products of weight $\geq n$.

It is shown in [6] that, for all $n \geq 1$, Δ^n is a free abelian group of rank $|G| - 1$. Thus if we are looking for a free Z-basis for Δ^n it suffices to find $|G| - 1$ proper products of weight $\geq n$ which span Δ^n.

In the following sections C_n will denote the cyclic group of order n and $C_n^{(k)}$ will denote the direct sum of k copies of C_n.

3. <u>The Groups</u> G_1, G_2 <u>and</u> G_3

In this section we consider the three groups

$$G_1 = \langle a,b,c\colon a^4 = b^2 = c^2 = [a,b] = 1,\ [b,c] = a^2\rangle$$

$$G_2 = \langle a,b,c\colon a^4 = b^2 = c^2 = [a,b] = [b,c] = 1,\ [a,c] = a^2\rangle$$

$$G_3 = \langle a,b,c\colon a^4 = b^4 = c^2 = [a,c] = [b,c] = 1,\ a^2 = b^2 = [a,b]\rangle .$$

The argument used to prove Proposition 2.7 of [6] actually proves the stronger result

<u>Proposition 2</u>. Let G be a finite p-group in which $\gamma_i(G)^p \leq \gamma_{ip}(G)$ for all $i \geq 1$. Let X be a fixed uniqueness basis for G with respect to the integer sequence $(p,p,\ldots,p)$. Then for any $n \geq 1$,

$$Q_n(G) \cong C_p^{(s(n))}$$

where $s(n)$ is the number of non-zero basic sequences α with the property that $n - W(\alpha)/(p-1)i_\alpha$ is a non-negative integer.

It is further proved that if X contains t_i elements of weight i, $1 \leq i \leq c =$ class of G, then there exists π dividing l.c.m.$\{1,2,\ldots,c\}$ such that $Q_n(G) = Q_{n+\pi}(G)$ for all $n \geq n_o$ where

$$n_o = (p-1)(t_1 + 2t_2 + \ldots + ct_c - 1) .$$

Each of the groups G_1, G_2 and G_3 satisfy the hypothesis of Proposition 2. We take $X = (a,b,c,a^2)$ and $R = (2,2,2,2)$ where $w(a) = w(b) = w(c) = 1$, $w(a^2) = 2$ in each case. If we now apply Proposition 2 we find

Proposition 3. Let G be any one of the groups G_1, G_2 or G_3. Then

$$Q_n(G) \cong C_n^{(s(n))}$$

where $s(1) = 3$, $s(2) = 7$, $s(3) = 10$ and $s(n) = 11$ for all $n \geq 4$.

4. The Groups G_4, G_5 and G_6

Throughout this section G will denote one of the groups

$G_4 = \langle a,b\colon a^2 = b^8 = 1,\ [a,b] = b^2\rangle$

$G_5 = \langle a,b\colon a^2 = b^8 = 1,\ [b,a] = b^2\rangle$

$G_6 = \langle a,b\colon a^4 = b^8 = 1,\ a^2 = b^4,\ [a,b] = b^2\rangle$.

Each of these groups has a uniqueness basis $X = (a,b,b^2,b^4)$ with respect to the integer sequence $(2,2,2,2)$ where $w(a) = w(b) = 1$, $w(b^2) = 2$, $w(b^4) = 3$. It will be convenient to define

$$\epsilon_G = \epsilon = \begin{cases} 0 & \text{if } a^2 = 1 \\ 1 & \text{if } a^2 = b^4 \end{cases}.$$

For any real number r we set

$$[r] = \begin{cases} 0 & \text{if } r < 0 \\ \text{the greatest integer } \leq r & \text{if } r \geq 0 \end{cases}.$$

Proposition 4. Let G be one of the groups G_5, G_5 or G_6 and let X be the uniqueness basis for G as above. For all $n \geq 1$, Δ^n has a free Z-basis $\mathcal{B}_n$ consisting of the elements

$U_1^{(n)} = 2^{n-1}(a - 1)$

$U_2^{(n)} = 2^{n-1}(b - 1)$

$U_3^{(n)} = 2^{[n-2]}(a - 1)(b - 1)$

$U_4^{(n)} = A_n$

$$U_5^{(n)} = (a-1)A_{n-1}$$
$$U_6^{(n)} = (b-1)A_{n-1}$$
$$U_7^{(n)} = (a-1)(b-1)A_{n-2}$$
$$U_8^{(n)} = 2^{[n/4]}(b^4-1)$$
$$U_9^{(n)} = 2^{[(n-1)/4]}(a-1)(b^4-1)$$
$$U_{10}^{(n)} = 2^{[(n-1)/4]}(b-1)(b^4-1)$$
$$U_{11}^{(n)} = 2^{[(n-2)/4]}(a-1)(b-1)(b^4-1)$$
$$U_{12}^{(n)} = 2^{[(n-2)/4]}(b^2-1)(b^4-1)$$
$$U_{13}^{(n)} = 2^{[(n-3)/4]}(a-1)(b^2-1)(b^4-1)$$
$$U_{14}^{(n)} = 2^{[(n-3)/4]}(b-1)(b^2-1)(b^4-1)$$
$$U_{15}^{(n)} = 2^{[(n-4)/4]}(a-1)(b-1)(b^2-1)(b^4-1)$$

where, for all $m \geq -1$,

$$A_m = 2^{[(m-1)/2]}(b^2-1) - 2^{[(m-3)/4]}(b^2-1)^{\tau_m}(b^4-1)$$

$$\tau_m = \begin{cases} 1 & \text{if } m \equiv 1 \text{ or } 2 \bmod 4 \\ 0 & \text{if } m \equiv 3 \text{ or } 4 \bmod 4 \end{cases} .$$

Proof. Free bases for Δ and Δ^2 are given in [5]. A simple transformation shows that $\mathcal{B}_1$ and $\mathcal{B}_2$ are equivalent to the bases given there. Thus we may assume $n \geq 3$.

Let S_n denote the additive subgroup of Δ spanned by $\mathcal{B}_n$. Since Δ^n is free abelian of rank 15 it suffices to prove that $S_n = \Delta^n$. To show that $S_n \subseteq \Delta^n$ we need only show that $\mathcal{B}_n \subseteq \Delta^n$. To show that $S_n \supseteq \Delta^n$ we will show that every proper product of weight $\geq n$ lies in S_n.

1) $2^k(b^4-1) = (-1)^k(b^4-1)^{k+1} \in \Delta^{4k+3}$ for all $k \geq 0$.

The equality follows by induction of k by means of the identity: $(b^4-1)^2 = -2(b^4-1)$. So it remains to prove $2^k(b^4-1) \in \Delta^{4k+3}$. This is clearly true for $k = 0$. Let $k > 0$ and assume $2^{k-1}(b^4-1) \in \Delta^{4k-1}$. Then

$$2^{k-1}(b^2-1)^2(b^4-1) = -2^k(b^2-1)(b^4-1) + 2^{k-1}(b^4-1)^2$$
$$= -2^k(b^2-1)(b^4-1) - 2^k(b^4-1)$$
$$= -2^k b^2(b^4-1) \in \Delta^{4k+3}.$$

Since b^2 is a unit in ZG, $2^k(b^4-1) \in \Delta^{4k+3}$. Thus the assertion follows by induction on k.

Remark: In the proof of assertion (1) and in what follows we make frequent use of the identity

$$(x-1)^2 = -2(x-1) + (x^2-1).$$

2) $U_r^{(n)} \in \Delta^n$ for $8 \leq r \leq 15$.

If $8 \leq r \leq 15$ then

$$U_r^{(n)} = 2^{[(n-m)/4]}(a-1)^i(b-1)^j(b^2-1)^k(b^4-1)$$

where $m = i + j + 2k$, $0 \leq i,j,k \leq 1$. By (1)

$$2^{[(n-m)/4]}(b^4-1) \in \Delta^{4[(n-m)/4]+3}$$

and so $U_r^{(n)} \in \Delta^{4[(n-m)/4]+3+m}$. Now

$$4[(n-m)/4] + 3 + m > 4((n-m)/4 - 1) + 3 + m = n - 1$$

and so $4[(n-m)/4] + 3 + m \geq n$. Thus $U_r^{(n)} \in \Delta^n$.

3) For $n \geq 1$ there exist integers s and t such that

$$A_n(b^2-1) = -A_{n+2} + sU_8^{(n+2)} + tU_{12}^{(n+2)}.$$

Suppose $n \equiv 0 \bmod 4$, say $n = 4k$, $k \geq 1$. Then

$$A_n(b^2-1) = 2^{2k-1}(b^2-1)^2 - 2^{k-1}(b^2-1)(b^4-1)$$
$$= 2^{2k}(b^2-1) + 2^{2k-1}(b^4-1) - 2^{k-1}(b^2-1)(b^4-1)$$
$$= (2^{2k}(b^2-1) - 2^k(b^2-1)(b^4-1))$$
$$+ 2^{2k-1}(b^4-1) + 2^{k-1}(b^2-1)(b^4-1)$$
$$= -A_{n+2} + 2^{k-1}U_8^{(n+2)} + U_{12}^{(n+2)}.$$

Thus assertion (3) holds in this case.

Now suppose $n \equiv 1 \mod 4$. If $n = 1$ then

$$\begin{aligned} A_1(b^2-1) &= (b^2-1)^2 - (b^2-1)^2(b^4-1) \\ &= -2(b^2-1) + (b^4-1) + 2(b^2-1)(b^4-1) + 2(b^4-1) \\ &= -A_3 + 2U_8^{(3)} + 2U_{12}^{(3)} . \end{aligned}$$

If $n = 4k + 1$, $k \geq 1$, then

$$\begin{aligned} A_n(b^2-1) &= 2^{2k}(b^2-1)^2 - 2^{k-1}(b^2-1)^2(b^4-1) \\ &= -2^{2k+1}(b^2-1) + 2^{2k}(b^4-1) + 2^k(b^2-1)(b^4-1) + 2^k(b^4-1) \\ &= -(2^{2k+1}(b^2-1) - 2^k(b^4-1)) \\ &\quad + 2^{2k}(b^4-1) + 2^k(b^2-1)(b^4-1) \\ &= -A_{n+2} + 2^kU_8^{(n+2)} + U_{12}^{(n+2)} . \end{aligned}$$

The cases $n \equiv 2 \mod 4$ and $n \equiv 3 \mod 4$ are handled in a similar manner.

4) $A_n \in \Delta^n$.

Since $A_1 = A_2 = (b^2-1) - (b^2-1)(b^4-1) \in \Delta^2$ and $A_3 = A_4 = 2(b^2-1) - (b^4-1) = (b^2-1)^2 \in \Delta^4 \subseteq \Delta^3$, the assertion is true for $n \leq 4$. Let $n > 4$ and assume (4) holds for $m < n$. Then $A_{n-2} \in \Delta^{n-2}$ and by (3) $A_{n-2} = -A_n + sU_8^{(n)} + tU_{12}^{(n)}$ for integers s and t. Since $U_8^{(n)}$ and $U_{12}^{(n)}$ belong to Δ^n by (2) it follows that $A_n \in \Delta^n$.

5) $U_r^{(n)} \in \Delta^n$ for $4 \leq r \leq 7$.

This is clearly true for $n = 1,2$. So we may assume $n \geq 3$. By (4), $U_4^{(n)} = A_n \in \Delta^n$ and, since $A_{n-1} \in \Delta^{n-1}$, $U_5^{(n)} = (a-1)A_{n-1} \in \Delta^n$. Similarly, $U_6^{(n)}, U_7^{(n)} \in \Delta^n$.

6) $2^k(b^2 - 1) \in \Delta^{2k+1}$ for all $k \geq 0$.

The assertion is true for $k = 0$. Let $k > 0$ and assume $2^k(b^2-1) \in \Delta^{2k+1}$. Then

$$2^k(b^2-1) = -2^{k+1}(b^2-1) + 2^k(b^4-1) \in \Delta^{2k+3} .$$

By (1), $2^k(b^4-1) \in \Delta^{4k+3} \subseteq \Delta^{2k+3}$. Therefore $2^{k+1}(b^2-1) \in \Delta^{2k+3}$ and the assertion follows by induction on k.

7) $U_2^{(n)} = 2^{n-1}(b - 1) \in \Delta^n$.

This is true for $n = 1$. Since $(b-1)^2 = -2(b-1) + (b^2-1) \in \Delta^2$, it follows that $2(b-1) \in \Delta^2$ and that the assertion is true for $n = 2$. Let $n \geq 2$ and assume the assertion holds for n. Then

$$2^{n-1}(b-1)^2 = -2^n(b-1) + 2^{n-1}(b^2-1) \in \Delta^{n+1} .$$

By (6), $2^{n-1}(b^{2}-1) \in \Delta^{2n-1} \subseteq \Delta^{n+1}$. Thus $2^n(b-1) \in \Delta^{n+1}$ and the assertion follows by induction on n.

8) $U_3^{(n)} = 2^{[n-2]}(a-1)(b-1) \in \Delta^n$.

This is true for $n = 1$. If $n \geq 2$ then by (7) $2^{n-2}(b-1) \in \Delta^{n-1}$ and hence $2^{n-2}(a-1)(b-1) \in \Delta^n$.

9) $U_1^{(n)} = 2^{n-1}(a - 1) \in \Delta^n$.

This is true for $n = 1$. Let $n \geq 1$ and assume $2^{n-1}(a-1) \in \Delta^n$. Then

$$2^{n-1}(a-1)^2 = -2^n(a-1) + \epsilon_G \cdot 2^{n-1}(b^4-1) \in \Delta^{n+1} .$$

By (1), $2^{n-1}(b^4-1) \in \Delta^{4n-1} \subseteq \Delta^{n+1}$ and hence $2^n(a-1) \in \Delta^{n+1}$.

10) $S_n \subseteq \Delta^n$.

This follows from (2), (5), (7), (8) and (9).

Henceforth we will assume that $n \geq 3$ and assume that $S_m = \Delta^m$ for all $m < n$.

11) $2\Delta^{n-1} \subseteq S_n$.

It is sufficient to show that $2U_r^{(n-1)} \in S_n$, $1 \leq r \leq 15$. This is easily seen to be the case for $r = 1, 2, 3$ and $8 \leq r \leq 15$. Now let us look at $2U_4^{(n-1)}$. If $n \equiv 0$ or $2 \bmod 4$ then $A_{n-1} = A_n$ and so $2U_4^{(n-1)} = 2U_4^{(n)} \in \Delta^n$. Suppose $n \equiv 1 \bmod 4$, say $n = 4k + 1, k \geq 1$. Then

$$\begin{aligned} 2A_{n-1} &= 2^{2k}(b^2-1) - 2^k(b^4-1) \\ &= (2^{2k}(b^2-1) - 2^k(b^4-1)) \\ &\qquad - 2^k(b^4-1) + 2^{k-1}(b^2-1)(b^4-1) \\ &= A_n - U_8^{(n)} + U_{12}^{(n)} \in S_n . \end{aligned}$$

If $n \equiv 3 \bmod 4$ then, similarly,

$$2A_{n-1} = A_n + U_8^{(n)} - U_{12}^{(n)} \in S_n .$$

Hence, in any case, $2U_4^{(n-1)} \in S_n$.

Similar arguments show that $2U_r^{(n-1)} \in S_n$, $r = 5, 6, 7$.

12) Let $\alpha = (i,j,k,t)$ be a proper sequence. If $t \geq 1$ and $i + j + 2k + 4t - 1 \geq n$ then

$$P(\alpha) = (a-1)^i(b-1)^j(b^2-1)^k(b^4-1)^k \in S_n .$$

We prove (12) by induction over the well ordered set $\mathcal{P}$ of proper sequences. Assume (12) holds for all sequences $\beta < \alpha$.

If $k \geq 2$ then

$$\begin{aligned} P(\alpha) &= -2(a-1)^i(b-1)^j(b^2-1)^{k-1}(b^4-1)^t \\ &\quad + (a-1)^t(b-1)^j(b^2-1)^{k-2}(b^4-1)^{t+1} \\ &= (a-1)^i(b-1)^j(b^2-1)^{k-1}(b^4-1)^{t+1} \\ &\quad + (a-1)^i(b-1)^j(b^2-1)^{k-2}(b^4-1)^{t+1} \\ &= P(\beta_1) + P(\beta_2) \end{aligned}$$

where $\beta_1 = (i,j,k-1,t+1)$ and $\beta_2 = (i,j,k-2,t+1)$. Now both $\beta_1 < \alpha$ and $\beta_2 < \alpha$ and $i+j+2(k-1)+4(t+1)-1 > i+j+2(k-2)+4(t+1)-1 = i+j+2k+4t-1 \geq n$. Hence, by the induction hypothesis, $P(\beta_1)$, $P(\beta_2)$ and, therefore, $P(\alpha)$ belong to S_n.

Similar arguments show that $P(\alpha) \in S_n$ if $i \geq 2$ or $j \geq 2$. Thus we may assume that $0 \leq i,j,k \leq 1$. Suppose $i+j+2k+4(t-1)-1 \geq n$. Since $n \geq 3$ we must have $t \geq 2$ and so

$$P(\alpha) = -2(a-1)^i(b-1)^j(b^2-1)^k(b^4-1)^{t-1} = -2P(\beta)$$

where $\beta = (i,j,k,t-1) < \alpha$. Since $i+j+2k+4(t-1)-1 \geq n$ it follows from the induction hypothesis that $P(\beta)$ and, hence, $P(\alpha)$ lie in S_n. Thus we may assume

$$i+j+2k+4t-1 \geq n > i+j+2k+4(t-1)-1$$

that is

$$t-1 = [(n-m)/4] \quad \text{where} \quad 0 \le m = i+j+2k \le 4 .$$

Hence, by (1),

$$P(\alpha) = 2^{[(n-m)/4]}(a-1)^i(b-1)^j(b^2-1)^k(b^4-1) = U_r^{(n)}$$

for some r, $8 \le r \le 15$. Thus $P(\alpha) \in S_n$.

13) For all $n \ge 3$, there exist integers s and t such that $(b^2-1)^{[(n+1)/2]} = \pm A_n + sU_8^{(n)} + tU_{12}^{(n)}$.

Since $(b^2-1)^2 = -2(b^2-1) + (b^2-1)(b^4-1) = -A_3 = -A_4$, the assertion is true for $n = 3$ and $n = 4$. Let $n \ge 5$ and assume (13) holds for all m, $3 \le m < n$. Then, taking $m = n - 2$,

$$(b^2-1)^{[(n-1)/2]} = \pm A_{n-2} + s_1U_8^{(n-2)} + t_1U_{12}^{(n-2)}$$

and hence

$$(b^2-1)^{[(n+1)/2]} = \pm A_{n-2}(b^2-1) + s_1U_8^{(n-2)}(b^2-1) + t_1U_{12}^{(n-2)}(b^2-1) .$$

Thus, by (3),

$$\begin{aligned}(b^2-1)^{[(n+1)/2]} &= \pm(-A_n + s_2U_8^{(n)} + t_2U_{12}^{(n)}) \\ &\quad + s2^{[(n-2)/4]}(b^2-1)(b^4-1) \\ &\quad + t2^{[(n-4)/4]}(b^2-1)^2(b^4-1) \\ &= \pm(-A_n + s'U_8^{(n)} + t'U_{12}^{(n)} + sU_{12}^{(n)} \\ &\quad - t2^{[n/4]}(b^2-1)(b^4-1) - t2^{[n/4]}(b^4-1) \\ &= \pm A_n + s''U_8^{(n)} + t''U_{12}^{(n)}\end{aligned}$$

for integers s'' and t''. Thus the assertion follows by induction on n.

14) Let $\alpha = (i,j,k,t)$ be a proper sequence. If $t = 0$ and $i + j + 2k \ge n$ then

$$P(\alpha) = (a-1)^i(b-1)^j(b^2-1)^k$$

belongs to S_n.

If $i \ge 2$ then

$$P(\alpha) = -2(a-1)^{i-1}(b-1)^j(b^2-1)^k + \epsilon_G\cdot(a-1)^{i-2}(b-1)^j(b^2-1)^k(b^4-1)$$
$$= -2P(\beta_1) + \epsilon_G\cdot P(\beta_2)$$

where $\beta_1 = (i-1,j,k,0)$ and $\beta_2 = (i-2,j,k,1)$. Since β_1 has weight $(i-1) + j + 2k \geq n-1$, $P(\beta_1) \in \Delta^{n-1}$ and hence, by (11), $2P(\beta_1) \in S_n$. Since $(i-2) + j + 2k + 4\cdot 1 - 1 > i + j + 2k \geq n$ it follows from (12) that $P(\beta_2) \in S_n$. Thus $P(\alpha) \in S_n$.

Similarly, $j \geq 2$ implies $P(\alpha) \in S_n$. Thus we may assume $0 \leq i,j \leq 1$. Suppose $i + j + 2(k-1) \geq n$. Then $k \geq 2$ and

$$P(\alpha) = -2(a-1)^i(b-1)^j(b^2-1)^{k-1} + (a-1)^i(b-1)^j(b^2-1)^{k-2}(b^4-1)$$
$$= -2P(\beta_1) + P(\beta_2)$$

where $\beta_1 = (i,j,k-1,0)$ and $\beta_2 = (i,j,k-2,1)$. Since $\beta_1 < \alpha$ and $i + j + 2(k-1) \geq n$, $P(\beta_1) \in S_n$ by the induction hypothesis. Since $i + j + 2(k-2) + 4\cdot 1 - 1 = i + j + 2k - 1 > i + j + 2(k-1) \geq n$, $P(\beta_2) \in S_n$ by (12). Hence $P(\alpha) \in S_n$.

Thus we may assume $i + j + 2k \geq n > i + j + 2(k-1)$, that is, $k = [((n + 1 - (i + j))/2]$. Let $m = i + j$. Then $0 \leq m \leq 2$ and, by (13),

$$P(\alpha) = (a-1)^i(b-1)^j(b^2-1)^{[((n-m)+1)/2]}$$
$$= (a-1)^i(b-1)^j(\pm A_{n-m} + sU_8^{(n-m)} + tU_{12}^{(n-m)})$$
$$= \pm U_x^{(n)} + sU_y^{(n)} + tU_z^{(n)} \in S_n$$

where $(x,y,z) = (4,8,12)$, $(5,9,13)$, $(6,10,14)$ or $(7,11,15)$ depending on whether $(i,j) = (0,0)$, $(1,0)$, $(0,1)$ or $(1,1)$.

This completes the proof of (14).

15) If α is a proper sequence of weight $\geq n$ then $P(\alpha) \in S_n$.

Let $\alpha = (i,j,k,t)$. If $t > 0$ then $1 + j + 2k + 4t - 1 \geq i + j + 2k + 3t = W(\alpha) \geq n$ and so, by (12), $P(\alpha) \in S_n$. If $t = 0$ then $i + j + 2k = W(\alpha) \geq n$ and, by (14), $P(\alpha) \in S_n$.

It follows from (15) that $\Delta^n \subseteq S_n$ and, therefore, that $\Delta^n = S_n$. Hence Proposition 4 is proved.

Proposition 5. Let G be one of the groups G_4, G_5 or G_6. Then

$$Q_n(G) \cong C_2^{(s(n))}$$

where $s(1) = 2$, $s(2) = 4$, $s(3) = 6$ and $s(n) = 7$ for all $n \geq 4$.

Proof. Let $n \geq 4$. By direct calculation the elements of the basis $\mathcal{B}_{n+1}$ for Δ^{n+1} can be expressed in terms of the basis $\mathcal{B}_n$ for Δ^n as in the following table (Table I). From this table we can immediately read off that

$$Q_n(G) = \Delta^n/\Delta^{n+1} \cong C_2^{(7)}$$

for all $n \geq 4$.

If $n = 3$ then the only change in the last column of the table is: $U_{15}^{(4)} = U_{15}^{(3)}$ (instead of $2U_{15}^{(3)}$). Thus

$$Q_3(G) = \Delta^3/\Delta^4 = C_2^{(6)} .$$

If $n = 1$ then, by the well known isomorphism $\Delta/\Delta^2 \cong G/G$, we find that $Q_1(G) \cong C_2^{(2)}$. If $n = 2$ then we can either calculate from the bases given in Proposition 4 or use the isomorphism

$$Q_2(G) \cong \gamma_2(G)/\gamma_3(G) \oplus Sp^2(G/\gamma_2(G))$$

(cf. [4]) to obtain $Q_2(G) \cong C_2^{(4)}$.

4. The Groups G_7 and G_8

In this section G will denote one of the groups

$$G_7 = \langle a,b,c\colon a^2 = b^4 = c^2 = [a,c] = [b,c] = 1,\ [a,b] = c\rangle$$

or

$$G_8 = \langle a,b,c\colon a^4 = b^4 = c^2 = [a,c] = [b,c] = 1,\ [a,b] = a^2 = c\rangle .$$

Each of these groups have a uniqueness basis $X = (a,b,b^2,c)$ with respect to the integer sequence $(2,2,2,2)$ where $w(a) = w(b) = w(b^2) = 1$, $w(c) = 2$. We set

TABLE I

BASIS FOR Δ^{n+1}, $n \geq 4$

	$n \equiv 0 \bmod 4$	$n \equiv 1 \bmod 4$	$n \equiv 2 \bmod 4$	$n \equiv 3 \bmod 4$
$U_1^{(n+1)} =$	$2U_1^{(n)}$	$2U_1^{(n)}$	$2U_1^{(n)}$	$2U_1^{(n)}$
$U_2^{(n+1)} =$	$2U_2^{(n)}$	$2U_2^{(n)}$	$2U_2^{(n)}$	$2U_2^{(n)}$
$U_3^{(n+1)} =$	$2U_3^{(n)}$	$2U_3^{(n)}$	$2U_3^{(n)}$	$2U_3^{(n)}$
$U_4^{(n+1)} =$	$2U_4^{(n)}+U_8^{(n)}-U_{12}^{(n)}$	$U_4^{(n)}$	$2U_4^{(n)}-U_8^{(n)}+U_{12}^{(n)}$	$U_4^{(n)}$
$U_5^{(n+1)} =$	$U_5^{(n)}$	$2U_5^{(n)}+U_9^{(n)}-U_{13}^{(n)}$	$U_5^{(n)}$	$2U_5^{(n)}-U_9^{(n)}+U_{13}^{(n)}$
$U_6^{(n+1)} =$	$U_6^{(n)}$	$2U_6^{(n)}+U_{10}^{(n)}-U_{14}^{(n)}$	$U_6^{(n)}$	$2U_6^{(n)}-U_{10}^{(n)}+U_{14}^{(n)}$
$U_7^{(n+1)} =$	$2U_7^{(n)}-U_{11}^{(n)}+U_{15}^{(n)}$	$U_7^{(n)}$	$2U_7^{(n)}+U_{11}^{(n)}-U_{15}^{(n)}$	$U_7^{(n)}$
$U_8^{(n+1)} =$	$U_8^{(n)}$	$U_8^{(n)}$	$U_8^{(n)}$	$2U_8^{(n)}$
$U_9^{(n+1)} =$	$2U_9^{(n)}$	$U_9^{(n)}$	$U_9^{(n)}$	$U_9^{(n)}$
$U_{10}^{(n+1)} =$	$2U_{10}^{(n)}$	$U_{10}^{(n)}$	$U_{10}^{(n)}$	$U_{10}^{(n)}$
$U_{11}^{(n+1)} =$	$U_{11}^{(n)}$	$2U_{11}^{(n)}$	$U_{11}^{(n)}$	$U_{11}^{(n)}$
$U_{12}^{(n+1)} =$	$U_{12}^{(n)}$	$2U_{12}^{(n)}$	$U_{12}^{(n)}$	$U_{12}^{(n)}$
$U_{13}^{(n+1)} =$	$U_{13}^{(n)}$	$U_{12}^{(n)}$	$2U_{13}^{(n)}$	$U_{13}^{(n)}$
$U_{14}^{(n+1)} =$	$U_{14}^{(n)}$	$U_{14}^{(n)}$	$2U_{14}^{(n)}$	$U_{14}^{(n)}$
$U_{15}^{(n+1)} =$	$U_{15}^{(n)}$	$U_{15}^{(n)}$	$U_{15}^{(n)}$	$2U_{15}^{(n)}$

$$\epsilon = \begin{cases} 0 & \text{if } a^2 = 1 \\ 1 & \text{if } a^2 = c. \end{cases}$$

Proposition 6. Let G be one of the groups G_7 or G_8 and let X be a uniqueness basis for G as above. For all $n \geq 1$, Δ^n has a free Z-basis $\mathcal{B}_n$ consisting of the elements

$$V_1^{(n)} = 2^{n-1}(a-1)$$
$$V_2^{(n)} = (b-1)^n$$
$$V_3^{(n)} = 2^{[n/2]}(b^2-1)$$
$$V_4^{(n)} = 2^{[n-2]}(a-1)(b-1)$$
$$V_5^{(n)} = 2^{[(n-2)/2]}(a-1)(b^2-1)$$
$$V_6^{(n)} = 2^{[(n-1)/2]}(b-1)(b^2-1)$$
$$V_7^{(n)} = 2^{[(n-1)/2]}(c-1)$$
$$V_8^{(n)} = 2^{[(n-3)/2]}(a-1)(b-1)(b^2-1)$$
$$V_9^{(n)} = 2^{[(n-2)/2]}(a-1)(c-1)$$
$$V_{10}^{(n)} = 2^{[(n-2)/2]}(b-1)(c-1)$$
$$V_{11}^{(n)} = 2^{[(n-2)/2]}(b^2-1)(c-1)$$
$$V_{12}^{(n)} = 2^{[(n-3)/2]}(a-1)(b-1)(c-1)$$
$$V_{13}^{(n)} = 2^{[(n-4)/2]}(a-1)(b^2-1)(c-1)$$
$$V_{14}^{(n)} = 2^{[(n-4)/2]}(b-1)(b^2-1)(c-1)$$
$$V_{15}^{(n)} = 2^{[(n-5)/2]}(a-1)(b-1)(b^2-1)(c-1).$$

Proof. Let S_n be the subgroup of Δ spanned by the $V_i^{(n)}$, $1 \leq i \leq 15$. As in the proof of Proposition 3.1 it is sufficient to show that $S_n = \Delta^n$. The proof is similar to (but simpler than) the proof of Proposition 4, and so we will only sketch it here. We begin by establishing

(1) For all $k \geq 0$,

(a) $2^k(b^2-1) \in \Delta^{2k+1}$

(b) $2^k(a-1)(b^2-1) \in \Delta^{2k+3}$

(c) $2^k(c-1) \in \Delta^{2k+2}$

(d) $2^k(a-1) \in \Delta^{k+1}$

(e) $2^k(b^2-1)(c-1) \in \Delta^{2k+4}$.

Each part of (1) is proved by induction on k. It follows from (1) that

(2) $v_i^{(n)} \in \Delta^n$, $1 \le i \le 15$

and, therefore, that

(3) $S_n \subseteq \Delta^n$ for all $n \ge 1$.

To get the opposite inclusion we show

(4) Let $\alpha = (i,j,k,t)$ be a non-zero proper sequence and let $m \ge 0$ be an integer.

(a) If $t \ge 1$ and $i+j+2k+2t+2m \ge n$ then
$2^m P(\alpha) = 2^m(a-1)^i(b-1)^j(b^2-1)^k(c-1)^t \in S_n$.

(b) If $t = 0$, $i \ge 1$, $k \ge 1$ and $i+j+2k+2m \ge n$ then
$2^m P(\alpha) = 2^m(a-1)^i(b-1)^j(b^2-1)^k \in S_n$.

(c) If $i = t = 0$, $k \ge 1$ and $j+2k+2m-1 \ge n$ then
$2^m P(\alpha) = 2^m(b-1)^j(b^2-1)^k \in S_n$.

(d) If $k = t = 0$, $i \ge 1$ and $i+j+m \ge n$ then
$2^m P(\alpha) = 2^m(a-1)^i(b-1)^j \in S_n$.

(e) If $i = k = t = 0$ and $j \ge n$ then
$(b-1)^j \in S_n$.

Assertions (a) - (d) are proved by induction over the well ordered set $\mathcal{P}$ of proper sequences; assertion (e) is proved by induction on j.

It is now easily seen that if $\alpha = (i,j,k,t)$ is a proper sequence of weight $W(\alpha) = i+j+k+2t \ge n$ then $P(\alpha) \in S_n$. Therefore, by Proposition 1, $\Delta^n \subseteq S_n$. Thus $\Delta^n = S_n$ and the proposition is proved.

It is clear that the elements $v_2^{(n)}$, $v_3^{(n)}$ and $v_6^{(n)}$ span the nth power $\Delta_B{}^n$ of the augmentation ideal Δ_B of ZB where $B = \langle b \rangle$, the cyclic subgroup of order 4 generated by b. But $\Delta_B{}^n$ has a basis consisting of the elements $(b-1)^n$, $(b-1)^{n+1}$ and $(b-1)^{n+2}$ (cf. [7]). Thus we can replace $v_2^{(n)}$, $v_3^{(n)}$ and $v_6^{(n)}$ in $\mathcal{B}_n$ by these elements to obtain:

Proposition 7. Let G be one of the groups G_7 or G_8 and let X be a uniqueness basis for G as in Proposition 4.1. For all $n \geq 1$, Δ^n has a free basis

$$\mathcal{B}_n^* = \{(b-1)^n, (b-1)^{n+1}, (b-1)^{n+2}, V_i^{(n)} \mid i \neq 2,3,6\}.$$

Now consider the basis elements $(b-1)^{n+1}$, $(b-1)^{n+2}$ and $(b-1)^{n+3}$ in $\mathcal{B}_{n+1}^*$. We have

$$(b-1)^{n+3} = -4(b-1)^n - 6(b-1)^{n+1} - 4(b-1)^{n+2} .$$

Thus we may replace the basis element $(b-1)^{n+3}$ in $\mathcal{B}_{n+1}^*$ by the element $4(b-1)^n$ to obtain a new basis $\mathcal{B}_{n+1}^{**}$ for Δ^{n+1}. In Table II we list the bases $\mathcal{B}_n^*$ and $\mathcal{B}_{n+1}^{**}$ for Δ^n and Δ^{n+1}, respectively. A comparison of these bases gives us the structure of $Q_n(G)$.

Proposition 8. Let G be one of the groups G_7 or G_8. Then

$$Q_n(G) \cong C_4 \times C_2^{(s(n))}$$

where $s(1) = 1$, $s(2) = 3$, $s(3) = 5$, $s(4) = 6$ and $s(n) = 7$ for all $n \geq 5$.

5. The Group G_9

Throughout this section G will denote the group

$$G_9 = \langle a,b \mid a^2 = b^8 = 1, [a,b] = b^4 \rangle .$$

This group has a uniqueness basis $X = (a,b,b^4)$ with respect to the integer sequence $(2,4,2)$ where $w(a) = w(b) = 1$ and $w(b^4) = 2$.

Proposition 9. Let $G = G_9$ and let $X = (a,b,b^4)$ be a uniqueness basis for G as above. Then, for all $n \geq 3$, Δ^n has a free Z-basis consisting of the elements

$$W_1^{(n)} = 2^{n-1}(a-1)$$
$$W_2^{(n)} = (b-1)^n$$
$$W_3^{(n)} = (a-1)(b-1)^{n-1}$$
$$W_4^{(n)} = (b-1)^{n+1}$$

TABLE II

$\mathcal{B}_n^{\ *}$	$\mathcal{B}_{n+1}^{\ **}$
$2^{(n-1)}(a-1)$	$2^n(a-1)$
$(b-1)^n$	$4(b-1)^n$
$(b-1)^{n+1}$	$(b-1)^{n+1}$
$2^{[n-2]}(a-1)(b-1)$	$2^{n-1}(a-1)(b-1)$
$2^{[(n-2)/2]}(a-1)(b^2-1)$	$2^{[(n-1)/2]}(a-1)(b^2-1)$
$(b-1)^{n+2}$	$(b-1)^{n+2}$
$2^{[(n-1)/2]}(c-1)$	$2^{[n/2]}(c-1)$
$2^{[(n-3)/2]}(a-1)(b-1)(b^2-1)$	$2^{[(n-2)/2]}(a-1)(b-1)(b^2-1)$
$2^{[(n-2)/2]}(a-1)(c-1)$	$2^{[(n-1)/2]}(a-1)(c-1)$
$2^{[(n-2)/2]}(b-1)(c-1)$	$2^{[(n-1)/2]}(b-1)(c-1)$
$2^{[(n-3)/2]}(b^2-1)(c-1)$	$2^{[(n-2)/2]}(b^2-1)(c-1)$
$2^{[(n-3)/2]}(a-1)(b-1)(c-1)$	$2^{[(n-2)/2]}(a-1)(b-1)(c-1)$
$2^{[(n-4)/2]}(a-1)(b^2-1)(c-1)$	$2^{[(n-3)/2]}(a-1)(b^2-1)(c-1)$
$2^{[(n-4)/2]}(b-1)(b^2-1)(c-1)$	$2^{[(n-3)/2]}(b-1)(b^2-1)(c-1)$
$2^{[(n-5)/2]}(a-1)(b-1)(b^2-1)(c-1)$	$2^{[(n-4)/2]}(a-1)(b-1)(b^2-1)(c-1)$

$$
\begin{aligned}
W_5^{(n)} &= 2^{[(n+1)/4]}(b^4-1)\\
W_6^{(n)} &= (a-1)(b-1)^n\\
W_7^{(n)} &= (b-1)^{n+2}\\
W_8^{(n)} &= 2^{[(n-1)/4]}(a-1)(b^4-1)\\
W_9^{(n)} &= 2^{[n/4]}(b-1)(b^4-1)\\
W_{10}^{(n)} &= 2(a-1)(b-1)^{n-2}\\
W_{11}^{(n)} &= 2^{[(n-2)/4]}(a-1)(b-1)(b^4-1)\\
W_{12}^{(n)} &= 2^{[(n-1)/4]}(b-1)^2(b^4-1)\\
W_{13}^{(n)} &= 2^{[(n-3)/4]}(a-1)(b-1)^2(b^4-1)\\
W_{14}^{(n)} &= 2^{[(n-2)/4]}(b-1)^3(b^4-1)\\
W_{15}^{(n)} &= 2^{[(n-4)/4]}(a-1)(b-1)^3(b^4-1)\ .
\end{aligned}
$$

Proof. Let S_n be the subgroup of Δ spanned by the $W_i^{(n)}$, $1 \leq i \leq 15$. As before it suffices to show that $S_n = \Delta^n$. Again we will merely sketch the proof. We begin by showing

(1) For all $k \geq 0$,

(a) $2^k(b^4-1) \in \Delta^{4k+2}$

(b) $2^k(a-1)(b^4-1) \in \Delta^{4k+4}$

(c) $2^k(a-1) \in \Delta^{k+1}$.

Each of these assertions is proved by induction on k.

It follows from (1) that

(2) $W_i^{(n)} \in \Delta^n$, $1 \leq i \leq 15$

and, thus, that

(3) $S_n \subseteq \Delta^n$.

To establish the opposite inclusion we show

(4) Let $\alpha = (i,j,k)$ be a non-zero proper sequence and $m \geq 0$ an integer.

(a) If $i = 0$, $k \geq 1$ and $j + 4k + 4m \geq n$ then
$2^m P(\alpha) = 2^m(b-1)^j(b^4-1)^k \in S_n$.

(b) If $i \geq 1$, $k \geq 1$ and $i + j + 4k + 4m - 1 \geq n$ then
$2^m P(\alpha) = 2^m(a-1)^i(b-1)^j(b^4-1)^k \in S_n$.

(c) If $k = 0$ and $i + j \geq n$ then
$P(\alpha) = (a-1)^i(b-1)^j \in S_n$.

Assertions (a) and (b) are proved by induction over the well ordered set $\mathcal{P}$ of proper sequences. Assertion (c) is proved by a double induction on i and j.

Now if $\alpha = (i,j,k)$ is any proper sequence of weight $W(\alpha) = i + j + 2k \geq n$ it follows from (4) that $P(\alpha) \in S_n$. Therefore, by Proposition 1, $\Delta^n \subseteq S_n$. Hence $\Delta^n = S_n$ and the proposition is proved.

Let S_n be the subgroup of Δ^n spanned by all those basic elements $W_i^{(n)}$ which contain a factor b^4-1, that is,

$$S_n = \langle W_i^{(n)} \mid i = 5,8,9,11,12,13,14,15 \rangle .$$

The arguments used in proving (4) in the preceding proof actually yield

<u>Lemma</u>. Let $n \geq 3$ and let $\alpha = (i,j,k)$ be a proper sequence.

i) If $i = 0$, $k \geq 1$, $m \geq 0$ and $j + 4k + 4m - 2 \geq n$ then

$$2^m P(\alpha) = 2^n (b-1)^j (b^4-1)^k \in S_n .$$

ii) If $i \geq 1$, $k \geq 1$, $m \geq 0$ and $i+j+4k+4m-1 \geq n$ then

$$2^m P(\alpha) = 2^m (a-1)^i (b-1)^j (b^4-1)^k \in S_n .$$

Now consider the basis $\mathcal{B}_n$ for Δ^n, $n \geq 4$. The basis element $W_7^{(n)} = (b-1)^{n+2}$ can be expressed as

$$(b-1)^{n+2} = -4(b-1)^{n-1} - 6(b-1)^m - 4(b-1)^{n+1} + (b-1)^{n-2}(b^4-1) .$$

Since $(b-1)^n$ and $(b-1)^{n+1}$ are in $\mathcal{B}_4$ and, by Lemma 5.2, $(b-1)^{n-2}(b^4-1)$ is a linear combination of elements of $\mathcal{B}_n$ different from $(b-1)^{n+2}$, it follows that by replacing $(b-1)^{n+2}$ in $\mathcal{B}_n$ by the element $4(b-1)^{n-1}$ we obtain a new basis $\mathcal{C}_n$ for Δ^n.

Further, the element $W_6^{(n)} = (a-1)(b-1)^n$ in $\mathcal{C}_n$ can be expressed as

$$(a-1)(b-1)^n = -4(a-1)(b-1)^{n-3} - 6(a-1)(b-1)^{n-2}$$
$$- 4(a-1)(b-1)^{n-1} + (a-1)(b-1)^{n-4}(b^4-1) .$$

Since $2(a-1)(b-1)^{n-2}$ and $(a-1)(b-1)^{n-1}$ are in $\mathcal{C}_n$ and, by the lemma, $(a-1)(b-1)^{n-4}(b^4-1)$ is a linear combination of elements of $\mathcal{C}_n$ different from $(a-1)(b-1)^n$ it follows that by replacing $(a-1)(b-1)^n$ in $\mathcal{C}_n$ by the element $4(a-1)(b-1)^{n-3}$ we obtain a new basis $\mathcal{D}_n$ for Δ^n.

In Table III we list the bases $\mathcal{B}_n$ and $\mathcal{D}_{n+1}$ for Δ^n and Δ^{n+1}, respectively, for $n \geq 3$. A comparison of these bases gives us the structure of $Q_n(G)$ for $n \geq 3$. The structures of $Q_1(G)$ and $Q_2(G)$ are known from previous results (cf. [4]).

<u>Proposition 10</u>. Let $G = G_9$. Then

$$Q_n(G) \cong C_4 \times C_2^{(s(n))}$$

where $s(1) = 1$, $s(2) = 3$, $s(3) = 4$ and $s(n) = 5$ for all $n \geq 4$.

TABLE III

BASES FOR Δ^n AND Δ^{n+1}

$\mathcal{B}_n$	$\mathcal{D}_{n+1}$
$2^{n-1}(a-1)$	$2^n(a-1)$
$(b-1)^n$	$4(b-1)^n$
$(a-1)(b-1)^{n-1}$	$2(a-1)(b-1)^{n-1}$
$(b-1)^{n+1}$	$(b-1)^{n+1}$
$2^{[(n+1)/4]}(b^4-1)$	$2^{[(n+2)/4]}(b^4-1)$
$(a-1)(b-1)^n$	$(a-1)(b-1)^n$
$(b-1)^{n+2}$	$(b-1)^{n+2}$
$2^{[(n-1)/4]}(a-1)(b^4-1)$	$2^{[n/4]}(a-1)(b^4-1)$
$2^{[n/4]}(b-1)(b^4-1)$	$2^{[(n+1)/4]}(b-1)(b^4-1)$
$2(a-1)(b-1)^{n-2}$	$4(a-1)(b-1)^{n-2}$
$2^{[(n-2)/4]}(a-1)(b-1)(b^4-1)$	$2^{[(n-1)/4]}(a-1)(b-1)(b^4-1)$
$2^{[(n-1)/4]}(b-1)^2(b^4-1)$	$2^{[n/4]}(b-1)^2(b^4-1)$
$2^{[(n-3)/4]}(a-1)(b-1)^2(b^4-1)$	$2^{[(n-2)/4]}(a-1)(b-1)^2(b^4-1)$
$2^{[(n-2)/4]}(b-1)^3(b^4-1)$	$2^{[(n-1)/4]}(b-1)^3(b^4-1)$
$2^{[(n-4)/4]}(a-1)(b-1)^3(b^4-1)$	$2^{[(n-3)/4]}(a-1)(b-1)^3(b^4-1)$

6. Summary

In this section we will summarize the results obtained in this paper and, for the sake of completeness, give the structure of $Q_n(G)$ for G an abelian group of order 2^4.

1. If $G = C_{16}$ then, for all $n \geq 1$,

$$Q_n(G) \cong C_{16} .$$

2. If $G = C_8 \times C_2$ then

$$Q_n(G) = C_8 \times C_2^{(s(n))}$$

where $s(1) = 1$, $s(2) = 2$, $s(3) = 3$ and $s(n) = 4$ for all $n \geq 4$.

3. If $G = C_4^{(2)}$ then

$$Q_1(G) \cong C_4^{(2)}$$
$$Q_2(G) \cong C_4^{(3)}$$
$$Q_3(G) \cong C_4^{(3)} \times C_2$$
$$Q_4(G) \cong C_4^{(3)} \times C_2^{(2)}$$
$$Q_n(G) \cong C_4^{(3)} \times C_2^{(3)}$$

for all $n \geq 5$.

4. If $G = C_4 \times C_2^{(2)}$ then

$$Q_n(G) \cong C_4 \times C_2^{(s(n))}$$

where $s(1) = 2$, $s(2) = 5$, $s(3) = 8$ and $s(n) = 9$ for all $n \geq 4$.

5. If $G = C_2^{(4)}$ then

$$Q_n(G) = C_2^{(s(n))}$$

where $s(1) = 4$, $s(2) = 10$, $s(3) = 14$ and $s(n) = 15$ for all $n \geq 4$.

6. If G is one of the groups

$$G_1 = \langle a,b,c \mid a^4 = b^2 = c^2 = [a,b] = 1,\ [b,c] = a^2 \rangle$$
$$G_2 = \langle a,b,c \mid a^4 = b^2 = c^2 = [a,b] = [b,c] = 1,\ [a,c] = a^2 \rangle$$
$$G_3 = \langle a,b,c \mid a^4 = b^4 = c^2 = [a,c] = [b,c] = 1,\ a^2 = b^2 = [a,b] \rangle$$

then

$$Q_n(G) = C_2^{(s(n))}$$

where $s(1) = 3$, $s(2) = 7$, $s(3) = 10$ and $s(n) = 1$ for $n \geq 4$.

7. If G is one of the groups

$G_4 = \langle a,b \mid a^2 = b^8 = 1, [a,b] = b^2 \rangle$

$G_5 = \langle a,b \mid a^2 = b^8 = 1, [b,a] = b^2 \rangle$

$G_6 = \langle a,b \mid a^4 = b^8 = 1, a^2 = b^4, [a,b] = b^2 \rangle$

then

$$Q_n(G) \cong C_2^{(s(n))}$$

where $s(1) = 2$, $s(2) = 4$, $s(3) = 6$ and $s(n) = 7$ for all $n \geq 4$.

8. If G is one of the groups

$G_7 = \langle a,b,c \mid a^2 = b^4 = c^2 = [a,c] = [b,c] = 1, [a,b] = c \rangle$

$G_8 = \langle a,b,c \mid a^4 = b^4 = c^2 = [a,c] = [b,c] = 1, [a,b] = a^2 = c \rangle$

then

$$Q_n(G) \cong C_4 \times C_2^{(s(n))}$$

where $s(1) = 1$, $s(2) = 3$, $s(3) = 5$, $s(4) = 6$ and $s(n) = 7$ for all $n \geq 5$.

9. If $G = G_9 = \langle a,b \mid a^2 = b^8 = 1, [a,b] = b^4 \rangle$ then

$$Q_n(G) \cong C_4 \times C_2^{(s(n))}$$

where $s(1) = 1$, $s(2) = 3$, $s(3) = 4$ and $s(n) = 5$ for all $n \geq 4$.

Parts (1) and (5) are due to Passi ([8]). The terminal values in parts (2), (3) and (4) were found by Singer ([9], [10]); the structure of $Q_n(G)$ for small values of n were calculated by methods similar to those used in the preceding sections.

REFERENCES

1. Bachmann, F. and Grünenfelder, L., Homological methods and the third dimension subgroup, Comm. Helv., 47 (1972), 526-531.

2. Bachmann, F. and Grünenfelder, L., The periodicity in the graded ring associated with an integral group ring, J. Pure and Appl. Agebra, 5 (1974), 253-264.

3. Burnside, W., Theory of Groups of Finite Order, 2 ed., Cambridge Univ. Press, 1911.

4. Losey, G., On the structure of $Q_2(G)$ for finitely generated groups, Canad. J. Math., 25 (1973), 353-359.

5. Losey, G., N-series and filtrations of the augmentation ideal, Canad. J. Math., 26 (1974), 962-977.

6. Losey, G. and Losey, N., Augmentation quotients of some nonabelian finite groups, Math. Proc. Cambridge Phil. Soc., 85 (1979), 261-270.

7. Losey, G. and Losey, N., The stable behaviour of the augmentation quotients of the groups of order p^3, J. Algebra, 60 (1979), 337-351.

8. Passi, I. B. S., Polynomial functors, Proc. Cambridge Phil. Soc., 66 (1969), 505-512.

9. Singer, M., On the augmentation terminal of a finite abelian group, J. Algebra, 41 (1976), 196-201.

10. Singer, M., Determination of the augmentation terminal for all abelian groups of exponent 8, Comm. Algebra, 5 (1977), 87-100.

11. Singer, M., Determination of the augmentation terminal for finite abelian groups, Bull. Amer. Math. Soc., 83 (1977), 1321-1322.

12. Tahara, K., On the structure of $Q_3(G)$ and the fourth dimension subgroups, Japan J. Math. (new series), 3 (1977), 381-394.

Supplementary Bibliography

The following items are relevant to the general problem of determining the augmentation quotients but were not directly referred to in this paper.

A. Ford, D. and Singer, M., Relations in $Q_n(Z_4 \times Z_8)$ and $Q_n(Z_8 \times Z_8)$, comm. Alg., 5 (1977), 83-86.

B. Hales, A. W., Augmentation terminals of finite abelian groups (to appear).

C. Hales, A. W. and Passi, I. B. S., The second augmentation quotient of an integral group ring, Arch. Math. (Basel), 31 (1978-79), 259-265.

D. Horibe, K., The stable behavior of the augmentation quotients of some groups of order p^4 (to appear).

E. Passi, I. B. S., The associated graded ring of a group ring, Bull. London Math. Soc., 10 (1978), 241-255.

F. Passi, I. B. S., Group Rings and their Augmentation Ideals, Lecture Notes in Mathematics, No. 715, Springer-Verlag, Berlin-New York, 1979.

G. Passi, I. B. S. and Vermani, L. R., The associated graded ring of an integral group ring, Proc. Cambridge Phil. Soc., 82 (1977), 25-33.

H. Sandling, R. and Tahara, K., Augmentation quotients of group rings and symmetric powers, Math. Proc. Cambridge Phil. Soc., 85 (1979), 247-252.

I. Singer, M., On the graded ring associated with an integral group ring, Comm. Alg., 3 (1975), 1037-1049.

J. Stallings, J. R., Quotients of the powers of the augmentation ideal in a group ring, Knots, Groups and 3-Manifolds, Ann. of Math. Studies, 84 (1975), 101-118.

K. Tahara, K., The augmentation quotients of group rings and the fifth dimension subgroup, J. Algebra, 71 (1981), 141-173.

L. Tahara, K., Augmentation quotients and dimension subgroups of semidirect products of groups, Math. Proc. Cambridge Phil. Soc., 91 (1982), 39-49.

THE UNIVERSITY OF MANITOBA
WINNIPEG, MANITOBA, CANADA

Received July 13, 1981

Contemporary Mathematics
Volume 33, 1984

FINITE-DIMENSIONAL ALGEBRAS AND k-GROUPS OF FINITE RANK

A. G. Myasnikov and V. N. Remeslennikov

The present paper accumulates the results obtained by the authors during the last two years, most of them are summarized in [1], [2], [11] and [12].[1] Some of the results are new; certain so far unsolved problems are listed at the end of this article. Principal achievements of the series of works under consideration can be stated as follows. First of all, certain new methods of studying unipotent algebraic groups over fields of characteristic zero have been proposed. The methods include: an extension of the method found by A. I. Mal'cev [3] which establishes a connection between algebras without unit and nilpotent groups of class 2. Second, theorems concerning relations between abstract and semilinear homomorphisms have been obtained. Third, nilpotent k-groups and finite-dimensional k-algebras over regularly definable fields k (in particular, over Q) have been classified according to their elementary properties, and complete theories of such groups and algebras have been described.

In the present paper we confine ourselves, as a rule, to giving brief summaries instead of complete proofs.

1. Preliminary Information

This section comprises some definitions and facts regarding nilpotent k-groups. For k-groups which are not necessarily nilpotent see R. Lyndon [4]. For the model theoretic concepts see [5].

[1]This paper was edited for publication by Kenneth Weston of the Department of Mathematics, University of Wisconsin-Parkside, Kenosha, WI 53141.

A binomial ring is an integral domain having Z as a subring, and containing, along with element λ, all the binomial coefficients

$$\binom{\lambda}{n} = \frac{\lambda(\lambda-1)\ \dots\ (\lambda-n+1)}{n!}\ ,\quad n \in N\ .$$

Throughout this paper the symbol σ will denote an arbitrary binomial ring, and k will stand for a field of characteristic zero.

Definition. A nilpotent group G of nilpotence class m is a group admitting exponent in a ring σ (or a σ-group), if for any $x \in G$ and $\lambda \in \sigma$ the element $x^\lambda \in G$ is defined univalently and, in addition, the following axioms hold ($x, y, x_1, \dots, x_n$ are arbitrary elements of G, and λ, μ are elements of σ):

1. $x^1 = x, x^\lambda x^\mu = x^{\lambda+\mu}$, $(x^\lambda)^\mu = x^{\lambda\mu}$,
2. $y^{-1}x^\lambda y = (y^{-1}xy)^\lambda$,
3. $x_1^\lambda x_2^\lambda \dots x_n^\lambda = (x_1 \dots x_n)^\lambda \tau_2(X)^{\binom{\lambda}{2}} \dots \tau_m(X)^{\binom{\lambda}{m}}$,

where $\tau_i(X)$ is the i-th Petrescu word of $x_1, \dots, x_n$. We remind the reader that for every $i \in N$ the i-th Petrescu word of $\tau_i(x_1, \dots, x_n) = \tau_i(X)$ is defined recursively by the following relation

$$x_1^i x_2^i \dots x_n^i = \tau_1(X)^i \tau_2(X)^{\binom{i}{2}} \dots \tau_{i-1}(X)^{\binom{i}{i-1}} \tau_i(X)$$

in a free group F with generators $x_1, x_2, \dots, x_n$. It is well known (see, e.g., [6]) that for any i of N the word $\tau_i(X)$ belongs to the subgroup $\gamma_i(F)$ which is the i-th term of the lower central series of the group F.

Let G be a σ-group. A subgroup H of the group G closed with respect to raising to the power λ, $\lambda \in \sigma$, is called a σ-subgroup. If for some elements $g_1, \dots, g_n$ of the group G the smallest σ-subgroup $(g_1, \dots, g_n)\sigma$ containing them equals G, then $g_1, \dots, g_n$ is a system of generators of G, and G is a σ-group of finite rank.

The group G is a σ-torsion-free group if for every $\lambda \in \sigma$, $g \in G$ if $g^\lambda = 1$ then either $\lambda = 0$, or $g = 1$.

Definition. An ordered set of elements $u_1, \ldots, u_n$ is called a Mal'cev basis of the σ-group G if:

1) every element x from G can be uniquely represented as

$$x = u_1^{t_1(x)} u_2^{t_1(x)} \ldots u_n^{t_n(x)}, \quad t_i(x) \in \sigma$$

and

2) if $G_1 = (u_i, \ldots, u_n)_\sigma$, then the chain of σ-subgroups $G = G_1 \geq G_2 \geq \ldots \geq G_n \geq 1$ forms a central series of the group G.

Elements $t_1(x), t_2(x), \ldots, t_n(x)$ are coordinates of the element x in the basis $u_1, \ldots, u_n$. Coordinates $t_i(xy)$ of products of elements x and y in a fixed Mal'cev basis $u_1, \ldots, u_n$ are polynomials over a ring σ_z of $2n$ variables $t_j(x)$, $t_j(y)$ $j = 1, \ldots, n$. Similarly, if $\lambda \in \sigma$, then coordinates $t_i(x^\lambda)$ are polynomials over the ring σ_z of $n + 1$ variables $t_j(x)$, λ, $j = 1, \ldots, n$.

Let G be a k-group of finite rank, and $u_1, \ldots, u_n$ be a Mal'cev basis for the group G (finite-rank groups over principal ideal rings, in particular, over a field k, always possess Mal'cev basis). Denote by τ_i, ω_i, $i = 1, 2, \ldots, n$ the multiplication polynomials and raising to the power polynomials in the basis $u_1, \ldots, u_n$. If a field K is the extension of the field k then one can construct a new group G^K, where G^K is the K-completion of G. All possible formal products $u_1^{\xi} u_2^{\xi^2} \ldots u_n^{\xi^n}$, $\xi_i \in K$, are used as elements of G^K, and multiplication and raising to the power $\lambda \in K$ are assigned by means of polynomials τ_i, ω_i, $i = 1, \ldots, n$. At such a definition G^K becomes a K-group, and G^K contains the group G as a subgroup.

Definition. Group G is called Q-determined if there exists a Mal'cev basis of the group G in which multiplication and raising to the power are assigned by polynomials with rational coefficients.

Proposition 1. (Criterion of Q-determination). A k-group G is Q-determined iff G is a k-completion of some Q-group of finite rank.

Definition. k-group $\overline{G} \leq G$, such that $G = \overline{G} \times A$ and, where $Z(\overline{G}) \leq \overline{G}'$, $A \leq Z(G)$, is called the stem of the k-group G. (Here and further G' is the commutator subgroup of the group G). Due to the category isomorphism between k-groups and Lie k-algebras, which is described in Section 6, the following duality of definitions in k-groups and k-algebras seems quite natural.

A finite-dimensional k-algebra P is Q-determined if there exists k-basis of P with rational structural constants. The algebra P is Q-determined iff $P = N \otimes k$ is a tensor product over Q of some finite-dimensional Q-algebra N and the field k.

Let P be an arbitrary k-algebra. By $\mathrm{Ann}_l P$, $\mathrm{Ann}_r P$, $\mathrm{Ann}\, P$ we denote the left, the right and two-sided annihilators of P, respectively, and P^2 is a subalgebra generated by all products of two elements.

Definition. A subalgebra $\overline{P}$ such that $P = \overline{P} \times C$ where $\mathrm{Ann}\, \overline{P} \leq \overline{P}^2$, and $C \leq \mathrm{Ann}\, P$ is the stem of the k-algebra P.

Theorem 1. Let G be a k-group of finite rank (k-algebra of finite dimension) then

1) the stem of the group (algebra) G does exist;

2) let B be a k-subgroup (subalgebra) of G, and $Z(B) \leq B'$ ($\mathrm{Ann}\, B \leq B^2$), then for any stem $\overline{G}$ of the group (algebra) G there exists a monomorphism of k-groups (k-algebras) $B \to \overline{G}$;

3) any two stems of the group (algebra) G are isomorphic to each other.

Let A be an algebraic system of signature Ω. A subset $M \subset A^n$ is called first-order definable with constants if there exists a formula $\phi(x_1,\ldots,x_n, y_n,\ldots,y_m)$ of the restricted predicate calculus with signature Ω, and a set of constants $b_1,\ldots,b_m$ from A such that

$$M = \{(a_1,\ldots,a_n) \in A^n \mid A \models \phi(a_1,\ldots,a_n, b_1,\ldots,b_m)\} .$$

In this case we say that M is determined by the formula ϕ with the constants $b_1,\ldots,b_m$.

If $m = 0$ we say the set M is first-order definable.

We say that a mapping $f: A^n \to A$ is first-order definable with constants if the graph of f is first-order definable.

Below, all the groups (including k-groups) are considered with signature $\langle \cdot^{(2)}, (-1)^{(1)}, e \rangle$, and all the rings (including k-algebras) - with ring signature $\langle +^{(2)}, \cdot^{(2)}, 0 \rangle$.

Verbal subgroups of finite width serve as examples of the first-order definable subgroups of the group G. A verbal subgroup V of G determined by a word v is said to have a width s if any element of V can be represented as products of no more than s values of the word v and its inverses, the number s being the minimal one with this property.

Proposition 2. Every term of the lower central series of a k-group of finite rank is a verbal subgroup of finite width.

Let A and B be algebraic systems with signature Ω. Notation $A \equiv B$ means that algebraic systems A and B are elementary equivalent, i.e. they both satisfy the same closed formulae with signature Ω.

2. Model and Quasimodel Groups

Model and quasimodel groups are the simplest examples of a non-commutative k-group. The set of all such factors of a non-commutative k-group G and their inner relationships determine the structure of G.

Definition. A nilpotent, of class 2 non-Abelian 2-generated σ-group which is σ-torsion-free is a model σ-group $M(\sigma)$.

Let $UT_3(\sigma)$ be a group of unitriangle 3×3 matrices over the ring σ. Define raising to the power $\lambda \in \sigma$ in the group $UT_3(\sigma)$ by the formula:

$$g^{\lambda} = \sum_{i=0}^{2} \binom{\lambda}{i} (g - e)^{i}$$

where $g \in UT_3(\sigma)$, and e is the matrix identity.

Let a,b generate an σ-group $M(\sigma)$ and $c = [b,a]$. By direct verification one can easily see that the mapping

$$a^{\alpha} b^{\beta} c^{\gamma} \text{ ---- } \begin{bmatrix} 1 & \alpha & \gamma \\ 0 & 1 & \beta \\ 0 & 0 & 1 \end{bmatrix}$$

is a σ-isomorphism of the group $M(\sigma)$ onto the group $UT_3(\sigma)$.

Definition. A nilpotent σ-group G of nilpotence class 2 is called a quasimodel group and is denoted by $K(\sigma)$, if G contains a model subgroup $M(\sigma)$ and $G = M(\sigma)Z(G)$.

In particular, if σ is an algebra over a field, then G is a quasimodel group $K(\sigma)$ iff $G = M(\sigma) \times A$, where $A \leq Z(G)$.

It follows easily that a quasimodel group $K(\sigma)$ satisfies the following conditions:

K.1. G is nilpotent of class 2 and contains elements a and b where $[a,b] \neq 1$. Denote as A the centralizer of the element a, and B is the centralizer of the element b.

K.2. G is generated by subgroups A and B. The subgroups A and B are Abelian and $A \cap B = Z(G)$.

K.3. $gp\{[a,b]\} = G'$, and besides, $[a,b_1] = [a,b_2]$ iff $b_1 \equiv b_2 \bmod Z(G)$.

K.4. $[A,b] = G'$, and besides, $[a_1 b] = [a_2,b]$ iff $a_1 \equiv a_2 \bmod Z(G)$.

On the other hand, if the conditions K.1 - K.4 hold in a group G, then one can define the structure of the ring on the commutator subgroup G' where the role of the addition operation is played by multiplication in the group (the group G' is Abelian), and multiplication $\times$ in a ring is assigned by the formula, for $c_1, c_2 \in G'$;

$$c_1 \times c_2 = [a_1 b_1]$$

where a_1, b_1 are defined by equalities

$$c_1 = [a_1,b] , \quad c_2 = [a,b_1] .$$

That the operation x is well defined and the realizability of the axioms of a ring are guaranteed by the conditions K.1 - K.4.

It is easy to see that the conditions K.1 - K.4 can be written by formulae in group signature with the constants a and b. In particular, using the operation x one can write (by means of group signature formulae) the associativity, commutativity and invertibility of every non-zero element in the ring $\langle G', \cdot, x\rangle$.

Proposition 3. A ring σ is relatively definable in a quasimodel group $K(\sigma)$.

Proposition 4. A class of all quasimodel (model) groups over a field is axiomatizable in the class of all groups and is, also, axiomatizable in the class of all Q-groups.

The leading idea of proofs of all the main results of the article is either reduction of the proof to the case of a quasimodel group, or induction on the group (algebra) based on the quasimodel factors of the group (algebra).

3. Regularly Definable Fields, First-Order Definable Bundles and Groups from the Class of k-Groups

The theory of k-groups turns out to be most constructive for a special type of field, the so-called regularly definable field.

Let k be a field of characteristic zero. By $R(k)$ we denote a class of all 2-nilpotent k-groups, in which for any pair of non-commuting elements a, b there exists a quasimodel subgroup over k which is first-order definable with constants, and contains the elements a,b.

Definition. A field k is regularly definable if for any natural number n there exists a formula $\phi_n(x)$ in ring signature which defines the field k in any of its algebraic power extensions of power not exceeding n.

An algebraically closed field is obviously a regularly definable one. The field of rational numbers is a less trivial example - its regular definability directly follows from the results of J. Robinson [7]. The only regularly definable algebraic number fields are those in which polynomials can be decomposed into polynomials with rational coefficients. Note also that a class of regularly definable fields is closed with respect to the operation of taking an ultrapower.

Theorem 2. Let k be a regularly definable field. Then every non-Abelian k-group of finite rank contains a first-order definable with constants non-Abelian k-group from the class $R(k)$.

Let $E = \{e_1,\ldots,e_n\}$ be a finite system of elements of the group G.

The system E is complete if the centralizer of E equals the centre of the group G.

Illustrative examples of groups having a complete system are σ-groups of finite rank, and groups with the minimality condition upon centralizers (in particular, ω-categorical groups).

A graph $\Gamma(E)$ of the system E is a non-oriented graph whose vertices are elements from E and vertices e_i, e_j are connected by an edge iff $[e_i,e_j] \neq 1$.

The system E is connected if the graph $\Gamma(E)$ is connected.

Definition. Let $E = \{e_1,\ldots,e_n\}$ be a connected complete system of elements of the group G. The graph $\Gamma(E)$, to each edge (e_i,e_j) of which is associated a quasimodel subgroup $K(e_i,e_j)$ over σ from G, containing elements e_i, e_j as σ-generators of the model subgroup $M(\sigma)$, is called a σ-bundle $\Gamma(E,\sigma)$ of the group G.

The bundle $\Gamma(E,\sigma)$ of the group G is first-order definable if all quasimodel subgroups, corresponding to the edges of the graph $\Gamma(E)$, are first-order definable with constants in the group G.

Proposition 5. Let G be a group from the class $R(k)$ possessing a complete system. Then first-order definable k-bundles of the group G do exist.

Up to the end of this section we fix an arbitrary non-Abelian group G from the class $R(k)$, possessing a first-order definable bundle T.

All the subsequent assertions of this section state first-order definability with constants for some subgroup or mapping. The defining formula is fully determined by the quasimodel subgroups of the bundle T. This fact will be marked by adding a symbol T to the defining formula.

Definition. Let $K(a,b)$ and $K(c,d)$ be quasimodel subgroups over k which contain non-permutational elements a, b and c, d respectively. Then the mapping $\delta\colon [a,b]^{\alpha} \to [c,d]^{\alpha}$, $\alpha \in k$, is the connecting isomorphism of the groups $K(a,b)$ and $K(c,d)$.

Note that δ yields an isomorphism between the interpreted field k in the commutator subgroups of the groups $K(a,b)$ and $K(c,d)$ respectively (Proposition 3).

Proposition 6. There exists a formula with group signature $\psi(x_1,y_1,y_2,T)$, such that for any non-commutating elements a, b of the group G the set $K(a,b) = \{g \in G \mid G \models \psi(g,a,b,T)\}$ is a quasimodel subgroup over k containing elements a, b.

Proposition 7. There exists a formula $\Delta(z,w,x,y_1,x_2,y_2,T)$ such that for any elements a, b, u, v, of the group G, $[a,b] \neq 1$, $[u,v] \neq 1$ the connecting isomorphism $\delta\colon [a,b]^{\alpha} \to [u,v]^{\beta}$, $\alpha \in k$ is defined by the formula $\Delta(z,w,a,b,u,v,T)$.

4. Basic Theorems (The Case of the Class of Nilpotence 2)

Let G be a group from the class $R(k)$ possessing a complete system.

The results of the previous section allow us to define k-linearly independent systems for groups $G/Z(G)$ and G'. Let us explain it in detail.

According to Proposition 5 there exists a first-order definable k-bundle $T = \Gamma(E,k)$ of the group G.

Let $c_1,\dots,c_n$ be elements of the commutator subgroup of G where $c_i = \prod_{j=1}^{s} [a_{ij},b_{ij}]$ is the decomposition of the element c_i, $i = 1,\dots,n$, into the product of no more than S commutators (it is clear that for an arbitrary finite system $c_1,\dots,c_n$ there always exists such an S). Then k-linear independent of the system $c_1,\dots,c_n$ is equivalent to the following condition:

$$\phi_1 = \forall \alpha_1,\dots,\alpha_n \in k \left[\prod_{i=1}^{n} \prod_{j=1}^{s} [a_{ij},b_{ij}]^{\alpha_i} = 1 \leftrightarrow \alpha_1 = \dots = \alpha_n = 0 \right].$$

Fix an arbitrary edge (e,f) of the graph $\Gamma(E)$ of the k-bundle T. Using the interpretation of the field k in the group $K(e,f)$ (Proposition 3), and the connecting isomorphisms (Proposition 7) $\delta_{ij}: [e,f]^{\alpha} \to [a_{ij},b_{ij}]^{\alpha}$, $\alpha \in k$ the condition ϕ_1 can be written by formulae with group signature with constants depending on the bundle T.

Let $c_1,\dots,c_n$ be elements of the group G. Then by the k-linear independence of the system $c_1,\dots,c_n$ by the module of the centre $Z(G)$ we mean the following condition is satisfied

$$\forall\alpha_1,\dots,\alpha_n \in k \; \Big(\prod_{i=1}^{n} c_i^{\alpha i} \in Z(G) \leftrightarrow \alpha_1 = \dots = \alpha_n = 0\Big) .$$

By the definition of the k-bundle $T = \Gamma(E,k)$ the system E is complete, i.e. $Z(G) = C_G(E)$, therefore, the k-independence condition of the system $c_1,\dots,c_n$ by the module $Z(G)$ is equivalent to the condition

$$\phi_2 = \forall\alpha_1,\dots,\alpha_n \in k, \forall e \in E \; \Big(\prod_{i=1}^{n} [c_i,e]^{\alpha i} = 1\Big) \leftrightarrow \alpha_1 = \dots = \alpha_n = 0)$$

which, like the condition ϕ_1, can be written by formulae with group signature with constants depending on the bundle T.

<u>Theorem 3</u>. Let G be a group of finite rank from the class $R(k)$, and $Z(G) = G'$. Then the set of Mal'cev bases of the group G associated to the upper central series is first-order definable with constants.

The theorem makes use of the fact that the Mal'cev bases serve as bases for the terms of the upper central as well as the lower central series.

<u>Theorem 4</u>. Let G and H be 2-nilpotent k-groups of finite ranks, G being a Q-determined group from the class $R(k)$. Then $G \equiv H$ if their stems $\overline{G}$ and $\overline{H}$ are k-isomorphic, the groups G and H either coinciding with their stems or being simultaneously unequal to them.

Theorem 4 in fact reduces the problem of elementary classification of groups from the class $R(k)$ to that of their classification to within an isomorphism. A few words about the proof of Theorem 4. Let $\overline{G}$ be the stem of the group G. Then $Z(\overline{G}) \leq G'$ and due to Theorem 3 one can first-order define some Mal'cev base $\overline{u} = \{u_1,\dots,u_n\}$ in the group G, and because the group G is Q-determined we can assume that the polynomials of multiplication and raising to the power in the base $\overline{u}$ have rational coefficients

(Proposition 1). Consequently, multiplication and raising to the power in the base $\bar{u}$ can also be written by formulae with group signature. Thus, in H there exists a system of elements $\bar{v} = \{v_1,\dots,v_n\}$ satisfying the conditions ϕ_1 and ϕ_2, the polynomials of multiplication and raising to the power in the systems $\bar{u}$ and $\bar{v}$ being the same. It remains to show that the formulae ϕ_1 and ϕ_2 in the group H also define a Mal'cev base for some stem H.

G is a Q-group (since $k \geq Q$), hence H is also a Q-group. Due to finite axiomatizability of quasimodel groups in the class of all Q-groups (Proposition 4), every formula defining a quasimodel subgroup G, defines a quasimodel subgroup in H. The formula Δ from Proposition 7 can be written by formulae with group signature. Consequently, the formula Δ also defines connecting isomorphisms for the subgroup H. Therefore, formulae ϕ_1 and ϕ_2 define in H, k-linearly independent systems in G' and G/Z(G), respectively.

We reduce the study of the theories of groups from the class R(k) to the study of the class of k-bundles.

<u>Definition</u>. Let G be a group from the class R(k), where the commutator subgroup has a finite width S. Then the bundle $T = \Gamma(E,k)$ is a k-map if for any $\alpha \in k$ the bundle T satisfies the following conditions:

1) $\forall x\ \exists y\ \forall e \in E\ ([x,e^{\alpha}] = [y,e])$. An element y satisfying condition 1) for the given x and α is denoted by $x^{(\alpha)}$;

2) $\forall x_1,x_2([x_1{}^{(\alpha)},x_2] = [x_1,x_2{}^{(\alpha)}])$;

3) $$\forall x,y_i,u_i,v_i\Big(\prod_{i=1}^{2s}[x_i,y_i] = \prod_{i=1}^{2s}[u_i,v_i]\Big) \to$$

$$\Big(\prod_{i=1}^{2s}[x_i{}^{(\alpha)}y_i] = \prod_{i=1}^{2s}[u_i{}^{(\alpha)},v_i]\Big) .$$

<u>Theorem 5</u>. Let G be a 2-nilpotent Q-group with a commutator subgroup of finite width; and T is a k-map of the group G. Then any stem of the group G is a k-group.

<u>Theorem 6</u>. Let G be a group of finite rank from the class R(k). If $G \equiv H$, then there exists a field k_1 which is elementary equivalent to k,

such that any stem of the group H is a k_1-group of finite rank. By means of Theorems 4 and 6 we have:

Theorem 7. Let G be a Q-determined group from the class $R(k)$. H is an arbitrary group. Then the following conditions are equivalent:

1) $G \equiv H$;

2) H is a Q-group;

a) there exists a field k, such that $k \equiv k_1$,

b) the stems $\overline{G}$ and $\overline{H}$ are k and k_1-completions, respectively, of some Q-group N of finite rank;

c) the groups G and H simultaneously either coincide with their stems or do not equal them.

5. Elementary Theories of Finite-Dimensional Algebras

Let P be an arbitrary ring (not necessarily associative, perhaps without unit).

To the ring P there is associated a group $M(P)$ defined as follows.

On a set of triples $\{(a,b,c) | a,b,c \in P\}$ introduced is a multiplication operation: $(a,b,c((x,y,z) = (a + x, b + y, c + z + bx)$. With respect to it a set of triples of elements from P forms a 2-nilpotent group.

If P is an algebra over a field k, then in the group $M(P)$ raising to the power λ from k is defined by the formula:

$$(a,b,c)^{\lambda} = (\lambda a, \lambda e, \lambda c + \frac{\lambda(\lambda-1)}{2} ba)$$

and $M(P)$ turns to a k-group.

The correspondence $P \to M(P)$ is the natural generalization of the correspondence $P \to UT_3(P)$ for algebras with a unit. The latter correspondence was investigated by A. I. Mal'cev in [3].

Note some properties of the correspondence $P \to M(P)$.

Proposition 8. Let P_1, P_2 be k-algebras with a unit (not necessary associative ones). Then the algebras P_1 and P_2 are isomorphic (elementary equivalent) iff the groups $M(P_1)$ and $M(P_2)$ are isomorphic (elementary equivalent).

Let us show a process X of changing a formula ϕ with group signature into a formula $X(\phi)$ with ring signature such that ϕ is true for $M(P)$ iff $X(\phi)$ is true for P.

We assume that ϕ is in the prenex normal form: $\phi = Q_1x_1 \dots Q_nx_n\phi_0$ where Q_i are quantifiers and ϕ_0 is the quantifierless part. Then the formula $X(\phi)$ is obtained by substitution of every quantifier Q_ix_i the three quantifiers $(Q_ix_i^{(1)})$ $(Q_ix_i^{(2)})$ $(Q_ix_i^{(3)})$ and every atomic subformula from ϕ_0 of the form $x_ix_j = x_m$ is replaced by the formula

$$x_i^{(1)} + x_j^{(1)} = x_m^{(1)} \quad \text{and} \quad x_1^{(2)} + x_j^{(2)} = x_m^{(2)}$$

$$\text{and} \quad x_i^{(3)} + x_j^{(3)} + x_i^{(2)} \times x_j^{(1)} = x_m^{(3)} .$$

It is obvious that ϕ is true in $M(P)$ iff $X(\phi)$ is true in P.

The idea of investigating elementary theories of rings consists in the transition from a ring P to a 2-nilpotent group $M(P)$, and applying to the latter with the help of Theorem 2 theorems like 4, 6 and 7. Then, information on the group $M(P)$ is "transferred" onto the ring P. Thus, the following theorems have been obtained.

Let k be an arbitrary regularly definable field of characteristic zero. We assume that k-algebras are not necessarily associative, and they don't necessarily possess a unit.

Theorem 8. Let P be a finite-dimensional k-algebra with non-zero multiplication. Then the field k is relatively definable in the algebra P.

Theorem 9. Let P be a finite-dimensional k-algebra, and $\operatorname{Ann} P \leq P^2$. Then a set of k-bases of the algebra P, which are associated to some series of subspaces $1 = P_1 \leq \dots \leq P_n = P$ is the first-order definable with constants.

Theorem 10. Two finite-dimensional Q-determined k-algebras P and P_1 are elementary equivalent iff their stems are isomorphic, P and P_1 simultaneously either coinciding with their stems or being unequal to them.

Theorem 11. Let P be a finite-dimensional direct indecomposable k-algebra; P_1 is an arbitrary ring. If $P \equiv P_1$, then there exists a field k_1, such that $k \equiv k_1$, and P_1 is a finite-dimensional k_1-algebra.

Theorem 12. Let P be a Q-determined finite dimensional k-algebra; P_1 is an arbitrary ring. Then the following conditions are equivalent:

1) $P \equiv P_1$

2) P_1 is a Q-algebra and there exists decompositions

$$\overline{P} = \prod_{i=1}^{n} S_i , \quad \overline{P}_1 = \prod_{i=1}^{n} T_i$$

of the stems of P and P_1 into direct products of k-algebras such that for any i, $1 \leq i \leq n-1$, there exist a field k_i for which the following conditions hold:

a) $k \equiv k_i$,

b) S_i and T_i are tensor products $N_i \otimes k$ and $N_i \otimes k_i$, respectively, for some N_i,

c) the algebras P and P_1 simultaneously either coincide with their stems or are not equal to them.

6. Elementary Theories of Nilpotent k-Groups (the General Case)

In this section we show how one can, using Theorems 9, 10, and 11 obtain classification theorems for nilpotent k-groups of arbitrary class of nilpotence.

In paper [8] A. I. Mal'cev proved that there exists categorical isomorphism between categories of nilpotent Lie Q-algebras and nilpotent Q-groups. The above isomorphism we denote by g. Thus, $g(L)$ is a Q-group corresponding to a Lie Q-algebra L. Let us remind the reader that the group $g(L)$ is constructed by the algebra L by means of introducing on L a multiplication operation due to the Campbell-Hausdorff formula

$$a * b = a + b + \frac{1}{2} [a,b] + \ldots$$

where the brackets denote multiplication in the Lie algebra.

Quillen showed [9] that there exists a categorical isomorphism between the categories of nilpotent Lie k-algebras and nilpotent k-groups where k has characteristic zero.

Note some properties of the correspondence g.

Proposition 9. Let $G = g(L)$. Then between k-subgroups of G and subalgebras of L there exists a one-to-one correspondence for which:

1) ideals of the algebra L correspond to normal k-subgroups of G;
2) powers L^i of the algebra L correspond to the terms of the lower central series G;
3) the terms of the upper central series of L correspond to the terms of the upper central series G;
4) stems of the algebra L correspond to those of the group G.

Proposition 10. Let L be a finite-dimensional rational Lie algebra. Then the following conditions are equivalent:

1) $L = N \otimes k$;
2) $G = g(N)^k$ is as k-completion of the subgroup $g(N)$.

Proposition 11. Let $G_i = g(L_i)$, $i = 1,2$. Then the following conditions are equivalent:

1) $G_1 \equiv G_2$,
2) $L_1 \equiv L_2$.

Using the categorical isomorphism g and Propositions 9, 10, 11, we deduce the following results from Theorems 9, 10, 11 and 12.

Let k be a regularly definable field of characteristic zero.

Theorem 13. Let G and H be k-groups of finite rank, G being Q-determined. Then $G \equiv H$ iff their stems $\overline{G}$ and $\overline{H}$ are k-isomorphic, the groups G and H either coinciding with their stems or simultaneously being unequal to them.

Theorem 14. Let G be a direct indecomposable k-group of finite rank; H is an arbitrary group. If $G \equiv H$, then there exists a field k_1, elementary equivalent to k, such that the group H is a k-group of finite rank.

Theorem 15. Let G be a Q-determined k-group of finite rank; H is an arbitrary group. Then the following conditions are equivalent:

1) $G \equiv H$;

2) H is a Q-group, and there exist decompositions

$$\overline{G} = \prod_{i=1}^{n} G_i, \quad \overline{H} = \prod_{i=1}^{n} H_i$$

of the stems of G and H into direct products of k-groups such that for any i, $1 \leq i \leq n$, there exist a field k_i for which the following conditions hold:

a) $k \equiv k_i$;

b) G_i and H_i are k and k_1 -completions, respectively, of some group N_i of finite rank;

c) the groups G and H simultaneously either coincide with their stems or are not equal to them.

7. Isomorphisms of k-Groups

In the course of proving some of the above theorems a number of results about connections between abstract and k-isomorphisms of k-groups were obtained.

Let G and H be σ-groups. The structure of k-groups allows us to point out the following subclasses of homomorphisms from all homomorphisms of G into H.

1. A homomorphism $\phi: G \to H$ is called a σ-homomorphism if $(g^{\alpha})^{\phi} = (g^{\phi})^{\alpha}$ for any $G \in G$, $\alpha \in \sigma$.

2. A homomorphism $\phi: G \to H$ is a ring or semilinear homomorphism if there exists an endomorphism θ of the ring σ, such that $(g^{\alpha})^{\phi} = (g^{\phi})^{\alpha\theta}$ for any $g \in G$, $\alpha \in \sigma$.

3. A homomorphism $\phi: G \to H$ is geometric if $(g^{\alpha})^{\phi} \in \mathrm{gp}(g^{\phi})\sigma$ for any $g \in G$, $\alpha \in \sigma$. It is σ-homomorphisms that are "actual" homomorphisms of σ-groups.

Among homomorphisms of every type one naturally defines isomorphisms. For instance, a semilinear homomorphism $\phi: G \to H$ is a semilinear isomorphism if there exists a semilinear homomorphism ψ such that

$$\psi\phi = id_G, \quad \psi\phi = id_H .$$

Note the following evident properties of the above homomorphism:

a) compositions of homomorphisms of one type is a homomorphism of the same type;

b) let F be a normal σ-subgroup of the group H, then the natural homomorphism: $H \to H/F$ is a σ-homomorphism;

c) if the groups G and H are torsion-free, then the isomorphism $\phi: G \to H$ is a semilinear isomorphism iff there exists $\theta \in \mathrm{Aut}\,\sigma$ such that $(g^{\alpha})^{\phi} = (g^{\phi})^{\alpha\theta}$, $\alpha \in \sigma$, $g \in G$.

Theorem 16. Let G and H be k-groups of finite rank, G being a k-completion of the Q-group G_0. Then a monomorphism $\phi: G_0 \to H$ can be raised up to a k-isomorphism $\overline{\phi}: G \to H$ iff the following conditions hold:

1) Mal'cev bases of the groups G_0 and H have an equal number of elements;

2) $gp(G_0^{\lambda})_k = H$.

The following theorem is the "group" generalization of the basic theorem of projective geometry, see, for instance, [10].

Theorem 17. Let G and H be non-Abelian σ-torsion-free σ-groups. Then any geometric isomorphism G onto H is semilinear.

Methods used to prove Theorems 14 and 15 allow us to prove the following results.

Theorem 18. Let G be a k-group of finite rank over a regularly definable field k. Then there exist a central series first-order definable with a constants k-subgroups $Z(G) = R_0 \leq \dots \leq R_n = G$ such that for any isomorphism of the k-groups $\psi: G \to H$, for every i, $1 \leq i \leq n-1$, the restriction ψ on R_{i+1} is semilinear on the module R_i.

Problems

1. What local fields are regularly definable? In particular will the fields of $\wp$-adic numbers be regularly definable?

2. To classify algebraic groups over regularly definable fields by elementary properties.

3. Is Theorem 13 true if the field k is not regularly definable?

4. To describe binomial rings for which Theorem 15 is valid.

REFERENCES

1. Мясников А.Г., Ремесленников В.Н. Элементарные свойства степенных нильпотентных групп. - ДАН СССР, 1981, т. 258, № 5, стр. 1056-1059.

2. Мясников А.Г., Ремесленников В.Н. Формульность множества мальцевских баз, 16 Всесоюзная алгебраическая конференция. Тезисы докладов, т. 1, Ленинград, 1981, стр. 113-114.

3. Мальцев А.И. Об одном соответствии между кольцами и группами. - Мат.сб., 1960, т. 50, № 3, стр. 257-266.

4. Lyndon R. C., Groups with parametric exponents. - Trans. Amer. Math. Soc. 1960, 96, 518-533.

5. Chang C. C., Keisler H. J. Model Theory . - North Holland, Amsterdam, 1972.

6. Hall P. Nilpotent groups, Canad. Math. Congress, Edmonton, 1957.

7. Robinson J. The undecidability of algebraic rings and fields. - Proc. Amer. Math. Soc., 10 (1959), 950-957.

8. Мальцев А.И. Нильпотентные группы без кручения. - Изв. АН СССР, сер. матем., 1949, 13, № 3, 201-212.

9. Quillen D. Rational Homotopy Theory, Ann. Math. 1969, 90, 205-295.

10. Artin, E., Geometric algebra, New York, 1957.

11. Myasnikov, A. G., Remeslennikov, V. N., Isomorphism and elementary properties of power nilpotent groups. Trudi Instituta Matematiki, v. 2, Mathematical Logic and the theory of algorithms; Nauka, Novosibirsk, 1981, 56-87.

12. Myasnikov, A. G., Remeslennikov, V. N., The formulation of sets of Mal'cev bases and the elementary theory of finite-dimensional algebras I, Sibusku Mat. Zhurnal, 5 (1982), 125-167.

INSTITUTE OF MATHEMATICS
OMSK

Received October 8, 1982

Contemporary Mathematics
Volume 33, 1984

THE COHOMOLOGY RING OF A ONE-RELATOR GROUP

John G. Ratcliffe

Dedicated to Roger C. Lyndon
on the occasion of his 65th birthday

In this paper, the cohomology ring of a one-relator group is calculated. As pointed out by Shapiro and Sonn in [8], this ring was essentially already determined by Labute [5] in 1965 when he calculated the cup product form of a one-relator pro-p-group. Not knowing of Labute's work, I made the calculation in this paper. Although my calculation is slightly longer than Labute's, it has the advantage that it reveals that the cup product form of a one-relator group can be computed in terms of second order Fox derivatives. This brings the cohomology ring of a one-relator group more in line with Lyndon's calculation of the cohomology of a one-relator group in [6]. One should also compare the calculation here with Horadam's calculation of homology cup co-products in [4]. For an application of the cohomology ring of a one-relator group, see [7].

1. The homology and cohomology of a one-relator group.

Let $G = (x_1,\ldots,x_n;r)$ be a finitely generated one-relator group, and R a commutative ring with identity. We start by computing the homology of G with trivial coefficients in R. Let F be the free group with generators $x_1,\ldots,x_n$, and let $r = s^e$ in F with $e \geq 1$ and as large as possible. Let σ_i be the exponent sum of r with respect to x_i. The Lyndon resolution [6] determines the homology chain complex

© 1984 American Mathematical Society
0271-4132/84 $1.00 + $.25 per page

$$\cdots \xrightarrow{\partial_4} R \xrightarrow{\partial_3} R \xrightarrow{\partial_2} R^n \xrightarrow{\partial_1} R \xrightarrow{\partial_0} 0$$

where $\partial_{odd} = 0$, $\partial_2(1) = (\sigma_1,\ldots,\sigma_n)$ and $\partial_{2k}(1) = e$ for $k \geq 2$.

Let d be the greatest common divisor of $\sigma_1,\ldots,\sigma_n$. Here it is understood that $d = 0$ if $\sigma_i = 0$ for each i. If $d = 0$, let $k_i = 0$ for each i; otherwise, let $k_i = \sigma_i/d$. Then $\partial_2(1) = d(k_1,\ldots,k_n)$. It follows that $H_2(G,R) \cong {}^dR$ where ${}^dR = \{u \in R \mid du = 0\}$. As $(k_1,\ldots,k_n)$ generates a direct summand of R^n, we have that $H_1(G,R) \cong R_d \oplus R^{n-1}$ where $R_d = R/dR$. Thus

$$H_i(G,R) \cong \begin{cases} R & \text{if } i = 0 \\ R_d \oplus R^{n-1} & \text{if } i = 1 \\ {}^dR & \text{if } i = 2 \\ R_e & \text{if } i = 3,5,\ldots \\ {}^eR & \text{if } i = 4,6,\ldots \end{cases}$$

Next we compute the cohomology of G. The Lyndon resolution determines the cohomology chain complex

$$0 \longrightarrow R \xrightarrow{\delta^0} R^n \xrightarrow{\delta^1} R \xrightarrow{\delta^2} R \xrightarrow{\delta^3} \cdots$$

where $\delta^{even} = 0$, the matrix for δ^1 is $(\sigma_1\ldots\sigma_n)$, and $\delta^{2k+1}(1) = e$ for $k \geq 2$. If $d \neq 0$, then $(k_1\ldots k_n)$ reduces by column operations to $(1\ 0\ \ldots\ 0)$. Therefore, $H^1(G,R) \cong {}^dR \oplus R^{n-1}$. Thus

$$H^i(G,R) \cong \begin{cases} R & \text{if } i = 0 \\ {}^dR \oplus R^{n-1} & \text{if } i = 1 \\ R_d & \text{if } i = 2 \\ {}^eR & \text{if } i = 3,5,\ldots \\ R_e & \text{if } i = 4,6,\ldots \end{cases}$$

The homology and cohomology of a one-relator group was first computed by Lyndon [6].

2. On the cup product form.

We now assume that $dR = 0$. If $d = 0$, this is no restriction at all. Note that $d = 0$ if and only if r is in the commutator subgroup of F. The condition $dR = 0$ implies that $\partial_2 = 0$ in the homology chain complex for G.

Hence,

$$H_i(G,R) \cong \begin{cases} R & \text{if } i = 0 \\ R^n & \text{if } i = 1 \\ R & \text{if } i = 2 \end{cases}$$

and $H^i(G,R) \cong \mathrm{Hom}_R(H_i(G,R),R)$ for $i = 0,1,2$, so that

$$H^i(G,R) \cong \begin{cases} R & \text{if } i = 0 \\ R^n & \text{if } i = 1 \\ R & \text{if } i = 2 \end{cases} .$$

Let x_i also denote the class in $H_1(G,R)$ corresponding to x_i. Then $\{x_i\}_{i=1}^n$ is a basis for $H_1(G,R)$. Let x_j^* be the class in $H^1(G,R)$ which is dual to $\{x_i\}_{i=1}^n$ with respect to the Kronecker product [3, p. 132]

$$H_1(G,R) \times H^1(G,R) \to R ,$$

that is, $[x_i,x_j^*] = \delta_{ij}$. Then $\{x_j^*\}_{j=1}^n$ is a basis for $H^1(G,R)$. We say that $\{x_j^*\}_{j=1}^n$ is the basis dual to $\{x_i\}_{i=1}^n$.

Let ρ be the class in $H_2(G,R)$ corresponding to r. Then evaluating at ρ by the Kronecker product

$$H_2(G,R) \times H^2(G,R) \to R$$

gives an isomorphism $H^2(G,R) \cong R$. Composing the cup product

$$H^1(G,R) \times H^1(G,R) \to H^2(G,R)$$

with this isomorphism gives a skew-symmetric bilinear form

$$H^1(G,R) \times H^1(G,R) \to R$$

which we call the <u>cup product form</u> of G. If $e = 1$, then $H^i(G,R) \cong 0$ for

$i > 2$, so the cohomology ring of G is completely determined by its cup product form in this case.

We consider the Fox derivatives [2] $\frac{\partial}{\partial x_i}$ to take values in the group ring $R(F)$. The symbol $^\circ$ will denote the augmentation $R(F) \to R$.

<u>Theorem 1</u>. Let $G = (x_1,\ldots,x_n;r)$ be a one-relator group, and R a commutative ring with identity such that $dR = 0$, where d is the greatest common divisor of the exponent sums of r. Then the ijth term of the matrix M_G for the cup product form

$$H^1(G,R) \times H^1(G,R) \to R$$

with respect to the basis $\{x_i^*\}$ is $\left(\frac{\partial^2 r}{\partial x_i \partial x_j}\right)^\circ$.

<u>Proof</u>. We follow the example of Greenberg [3, p. 150]. Write

$$r = x_{i_1}^{\epsilon_1} \cdots x_{i_m}^{\epsilon_m} \text{ with } \epsilon_j = \pm 1 \text{ and } m \geq 4 .$$

Consider a regular m-gon D triangulated as indicated by the following diagram

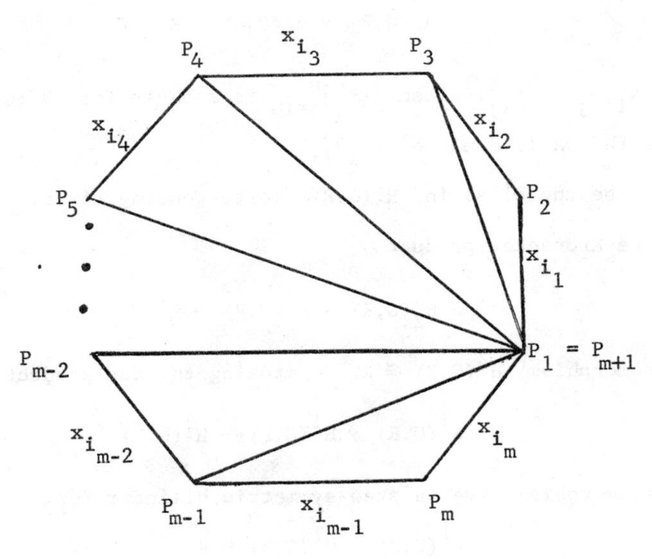

We orient the edge (P_jP_{j+1}) by ϵ_j. Let K be the 2-complex obtained from D by (1) identifying the P_j to one point P, (2) identifying the

edges corresponding to the same generator according to the orientations, and (3) wedging on circles corresponding to the generators not appearing in r. Then K is a model for the presentation.

Consider the singular 2-chain in D

$$\sigma = \sum_{j=2}^{m-2} (P_1 P_j P_{j+1}) - \sum_{\substack{j=1 \\ j:\ \epsilon_j = -1}}^{m-2} (P_j P_{j+1} P_j)$$

$$- (P_1 P_m P_{m-1}) + \sum_{\substack{j=m-1 \\ j:\ \epsilon_j = 1}}^{m} (P_j P_{j+1} P_j) .$$

Observe that

$$\partial\sigma = \sum_{j=1}^{m-2} (P_j P_{j+1}) - (P_1 P_{m-1})$$

$$- \sum_{\substack{j=1 \\ j:\ \epsilon_j = -1}}^{m-2} [(P_{j+1} P_j) - (P_j P_j) + (P_j P_{j+1})]$$

$$- (P_m P_{m-1}) + (P_1 P_{m-1}) - (P_1 P_m)$$

$$+ \sum_{\substack{j=m-1 \\ j:\ \epsilon_j = 1}}^{m} [(P_{j+1} P_j) - (P_j P_j) + (P_j P_{j+1})] ,$$

so that

$$\partial\sigma = \sum_{\substack{j=1 \\ j:\ \epsilon_j = 1}}^{m} (P_j P_{j+1}) - \sum_{\substack{j=1 \\ j:\ \epsilon_j = -1}}^{m} (P_{j+1} P_j)$$

$$+ \sum_{\substack{j=1 \\ j:\ \epsilon_j = -1}}^{m-2} (P_j P_j) - \sum_{\substack{j=m-1 \\ j:\ \epsilon_j = 1}}^{m} (P_j P_j) .$$

Hence, σ is a relative 2-cycle on $(D, \partial D)$ representing the generator of $H_2(D, \partial D; R)$ which determines the generator of $H_2(K, P; R) \cong H_2(G, R)$ corresponding to ρ.

Suppose $p \neq q$. Then

$$[\rho, x_p^* \cup x_q^*] = [\epsilon_1 x_{i_1}, x_p^*][\epsilon_2 x_{i_2}, x_q^*]$$

$$+ \ldots + [\sum_{j=1}^{m-3} \epsilon_j x_{i_j}, x_p^*][\epsilon_{m-2} x_{i_{m-2}}, x_q^*]$$

$$- [-\epsilon_m x_{i_m}, x_p^*][-\epsilon_{m-1} x_{i_{m-1}}, x_q^*] .$$

Now replace $-\epsilon_m x_{i_m}$ by $\sum_{j=1}^{m-1} \epsilon_j x_{i_j}$. As $p \neq q$, we have that

$$[\epsilon_{m-1} x_{i_{m-1}}, x_p^*][\epsilon_{m-1} x_{i_{m-1}}, x_q^*] = 0 .$$

Hence,

$$[\rho, x_p^* \cup x_q^*] = \sum_{k=2}^{m-1} [\sum_{j=1}^{k-1} \epsilon_j x_{i_j}, x_p^*][\epsilon_k x_{i_k}, x_q^*]$$

$$= \sum_{k:\, i_k = q}^{m-1} \epsilon_k (\sum_{j:\, i_j = p}^{k-1} \epsilon_j) .$$

If $i_m = q$, then $\sum_{j:\, i_j = p}^{m-1} \epsilon_j = \sigma_p$ which is zero in R. Thus

$$[\rho, x_p^* \cup x_q^*] = \sum_{k:\, i_k = q}^{m} \epsilon_k (\sum_{j:\, i_j = p}^{k-1} \epsilon_j) .$$

The kth <u>initial section</u> of

$$r = \prod_{j=1}^{m} x_{i_j}^{\epsilon_j} \quad \text{is} \quad \prod_{j=1}^{k-1} x_{i_j}^{\epsilon_j} \quad \text{or} \quad \prod_{j=1}^{k} x_{i_j}^{\epsilon_j}$$

according as $\epsilon_k = 1$ or $\epsilon_k = -1$. By Formula (2.8) of Fox [2],

$$\frac{\partial r}{\partial x_q} = \sum_{k:\, i_k = q}^{m} \epsilon_k r_{(k)}$$

where $r_{(k)}$ is the kth section of r. Hence

$$\frac{\partial^2 r}{\partial x_p \partial x_q} = \sum_{k:\, i_k = q}^{m} \epsilon_k (\sum_{j:\, i_j = p}^{k-\frac{1}{2}(\epsilon_k+1)} \epsilon_j r_{(j)}) .$$

Thus

$$\left(\frac{\partial^2 r}{\partial x_p \partial x_q}\right)^\circ = \sum_{k:\ i_k = q}^{m} \epsilon_k \Big(\sum_{j:\ i_j = p}^{k-1} \epsilon_j \Big) .$$

Therefore,

$$[\rho, x_p^* \cup x_q^*] = \left(\frac{\partial^2 r}{\partial x_p \partial x_q}\right)^\circ .$$

Now suppose $p = q$. Then

$$[\rho, x_p^* \cup x_p^*] = \sum_{k=2}^{m-2} \Big[\sum_{j=1}^{k-1} \epsilon_j x_{i_j}, x_p^* \Big][\epsilon_k x_{i_k}, x_p^*]$$

$$- \sum_{\substack{j=1 \\ j:\ \epsilon_j = -1}}^{m-2} [\epsilon_j x_{i_j}, x_p^*][-\epsilon_j x_{i_j}, x_p^*]$$

$$- [-\epsilon_m x_{i_m}, x_p^*][-\epsilon_{m-1} x_{i_{m-1}}, x_p^*]$$

$$+ \sum_{\substack{j=m-1 \\ j:\ \epsilon_j = 1}}^{m} [\epsilon_j x_{i_j}, x_p^*][-\epsilon_j x_{i_j}, x_p^*] .$$

As before replace $-\epsilon_m x_{i_m}$ by $\sum_{j=1}^{m-1} \epsilon_j x_{i_j}$. Then

$$[\rho, x_p^* \cup x_p^*] = \sum_{k=2}^{m-1} \Big[\sum_{j=1}^{k-1} \epsilon_j x_{i_j}, x_p^* \Big][\epsilon_k x_{i_k}, x_p^*]$$

$$+ \sum_{\substack{j=1 \\ j:\ \epsilon_j = -1}}^{m-2} [\epsilon_j x_{i_j}, x_p^*][\epsilon_j x_{i_j}, x_p^*]$$

$$+ [\epsilon_{m-1} x_{i_{m-1}}, x_p^*][\epsilon_{m-1} x_{i_{m-1}}, x_p^*]$$

$$- \sum_{\substack{j=m-1 \\ j:\ \epsilon_j = 1}}^{m} [\epsilon_j x_{i_j}, x_p^*][\epsilon_j x_{i_j}, x_p^*]$$

$$+ \sum_{j=1}^{m} [\epsilon_j x_{i_j}, x_p^*][\epsilon_m x_{i_m}, x_p^*] .$$

Therefore,

$$[\rho, x_p^* \cup x_p^*] = \sum_{k=2}^{m} [\sum_{j=1}^{k-1} \epsilon_j x_{i_j}, x_p^*][\epsilon_k x_{i_k}, x_p^*]$$

$$+ \sum_{j:\ \epsilon_j = -1}^{m} [\epsilon_j x_{i_j}, x_p^*][\epsilon_j x_{i_j}, x_p^*]$$

$$= \sum_{k:\ i_k = p}^{m} \epsilon_k (\sum_{j:\ i_j = p}^{k-\frac{1}{2}(\epsilon_k+1)} \epsilon_j)$$

$$= (\frac{\partial^2 r}{\partial x_p^2})^\circ \quad . \qquad \square$$

The next theorem gives an interpretation for the entries of the cup product form matrix in Theorem 1.

Theorem 2. Write

$$r = x_1^{\sigma_1} \ldots x_n^{\sigma_n} \prod_{j=1}^{k} t_j [x_{i_j}, x_{i'_j}]^{\epsilon_j} t_j^{-1}$$

where $\epsilon_j = \pm 1$ and $i_j < i'_j$ for each j. If $p < q$, then

$$(\frac{\partial^2 r}{\partial x_p \partial x_q})^\circ = \Sigma \{\epsilon_j \mid (i_j, i'_j) = (p,q)\} .$$

If $p = q$, then

$$(\frac{\partial^2 r}{\partial x_p^2})^\circ = \begin{cases} \sigma_p/2 & \text{if } d \text{ is even} \\ 0 & \text{if } d \text{ is odd} \end{cases} .$$

Proof. Let $u = x_1^{\sigma_1} \ldots x_n^{\sigma_n}$ and

$$v = \prod_{j=1}^{k} t_j [x_{i_j}, x_{i'_j}]^{\epsilon_j} t_j^{-1} .$$

Then

$$\frac{\partial^2}{\partial x_p \partial x_q} (uv) = \frac{\partial}{\partial x_p} (\frac{\partial u}{\partial x_q} + u \frac{\partial v}{\partial x_q})$$

$$= \frac{\partial^2 u}{\partial x_p \partial x_q} + \frac{\partial u}{\partial x_p} + \frac{\partial^2 v}{\partial x_p \partial x_q} .$$

Observe that $(\frac{\partial u}{\partial x_p})^\circ = \sigma_p$ which is zero in R. Therefore,

$$(\frac{\partial^2 r}{\partial x_p \partial x_q})^\circ = (\frac{\partial^2 u}{\partial x_p \partial x_q})^\circ + (\frac{\partial^2 v}{\partial x_p \partial x_q})^\circ .$$

If $p \neq q$, then $(\frac{\partial^2 u}{\partial x_p \partial x_q})^\circ = 0$, since $\frac{\partial^2 u}{\partial x_p \partial x_q} = 0$ when $p > q$. Therefore, if $p < q$, we have that

$$(\frac{\partial^2 r}{\partial x_p \partial x_q})^\circ = (\frac{\partial^2 v}{\partial x_p \partial x_q})^\circ .$$

Let $v_j = t_j [x_{i_j}, x_{i'_j}]^{\epsilon_j} t_j^{-1}$. By Formulas (5.1) and (5.2) of Fox [2],

$$(\frac{\partial^2 v}{\partial x_p \partial x_q})^\circ = \sum_{j=1}^{k} (\frac{\partial^2 v_j}{\partial x_p \partial x_q})^\circ$$

$$= \sum_{j=1}^{k} \left\{ \frac{\partial^2 [x_{i_j}, x_{i'_j}]^{\epsilon_j}}{\partial x_p \partial x_q} \right\}^\circ$$

$$= \Sigma \{\epsilon_j \mid (i_j, i'_j) = (p,q)\} .$$

Suppose $p = q$. By Formula (3.9) of Fox [2], we have that $(\frac{\partial^2 r}{\partial x_p^2})^\circ = \binom{\sigma_p}{2}$.

Now $\binom{\sigma_p}{2} = \sigma_p(\sigma_p - 1)/2$, even if σ_p is negative. If d is even, then σ_p is even; and so $\sigma_p(\sigma_p-1)/2 = -\sigma_p/2 = \sigma_p/2$. If d is odd, then the characteristic of R is odd, since $dR = 0$. This implies that 2 is invertible in R. Therefore, $\sigma_p(\sigma_p - 1) = 0$ implies that $\sigma_p(\sigma_p-1)/2 = 0$. This shows that

$$(\frac{\partial^2 r}{\partial x_p^2})^\circ = \begin{cases} \sigma_p/2 & \text{if } d \text{ is even} \\ 0 & \text{if } d \text{ is odd} \end{cases} . \quad \square$$

Some examples:

(1) Let G be an orientable surface group with presentation

$$(x_1, \ldots, x_{2m}; [x_1, x_2][x_3, x_4] \ldots [x_{2m-1}, x_{2m}]) .$$

Then M_G is the matrix

$$\begin{bmatrix} 0 & 1 & & & & \\ -1 & 0 & & & & \\ & & \cdot & & & \\ & & & \cdot & & \\ & & & & \cdot & \\ & & & & 0 & 1 \\ & & & & -1 & 0 \end{bmatrix}.$$

(2) Let G be a nonorientable surface group with presentation $(x_1,\ldots,x_n;x_1^2\ldots x_n^2)$. Then R has characteristic two and M_G is the $n \times n$ identity matrix.

(3) Let G be the Klein bottle group with presentation $(x,y;xyx^{-1}y)$ then $M_G = \begin{bmatrix} 0 & 1 \\ 1 & 1 \end{bmatrix}$. Here it is easier to compute M_G directly with Fox derivatives. Notice that M_G depends on the presentation for G.

3. The higher ring structure.

Suppose $e > 1$. Let $h\colon \mathbb{Z}_e \to G$ be the homomorphism defined by $h(1) = s$. Then h induces a chain transformation from the Lyndon resolution of $\mathbb{Z}_e = (x;x^e)$ to the Lyndon resolution of $G = (x_1,\ldots,x_n;s^e)$. This chain transformation induces a cohomology chain transformation

$$\begin{array}{lccccccccl} G\colon 0 \longrightarrow & R & \xrightarrow{\delta^0} & R^n & \xrightarrow{\delta^1} & R & \xrightarrow{\delta^2} & R & \longrightarrow & \ldots \\ & \downarrow h^0 & & \downarrow h^1 & & \downarrow h^2 & & \downarrow h^3 & & \\ \mathbb{Z}_e\colon 0 \longrightarrow & R & \xrightarrow{\delta^0} & R & \xrightarrow{\delta^1} & R & \xrightarrow{\delta^2} & R & \longrightarrow & \ldots \end{array}$$

One checks easily that $h^p = id_R$ if $p \neq 1$, and h^1 has the matrix $(\tau_1\cdots\tau_n)$ where τ_i is the exponent sum of s with respect to x_i.

Let us now assume that $eR = 0$. This implies that all the coboundaries are trivial. Therefore,

$$h^*\colon H^p(G,R) \to H^p(\mathbb{Z}_e, R)$$

is an isomorphism for $p \neq 1$, and $h^*\colon H^1(G,R) \to H^1(\mathbb{Z}_e,R)$ has the matrix $(\tau_1\cdots\tau_n)$.

Recall what the cohomology ring of $\mathbb{Z}_e$ looks like [1, p. 252]. There is a generator $y^p \in H^p(\mathbb{Z}_e, R) \cong R$ such that

$$y^p \cup y^q = \begin{cases} y^{p+q} & \text{if } p \text{ or } q \text{ is even}, \\ \frac{e}{2} y^{p+q} & \text{if } p \text{ and } q \text{ are odd} . \end{cases}$$

Note that if e is odd, then the characteristic of R is odd, since $eR = 0$. Therefore, 2 is invertible in R. Thus, $e/2$ makes sense and is zero.

Let x^p be the class in $H^p(G,R)$, $p > 1$, such that $h^*(x^p) = y^p$. Then x^p is a generator of $H^p(G,R) \cong R$; and $h^*(x^p) \cup h^*(x^q) = h^*(x^p \cup x^q)$ implies that

$$x^p \cup x^q = \begin{cases} x^{p+q} & \text{if } p \text{ or } q \text{ is even}, \\ \frac{e}{2} x^{p+q} & \text{if } p \text{ and } q \text{ are odd} . \end{cases}$$

It remains only to compute $x_i^* \cup x^q$ for $i = 1,\ldots,n$ and $q > 1$. Observe that

$$\begin{aligned} h^*(x_i^* \cup x^q) &= h^*(x_i^*) \cup h^*(x^q) \\ &= \tau_i y^1 \cup y^q . \\ &= \begin{cases} \tau_i y^{q+1} & \text{if } q \text{ is even}, \\ \frac{\sigma_i}{2} y^{q+1} & \text{if } q \text{ is odd} . \end{cases} \end{aligned}$$

Therefore,

$$x_i^* \cup x^q = \begin{cases} \tau_i x^{q+1} & \text{if } q \text{ is even} \\ \frac{\sigma_i}{2} x^{q+1} & \text{if } q \text{ is odd} . \end{cases}$$

REFERENCES

1. H. Cartan and S. Eilenberg, Homological Algebra, Princeton University Press, Princeton, 1956.

2. R. H. Fox, Free differential calculus I, Ann. Math. 57 (1953), 547-560.

3. M. J. Greenberg, Lectures on Algebraic Topology, W. A. Benjamin, Reading, Mass., 1967.

4. K. J. Horadam, The diagonal comultiplication on homology, J. Pure Appl. Algebra 20 (1981), 165-172.

5. J. P. Labute, Classification of Demushkin groups, Canad. J. Math. 19 (1967), 106-132.

6. R. C. Lyndon, Cohomology theory of groups with a single defining relation, Ann. Math. 52 (1950), 650-665.

7. J. G. Ratcliffe, On one-relator groups which satisfy Poincaré duality, Math. Z. 177 (1981), 425-438.

8. J. Shapiro and J. Sonn, Free factor groups of one-relator groups, Duke Math. J. 41 (1974), 83-88.

DEPARTMENT OF MATHEMATICS
UNIVERSITY OF ILLINOIS
URBANA, IL 61801

Received March 1, 1982

Contemporary Mathematics
Volume 33, 1984

COMBINATORIAL GROUP THEORY, RIEMANN SURFACES AND DIFFERENTIAL EQUATIONS

C. L. Tretkoff* and M. D. Tretkoff**

Group theory has several points of contact with the study of Riemann surfaces and differential equations. In this article we shall discuss some of the interactions between these disciplines resulting from the notion of the monodromy group. The actual contents are described in the next paragraph. Here, we wish to state that we have written neither a survey article nor a historical account of monodromy groups. Our goals are more modest. First of all, a portion of our paper represents new results: the introduction and application of a combinatorial algorithm. Second, we report selectively on recent results. We hope that our choice of material will be appealing and will provide a view which is orthogonal to that in other sources. Finally, because we have supposed that most of our readers will be group theorists, we have provided more background material about Riemann surfaces and differential equations than about group theory.

Section 1 contains the definitions of the monodromy group of a branched covering, X, of the Riemann sphere $\hat{\mathbb{C}}$ and of the Hurwitz system used to obtain a cellular decomposition of it. We also show how to obtain the intersection matrix for the 1-cycles on X from the Hurwitz system. Oddly enough, despite the importance of intersection numbers, this material does not appear in the standard sources.(1) In Section 2 we given an algorithm for

*This research was supported in part by a grant from The City University of New York PSC-CUNY Research Award Program.
**Supported in part by National Science Foundation Grant MCS-8103453.

© 1984 American Mathematical Society
0271-4132/84 $1.00 + $.25 per page

calculating these intersection matrices and we show how to determine a homology basis from it. Usually, the homology basis is found by puncturing X and $\hat{\mathbb{C}}$ and applying the Reidemeister-Schreier process to the fundamental groups; additional relations must be introduced to account for the punctures. This procedure does not give the intersection matrix of the resulting basis. Section 3 contains a brief survey of results about monodromy groups and groups of conformal self mappings, as well as a combinatorial proof of a theorem of Hurwitz. We especially hope the latter will appeal to Roger Lyndon; indeed he has used combinatorial methods to prove the Riemann-Hurwitz formula [31]. In Section 4 we apply the results of Section 2 to determine the periods and quadratic periods of Abelian integrals on the Klein-Hurwitz curve and on the Fermat curve. We also see that these curves provide an answer to a question of H. Martens and C. Earle. Section 5 is devoted to the monodromy group of a homogeneous linear differential equation. Our discussion of this vast subject is necessarily limited. We give a brief report on some modern developments whose "classical" origins are related to the theory of Riemann surfaces. A few applications of modern results in group theory are also mentioned. Readers wishing to learn more about the "classical" period should consult the following articles:

Hilb, E., <u>Lineare Differentialgleichungen im komplexen Gebiet</u>, (1916), Encyklopadie der math. Wiss. II, B(5), 471-562.

Vessiot, E., <u>Gewohnliche Differentialgleichungen: elementare Integrationsmethoden</u> (1900), Encyklopadie der math. Wiss. II, A(4b), 230-293.

There is also a very useful bibliography in Hejhal's article [23].

Finally, we remark that because there is no longer a shortage of books about Riemann surfaces and differential equations, we have simply referred to our favorites. In order to keep our bibliography manageable, we have listed relatively few papers.

Section 1: Monodromy groups and cell decompositions

Let $p: X \to \hat{\mathbb{C}}$ be a branched covering of the z-sphere, $\hat{\mathbb{C}}$, by a compact Riemann surface, X, of genus g. This covering determines the following data whose precise meaning will be recalled below:

i) branch points $z = b_1,\ldots,b_t$; $t \geq 2$

ii) the sheet number, n, of p

iii) the conjugacy class of a transitive permutation group, M, on n symbols which is called the monodromy group of the covering.

It is conventional to call a point $z = b$ a branch point if and only if p is not one to one in a neighborhood of some point of the fibre $p^{-1}(b)$. Applying the compactness of $\hat{\mathbb{C}}$ and the analyticity of p, we easily see that the totality of branch points is finite. Setting $Z = \hat{\mathbb{C}} - \{b_1,\ldots,b_t\}$ and noting that $\tilde{Z} = p^{-1}(Z)$ is connected, we find that the restriction of p to $\tilde{Z}$ is a covering projection in the sense customarily adopted in algebraic topology texts. Thus, the cardinality of $p^{-1}(z)$, $z \in Z$, is independent of z. This number is usually called the "number of sheets" of $\tilde{Z}$ over Z; it is easily seen to equal the degree of the branched covering map p.

We recall that the monodromy representation of a covering is defined by selecting a base point, say $z = a$, and letting $\pi_1(Z,a)$ act as a group of permutations of the fibre $p^{-1}(a) = \{a_1,\ldots,a_n\}$. More precisely, if γ is a loop in Z based at $z = a$, then there is a unique loop, $\tilde{\gamma}$, in $\tilde{Z}$ beginning at a_j such that $p\tilde{\gamma} = \gamma$. The end point of $\tilde{\gamma}$ depends only on the homotopy class, $[\gamma]$, of γ and belongs to $p^{-1}(a)$. Thus, it is of the form a_k, $1 \leq k \leq n$, and setting $k = m([\gamma])(j)$ we obtain a permutation, $m([\gamma])$, of the set of integers $\{1,\ldots,n\}$. The representation, m, of $\pi_1(Z,a)$ in the symmetric group on n symbols, $\Sigma(n)$, is called the monodromy representation and the image, M, of m is called the monodromy group. Since the fibre $p^{-1}(a) = Z_a$ is an unordered set of n points, rearranging their indices gives all representations of $\pi_1(Z,a)$ in $\Sigma(n)$ which are equivalent to m.

Thus we see that the covering, in fact, only determines the <u>conjugacy class</u> of M.

If $q: Y \to \hat{\mathbb{C}}$ is a branched covering with the same data (i), (ii), (iii) as $p: X \to \hat{\mathbb{C}}$, then there is a one to one conformal mapping, f, of X onto Y such that $qf = p$. This is a consequence of the fact that the combinatorial data (ii), (iii) determine topologically equivalent coverings $\tilde{Z} = p^{-1}(Z)$ and $\tilde{W} = q^{-1}(Z)$ whose complex structures are induced from Z. Now, it is a standard fact that unbranched coverings are determined up to equivalence by conjugacy classes of subgroups of the fundamental group of the base. This is so because the construction of a covering space requires the choice of a base point, although it plays no role in the definition of a covering space. All of this is a topological manifestation of the fact that any representation of a group, say $\pi_1(Z,a)$, as a transitive group on n symbols can be realized as a group of permutations on the cosets of a subgroup, H_j, of index n. Namely, we take for H_j the stabilizer of the jth symbol. The kernel of this representation is obviously the intersection of all the conjugates of H_j. Once again, because the choice of j is arbitrary, we only obtain an equivalence class of representations. This is tantamount to rearranging the names of the n symbols, that is, following the representation by an inner automorphism $\Sigma(n)$.

In order to decompose X as a CW-complex we pick a base point $a \neq b_i$ $i = 1,\ldots,t$, and join a to b_i by simple directed arcs ℓ_i that meet only at $z = a$. The union, Γ, of all the $p^{-1}(\ell_i)$ obviously forms a 1-dimensional CW-complex. Now, the complement, D, in $\hat{\mathbb{C}}$ of the union of the ℓ_i is homeomorphic to the interior of a 2-cell, so the same is true of each component, D_j, $j = 1,\ldots,n$, of $p^{-1}(D)$. Thus, X is a 2-dimensional CW-complex obtained from Γ by attaching 2-cells E_j along the boundaries of $\partial\bar{D}_j$ of the closures $\bar{D}_j$ of the D_j.

The cellular decomposition of X enables us to determine $H_1(X,\mathbb{Z})$ as the quotient of $H_1(\Gamma,\mathbb{Z})$ by the subgroup generated by $\partial\bar{D}_1,\ldots,\partial\bar{D}_n$. Since

$\partial\overline{D}_1+\ldots+\partial\overline{D}_n = 0$, the subgroup in question is generated by $\partial\overline{D}_1,\ldots,\partial\overline{D}_{n-1}$. For our purposes, it is also necessary to determine the intersection matrix of the resulting homology basis. We shall now indicate in general terms how this information may be obtained with the aid of the cell decomposition; a more detailed account, taking the monodromy transformations into consideration, is given in Section 2. In fact, it seems simplest to first determine the intersection matrix of certain 1-cycles and then to observe that a subset of them forms a homology basis.

Thus, let $T \subset \Gamma$ be a maximal tree and let $e_1,\ldots,e_r$ be the edges of Γ not in T. Since T contains a unique reduced path joining any pair of vertices of Γ, we obtain a 1-cycle c_i by traversing e_i and returning from its terminal point to its initial point by a reduced path in T. The cycles $c_1,\ldots,c_r$ represent a basis for $H_1(\Gamma, \mathbb{Z})$; this is easily seen by contracting T to a single vertex and noting that the resulting complex is a bouquet of r circles. It is also clear that there is an open set $U \subset X$ containing T which may be mapped homeomorphically, preserving orientation, onto the unit disk and such that $\overline{U}$ meets e_i in distinct points p_i and q_i whose images on the unit circle are P_i and Q_i. Now, the element of $H_1(X, \mathbb{Z})$ represented by c_i is unchanged if we replace the subarc of c_i belonging to $\overline{U}$ by the image in $\overline{U}$ of the chord P_iQ_i; the intersection number $c_i \cdot c_j$ is therefore the intersection number of P_iQ_i with P_jQ_j, so its value is one of the integers $-1,0,1$.

It is easy to see how to express $n-1$ of the cycles c_i in terms of the others and thereby obtain a basis for $H_1(X, \mathbb{Z})$. Namely, we note that each e_i appears once with coefficient 1 and once with coefficient -1 in the system of equations $\partial\overline{D}_j = 0$, $j = 1,\ldots,n$. An easy inductive argument shows that successive addition of appropriate pairs of these equations yields $n-1$ equations and distinguishes $n-1$ of the e_i so that each of them appears in precisely one of the equations and has coefficient 1. Eliminating these

e_i we obtain a basis for $H_1(X, \mathbb{Z})$, and it is a simple matter to determine the corresponding intersection matrix K.

Denoting the genus of X by g, we see that K is an integral skew symmetric matrix of degree $2g = r-n+1$, so we may apply the algorithm of Frobenius[2] to obtain a unimodular matrix T of degree $2g$ such that ${}^tTKT = J$. Here,

$$J = \begin{pmatrix} 0 & I \\ -I & 0 \end{pmatrix}$$

and I is the identity matrix of degree g. Clearly, the determinant of K is 1. Conversely, if the principal minor of the intersection matrix $(c_i \cdot c_j)$ given by the sequence $i_1 = j_1, \ldots, i_{r-n+1} = j_{r-n+1}$ has determinant 1, then the 1-cycles $c_{i_1}, \ldots, c_{i_{r-n+1}}$ represents a basis for $H_1(X, \mathbb{Z})$. A basis for $H_1(X, \mathbb{Z})$ with intersection matrix J is called canonical.

Finally, we recall the traditional method of choosing generators for $\pi_1(Z,a)$. Let Γ_j, $j = 1,\ldots,t$, be pairwise disjoint circles centered about b_j which meet ℓ_j in a single point, r_j, and do not meet ℓ_k, $k \neq j$. Then $\pi_1(Z,a)$ is generated by the homotopy classes of the loops, γ_j, obtained by running along ℓ_j from a to r_j, traversing Γ_j once in the positive direction, and returning from r_j to a along ℓ_j. Supposing that the ℓ_j are indexed so that sufficiently small circles centered about $z = a$ meet them in the order $\ell_1, \ldots, \ell_t$ or in a cyclic permutation thereof, we see that $\gamma_1 \cdots \gamma_t$ is null homotopic. Thus, $\pi_1(Z,a)$ is freely generated by $[\gamma_1], \ldots, [\gamma_{t-1}]$. It follows that the monodromy group M is a transitive subgroup of $\Sigma(n)$ generated by permutations $\pi_j = m([\gamma_j])$ which satisfy the relation $\pi_1 \cdots \pi_t = 1$. The data $(z = a, b_1, \ldots, b_t, \ell_1, \ldots, \ell_t, \pi_1, \ldots, \pi_t)$ is often called a Hurwitz system; it yields cell decompositions for X and $\hat{\mathbb{C}}$.

With the following simple modifications, our discussion also applies to branched coverings, $p: X \to S$, of a Riemann surface, S, of genus $\hat{g} > 0$. We begin by selecting $2\hat{g}$ simple closed curves α_j, β_j, $j = 1, \ldots, \hat{g}$, on S based at a, with intersection matrix J, and supposed not to meet the branch points. The ℓ_i are then required to meet the α_j, β_j only at a. Finally, the complement in S of the union of the ℓ_i, α_j, β_j is the open 2-cell D, and Γ is the pre-image in X of this union. Thus, we obtain cell decompositions of X and S such that $X^{(k)} = p^{-1}(S^{(k)})$, where $X^{(k)}$ and $S^{(k)}$ denote the k-skeletons of X and S respectively. Once again, the data $(a,b_i,\ell_i,\pi_i,\alpha_j,\beta_j,\mu_j,\nu_j)$, $1 \le i \le t$, $1 \le j \le \hat{g}$, is called a Hurwitz system. Here μ_j and ν_j are the permutations of $p^{-1}(a)$ obtained by lifting α_j and β_j respectively to X. The monodromy group is generated by the π_i, μ_j, ν_j; the paths ℓ_i, α_j, β_j are selected so that

$$[\alpha_1]^{-1}[\beta_1]^{-1}[\alpha_1][\beta_1]\ldots[\alpha_g]^{-1}[\beta_g]^{-1}[\alpha_g][\beta_g][\ell_1]\ldots[\ell_t] = 1 .$$

Hurwitz systems play an important role in algebraic geometry, topology and number theory. For example, the reader may consult [16], [15], and [7]. The papers of M. Fried will especially interest readers of the present article because of their interaction with combinatorial group theory[3].

Finally, we recall that the Riemann-Hurwitz formula for the branched covering $p: X \to S$ is

$$2 - 2g = (2-2\hat{g})n - (v(\pi_1) + \ldots + v(\pi_t)) ,$$

where $v(\pi_j) = n - z_j$ when π_j is the product of z_j disjoint cycles.

Section II: Algorithm for the intersection form

In this section we solve the following problem: given a branched covering of $\hat{\mathbb{C}}$, find a collection of curves on $\hat{\mathbb{C}}$ which lift to a canonical homology basis for X. Our motivation for this investigation has its roots in the study of periods of Abelian integrals, so we defer comments about its background to Section 4. Here, we use a Hurwitz system to construct a tree, Δ, embedded in the plane with the following properties:

(i) the 1-skeleton, Γ, of X is a quotient of Δ

(ii) deleting certain edges from Δ, we obtain a tree, T', which is homeomorphic to a maximal tree T in Γ

(iii) there is a planar neighborhood, U', of T' which is homeomorphic to the neighborhood U of T introduced in Section I .

This enables us to effectively carry out the general procedure introduced in Section 1 for computing intersection numbers. We illustrate the process with several examples which play a role in the sequel. Finally, we note that our procedure is an algorithm, suitable for implementation on a computer, which can be effected without drawing pictures. This has, in fact, been carried out by the first author; but the following exposition seems preferable in the present context.

We begin with a general description of our construction of Δ. Full details and some illustrative examples are provided in the following paragraphs. We shall use the same notation for corresponding vertices of Γ and Δ. In particular, we label the elements of $p^{-1}(a)$ by the integers $1,\ldots,n$ and the elements of $p^{-1}(b_i)$ by the cycles occurring in the disjoint cycle decomposition of π_i. It is also convenient to write $c_i(j)$ for the cycle of π_i containing the integer j. Now, we select a point labeled 1 as the center of a family C_m, $m = 1,2,\ldots,$ of positively oriented concentric circles of increasing radii. The graph Δ will be constructed in stages by selecting vertices on C_m and joining them to vertices previously constructed on C_{m-1}. We shall say that a sequence of vertices or disjoint closed arcs on C_m is "successive" if, beginning with the first member, we encounter its members in the prescribed sequence when traversing C_m in the positive direction. Now, we select arbitrary successive points $c_1(1),c_2(1),\ldots,c_t(1)$ on C_1 and join them to 1 by oriented segments beginning at 1 which lie inside C_1 and meet only at 1. Our construction proceeds by first selecting a sequence of successive disjoint closed arcs on C_m corresponding to the sequence of successive vertices of Δ on C_{m-1}. Next, we choose a sequence of successive vertices on each of these arcs. Finally, we join each vertex of Δ on C_{m-1} to the vertices on the arc it determines on C_m by oriented edges in the region between C_m and C_{m-1}. These edges are oriented so that their initial vertices lie on C_m, m even; each pair of these edges is either disjoint or meets on C_{m-1}. Whenever our construction yields a vertex whose label has previously appeared, we affix a prime to that label and call the vertex final. If a final vertex appears on C_m, we do not pick an arc corresponding to it on C_{m+1}.

Now, we describe the labeling of the vertices in detail. It may be helpful to examine Figures 1, 2, 3, 4 and 5, which illustrate our construction for several interesting Riemann surfaces. The vertices on the closed arc on C_2 corresponding to $c_i(1)$ will be denoted by $\pi_i(1), \pi_i^2(1), \ldots$. Since π_i has finite order, we obtain a finite set of vertices. Next, the arc on C_3 corresponding to $\pi_i^k(1)$ will contain vertices with the labels $c_{i+1}(\pi_i^k(1)), c_{i+2}(\pi_i^k(1)), \ldots, c_t(\pi_i^k(1)), c_1(\pi_i^k(1)), \ldots, c_{i-1}(\pi_i^k(1))$. The arc on C_4 corresponding to $c_j(\pi_i^k(1))$ will have vertices labeled $\pi_j(\pi_i^k(1)), \pi_j^2(\pi_i^k(1)), \ldots$. It should now be clear how to proceed.

It is apparent from our construction that Δ is finite and that there is a surjection, δ, of Δ onto Γ which carries vertices to vertices, edges to edges, and preserves orientation. Moreover, if e is an edge of Γ, then $\delta^{-1}(e)$ consists of one or two edges. The latter case occurs if and only if at least one of the edges of $\delta^{-1}(e)$ has a final vertex. Thus, we see that Γ is a quotient graph of Δ.

Next, we denote by Δ' the subgraph of Δ consisting of all edges with a final vertex. Observe that if $e' \in \Delta'$ and if there is an edge e in Δ but not in Δ' such that $\delta(e) = \delta(e')$ and $\{v\} = e \cap e' \neq \emptyset$, then $\delta(v)$ is a vertex of $\Gamma \subset X$ which projects to a branch point $z = b_i$ for which π_i has a fixed point. Now, denote by T' the closure of the complement of Δ' in Δ. Of course, δ maps T' homeomorphically onto $T = \delta(T')$, which is a maximal tree because it contains all the vertices of Γ. Thus, we have established properties (i) and (ii).

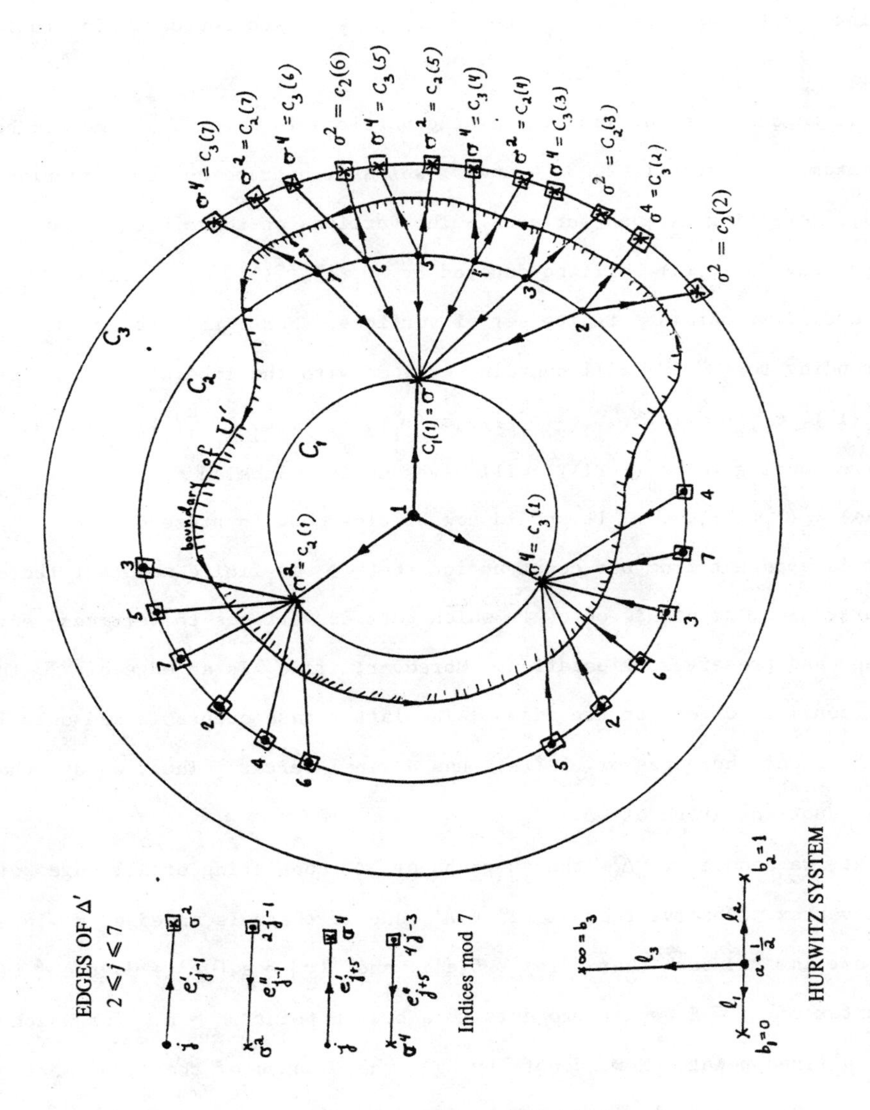

Figure 1

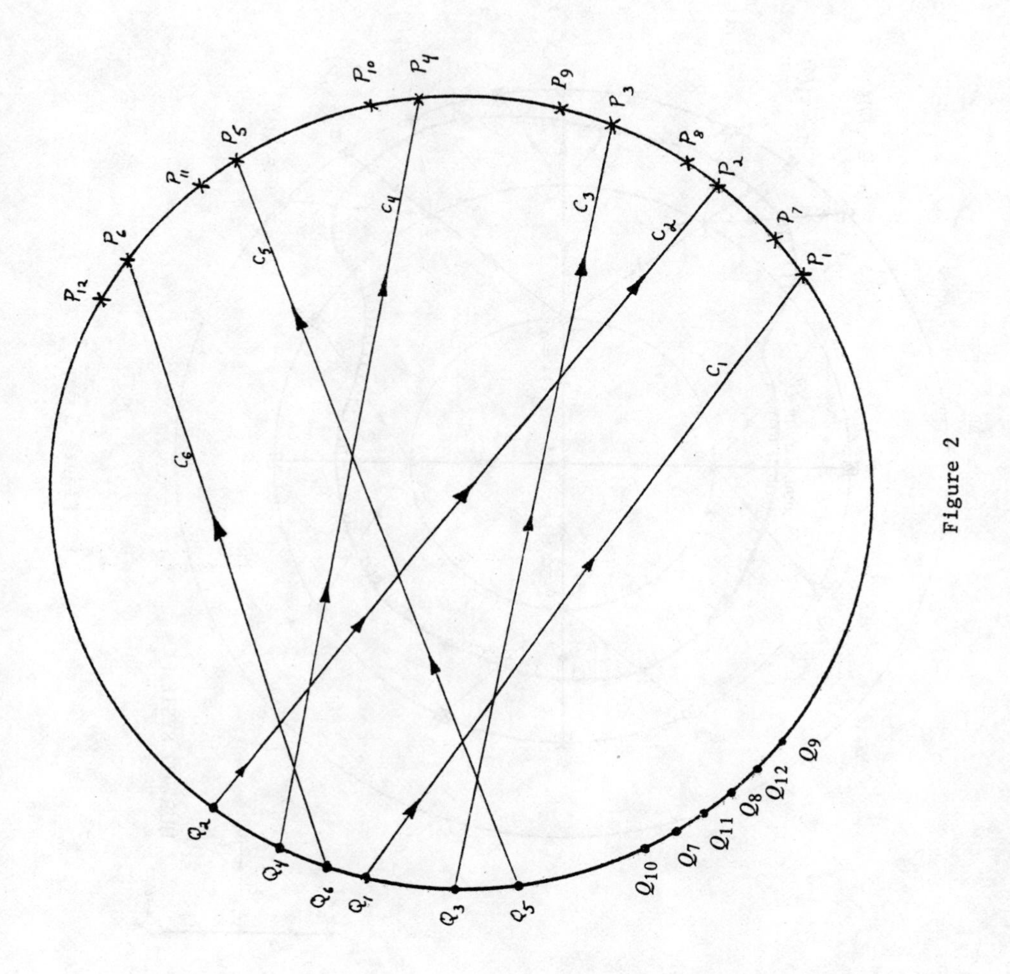
P12
P6
P11
P5
P10
P4
P9
P3
P8
P2
P7
P1
C1
C2
C3
C4
C5
C6
Q2
Q4
Q6
Q1
Q3
Q5
Q10
Q7
Q11
Q8
Q12
Q9

Figure 2

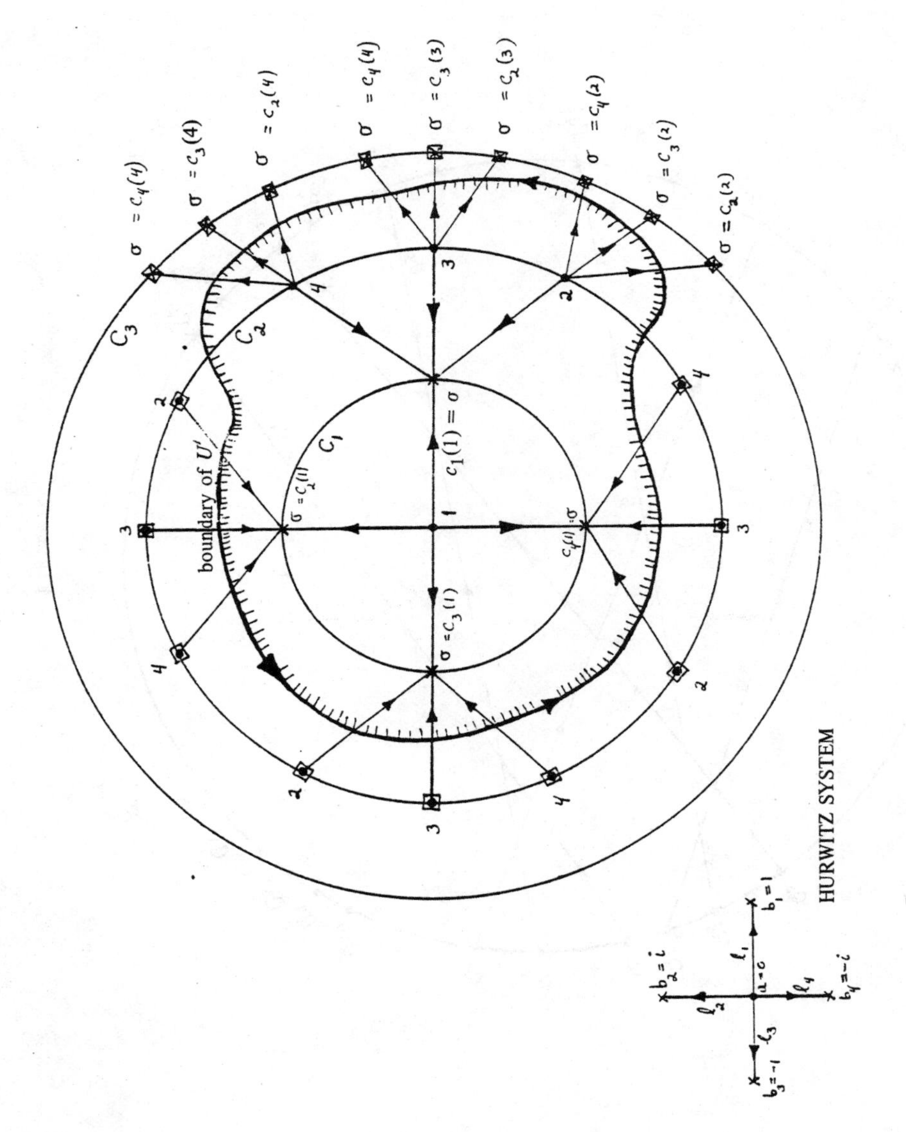

C1
C2
C3
boundary of U'
c1(1) = σ
σ = c2(1)
σ = c3(1)
c4(1) = σ
σ = c4(4)
σ = c3(4)
σ = c2(4)
σ = c4(4)
σ = c3(3)
σ = c2(3)
σ = c4(2)
σ = c3(2)
σ = c2(2)
HURWITZ SYSTEM
b1 = 1
b2 = i
b3 = -1
b4 = -i
a = 0

Figure 3

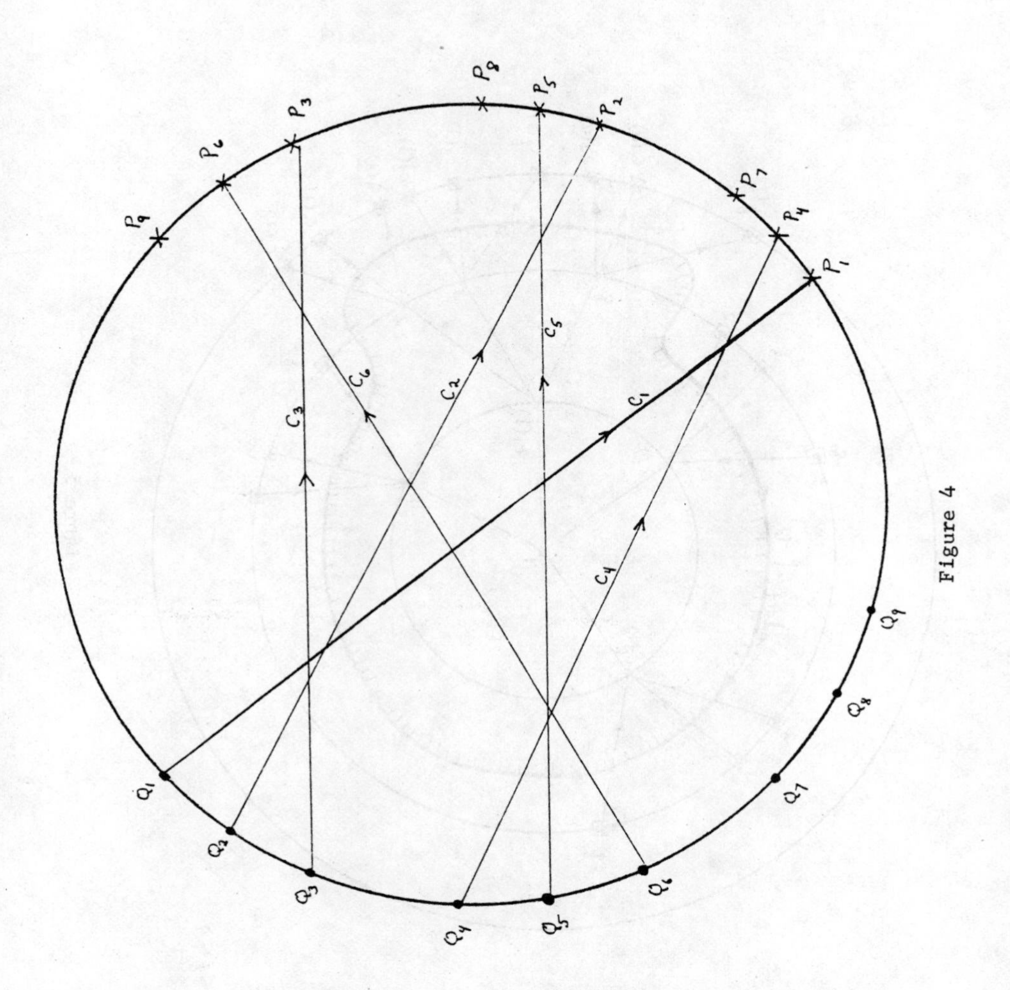

Figure 4

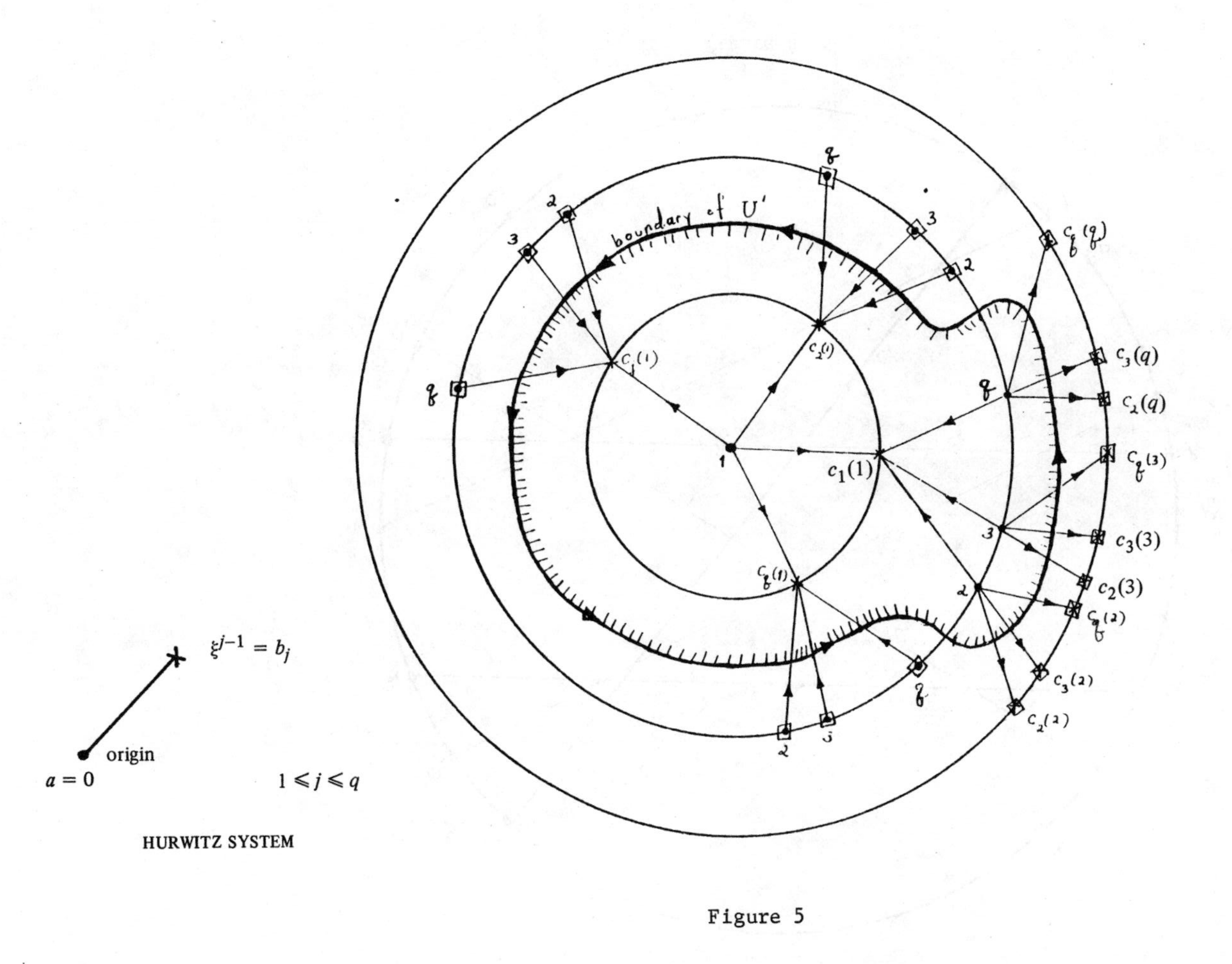

boundary of U'
$c_1(1)$
$c_3(q)$
$c_2(q)$
$c_3(3)$
$c_2(3)$
origin
$a = 0$
$\xi^{j-1} = b_j$
$1 \leqslant j \leqslant q$
HURWITZ SYSTEM

Figure 5

Next, let U' denote the interior of the region bounded by a positively oriented simple closed curve which winds once around the center of our family $\{C_m\}$ of circles, does not meet any edge of T', crosses each edge of Δ' once, and does not meet any vertices of Δ. Thus, $T' \subset U'$ and there is an orientation preserving homeomorphism of the closure, $\overline{U}'$, of U' onto the region $\overline{U} \subset X$, introduced in Section 1, which carries T' onto T. Thus (iii) is also established.

Of course, there is also an orientation preserving homeomorphism of $\overline{U}'$ onto the unit disk in the plane which carries T' into the interior of the disk. Now, suppose that e_i' and e_i'' are edges of Δ' with $\delta(e_i') = \delta(e_i'')$. Then e_i' and e_i'' meet the boundary of $\overline{U}'$ at points p_i and q_i whose images on the unit circle will be denoted by P_i and Q_i respectively. We draw the chord P_iQ_i and orient it so that P_i is its initial point. Here, we suppose the intersection number of e_i' and the boundary of U' is -1. As we explained in Section 1, the chord P_iQ_i determines a 1-cycle, c_i, on X and the intersection number of the chord P_iQ_i and P_jQ_j equals the intersection number, $c_i \cdot c_j$, of c_i with c_j. Finally, Frobenius' algorithm enables us to find appropriate integral linear combinations of the c_i which form a canonical homology basis.

The embedded tree Δ enables us to readily describe a loop in $\hat{\mathbb{C}}$ which lifts to a loop in X homotopic to c_i. First, we observe that traversing abutting edges e_1 and e_2 whose vertices are, successively, h, $c_i(h)$, $\pi_i^k(h)$ corrresponds to traversing the lift to X beginning at h of the loop γ_i traversed k times in the positive direction. Needless to say, this lifting does not lie in Γ, so it does not coincide with the edge path $\delta(e_1)$, $\delta(e_2)$. However, it does have the same end points as that edge path and it is homotopic to it. Now, suppose that $e_i \in \Gamma$, $e_i \notin T$ and that $\delta(e_i') = e_i = \delta(e_i'')$, e_i' and e_i'' in Δ but not in T'. Then, the 1-cycle c_i on X determined by e_i is the image under δ of the reduced path in Δ obtained by first traversing the reduced path in T' joining the vertex of e_i' in T'

to the vertex of e_i'' in T' and then traversing e_i''. Since Γ is a bipartite graph, this path has an even number of edges; so we may describe a sequence of loops on $\hat{\mathbb{C}}$ whose lifts to X combine to yield a loop homotopic to c_i.

Example I: Suppose that X is the Riemann surface of the Klein-Hurwitz curve defined by the equation $w^7 = z(1-z)^2$. The branch points may be labeled as follows: $b_1 = 0$, $b_2 = 1$, $b_3 = \infty$. Setting $a = \frac{1}{2}$ and taking the line segments $0 \leq z \leq \frac{1}{2}$, $\frac{1}{2} \leq z \leq 1$, and $\{z \mid \mathrm{Re}\, z = \frac{1}{2}, \mathrm{Im}\, z \geq 0\}$ as ℓ_1, ℓ_2, and ℓ_3 respectively, we obtain a Hurwitz system with $\pi_1 = \sigma$, $\pi_2 = \sigma^2$, $\pi_3 = \sigma^4$, where σ is the cycle $(1\ 2\ 3\ 4\ 5\ 6\ 7)$. Figure 1 shows the graph Δ and the region U'; final vertices are surrounded by boxes in the illustration. The reader will easily identify the twenty-four edges of Δ'; they form eight connected components. Figure 1 also illustrates our Hurwitz system.

In Figure 2, we see a sequence of successive points on the unit circle which correspond biuniquely to the edges of Δ'. They are labeled P_j and Q_j in accordance with the previous discussion. Figure 2 also illustrates the chords corresponding to the 1-cycles $c_1,\ldots,c_6$. Their intersection matrix is easily seen to be

$$K = \begin{pmatrix} 0 & 0 & 1 & 0 & 1 & 0 \\ 0 & 0 & 1 & 1 & 1 & 1 \\ -1 & -1 & 0 & 0 & 1 & 0 \\ 0 & -1 & 0 & 0 & 1 & 1 \\ -1 & -1 & -1 & -1 & 0 & 0 \\ 0 & -1 & 0 & -1 & 0 & 0 \end{pmatrix}.$$

Here, the intersection number, $c_i \cdot c_j$, of c_i with c_j is the entry k_{ij} of the matrix $K = (k_{ij})$.

Finally, applying Frobenius' algorithm [A3, Vol. III], we find that $J = {}^tTKT$, where

$$T = \begin{bmatrix} 0 & 0 & 0 & 0 & 1 & 0 \\ 1 & -1 & -1 & 0 & -1 & 0 \\ 0 & 0 & -1 & 1 & 0 & 0 \\ 0 & 1 & 0 & 0 & 0 & -1 \\ 0 & -1 & 1 & 0 & 0 & 0 \\ 0 & 0 & 0 & 0 & 0 & 1 \end{bmatrix}.$$

Example II: Let X be the Riemann surface defined by the Fermat curve $z^q + w^q = 1$. The branch points may be labeled $b_j = \xi^{j-1}$, where $\xi = e^{2\pi i/q}$. A useful Hurwitz system is obtained by selecting $a = 0$ and choosing the line segment from a to b_j for ℓ_j. Obviously, each $\pi_j = \sigma_j = (1\ 2\ .\ .\ .\ q)$. This Hurwitz system is shown in Figure 5, which also illustrates the corresponding embedded graph Δ. The special case $q = 4$ is shown in Figure 3. The chords corresponding to the 1-cycles $c_1, \ldots, c_6$ are shown in Figure 4. Their intersection matrix is

$$K = \begin{bmatrix} 0 & 1 & 1 & 1 & 1 & 1 \\ -1 & 0 & 1 & 0 & 1 & 1 \\ -1 & -1 & 0 & 0 & 0 & 1 \\ -1 & 0 & 0 & 0 & 1 & 1 \\ -1 & -1 & 0 & -1 & 0 & 1 \\ -1 & -1 & -1 & -1 & -1 & 0 \end{bmatrix}.$$

Applying Frobenius' algorithm, we obtain $J = {}^tTKT$, where

$$T = \begin{bmatrix} 1 & 1 & 1 & 0 & 0 & 1 \\ 0 & -1 & -1 & 1 & -1 & 0 \\ 0 & 1 & 0 & 0 & 0 & 0 \\ 0 & 0 & 0 & 0 & 1 & -1 \\ 0 & 0 & 1 & 0 & 0 & 0 \\ 0 & 0 & 0 & 0 & 0 & 1 \end{bmatrix}.$$

Section III: A combinatorial proof of a theorem of Hurwitz

Now, suppose that X is the Riemann surface of an algebraic function. That is, suppose that X is the domain of a n-valued meromorphic function, $w(z)$, defined by the vanishing of an irreducible polynomial, $f(z,w) = 0$, with complex coefficients and variables z and w. It is supposed that n is the highest power to which w appears in $f(z,w)$. Of course, such an equation defines an algebraic extension field, $\mathbb{C}(z,w)$, of the field of rational functions $\mathbb{C}(z)$. The degree of this extension is at least n; it equals n precisely when it is a Galois extension. In any event, we may form the normal or Galois closure, $\overline{\mathbb{C}(z,w)}$, of $\mathbb{C}(z,w)$ and refer to its Galois group as the Galois group of the equation $f(z,w) = 0$. This is a group of permutations of the n roots of the equation, and it is known (see, for example, [54], for a proof) to coincide with the monodromy group of the corresponding branched covering X of $\hat{\mathbb{C}}$.

Any transitive group of permutations of a finite set of symbols can occur as the monodromy group of an algebraic function. This may be proved, for example, by making simple modifications of the proof given in [54] for the case of regular permutation groups. It is important to realize that no purely algebraic proof of this fact is known; an existence theorem from analysis is used, at least implicitly, to show that the manifolds constructed in the proof are defined by algebraic functions. Thus, we cannot say anything about the number theoretic properties of the constants in the equation defining the algebraic function. Finally, we note that the existence theorem in question, usually called Riemann's existence theorem, can be proved by several methods. For example, see A, [3], Vol. I, page 168.

Given a finite group G, we may ask what are the genera of Riemann surfaces with monodromy group isomorphic to G. Now, Kuiken [33] has made a start at determining which groups can appear if the Riemann surface has genus zero or one. With some obvious exceptions, she conjectures that solvable groups cannot be monodromy groups of Riemann surfaces of genus zero.[(3)]

Rimhak Ree [44] observed that the following combinatorial fact may be proved by combining the existence of a branched covering of $\hat{\mathbb{C}}$ with given monodromy group and the inequality obtained from the Riemann-Hurwitz formula by deleting the term involving the genus: if $\pi_1, \ldots, \pi_t$ are permutations of n symbols which satisfy $\pi_1 \cdots \pi_t = 1$ and generate a group with s orbits, then $v(\pi_1) + \ldots + v(\pi_t) \geq 2(n-s)$, where $v(\pi) = n-z$ when π is the product of z disjoint cycles. Subsequently, W. Feit, R. C. Lyndon and L. L. Scott [14] and the present authors [53] gave proofs of Ree's theorem which avoid surface topology. J. McKay showed that Ree's theorem may be applied as a "Brauer trick" to prove the existence of conjectured subgroups of a given finite group. Subsequently, L. L. Scott [46] proved an analogue of Ree's theorem for matrix groups which yields a general "Brauer trick" and many interesting results centering around group cohomology. Finally, we note that the present authors [53] showed how Kuratowski's planarity criteria and its generalizations may be applied to strengthen Ree's inequalities when partial information about the permutations is available. For example, suppose that (123), (1324) and (1237) occur among the cycles in a set of generators of a monodromy group. Then, the corresponding Riemann surface has genus greater than zero because its 1-skeleton contains the complete bipartite graph $K_{3,3}$. This application may be of interest in view of the difficulty of determining monodromy groups.

The collection of all biholomorphic mappings of a Riemann surface, X, one to one onto itself obviously forms a group, Aut(X). In case X is compact and has genus $g > 1$, Hurwitz [25] proved that this group is finite and has order at most $84(g-1)$. Moreover, Hurwitz showed that Aut(X) admits faithful representations on both the g-dimensional complex vector space of Abelian differentials of the first kind and on the free Abelian group $H_1(X, \mathbb{Z})$.(5) Hurwitz proved his theorems by applying the Riemann-Hurwitz formula and information about the number of Weierstrass points on X. The latter results are usually derived with the aid of the Riemann Roch theorem. Thus, Hurwitz' proof depends in an essential fashion on function theoretic or, depending on one's viewpoint, algebraic geometric information.

We now present a simple combinatorial proof of one of Hurwitz' theorems. Namely, we prove the following:

Theorem (Hurwitz): Let X be a compact Riemann surface of genus $g > 1$ and let θ be a biholomorphic self-mapping of X whose order is n. If θ induces the identity automorphism on $H_1(X, \mathbb{Z})$, then θ is the identity.

Proof: First, we observe that the theorem follows from the special case in which θ has prime order. Indeed, if the order of θ is a composite number, n, then θ^m has prime order for some integer m with $1 < m < n$. Setting $\phi = \theta^m$, noting that ϕ also induces the identity on $H_1(X, \mathbb{Z})$, and applying the special case of the theorem, we obtain the contradiction $\phi = \theta^m = 1$.

Our proof proceeds by contradiction. Namely, if θ is not the identity, we shall exhibit a homology basis upon which it acts non-trivially.

Now, suppose that θ has prime order, q, and let G denote the cyclic group generated by θ. It is an elementary exercise to show that G acts properly discontinuously on X and yields a branched covering, $p: X \to S$, of the orbit space $S = X/G$. Picking a Hurwitz system for this covering and recalling that the k-skeletons are related by the equations $X^{(k)} = p^{-1}(S^{(k)})$, we see that we have a G-invariant cellular decomposition of X. Thus, we may suppose that the points belonging to $p^{-1}(a) = \{a_1, \ldots, a_q\}$ satisfy $\theta a_k = a_{k+1}$, where the indices are integers modulo q. Similarly, we have $\theta(\ell(j,k)) = \ell(j,k+1)$, $\theta(\alpha(j,k)) = \alpha(j,k+1)$, $\theta(\beta(j,k)) = \beta(j,k+1)$, where $\ell(j,k)$, $\alpha(j,k)$, $\beta(j,k)$ are the lifts to X beginning at a_k of $\ell(j)$, α_j, β_j respectively. We may also suppose that the 2-cells lying over D are indexed so that $\theta D_k = D_{k+1}$. Finally, since the stabilizer of a point fixed by a non-trivial element of G must be G itself, there is a single point, $\tilde{b}_j$, on X lying over the branch point b_j.

We shall now exhibit a basis for $H_1(X, \mathbb{Z})$ upon which θ induces a non-trivial action. We simplify the exposition by supposing at first that S has genus zero; the general case is only slightly more complicated. Thus, we suppose that $t > 1$; in fact, t must be greater than two, lest the genus of X be zero. The required 1-cycles are

$$\gamma(j,k) = -\ell(1,k) + \ell(j,k) - \ell(j,k+1) + \ell(1,k+1),\ 2 \leq j \leq t-1,\ 1 \leq k \leq q .$$

These cycles obviously satisfy

(i) $\theta(\ell(j,k)) = \ell(j,k+1)$

(ii) $\ell(j,q) = -[\ell(j,k) + \ldots + \ell(j,q-1)]$,

so the proof of Hurwitz' theorem will be complete if we prove that they represent a basis for $H_1(X, \mathbb{Z})$. In fact, we shall see that the $\gamma(j,k)$ represent $2g$ generators for $\pi_1(X,\tilde{b}_1)$ and, therefore, a basis for $H_1(X, \mathbb{Z})$. For this purpose, we let E' denote the complement of the $\gamma(j,k)$ in X and observe that E' is homeomorphic to the interior of the closed 2-cell, E, in Figure 6. Now, let Γ' be a graph homeomorphic to the subgraph of Γ consisting of all $\gamma(j,k)$, and give the vertices and edges of Γ' the same labels as the corresponding constituents of Γ. Attaching E to Γ', we obtain X. Thus, $\pi_1(X,\tilde{b}_1)$ is a one-relator group generated by the $\gamma(j,k)$; each $\gamma(j,k)$ appears once with exponent $+1$ and once with exponent -1 in the relator. This will not in general be the "standard" relator for the fundamental group of a surface of genus g because the $\gamma(j,k)$ do not have intersection matrix J.

Incidentally, another proof that the $\gamma(j,k)$ form a homology basis can be given by noting that their intersection matrix has determinant one.

Now, suppose that X is an unramified covering of S, so $t = 0$. Since $g \geq 2$ and unramified coverings of the torus all have genus one, we see that the genus, $\hat{g}$, of S is also at least two. Now, $p_*\pi_1(X,a_1)$ is a normal subgroup of $\pi_1(S,a)$ with cyclic quotient of order q, so every lifting of $\alpha_j^{-1}\beta_j^{-1}\alpha_j\beta_j$ to X is a loop. On the other hand, not all the $\alpha(j,k)$ and $\beta(j,k)$ can be loops; otherwise X would be a disjoint union of q copies of

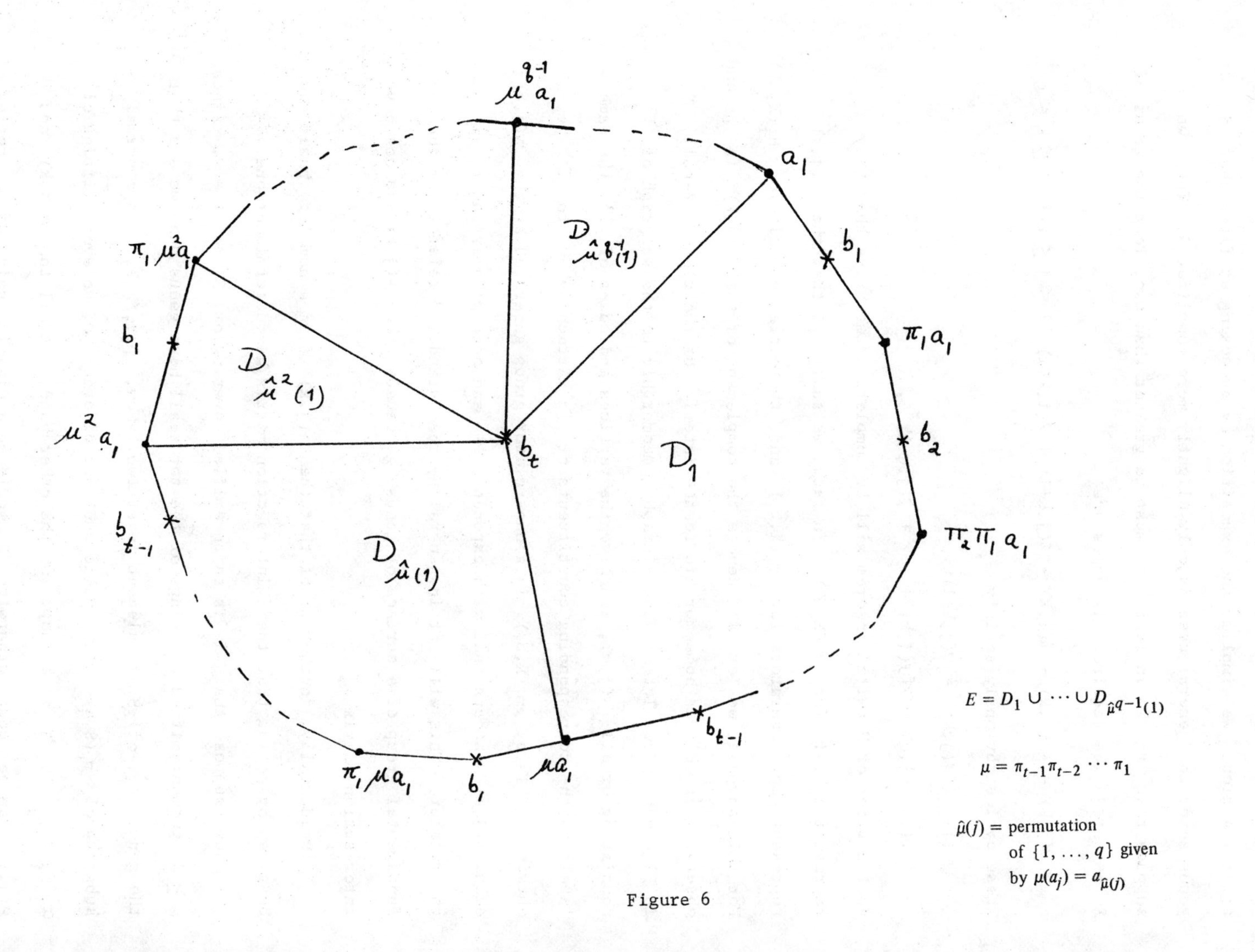
$\mu^{q-1} a_1$
a_1
b_1
$\pi_1 a_1$
b_2
$\pi_2\pi_1 a_1$
b_{t-1}
μa_1
b_1
$\pi_1 \mu a_1$
b_{t-1}
$\mu^2 a_1$
b_1
$\pi_1 \mu^2 a_1$
b_t
D_1
$D_{\hat{\mu}^{q-1}(1)}$
$D_{\hat{\mu}^2(1)}$
$D_{\hat{\mu}(1)}$
$E = D_1 \cup \cdots \cup D_{\hat{\mu}^{q-1}(1)}$
$\mu = \pi_{t-1}\pi_{t-2} \cdots \pi_1$
$\hat{\mu}(j)$ = permutation of $\{1, \ldots, q\}$ given by $\mu(a_j) = a_{\hat{\mu}(j)}$

Figure 6

S, a contradiction of its connectedness. Thus, we suppose that α_1 does not lift to a loop on X and observe that $\alpha(1,1),\ldots,\alpha(1,q-1)$ form a maximal tree in the 1-skeleton, Γ, of X. Each of the $2g-1$ remaining α_j and β_j lifts to yield q one-cycles which are permuted cyclically. Using the disks D_k, we may express q of these 1-cycles in terms of the others. Since $2g-1 \geq 3$, the action induced by θ on these remaining cycles is non-trivial.

A similar proof can be given in case $\hat{g} > 0$ and $\hat{t} > 0$.

Q.E.D.

The elegant idea of working directly with chains is due to Jane Gilman [17]. Her proof differs from ours because, as she observes, it is based on the Schreier-Reidemeister rewriting process, following a paper of J. Nielsen [62] (1937). Moreover, Gilman's paper is presented in the context of Fuchsian groups; so, in addition to the implicit use of the highly trascendental uniformization theorem, her proof has a different flavor than ours. We refer the reader to the final section of her paper for a number of interesting related applications of the "adapted" basis. Subsequently, Gilman and Patterson [18] determined the intersection matrix of this basis. Finally, we note that, in the case $S = \hat{\mathbb{C}}$, the adapted basis and its invariance appear in the famous paper of Lefschetz [61] (1921); thus Gilman's applications of it seem all the more remarkable.

Now, using the elementary theory of Jacobian varieties, H. Martens [41] has given yet another proof of Hurwitz' theorem and a generalization of it. In particular, he shows that a holomorphic surjection between compact Riemann surfaces of genus at least two is completely determined by the homeomorphism it induces on the homology. It would be interesting to know whether Marten's theorem can be proved purely on the homology level, in the spirit of our proof of Hurwitz' theorem.

Hurwitz' theorems can also be proved by the methods of algebraic topology. For this approach, we refer the reader to the paper of R. Kulkarni [35] which contains many interesting extensions of the original results.

In view of Hurwitz' theorem, it is natural to ask whether every finite group is the automorphism group of a Riemann surface. The answer to this question was given by L. Greenberg [19] who proved: given a non-negative integer g and a finite group G, there is a compact Riemann surface, X, whose group of conformal self-mappings, $\mathrm{Aut}(X)$, is isomorphic to G and whose orbit space X/G, has genus g. The monodromy group of the resulting branched covering is a regular permutation group because branched coverings defined by discontinuous group actions are regular. This remark should not obscure the depth of Greenberg's theorem, which is perhaps more appropriately viewed in the context of Fuchsian groups and is proved with the aid of the Teichmuller theory of families of Riemann surfaces. Apparently, no proof involving a single Riemann surface is known. Finally, we note that Greenberg's theorem does not tell us the genus of X.

Considerable effort has been devoted to the study of Riemann surfaces of genus g whose group of conformal self-mappings has order $84(g-1)$. Let $N(g)$ denote the maximal order of $\mathrm{Aut}(X)$, X a Riemann surface of genus g. A. M. Macbeath [38] showed that there is an infinite sequence $\{g_n\}$ for which there are Riemann surfaces, X_n, of genus g_n with $|\mathrm{Aut}(X_n)| = 84(g_n-1)$. Thus, $N(g_n) = 84(g_n-1)$. A related result of R. Accola [1] shows that $8(g+1) \leq N(g)$ for all g, and that equality holds for infinitely many values of g. In fact, $|\mathrm{Aut}(X)| = 84(g-1)$ if and only if X is uniformized by a normal subgroup Γ of finite index in the (2,3,7) triangle group, that is, if and only if X is the quotient of the unit disk by the discontinuous group Γ. For this reason, finite quotients of the (2,3,7) group are of interest and are often called Hurwitz groups.

Three is the smallest value of g for which $N(g) = 84(g-1)$. There is a famous example, commonly called the Klein-Hurwitz curve, of a Riemann surface, X, of genus three with $\mathrm{Aut}(X) \cong PSL(2,7)$. This Riemann surface, which may be defined by the equation $w^7 = z(1-z)^2$, plays a role in the construction of Macbeath's sequence $\{X_n\}$. The next value of g for which

$N(g) = 84(g-1)$ is seven, and Macbeath [39] has given an explicit description of such a Riemann surface with group of conformal self mappings $PSL(2,2^3)$. Both the Klein-Hurwtiz curve and Macbeath's curve of genus seven can be defined by a polynomial equation $f(z,w) = 0$ with <u>real</u> coefficients. In fact, D. Singerman [49] has shown that all the X_n in Macbeath's sequence correspond to real algebraic curves. Moreover, Singerman employed a non-split extension of $\mathbb{Z}_2^3$ by $PSL(2,7)$ discovered by A. Sinkov [50] to prove the existence of a Riemann surface of genus 17 admitting 1344 conformal self mappings which <u>cannot</u> be defined by a real algebraic curve. Finally, we note relevant recent papers by J. Cohen [10] and R. Kulkarni [34]; in particular, group theory plays an important role in Kulkarni's paper.

Additional information about automorphisms of Riemann surfaces can be found in [58] and [A,1].

Section IV: Periods and quadratic periods of Abelian integrals

We now turn our attention to certain transcendental invariants of compact Riemann surfaces, the periods and quadratic periods of Abelian integrals. The former have played a central role in our subject since its earliest days. Although they satisfy a well developed theory, important questions about them remain unanswered and they are still a subject of active investigation. The quadratic periods are of recent origin, so their ultimate role in the theory is less clear despite encouraging signs. It is important to understand that because of their transcendental nature these invariants can rarely be determined explicitly and that their values are often of number theoretical interest.(6) Here, with the aid of the combinatorial technique introduced in Sections 2 and 3, we determine the periods and quadratic periods explicitly on certain Riemann surfaces. Previously, the quadratic periods were only determined on hyperelliptic Riemann surfaces, that is, two-sheeted coverings of $\hat{\mathbb{C}}$. Our calculation of the ordinary periods on the Klein-Hurwitz curve and Fermat curves yields an answer to a question of C. Earle [13]. We begin this section with a thumbnail sketch of the relevant theory of Abelian integrals and other background remarks. Next, we make the indicated computations.

Once again, let X be a compact Riemann surface of genus $g > 1$ defined by an irreducible polynomial $f(z,w) = 0$ of degree n. It is a famous theorem (for example, see [A,2], page 225) that any compact Riemann surface admits at least one such representation. We are now interested in the integrals of those differential one-forms, called _Abelian differentials of the first kind_, which are holomorphic at each point of X. The totality of these differentials can be shown to form a g-dimensional complex vector space. Without going into the details, we note that these differentials generally have the form

$$du_j = \frac{Q_j(z,w)\,dz}{\frac{\partial f}{\partial w}}\ , \quad j = 1,\ldots,g\ ,$$

where the $Q_j(z,w)$ are g linearly independent polynomials of degree at most $n-3$ which pass through an r-fold singularity of $f(z,w) = 0$ at least $r-1$ times.

Denoting the integrals of the du_j along members α_k and β_k, $1 \leq k \leq g$, of a canonical homology basis by a_{jk} and b_{jk}, we obtain a period matrix $\Pi = (A\ B)$, $A = (a_{ij})$, $B = (b_{ij})$. Riemann proved that A is non-singular and that $Z = A^{-1}B$ is symmetric with positive definite imaginary part. We call $(I\ Z)$ a _normalized_ period matrix. Now, the totality of symmetric complex matrices of degree g with positive definite imaginary part may be viewed as an analogue of the upper half plane, called the _Siegel upper half space of degree_ g. It is an outstanding problem, known as the _Schottky problem_, to characterize the subspace of the Siegel space corresponding to normalized period matrices when $g \geq 4$.

The period matrix Π will change if we change either the basis for the Abelian differentials of first kind or the homology basis. More precisely, if a new basis for the differentials is given by $d\underline{v} = Md\underline{u}$, M in $GL(g,\mathbb{C})$, then the period matrix with respect to $d\underline{v}$ and α_k, β_k is $M\Pi$. Similarly, if N belongs to the symplectic group $Sp(g, \mathbb{Z})$, then ΠN is the period matrix of the du_j with respect to a new canonical homology basis. A famous theorem of Torelli asserts that if $\Pi = (I\ Z)$ and $\Pi' = (I\ Z')$ are normalized period matrices of Riemann surfaces X and X' and if there is a symplectic matrix N such that Z' is obtained by normalizing ΠN, then X and X' are conformally equivalent. A beautiful proof of Torelli's theorem by H. Martens also shows that Z cannot be a direct sum of submatrices. Finally, suppose that Π is a normalized period matrix for X. Then the $2g$ columns of Z are linearly independent over the real numbers and the free Abelian group generated by these vectors is a lattice, L, in $\mathbb{C}^g$. The quotient $\mathbb{C}^g/L$ is a compact, connected g-dimensional complex Abelian Lie group known as the Jacobian variety of X and denoted by $J(X)$. A famous result, Abel's theorem, allows us to conclude that there is a biholomorphic mapping of X onto a proper submanifold of $J(X)$.

Now, suppose that a period matrix $\Pi = (A\ B)$ for X is given with respect to a homology basis which is not canonical. It is conceivable that A is non-singular and that $Z = A^{-1}B$ is the direct sum of period matrices. In fact, examples of this phenomenon were found in genus two by Hayashida and Nishi [59] in genus four by A. Weil (unpublished) and for arbitrary even genus by C. Earle [13]. The latter asked whether examples exist if g is odd. For additional information about this curious phenomenon, we refer to H. Martens' article [40] and H. Lange's characterization [36] of Abelian varieties which are isomorphic to the product of elliptic curves.

The periods of an Abelian integral depend only on the homology class of the path of integration. These quantities have been studied for more than one hundred years. More recently, R. C. Gunning [20] introduced new transcendental invariants $Q_{ij}(\gamma)$, which he called quadratic periods, by selecting a base point, x_0, on X and integrating the product $(\int_{x_0}^{x} du_j)du_k$ along a loop representing the element γ of $\pi_1(X,x_0)$. Here, the multi-valued integrand is made meaningful by selecting the branch of the Abelian integral $\int_{x_0}^{x} du_j$

which vanishes at x_0. Of course, the quadratic periods reflect the non-Abelian nature of the fundamental group, yet they are related to it by the formula

$$Q_{jk}(\gamma_1\gamma_2) = Q_{jk}(\gamma_j) + Q_{jk}(\gamma_2) - \pi_j(\gamma_1)\pi_k(\gamma_2) ,$$

where $\pi_j(\gamma) = \int_\gamma du_j$. Moreover, certain combinations of quadratic periods appear in the classical theory of the Riemann theta function as the vector of Riemann constants. In fact, the Riemann constants are invariant under those automorphisms of the fundamental group which induce the identity on homology. E. Jablow [26] proved that these are essentially the only combinations of quadratic periods invariant under such automorphisms. Thus, the quadratic periods appear to be more delicate transcendental invariants than the ordinary periods; and it is hoped that their investigation will shed additional light on the Schottky problem.

The quadratic periods also play a role in the discovery by B. Harris [22] that the embedded Fermat quartic curve and its negative are homologous cycles on their Jacobian variety which are not algebraically equivalent. Roughly speaking, this means that there is no family of curves on the Jacobian which are defined by algebraic equations and contains both the embedded copy of the Fermat quartic and the curve obtained by taking the inverse of each of its points.

Finally, we note that Gunning's construction can be generalized by taking n-fold iterated integrals. Thus, it is a manifestation in the holomorphic category of the deRham homotopy theory developed by K. T. Chen [9] and D. Sullivan [51].

We shall determine the periods and quadratic periods on the Klein-Hurwitz curve and on the Fermat curves. We already mentioned that the quadratic periods were previously determined only on the hyperelliptic curves, where they proved to be products of ordinary periods. On the other hand, the periods of Abelian integrals on these curves have been determined by several authors. For example, Rauch and Lewittes [43] determined a point in the Siegel upper half space corresponding to the Klein-Hurwitz curve. Their work utilized the fact that this Riemann surface is uniformized by the principal congruence subgroup of level seven in the modular group. They also determined the action of $PSL(2,7)$ on the first homology group of this curve by essentially group-theoretic methods. Their paper contains interesting footnotes, pointing out errors both by Poincaré and by Hurwitz in determining the period matrix of the Klein-Hurwitz curve, and citing a discussion of the curve in a book by H. F. Baker [3]. In fact, Baker determined the periods of Abelian integrals by the same procedure we shall follow. However, he did not determine the intersection matrix of the cycles he employed, and he dutifully noted this fact. Baker's homology basis was not canonical, and the point he determined in Siegel space was a diagonal matrix. Thus, he answered the question of Martens and Earle many years before they asked it. Moreover, Baker obtained diagonal period matrices for the Fermat quartic and certain hyperelliptic curves of arbitrary even genus. Finally, we note that A. Adler [2] also observed that a period matrix of the Klein-Hurwitz curve can be diagonal. Adler's papers are a noteworthy application of integral representations to algebraic geometry, in the spirit of Hecke.

Example I: Let X denote the Riemann surface defined by the Klein-Hurwitz curve $w^7 = z(1-z)^2$. It is easy to verify that a basis for the Abelian differentials of the first kind on X is given by

$$du_1 = \frac{(1-z)dz}{w^6}, \; du_2 = \frac{(1-z)dz}{w^5}, \; du_3 = \frac{dz}{w^3} .$$

Now the six series solutions of $w^7 = z(1-z)^2$ in powers of $z - \frac{1}{2}$ may be labelled $w_k = w_k(z)$, $1 \leq k \leq 6$, and identified with the points a_k, $1 \leq k \leq 6$, on X. Moreover, it is apparent that any one of these solutions is obtained from any other by multiplication by a suitable power of $\xi = \exp(\frac{2\pi i}{7})$. Thus, we may also suppose that $w_k = \xi^{k-1} w_1$; and it is a simple matter to give explicit expressions for the du_j in a neighborhood of a_k in terms of the expressions for the du_j in a neighborhood of a_1. For example, in a neighborhood of a_k the differential du_1 is obtained from its expression at a_1 by multiplication by ξ^{k-1}. Similarly, du_2 and du_3 are defined in a neighborhood of a_k by multiplying their expressions at a_1 by ξ^{2k-2} and ξ^{4k-4} respectively.

Now, let $e_k = -\ell(1,k) + \ell(2,k)$, so $\gamma_k = e_k - e_{k+1}$, where $\gamma(2,k) = \gamma_k$, $1 \leq k \leq 6$, are the 1-cycles introduced in Section 3. The integrals of the du_j, $j = 1,2,3$, along e_k are obtained from the integrals of the du_j along e_1 by multiplication by $\xi^{k-1}, \xi^{2k-2}, \xi^{4k-4}$ respectively. Denoting the integral of du_j along γ_k by $u(j,k)$ we therefore find that

$$u(1,k) = \xi^{k-1}(1 - \xi)B(1/7,2/7)$$

$$u(2,k) = \xi^{2k-2}(1 - \xi^2)B(2/7,4/7)$$

$$u(3,k) = \xi^{4k-4}(1 - \xi^4)B(4/7,1/7) \quad ,$$

where $B(p,q) = \int_0^1 z^{p-1}(1-z)^{q-1}dz$ denotes the Beta function as usual.

Now, letting

$$(1 - \xi)B(1/7,2/7)dv_1 = du_1$$

$$(1 - \xi^2)B(2/7,4/7)dv_2 = du_2$$

$$(1 - \xi^4)B(4/7,1/7)dv_3 = du_3 \quad ,$$

we obtain another basis dv_1, dv_2, dv_3 for the differentials of the first kind on X. The period matrix with respect to this basis is

$$(A\ B) = \begin{bmatrix} 1 & \xi & \xi^2 & \xi^3 & \xi^4 & \xi^5 \\ 1 & \xi^2 & \xi^4 & \xi^6 & \xi & \xi^3 \\ 1 & \xi^4 & \xi & \xi^5 & \xi^2 & \xi^6 \end{bmatrix} .$$

The determinant of A is $(\xi^3 + \xi^5 + \xi^6) - (\xi + \xi^2 + \xi^4)$, which is not zero; in fact $\xi^3 + \xi^5 + \xi^6 = -2\tau$ and $\xi + \xi^2 + \xi^4 = 2\tau - 1$, $\tau = \frac{1 + \sqrt{-7}}{4}$. Thus, the matrix equation $A d\underline{t} = d\underline{v}$ yields a basis dt_1, dt_2, dt_3 for the Abelian differentials of first kind whose period matrix is

$$(C\ D) = \begin{bmatrix} 1 & 0 & 0 & 1 & 2\tau-1 & -1 \\ 0 & 1 & 0 & 2\tau & -1 & -1 \\ 0 & 0 & 1 & 2\tau-1 & -1 & -2\tau \end{bmatrix}.$$

Finally, letting $ds_1 = dt_2$, $ds_2 = dt_1$, $ds_3 = dt_3$ and replacing the homology basis $\gamma_1,\ldots,\gamma_6$ by the basis γ_2, γ_1, γ_3, $\gamma_3 + \gamma_4 + \gamma_6$, $\gamma_2 + \gamma_3 + \gamma_5$, $-\gamma_1 - \gamma_2 - \gamma_3 - \gamma_6$, we obtain the period matrix

$$\begin{bmatrix} 1 & 0 & 0 & 2\tau-1 & 0 & 0 \\ 0 & 1 & 0 & 0 & 2\tau-1 & 0 \\ 0 & 0 & 1 & 0 & 0 & 2\tau-1 \end{bmatrix}.$$

Thus, the Klein-Hurwitz curve provides an affirmative answer to Earle' question, as we asserted. Another way of stating this result is the following: the Jacobian variety of the Klein-Hurwitz curve is biholomorphically equivalent to the complex manifold $E \times E \times E$, where E is the elliptic curve obtained from $\mathbb{C}$ by factoring out the lattice generated by 1 and $2\tau-1$.

Next, we determine a point in the Siegel upper half space of degree three corresponding to the Klein-Hurwitz curve. First, we recall that in Section 2 we determined the intersection matrix for the homology basis given by $c_k = -e_1 + e_{k+1}$, $1 \le k \le 6$, and that a canonical basis on the Klein-Hurwitz curve was expressed in terms of the c_k by means of a matrix designated T. Since $c_1 = -\gamma_1$ and $c_k = -\gamma_1 - \ldots - \gamma_k$, $2 \le k \le 6$, we obtain the matrix equation $\underline{c} = N\underline{\gamma}$, where $N = (n_{ij})$ and $n_{ij} = -1$ for $j \ge 1$ and $n_{ij} = 0$ for $j < 1$. Thus, the period matrix for dt_1, dt_2, dt_3 with respect to the canonical homology basis under consideration is $(E\ F) = (C\ D)NT$. The reader may check that this matrix is

$$\begin{bmatrix} -1 & 2\tau & -1-2\tau & -1 & 0 & 2-2\tau \\ -1 & 0 & -2\tau & -1 & 1 & 2 \\ 0 & -1 & 2-2\tau & -1 & 0 & 1+2\tau \end{bmatrix}$$

and that $Z = E^{-1}F$ is

$$\begin{pmatrix} -4\tau^3 + \tau^2 & \tau^2 & 2\tau^3 \\ \tau^2 & -2\tau^3 + 2\tau^2 & \tau^2 \\ 2\tau^3 & \tau^2 & -4\tau^4 \end{pmatrix} = \begin{pmatrix} \frac{7 + 3\sqrt{-7}}{8} & \frac{-3 + \sqrt{-7}}{8} & \frac{-5 + 2\sqrt{-7}}{8} \\ \frac{-3 + \sqrt{-7}}{8} & \frac{-1 + 3\sqrt{-7}}{8} & \frac{-3 + \sqrt{-7}}{8} \\ \frac{-5 + 2\sqrt{-7}}{8} & \frac{-3 + \sqrt{-7}}{8} & \frac{-1 + 3\sqrt{-7}}{8} \end{pmatrix}.$$

Using Jacobi's criteria, we may easily verify that the imaginary part of Z is positive definite.

We now determine the quadratic periods on the Klein-Hurwitz curve along the generators $\gamma_1, \ldots, \gamma_6$ of $\pi_1(X, \tilde{0})$. Here, $\tilde{0}$ is the branch point on X above $z = 0$. We have already seen that

$$u(j,k) = \int_{\gamma_k} du_j = \xi^{\omega(j,k)}(1-\xi^{\mu(j)})B(\frac{\mu(j)}{7}, \frac{\nu(j)}{7}),$$

where $\omega(j,k) = \begin{matrix} k-1 \\ 2k-2 \\ 4k-4 \end{matrix}$, $\mu(j) = \begin{matrix} 1 \\ 2 \\ 4 \end{matrix}$, $\nu(j) = \begin{matrix} 2 \\ 4 \\ 1 \end{matrix}$ for $j = \begin{matrix} 1 \\ 2 \\ 3 \end{matrix}$.

Now, $\int_{\gamma_k} u_r du_s = \int_{e_k} u_r du_s - \int_{e_{k+1}} u_r du_s$, where the multi-valued factor in the integrand is given by $u_r = \int_{\tilde{0}}^{\tilde{z}(t)} du_r$ and $\tilde{z}(t)$ traverses the path γ_k parametrized by $0 \leq t \leq 1$. On e_k, we use the branch of u_r which vanishes at $\tilde{0}$, but on e_{k+1} we must add the constant $u(r,k)$ to that branch. Thus, we obtain

$$\begin{aligned} Q_{r,s}(\gamma_k) &= \int_{\gamma_k} u_r du_s \\ &= \int_{e_k} u_r du_s - \int_{e_{k+1}} u_r du_s - \int_{\gamma_k} du_r \int_{e_{k+1}} du_s \\ &= \xi^{\omega(r,k) + \omega(s,k)} \int_0^1 (\int_0^z du_r) du_s \\ &\qquad - \xi^{\omega(r,k+1) + \omega(s,k+1)} \int_0^1 (\int_0^z du_r) du_s \\ &\qquad - \xi^{\omega(r,k)}(1-\xi^{\mu(r)})B(\frac{\mu(r)}{7}, \frac{\nu(r)}{7}) \int_{e_{k+1}} du_s \\ &= [\xi^{\omega(r,k) + \omega(s,k)} - \xi^{\omega(r,k+1) + \omega(s,k+1)}] \int_0^1 (\int_0^z du_r) du_s \\ &\qquad - \xi^{\omega(r,k) + \omega(s,k+1)}(1-\xi^{\mu(r)})B(\frac{\mu(r)}{7}, \frac{\nu(r)}{7})B(\frac{\mu(s)}{7}, \frac{\nu(s)}{7}), \end{aligned}$$

where $1 \leq r,s \leq 3$.

Example II: Let X denote the Riemann surface of genus $\frac{(q-1)(q-2)}{2}$ defined by the Fermat curve $z^q + w^q = 1$. A basis for the Abelian differentials of the first kind on X is given by

$$du_{r,s} = z^r w^{s+1-q} dz \ , \ 0 \leq r,s \leq q-3, \ r+s \leq q-3 \ .$$

The power series solutions of $w^q = 1-z^q$ will be labeled by $w_k = w_k(z)$, $1 \leq k \leq q$, and the point a_k on X above $z = 0$ will be identified with w_k. As in Example I, we have $w_k = \xi^{k-1} w_1$, $\xi = \exp(\frac{2\pi i}{q})$, and in a neighborhood of a_k an expression for $du_{r,s}$ is obtained from its expression in a neighborhood of a_1 by multiplication by the constant $\xi^{(s-q+1)(k-1)} = \xi^{(s+1)(k-1)}$.

The automorphism of the z-sphere given by replacing z by ξz is easily seen to lift to an automorphism of X. The effect of this automorphism on the space of Abelian differentials is to replace $du_{r,s}$ by $\xi^{(r+1)} du_{r,s}$.

Using these substitutions, we find that

$$\int_{\ell(j,k)} du_{r,s} = \xi^{(s+1)(k-1)+(r+1)(j-1)} \int_0^1 z^r w^{s+1-q} dz \ ,$$

with $w = w_1(z)$. Since $\gamma(j,k) = -\ell(1,k) + \ell(j,k) - \ell(j,k+1) + \ell(1,k+1)$, we obtain

$$\int_{\gamma(j,k)} du_{r,s} = \xi^{(s+1)(k-1)} (\xi^{(s+1)}-1)(1-\xi^{(r+1)(j-1)}) \int_0^1 z^r w^{s+1-q} dz \ .$$

Letting $t = z^q$ and introducing a new basis by setting $du_{r,s} = (\xi^{(s+1)}-1) dv_{r,s}$, we find that

$$\int_{\gamma(j,k)} dv_{r,s} = \frac{1}{q} \xi^{(s+1)(k-1)} (1-\xi^{(r+1)(j-1)}) \int_0^1 t^{\frac{r+1}{q}-1} (1-t)^{\frac{s+1}{q}-1} dt$$

$$= \frac{1}{q} \xi^{(s+1)(k-1)} (1-\xi^{(r+1)(j-1)}) \frac{\Gamma(\frac{r+1}{q}) \Gamma(\frac{s+1}{q})}{\Gamma(\frac{r+s+2}{q})} \ .$$

Finally, we let

$$dv_{r,s} = \frac{1}{q}\,\frac{\Gamma(\frac{r+1}{q})\,\Gamma(\frac{s+1}{q})}{\Gamma(\frac{r+s+2}{q})}\,dt_{r,s}$$

and obtain

$$\int_{\gamma(j,k)} dt_{r,s} = \xi^{(s+1)(k-1)}(1-\xi^{(r+1)(j-1)}) \,. \tag{7}$$

The special case $q = 4$ provides another example of a Riemann surface whose Jacobian variety is biholomorphically equivalent to a product of elliptic curves. Namely, we find that the Jacobian variety of X is isomorphic qua complex manifold to $E_1 \times E_1 \times E_2$, where E_1 is the elliptic curve obtained from $\mathbb{C}$ by factoring out the lattice generated by 1 and i and E_2 is obtained from $\mathbb{C}$ by factoring out the lattice generated by 1 and $2i$. To see this, observe that we have the period matrix

$$\begin{array}{c} \\ dt_{0,0} \\ dt_{1,0} \\ dt_{0,1} \end{array} \begin{array}{c} \begin{array}{cccccc} \gamma(2,1) & \gamma(3,1) & \gamma(3,3) & \gamma(2,2) & \gamma(2,3) & \gamma(3,2) \end{array} \\ \begin{bmatrix} 1-i & 2 & -2 & 1+i & -1+i & 2i \\ 2 & 0 & 0 & 2i & -2 & 0 \\ 1-i & 2 & 2 & -1+i & 1-i & -2 \end{bmatrix} \end{array} .$$

Multiplying on the left by

$$\begin{bmatrix} 1 & 2 & 0 \\ 1 & 2+2i & 0 \\ 0 & 1 & 1 \end{bmatrix}$$

and on the right by

$$\begin{bmatrix} 1 & 0 & 0 & 1 & -2 & -1 \\ 0 & 0 & 0 & 0 & 1 & 0 \\ 0 & 0 & 1 & 1 & -1 & 0 \\ -1 & 1 & 0 & 1 & 0 & 0 \\ 0 & 0 & 0 & 0 & 0 & -1 \\ 1 & -1 & 0 & 0 & 0 & -1 \end{bmatrix}$$

we obtain the period matrix

$$\begin{bmatrix} 1 & 0 & 0 & i & 0 & 0 \\ 0 & 1 & 0 & 0 & 2i & 0 \\ 0 & 0 & 1 & 0 & 0 & i \end{bmatrix} .$$

Finally, we evaluate the quadratic periods on the Fermat curve along the generators $\gamma(j,k)$, $2 \leq j \leq q-1$, $1 \leq k \leq q-1$, of $\pi_1(X,\tilde{b}_1)$, where $\tilde{b}_1$ is the branch point on X above the point $z = b_1 = 1$. As in the case of the ordinary periods, we express $du_{r,s}$ and $du_{m,n}$ along $\ell(j,k)$ in terms of their expressions along the segment $0 \leq z \leq 1$, and we find that

$$\int_{-\ell(j,k)} u_{r,s}du_{m,n}$$

$$= \xi^{(s+n+2)(k-1)+(r+m+2)(j-1)} \int_1^0 (\int_1^z du_{r,s})du_{m,n} .$$

Therefore,

$$\int_{\gamma(j,k)} u_{r,s}du_{m,n}$$

$$= [\xi^{(s+n+2)(k-1)} - \xi^{(s+n+2)(k-1)+(r+m+2)(j-1)}$$

$$+ \xi^{(s+n+2)k+(r+m+2)(j-1)} - \xi^{(s+n+2)k}] \int_1^0 (\int_1^z du_{r,s})du_{m,n}$$

$$+ (\int_{\gamma(j,k)} du_{r,s})(\xi^{(n+1)k+(m+1)(j-1)} - \xi^{(n+1)k}) \int_1^0 du_{m,n}$$

$$= (\xi^{(s+n+2)}-1)(\xi^{(r+m+2)(j-1)}-1)\xi^{(s+n+2)(k-1)} \int_0^1 (\int_0^z du_{r,s})du_{m,n}$$

$$+ \xi^{(n+1)k}(\xi^{(m+1)(j-1)}-1) \int_1^0 du_{m,n} \int_{\gamma(j,k)} du_{r,s} .$$

Expressing $\int_1^0 du_{m,n}$ in terms of $\int_{\gamma(j,k)} du_{m,n}$, we obtain

$$Q_{(m,n),(r,s)}(\gamma(j,k))$$

$$= (\xi^{(s+n+2)}-1)(\xi^{(r+m+2)(j-1)}-1)\xi^{(s+n+2)(k-1)} \int_1^0 (\int_1^z du_{r,s})du_{m,n}$$

$$+ \frac{\xi^{(n+1)}}{\xi^{(n+1)}-1} \int_{\gamma(j,k)} du_{m,n} \int_{\gamma(j,k)} du_{r,s} .$$

Section V: Monodromy groups of differential equations

In this section we discuss the monodromy group of an analytic function and a few function theoretic implications of group theory resulting from it. The global structure of an analytic function, $w(z)$, with a finite set of branch points on $\hat{\mathbb{C}}$ is determined by its monodromy group. For example, $w(z)$ is single valued if and only if its monodromy group is the identity. In fact, the monodromy group of $w(z)$ may be identified with the monodromy group of the Riemann surface, possibly infinite sheeted, branched over $\hat{\mathbb{C}}$ upon which $w(z)$ becomes single valued. However, for present purposes, it is more useful to view the monodromy group as a group of linear transformations of a complex vector space, V, than to consider it as a group of permutations. By far the most important case occurs when V is finite dimensional; for then $w(z)$ satisfies a homogeneous linear differential equation, and it may be considered to be a "special function".

Recall that the study of the algebraic aspects of differential equations was actively pursued in the second half of the nineteenth century. Group theorists will recognize such familiar names as Frobenius and Jordan among the investigators of this topic. For example, Frobenius defined the notion of reducibility of a homogeneous linear differential equation in terms of its solutions, and Jordan interpreted this in terms of the reducibility of the monodromy representation. Here it is not our intention to present a historical account of our subject. Rather, we wish to observe that, from the beginning, group theoretic properties have been the key to successful applications of the monodromy group. In fact, our monodromy groups are always finitely generated, but the relators usually play an indirect role.

Suppose that $w = w(z-a)$ is a power series which converges in a neighborhood of $z = a$ and can be continued analytically to every point of $Z = \hat{\mathbb{C}} - \{b_1,\ldots,b_t\}$. Now, let $\mathcal{O}_a$ denote the totality of convergent power series $f = f(z-a)$ which may be obtained from w by analytic continuation along loops in Z based at $z = a$. According to the monodromy theorem, analytic continuation along homotopic loops yields the same series, so there is a homomorphism $m: \pi_1(Z,a) \to \Sigma(\mathcal{O}_a)$, where $\Sigma(\mathcal{O}_a)$ is the group of all permutations of the elements of $\mathcal{O}_a$. To be precise, $m([\gamma])(f)$ is the power series obtained by analytically continuing $f \in \mathcal{O}_a$ along the loop γ in Z based at $z = a$. We call m the monodromy representation and refer to its image, M, as the monodromy group of $w(z-a)$. An indication of the function theoretic significance of the monodromy group is given by the fact that M is finite if $w(z-a)$ is the branch of an algebraic function. Indeed, this is clear because $\mathcal{O}_a$ must be finite. Moreover, the converse is also valid if the function approaches a definite limit, possibly infinity, as z approaches b_j in every sector with vertex at b_j. For example, see A [3], Vol. I, page 168.

Now, suppose that V_a is the complex vector space spanned by the elements of $\mathcal{O}_a$. Since $\mathcal{O}_a \subset V_a$ and because analytic continuation commutes with algebraic operations, each element of $\Sigma(\mathcal{O}_a)$ induces an invertible linear transformation of V_a. Thus, we may view $\Sigma(\mathcal{O}_a)$ as a subgroup of $GL(V_a)$ and regard the monodromy representation as a homomorphism $m: \pi_1(Z,a) \to GL(V_a)$ whose image, M, is the monodromy group.

Basic to our considerations is the fact that: V_a is finite dimensional if and only if $w(z-a)$ is a solution of a homogeneous linear differental equation

$$(*) \qquad a_0(z)w^{(n)} + a_1(z)w^{(n-1)} + \dots + a_n(z)w = 0$$

whose coefficients, $a_j(z)$, are single-valued holomorphic functions on Z.

Obviously, the totality of solutions of (*) at $z = a$ forms an n-dimensional complex vector space. Moreover, differentiation and analytic continuation are commuting operations; so if $w(z-a)$ is a solution of (*), the same is true of all the elements of $\mathcal{O}_a$, and V_a has dimension at most n. The converse is only slightly more difficult to establish. Its proof involves the Wronskian determinant and may be found, for example, in Poole [B,2], page 47.

For brevity, we shall often say that a function "satisfies a linear differential equation" when we wish to indicate that it is a solution of an equation of the type given by (*). The points $z = b_1,\ldots,b_t$ are called the singular points of the equation, and a singularity, $z = b_j$, is a regular singular point if every solution of the differential equation approaches a limit, possibly infinity, as z approaches b_j in every sector formed by two rays with vertex at b_j. If all the singular points are regular singular points, the differential equation is said to be of Fuchsian class. The hypergeometric equation is an example of such an equation. On the other hand, Bessel's equation is not of Fuchsian class because it possesses an irregular singular point at infinity. Finally, we recall that there are simple criteria, essentially due to L. Fuchs, which allow us to decide if a point is a regular singular point. Using these, Fuchs showed that if (*) is of Fuchsian class then the $a_j(z)$ are polynomials of specified form.

If a group is isomorphic to a subgroup of $GL(n,\mathbb{C})$, we shall refer to it as a linear group. The term algebraic group will be used to indicate that the group in question is isomorphic to a subgroup of $GL(n,\mathbb{C})$ which is closed in the Zariski topology. Thus, we have seen that if $w(z-a)$ satisfies a linear differential equation, then its monodromy group is a finitely generated linear group. The converse assertion is commonly called the Riemann-Hilbert Problem or Hilbert's Twenty-First Problem, in recollection of its appearance in the famous collection of problems posed by Hilbert in 1900. In order to state the problem precisely, suppose we are given the data of a Hurwitz system, $(z = a, b_1,\ldots,b_t, \ell_1,\ldots,\ell_t, m_1,\ldots,m_t)$, where instead of permutations π_j we have matrices $m_j \in GL(n,\mathbb{C})$ with $m_1\ldots m_t = 1$. Then, we ask whether there is a differential equation of Fuchsian class of order n whose singularities are at $z = b_1,\ldots,b_t$ and whose solution space at $z = a$ admits a basis, $\underline{w}(z-a)$, which is replaced by $m_j \underline{w}(z-a)$ upon analytic continuation along the loop γ_j associated with ℓ_j.

The Riemann-Hilbert Problem, as we posed it, has an affirmative answer. In fact, Hilbert himself solved the problem for the case $n = 2$ with the aid of his then newly created theory of integral equations [24]. The case of arbitrary n was settled shortly thereafter by Plemelj [42]. Solutions were also given by G. D. Birkhoff [8] and, more recently, by H. Rohrl [45]. Here, we only wish to point out that Riemann's theory of the hypergeometric equation may be viewed as his solution of the problem when $n = 2$ and $t = 3$, and that H. Poincaré and L. Schlesinger attacked the general problem vigorously with the aid of Poincaré's theory of automorphic functions and Zetafuchsian systems.

The Riemann-Hilbert Problem has been vastly generalized in recent times. For example, the domain Z may now be an arbitrary smooth quasi-projective complex algebraic variety or certain types of complex manifolds. Here, we can do no more than refer the reader to the article by N. Katz [28], which explains the algebro-geometric viewpoint and Deligne's solution of the resulting problem. Naturally, Katz explains the role played by important earlier works of Grothendieck, Hironaka and Serre; and the reader cannot fail to sense the great accomplishment of modern algebraic geometers in extending Riemann's ideas from one to many variables. However, lest we convey the impression that the story is now complete, we wish to note that from one viewpoint, at least, the solution of the generalized Riemann-Hilbert Problem is not as satisfactory as that of the original version. Namely, although Deligne shows that every finite dimensional representation of the fundamental group of Z can be realized as the monodromy group of a system of differential equations, he does not tell us which groups may arise in this fashion. In fact, this question is more properly part of another story, that of determining the fundamental group of the complement of a variety in projective space ([12], [57]).

Finally, we note that the Riemann-Hilbert Problem plays a role in certain recent developments in mathematical physics and number theory. We refer the reader to the Chudnovsky brothers' seminar [11] for a report on these exciting developments. Here we only wish to remark that the problem of deforming differential equations, for instance by varying their singular points, while preserving their monodromy groups plays a key role in this work. The problem of determining all differential equations with a given monodromy group also plays a role in the theory of moduli of Riemann surfaces (see, for example, [23]).

The problem of determining the monodromy group of (*) is essentially unsolved. A notable exception occurs for second order equations of Fuchsian class with three singular points. Such equations can always be put in the following normal form, due to Gauss:

$$z(1-z)w'' + [c-(a+b+1)z]w' - abw = 0 ,$$

where a,b,c are parameters. This equation is called the <u>hypergeometric equation</u>, and its monodromy group was determined by Riemann and by Jordan. For example, the following two matrices may be taken as its generators:

$$\begin{pmatrix} 1 & e^{2\pi i(c-a-b)}-e^{2\pi ia} \\ 0 & e^{-2\pi ic} \end{pmatrix}$$

and

$$\begin{pmatrix} e^{2\pi i(c-a-b)} & 0 \\ e^{-2\pi ic}-e^{-2\pi ia} & 1 \end{pmatrix} .$$

Setting $a = \frac{1}{2}$, $b = \frac{1}{2}$, and $c = 1$, we obtain

$$\begin{pmatrix} 1 & 2 \\ 0 & 1 \end{pmatrix} \quad \text{and} \quad \begin{pmatrix} 1 & 0 \\ 2 & 1 \end{pmatrix} .$$

These matrices generate the principal congruence subgroup modulo two in the modular group, a subgroup which is easily seen to be free to rank two. Thus, the monodromy group of the hypergeometric equation corresponding to these special values of the parameters a,b,c is free of rank two. On the other hand, it is also possible to select values of a,b,c so that the monodromy group is finite; consequently the corresponding hypergeometric equation only has algebraic functions as its solutions. In fact, H. A. Schwarz (1873) determined all second order equations of Fuchsian class with three singularities whose only solutions are algebraic functions. Needless to say, this work is intimately connected with Schwarz' triangle functions, the non-linear differential operator now called the Schwarzian derivative, and the finite subgroups of $PSL(2,\mathbb{C})$. The latter also play a key role in all subsequent attempts to solve the following problem: given a second order differential equation (*) whose coefficients $a_j(z)$ belong to $\mathbb{C}(z)$, determine in a finite number of steps whether all its solutions are algebraic functions. Apparently, both F. Klein and L. Fuchs claimed to have solved this problem, but there are gaps in their work which seem to have gone unnoticed. In any event, Baldassarri and Dwork [4] reexamined the problem and provided a complete solution of it. Subsequently, Baldassarri [6] extended these results to include differential equations defined on Riemann surfaces of genus greater than zero and applied these results in an investigation of the algebraic solutions of Lamé's equation [5]. For our purposes, a differential equation on the Riemann surface X defined by $f(z,u) = 0$ is defined by replacing the coefficients $a_j(z)$ in (*) by rational functions $a_j(z,u)$ of z and u. In the case of Lamé's equation, X has genus one.

There is also an arithmetic approach to the problem of deciding whether all the solutions of (*) are algebraic functions. Let us suppose that all the constants appearing in the $a_j(z)$ are rational numbers. Then, for almost all primes p, we may reduce (*) modulo p. The resulting equation may then be viewed as defining the kernel of a linear operator, L_p, acting on $\mathbb{F}_p(z)$ viewed as a vector space over $\mathbb{F}_p(z^p)$, where $\mathbb{F}_p$ denotes the field with p elements. Now, it is a conjecture of Grothendieck that if the dimension of the kernel of L_p is n for almost all p, then all the solutions of (*) are algebraic functions. In fact, N. Katz [29] proved this conjecture for the cases where (*) is a suitable direct factor of a Picard-Fuchs equation. His proof uses many of the notions of modern algebraic geometry, and even the rigorous definition of Picard-Fuchs equations is out of place here. Nevertheless, we do wish to make a few remarks about the simplest case because it illustrates the connection between the periods of Abelian integrals and differential equations.

We begin by rewriting the hypergeometric equation in new variables t and u and writing the well known integral representation for its solutions. Thus, we have the equation

$$t(1-t)u'' + [c-(a+b+1)t]u' - abu = 0$$

and the solution

$$u(t) = \int_{\gamma} z^{a-c}(z-1)^{c-b-1}(z-t)^{-a}dz , \quad t \neq 0,1, \quad ,$$

where γ is a loop in the z-plane customarily called a double loop. In fact, γ is a commutator, $\gamma_j\gamma_k\gamma_j^{-1}\gamma_k^{-1}$, where $\gamma_1, \gamma_2, \gamma_3, \gamma_4$ are the simple loops determined by any Hurwitz system with $b_1 = 0$, $b_2 = 1$, $b_3 = t$, $b_4 = \infty$. In the special case $a = \frac{1}{2}$, $b = \frac{1}{2}$, $c = 1$, the integrand under consideration is an Abelian differential of first kind on the elliptic curve defined by the equation $w^2 = z(z-1)(z-t)$. Moreover, the loop γ lifts to a loop $\tilde{\gamma}$ on the Riemann surface $X(t)$ defined by this equation, and $\tilde{\gamma}$ is homotopic to one of the loops $\gamma(j,k)$ introduced in Section 3. Thus, we may view the periods of the elliptic integral of first kind as a function on the t-sphere which satisfies a differential equation. Of course, the t-sphere with $t = 0,1,\infty$ deleted is also a parameter space for the family of Riemann surfaces $\{X_t\}$ of genus one. The exceptional values $t = 0,1,\infty$ correspond to Riemann surfaces with singularities, as well as to singularities of the differential equation.

Finally, without giving any details, we note that the Picard-Fuchs equations arise by replacing the family $\{X_t\}$ of elliptic curves by any family $\{V_t\}$ of algebraic varieties defined by algebraic equations involving the coordinates of the parameter space and the ambient space of the V_t. Now, the periods of the Abelian integral on X_t may be interpreted in terms of the cohomology of X_t; and we have seen that they satisfy a differential equation. Similarly, elements in the cohomology of V_t may be viewed as functions of t and proved to satisfy a differential equation, called the Picard-Fuchs equation.

We now turn briefly to the Differential Galois Theory created by Picard and Vessiot at the end of the nineteenth century. Essentially, those authors achieved their goal of establishing an analogue of the Galois theory of algebraic equations for homogeneous linear differential equations. Perhaps the highlight of their theory was the theorem that such a differential equation is solvable by quadratures if and only if its differential Galois group is solvable. Apparently, from the present viewpoint, their work is somewhat flawed by a lack of rigour. In any event, their results were firmly established by the efforts of E. Kolchin, whose pioneering work in algebraic groups was directly linked to this topic. An admirably clear and concise presentation of Differential Galois Theory is given by Kaplansky in [27].

For our purposes, it suffices to know that a differential field, F, is a field together with a derivation, that is, an additive map of F into itself which satisfies Leibnitz' rule when applied to products. For example, $\mathbb{C}(z)$ with the usual differentiation is a differential field. The analogue of a splitting field of an algebraic equation is called a Picard-Vessiot extension and may be described as follows for the equation (*) with coefficients in $\mathbb{C}(z)$. Suppose that $w_1,\ldots,w_n$ form a basis over $\mathbb{C}$ for the solutions of (*). Then the field E consisting of quotients of polynomials in the w_j, the derivatives of the w_j with respect to z, and z is easily seen to be a differential field. It is the desired field, and it has the property that it is contained in every differential field containing $\mathbb{C}(z)$ and $w_1,\ldots,w_n$. The differential Galois group of $E/\mathbb{C}(z)$ consists of all automorphisms of E which fix each element of $\mathbb{C}(z)$ and commute with differentiation. Clearly, this group is a linear group; in fact, it is even an algebraic group. Finally, we note that there is also an analogue of the fundamental theorem of Galois theory.

As might be expected, most statements involving classical Galois theory have analogues in the differential case. For example, in his address to the International Congress of Mathematicians at Moscow in 1966, E. Kolchin [31] cites as unsolved the following inverse problem of differential Galois theory. Given an algebraic group $G \subset GL(n,\mathbb{C})$, does there exist a Picard-Vessoit extension E of $\mathbb{C}(z)$ with differential Galois group G? Of course, this problem reminds us of the classical case where a <u>finite</u> group is given and we are asked whether there is an extension of the <u>rational numbers</u> for which it is the Galois group. I. R. Shafarevitch [63] gave an affirmative reply to this question when the finite group in question is <u>solvable</u>. His results also apply if the rational numbers are replaced by an arbitrary algebraic number field.

Returning to the differential case, we note the Kovacic [32] showed that the inverse problem has an affirmative answer whenever G is solvable. Moreover, his treatment is purely algebraic and applies when $\mathbb{C}$ is replaced by any algebraically closed field of characteristic zero. Next, the present authors [52] showed that the <u>inverse problem of classical differential Galois theory has an affirmative solution</u>. Here we use the adjective "classical" to stress that the ground field is $\mathbb{C}(z)$ and that the tools of analysis are available to us. In fact, our proof is an immediate consequence of three statements and only the first belongs entirely to the domain of algebra. Namely, we have:

(i) If $G \subset GL(n,\mathbb{C})$ is an algebraic group, then there is a finitely generated Zariski dense subgroup $M \subset G$.

(ii) There is an equation (*) of Fuchsian class whose monodromy group is M.

(iii) The Zariski closure, $\overline{M}$, of the monodromy group of an equation (*) of Fuchsian class coincides with the differential Galois group of a Picard-Vessiot extension of $\mathbb{C}(z)$ defined by (*).

Finally, we note that (iii) need not hold if (*) is not of Fuchsian class and that statement (ii) is obviously the assertion of the Riemann-Hilbert Problem.

We now turn to the notable work of M. F. Singer ([47], [48]) which makes effective Picard and Vessiot's characterization of (*) as solvable by quadratures if and only if its differential Galois group is solvable. Previously, the situation was like that of classical Galois theory: despite the intrinsic beauty of Picard and Vessiot's theorem, its utility was limited by the difficulty of determining Galois groups. Indeed, it should be clear from the foregoing discussion of monodromy groups that the determination of the Galois group of (*) is generally a formidable task.

Now, suppose that the coefficients $a_j(z)$ of (*) belong to a finite algebraic extension, K, of the field $\mathbb{Q}(z)$. Then, Singer proves that <u>there is an algorithm which determines in a finite number of steps whether all the solutions of (*) can be obtained from K by quadratures</u>. In fact, the solutions of (*) which can be obtained by quadratures form a subspace of the solution space, and Singer's algorithm yields a basis for the subspace of solutions of (*) which are algebraic functions. Of course, it is possible that none of the solutions of (*) is an algebraic function and that none of them is obtainable by quadratures; but, in either case, the algorithm informs us of this fact.

Singer's use of the Picard-Vessiot theory allows him to bring the full power of group theory to bear on his problem without actually determining the monodromy group or the Galois group. By way of contrast, it seems to us that the main thrust of F. Klein and his followers was to handle the second order case by reducing their problem to the hypergeometric equation and Schwarz' list. However, we should note that Kimura [30] applied differential Galois theory to determine all second order equations (*) of Fuchsian class with three singularities whose solutions may all be obtained by quadratures. Returning to Singer's algorithms, we note that a key role is played by the theorem that: there is a computable integer valued function, $I(n)$, such that any solvable-by-finite subgroup $G \subset GL(n,\mathbb{C})$ contains a triangularizable subgroup H of index at most $I(n)$. Now, if (*) has a solution obtainable by quadratures, this theorem may be applied via the Galois theory to show that (*) has a solution, u, the common eigenvector for elements of H, such that u'/u is algebraic over K of degree at most $I(n)$. The problem is eventually reduced to elimination theory, but this a priori bound is the keystone of the algorithm.

The group-theoretic result cited above is obtained by combining a theorem of Platonov with Malcev's version of the Lie-Kolchin Theorem (see, for example [56], Corollary 10.11 and Theorem 3.6). In fact, Singer discovered and applied a somewhat weaker result, but both of them are related to a famous theorem of Jordan. Namely, there is a computable function, $b(n)$, such that every finite subgroup $G \subset GL(n,\mathbb{C})$ contains an Abelian normal subgroup H of index at most $b(n)$. In fact, Jordan was also occupied with the problem of determining when (*) possesses only algebraic functions as solutions, and his theorem appears in a paper [60] devoted to this topic. In particular, suppose that all the solutions of (*) are algebraic functions and that $w_1,\ldots,w_n$ is a basis for the solution space of that equation. Then, as shown by Jordan, there is a homogeneous polynomial $f(x_1,\ldots,x_n)$ of degree at most $b(n)$ such that $f(w_1,\ldots,w_n)$ is the radical of a rational function. Results of this type play a role in the paper of Baldassarri and Dwork (see [4], page 44), and the latter was led to inquire of J. G. Thompson as to the best possible bounds. Subsequently, Thompson [55] proved the following theorem: Let $H_d(\mathbb{C})$ denote the space of homogeneous polynomials of degree d in n variables with coefficients in $\mathbb{C}$ and let the finite group $G \subset GL(n,\mathbb{C})$ act on H_d via its usual action on $\mathbb{C}[x_1,\ldots,x_n]$. If $d(G)$ is the smallest integer for which H_d contains a 1-dimensional G-invariant subspace, then $d(G) \leq 4n^2$. Moreover, Thompson conjectures that $d(G) \leq Cn$, for some constant C, independent of n and G.

The Picard-Vessiot theory also plays a role in a forthcoming paper of M. F. Singer and M. D. Tretkoff [64]. Among other things, there it is shown that there are n^{th} order linear differential equations of Fuchsian class with three singular points whose solutions cannot all be written as rational functions of the solutions of second order linear equations of Fuchsian class with three singular points. Here, it is assumed that $n \geq 6$. The group-theoretical key to this result is Platonov's theorem (see [56], Lemma 10.10) which asserts that an algebraic group $G \subset GL(n,\mathbb{C})$ may be written in the form $G = HG^o$, where H is a finite group and G^o is the connected component of the identity. In addition, [64] contains an inductive argument based on the maximal order of non-Abelian composition factors of a finite group. Apparently, finite groups have not been studied from this viewpoint. Finally, we note that [64] contains a discussion of some function theoretic ramifications of Tits' theorem that finitely generated linear groups are either solvable-by-finite or contain a non-Abelian free subgroup. Roughly speaking, this yields a partition of the differential equations (*) into two classes: those which are solvable by quadratures and those whose solutions satisfy certain functional equations.

We close this section with the remark that combinatorial group theory has been applied to the discovery of a new class of transcendental functions. Namely, it is shown in [21] that there are many multi-valued transcendental functions with three singular points on $\hat{\mathbb{C}}$ which do not satisfy any linear differential equation (*). Such functions arise by combining the existence of two-generator groups which have no faithful matrix representations over $\mathbb{C}$ with the analogue of the Riemann existence theorem for non-compact Riemann surfaces. Some of these functions reflect their monodromy groups in interesting ways. For example, suppose we select a two-generator infinite group of exponent N as the monodromy group. Then, we obtain a transcendental function $T(z)$, with three singularities, say at $z = 0,1,\infty$, such that analytic continuation repeated N times along any loop avoiding $z = 0,1,\infty$ yields the original branch. Thus, $T(z)$ appears to be an algebraic function, but it really is transcendental. Moreover, $T(z)$ is not a solution of the hypergeometric equation. The present authors have recently obtained another appealing example. Namely, they observed the existence of a multi-valued function on $\hat{\mathbb{C}}$ whose Riemann surface has a single infinite spiral over each of its three singularities, $z = 0,1,\infty$. This should be contrasted with Log z which has a single infinite spiral over $z = 0,\infty$ and the inverse of Legendre's elliptic modular function $\lambda(\tau)$ which has infinitely many infinite spirals over its three branch points $z = 0,1,\infty$. It is of interest to note that each of the new "special functions" we have described corresponds to a subgroup of infinite index in the modular group. Finally, we remark that a function theoretic proof of the existence of $T(z)$ would yield an analytic solution of the Burnside problem.

REFERENCES

A. Books about Riemann surfaces

1. Farkas, H. M. and Kra, I., Riemann Surfaces, Graduate Texts in Mathematics, Vol. 71, Springer-Verlag (1980).
2. Gunning, R. C., Lectures on Riemann Surfaces, Princeton University Press (1967).
3. Siegel, C. L., Topics in Complex Function Theory, Vol. I (1969), Vol. II (1971), Vol. III (1973), Wiley-Interscience, N. Y.

B. Books about differential equations

1. Hille, E., Ordinary differential equations in the complex domain, Wiley-Interscience, New York (1971).
2. Poole, E. G. C., Introduction to the theory of linear differential equations, Dover reprint, New York (1960).

C. Papers

1. Accola, R., On the number of automorphisms of a closed Riemann surface, Trans. Amer. Math. Soc. 131 (1968), 398-407.
2. Adler, A., Some integral representations of $PSL_2(\mathbb{F}_p)$ and their applications. J. of Algebra, 72 (1981), 115-145.
3. Baker, H. F. Multiply Periodic Functions (1907), Cambridge University Press.
4. Baldassarri, F. and Dwork, B. On second order linear differential equations with algebraic solutions, Am. J. Math. 101 (1979), no. 1, 42-76.

5. Baldassarri, F. On algebraic solutions of Lamé's differential equation, J. of Diff. Eqns. 41 (1981), 44-58.

6. Baldassarri, F. On second order linear differential equations with algebraic solutions on algebraic curves, Am. J. Math. 102 (1980), 517-535.

7. Berstein, I. and Edmonds, A. L. On the construction of branched coverings of low dimensional manifolds, Trans. A.M.S. 247 (1979), 27-124.

8. Birkhoff, G. D. Collected Mathematical Papers, Volume I (1950), American Mathematical Society.

9. Chen, K. T. Iterated path integrals, Bull A.M.S., 83 (1977), 831-879.

10. Cohen, J. Families of compact Riemann surfaces with automorphism groups, J. London Math. Soc. (2), 24 (1981), 161-164.

11. Chudnovsky, D. and Chudnovsky, G. The Riemann Problem, Complete Integrability and Arithmetic Applications, Lecture Notes in Mathematics 925, Springer-Verlag.

12. Deligne, P. Le groupe fondamental du complément d'une courbe plane n'ayant que des points doubles ordinaries est abélien (d'après W. Fulton), Séminaire Bourbaki 1979/80, Lecture Notes in Mathematics 842 (1981), 1-10, Springer-Verlag.

13. Earle, C. J. Some Jacobian Varieties which split, Lecture Notes in Mathematics 747, 101-107, Springer-Verlag.

14. Feit, W., Lyndon, R. C. and Scott, L. L. A remark about permutations, J. Combinatorial Theory, A, 18 (1975), 234-235.

15. Fried, M. Exposition on an arithmetic-group theoretic connection via Riemann's existence theorem, Proceedings of Symposia in Pure Mathematics, Vol. 37 (1980), American Mathematical Society, 571-602.

16. Fulton, W. Hurwitz schemes and irreducibility of moduli of algebraic curves, Ann. of Math. 90 (1969), 542-575.

17. Gilman, J. A matrix representation for automorphisms of compact Riemann surfaces, Linear Algebra and its Applications 17 (1977), 139-147.

18. Gilman, J. and Patterson, D. Intersection matrices for bases adapted to automorphisms of a compact Riemann surface, Annals of Mathematics Studies 97 (1981), Princeton University Press, 149-166.

19. Greenberg, L. Maximal groups and signatures, Ann. of Math. Studies 79 (1974), Princeton University Press, 99-105.

20. Gunning, R. C. Quadratic Periods of Abelian Integrals, in Problems in Analysis, Princeton University Press.

21. Haimo, F., Singer, M. F. and Tretkoff, M. Remarks on analytic continuation, Bull. London Math. Soc. 12 (1980), 9-12.

22. Harris, B. Homological versus algebraic equivalence in a Jacobian, preprint.

23. Hejhal, D. Monodromy groups and linearly polymorphic functions, Acta Math. 135 (1975), 1-55.

24. Hilbert, D. Linearen Integralgleichungen, reprinted by Chelsea Publishing Co., Inc., New York, 1953.

25. Hurwitz, A. Mathematische Werke, Bd. I., Basel (1932), 415-417.

26. Jablow, E. Thesis, Princeton University (1982).

27. Kaplansky, I. An Introduction to Differential Algebra, Hermann, Paris (1957).

28. Katz, N. An overview of Deligne's work on Hilbert's Twenty-First Problem, Mathematical Developments Arising From Hilbert Problems, American Mathematical Society (1976), 537-557.

29. Katz, N. Algebraic Solutions of differential equations, Inventiones Math. 18 (1972), 1-118.

30. Kimura, T. On Riemann's equations which are solvable by quadratures, Funkcialaj Ekvacioj, 12 (1969), 269-281.

31. Kolchin, E. Some problems in differential algebra, Proc. Intl. Congress of Mathematicians (Moscow, 1966), Mir, Moscow, 1968.

32. Kovacic, J. On the inverse problem in the Galois theory of differential fields, I and II, Annals of Math. 89 (1969), 583-608 and 93 (1971), 269-284.

33. Kuiken, K. On the monodromy groups of Riemann surfaces of genus ≥ 1, Canadian J. of Math. XXXIII (1981), 1142-1156.

34. Kulkarni, R. Normal subgroups of Fuchsian groups, preprint.

35. Kulkarni, R. Pseudofree actions and Hurwitz's 84 (g-1) theorem, preprint.

36. Lange, H. Produkte elliptischer Kurven, Nachr. Akad. Wiss. Gottingen, II. Math.-phys. Kl. 8 (1975), 95-108.

37. Lyndon, R. C. On the combinatorial Riemann-Hurwitz formula, Symposia Mathematica XVII (1976), 435-439.

38. Macbeath, A. M. On a Theorem of Hurwitz, Proc. Glasgow Math. Assoc. 5 (1961), 90-96.

39. Macbeath, A. M. On a curve of genus 7, Proc. London Math. Soc. (3), 15 (1965), 527-542.

40. Martens, H. Riemann matrices with many polarizations, Complex Analysis and its applications, Vol. III (1976), International Atomic Energy Agency, Vienna, 35-48.

41. Martens, H. Observations on morphisms of closed Riemann surfaces, Bull. London Math. Soc. 10 (1978), 209-212.

42. Plemelj, J. Riemannsche Funktionenscharen mit gegebener Monodromiegruppe, Monatshefte für Math. und Physik, 19 (1908), 211-246.

43. Rauch, H. E. and Lewittes, J. The Riemann surface of Klein with 168 automorphisms, in Problems in Analysis, R. C. Gunning, ed., Princeton University Press (1970).

44. Ree, R. A theorem about permutations, J. Combinatorial Theory, A, 10 (1971), 174-175.

45. Rohrl, H. Das Riemann-Hilbertsche Problem der Theorie der linearen Differentialgleichungen, Math. Ann. 133 (1957), 1-25.

46. Scott, L. L. Matrices and cohomology, Ann. of Math. 105 (1977), 473-492.

47. Singer, M. F. Liouvillian solutions of the nth order homogeneous linear differential equations, Am. J. Math. 103 (1981), 661-682.

48. Singer, M. F. Algebraic solutions of the nth order linear differential equations, Queens Papers in Mathematics.

49. Singerman, D. Symmetries of Riemann Surfaces with Large Automorphism Group, Math. Ann. 210 (1974), 17-32.

50. Sinkov, A. On the group defining relations $(2,3,7,\rho)$, Ann. of Math. (2) 38 (1937), 577-584.

51. Sullivan, D. Differential forms and the topology of manifolds, in Manifolds-Tokyo 1973, A. Hattori, ed., University of Tokyo Press.

52. Tretkoff, C. L. and Tretkoff, M. D. The solution of the inverse problem of Differential Galois Theory in the classical case, Am. J. Math. 101 (1979), 1327-1332.

53. Tretkoff, C. L. and Tretkoff, M. D. On a theorem of Rimhak Ree about permutations, J. Combinatorial Theory, A 26 (1979), 84-86.

54. Tretkoff, M. Algebraic extensions of the field of rational functions, Communications on Pure and Applied Math. XXIV (1971), 491-497.

55. Thompson, J. G. Invariants of finite groups, J. of Algebra, 69 (1981), 143-145.

56. Wehrfritz, B. A. F. Infinite Linear Groups, Springer-Verlag (1973).

57. Zariski, O. Algebraic Surfaces, Second edition, Springer-Verlag (1971).

58. Greenberg, L. Finiteness Theorems for Fuchsian and Kleinian Groups, in Discrete Groups and Automorphic Functions, ed. W. J. Harvey, Academic Press (1977).

59. Hayashida, T. and Nishi, M. Existence of curves of genus two on a product of two elliptic curves, J. Math. Soc. Japan, 17 (1965), 1-16.

60. Jordan, C. Memoire sur les equationes differentielles lineaires a integrale algebrique, J. fur Math. 84 (1878), 89-215.

61. Lefschetz, S. On certain numerical invariants of algebraic varieties with application to Abelian varieties, Trans. Amer. Math. Soc. 22 (1921), 327-482.

62. Nielsen, J. Die Structur periodischer Transformationen von Flachen, D. K. Dan. Vidensk. Selsk. Math-fys. Medd. XV (1937), 1-77.

63. Shafarevitch, I. R. Construction of fields of algebraic numbers with given solvable Galois group, Amer. Math. Soc. Translations, Vol. 4 (1956), 185-237.

64. Singer, M. F. and Tretkoff, M. D. Some applications of linear groups to differential equations, Am. J. Math., to appear.

65. Gross, B., On the periods of Abelian integrals and a formula of Chowla and Selberg (with an appendix by D. E. Rohrlich), Inventiones math. 45 (1978), 193-211.

66. Deligne, P., Valeurs de fonctions L et périodes d'intégrales, Proc. Symp. Pure Math., A.M.S. 33 (1979), Part 2, 313-346.

67. Ritt, J. F., On algebraic functions which can be expressed in terms of radicals, Trans. Am. Math. Soc., 24 (1922), 21-30.

68. Siegel, C. L., Transcendental Numbers, Princeton University Press (1949).

Notes added in proof

1. (page 467) The intersection numbers can also be determined by calculating the cohomology ring of the manifold. Given a triangulation, it is well known how to do this. However, the advantage of the present method is its directness; everything is done in terms of the Hurwitz system.

2. (page 472) See, for example, A3, Vol. III, page 65.

3. (page 473) In particular, Fried uses Hurwitz systems in an attack on the inverse problem of Galois theory over $\mathbb{Q}$. Recently, there has been important progress in extending Shafarevitch's result (see page 510) from solvable groups to simple groups. For example, J. G. Thompson has shown that the Fisher-Greiss "Monster" is the Galois group of an extension of $\mathbb{Q}$.

4. (page 484) Also, see the old paper of Ritt [67]. There, the functions with solvable monodromy group that also define Riemann surfaces with a <u>prime</u> number of sheets are determined.

5. (page 485) In fact, Hurwitz only considered the representation on the space of Abelian differentials of the first kind. However, it can easily be seen that the faithfulness of this representation is equivalent to the faithfulness of the representation on $H_1(X, \mathbb{Z})$. Thus, the custom of crediting the latter fact to Hurwitz seems reasonable.

6. (page 491) See, for example, C. L. Siegel [68] Chapter IV, especially pages 99 and 100. More recent investigations of this topic include Deligne [65] and Gross [66].

7. (page 500) Our results differ from those obtained by Rohrlich [66] because we used a different homology basis. In particular, we use 1-cycles which also represent generators for the fundamental group. This facilitates computation of the quadratic periods, a topic which does not occur in [66].

DEPARTMENT OF COMPUTER AND INFORMATION SCIENCE
BROOKLYN COLLEGE-C.U.N.Y.
BROOKLYN, N.Y. 11210

Received September 20, 1982

ABCDEFGHIJ–AMS–8987654

22

/Statistics/Computer Science Library
642-3381
LIBRARY OF THE UNIVERSITY OF CALIFORNIA
Berkeley